Comparative Anatomy
of the

VERTEBRATES

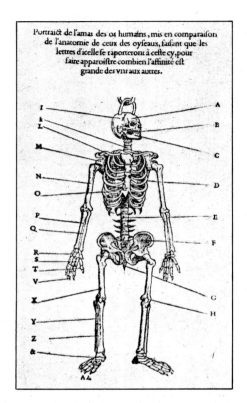

Portraict de l'amas des os humains, mis en comparaison de l'anatomie de ceux des oyseaux, faisant que les lettres d'icelle se raporteront à ceste cy, pour faire apparoistre combien l'affinité est grande des vns aux autres.

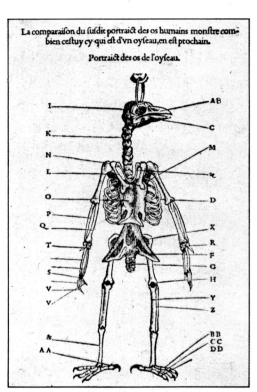

La comparaison du susdit portraict des os humains monstre combien cestuy cy qui est d'vn oyseau, en est prochain.

Portraict des os de l'oyseau.

Woodcut by Belon, 1555

Comparative Anatomy
of the
VERTEBRATES

George C. Kent

Alumni Professor Emeritus, Department of Zoology,
Louisiana State University, Baton Rouge

Sixth Edition

with 1057 illustrations including 469 in color

Original text artwork by Karen A. Westphal
Baton Rouge, Louisiana

Times Mirror/Mosby
College Publishing
ST. LOUIS • TORONTO • SANTA CLARA   1987

Executive Editor: Donald G. Mason
Editorial Assistant: June Schaffer Heath
Project Editor: Teri Merchant
Editing and Production: Top Graphics
Designer: John R. Rokusek

Illustrator: Karen A. Westphal

Cover photo: ''Frog/Colos Thethus bocagei''
        by Rosamond Wolff Purcell
        from *Illuminations/A Bestiary*,
        W.W. Norton & Co., Inc. 1986

Sixth Edition
Copyright © 1987 by Times Mirror/Mosby College Publishing

A division of The C.V. Mosby Company
11830 Westline Industrial Drive
St. Louis, Missouri 63146

Previous editions copyrighted 1965, 1969, 1973, 1978, 1983

Printed in the United States of America

Library of Congress Cataloging-in-Publication Data

Kent, George Cantine
   Comparative anatomy of the vertebrates.

   Bibliography: p.
   Includes index.
   1. Vertebrates   2. Anatomy, Comparative.   I. Title.
QL805.K43   1987   596'.04   86-23188
ISBN 0-8016-2657-9

TS/VH/VH   9 8 7 6 5 4 3 2 1   02/B/244

PREFACE

This is a textbook of functional and comparative morphology with a developmental and evolutionary perspective. It examines the primitive architectural patterns of vertebrate systems, speculates (when justified) on the survival value of successive modifications, and points out how functional combinations of recent characters contribute to survival in the contemporary environment. Repeatedly expressed or implied is the fact that existing structural patterns are modifications of ancestral ones, that the adult is a modification of the embryo, that individual differences as well as species differences exist, and that structure is broadly determined by heredity and adaptively modified through natural selection. The text assumes an elementary understanding of evolutionary theory. A reminder of what organic evolution implies and, equally important, what it does not imply is found in Chapter 1.

CHANGES IN THIS EDITION

Readers familiar with earlier editions will discover changes that, it is hoped, will meet with general approval. More than 100 new illustrations have been added, and more detailed coverage of many topics has been provided.

- Discussions of phylogeny, taxonomy and systematics, homology, convergent and parallel evolution, adaptation, Lamarckism, Darwinism (natural selection), and speciation have been combined, supplemented, and presented in Chapter 1, following an introduction to the phylum Chordata, along with an expanded discussion of the historical roots of comparative anatomy.

- Chapter 2, Protochordates and the Origin of Vertebrates, has been restructured and the discussion of hemichordates has been amplified.

- In Chapter 3, discussions of vertebrate phylogeny have been supplemented by the addition of new family trees and many new or redrawn illustrations of representatives of the vertebrate classes.

- The discussion of early chordate morphogenesis (Chapter 4) has been restructured, and the discussions of gastrulation, neurulation, ectodermal placodes, and neural crests have been expanded.

- The chapter on the integument (Chapter 5) has been expanded to include a recapitulation of the major features of vertebrate skin on a class-by-class basis.

• The discussion of mineralized tissues (Chapter 6) has been restructured, and discussions of the skeletal system (Chapters 7 to 9) have been expanded and supplemented with new or redrawn illustrations of fossil and living vertebrates.

• Chapters 10 through 16 (Muscles through Sense Organs) have undergone fewer changes, although some restructuring will be evident. Expanded topics include, among others, the appendicular muscles of tetrapods, mammalian dentition, air-breathing in bony fishes, respiratory mechanisms in bullfrogs and amniotes, the larynx as an organ of communication, blood as a tissue, capillary morphology, osmotic changes in the loop of Henle, the tetrapod ear complex, and the amniote retina.

• An informative section on hormonal control of biorhythms has been added to Chapter 17 (Endocrine Organs) along with several new histology drawings.

FEATURES RETAINED

The many additions to content join the well-received student-oriented features of earlier editions that have been retained in the present one: the overviews and outlines that precede each chapter; numbered chapter summaries that facilitate review; a variety of end-of-chapter readings (updated); and a popular list of relevant prefixes, suffixes, roots, and stems. The latter, which is not a glossary, is unconventional and lacks formidability in the hope that it will have maximal appeal.

ACKNOWLEDGMENTS

Prefaces to earlier editions have documented my ever-increasing indebtedness to the many colleagues and students who have offered suggestions and criticism in the past. I am especially indebted to Professors Larry Brown, University of South Florida; Margaret Haag, University of Alberta; Kenneth Kardong, Washington State University; William Muchmore, University of Rochester; Thane Robinson, University of Louisville; and Keith Standing, California State University—Fresno for in-depth critiques of the previous edition and valuable suggestions for the present one. To each, I extend my most sincere thanks. I am also indebted to John Wilson for his contribution on circadian rhythms, and to Karen A. Westphal, whose professional illustrations contribute immeasurably to this edition. The Frontispiece shows comparative anatomical wood cuts by Belon (1555).

In conclusion, I gratefully acknowledge the continuing aid and encouragement of my daughter, Carolyn Rovee-Collier, and of my wife, Lila, to both of whom the original work, commissioned by the Blakiston Company of Philadelphia, was affectionately dedicated.

George C. Kent

CONTENTS

vii

APPENDIX I

APPENDIX II

INTRODUCTION

This chapter is to the rest of this book like an overture is to an extended musical work. It is a foretaste of things to come—a preparation of the mind designed to minimize surprise. The first part of the chapter is orientation to the status of vertebrates in the animal kingdom and to the basic structure of animals with backbones. The rest of the chapter provides background in capsule form for understanding the concepts that have been woven into the fabric of vertebrate anatomy since the Renaissance. The Selected Readings provide a wealth of additional input from easy-to-read articles to in-depth conceptual discussions of some of the topics.

Comparative vertebrate anatomy is the study of the structure of vertebrates (descriptive morphology), of the functional significance of structure (functional morphology), and of the variation in structure and function in geological time. Since structure is the end-product of development of the individual (ontogenesis) and of the species (phylogenesis), the discipline embraces these areas of inquiry as well. Every area of biological inquiry provides relevant data to the discipline. Ecology, genetics, molecular biology, serology, immunology, biochemistry, and paleobiology are all sources of valuable data. Geology is an indispensable source. Because of the physical limitations of a single text, the contribution of some of these areas will be mentioned only in passing. The thrust of the chapters will be the organs and systems, their roles in survival, their embryogenesis, and their historical background in geological time. The latter entails consideration of the phylogenesis of vertebrates in general.

To the extent that comparative anatomy is concerned with phylogenesis, it is a study of history. It is a study of the species that preceded us on this planet, of the effects of mutations, of adaptations, of the struggle for compatibility with an ever-changing environment, of the invasion of new territory by those best equipped for survival, and of the extinction of aging species. It is a study of what vertebrates once were like and what they are like today. It is the study of history, just as are man's conquests, political fortunes, and social evolution. The generalizations and conclusions arrived at in the discipline, like those of science as a whole, add to the enlightenment of the human mind.

THE PHYLUM CHORDATA

It is conventional to think of animals as falling into two categories—those lacking vertebral columns, or invertebrates, and animals with vertebral columns, or vertebrates. Such a dichotomy, although valid, does not recognize a group of small marine animals who are transitional between invertebrates and vertebrates—the protochordates—whom we will be studying in the next chapter. Protochordates have no vertebral column, but they share with vertebrates and with no other animals a combination of three other morphological features—a **notochord,** a **dorsal hollow central nervous system,** and a **pharynx with paired pouches and clefts in the embryo stage at least.** These characteristics are so fundamental in the architecture of vertebrates that they are among the first to appear in vertebrate embryos. Indeed, without them no vertebrate could proceed beyond the earliest stages of embryonic development. Because of the primacy of these structures in protochordates and vertebrates alike, these two groups have been incorporated into a single taxon, or classification category, the **phylum Chordata.** The taxonomic relationship of protochordates (two subphyla) and vertebrates (one subphylum) as follows*:

> Kingdom Animalia
>   Phylum Chordata
>   Subphylum Urochordata
>   Subphylum Cephalochordata
>   Subphylum Vertebrata (Craniata)

Chordates are animals that have a notochord in the embryo stage at least. Vertebrates, except jawless ones, are chordates with vertebrae. Vertebrae appear during embryonic development after the notochord has formed. Subsequently, they reinforce the notochord or replace it functionally.

Protochordates are discussed in Chapter 2, and the major taxa (categories) of vertebrates are introduced in Chapter 3. A synoptic classification of vertebrates is provided in Appendix I.

THE VERTEBRATE BODY: GENERAL PLAN

All vertebrates conform to a generalized pattern of anatomical structure. This is revealed by dissection and is the result of common ancestry. Vertebrates also exhibit similar, but not identical, patterns of embryonic development. This, too, is a result of common ancestry. Both morphology and developmental processes have been altered during the passage of time which, as it lengthens, provides increasing opportunities for genetic change. Yet, despite changes, innumerable architectural similarities still exist. These along with any mutations are examined in detail in subsequent chapters. In this chapter we discuss only a few highlights that characterize vertebrates in general.

Regional Differentiation

The vertebrate body consists of four regional components—head, trunk, postanal tail, and paired locomotor appendages. Concentrated on or in the **head** are special sense organs for monitoring the external environment;

---

*A third subphylum of protochordates, of debatable status, is discussed in Chapter 2.

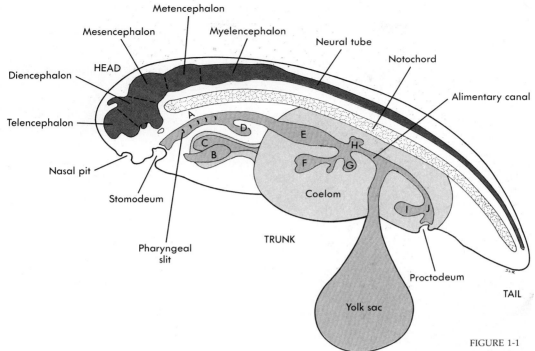

Metencephalon
Mesencephalon
Myelencephalon
Neural tube
Notochord
Diencephalon
HEAD
Telencephalon
Alimentary canal
Nasal pit
Stomodeum
Coelom
Pharyngeal slit
TRUNK
Proctodeum
TAIL
Yolk sac

FIGURE 1-1

Sagittal section of vertebrate embryo. *A,* Pharynx *(light red)* with pharyngeal slits; *B* and *C,* ventricle and atrium of heart; *D,* diverticulum that gives rise to the lung in tetrapods and swim bladder in fishes; *E,* stomach; *F,* liver bud and associated gallbladder; *G,* ventral pancreatic bud; *H,* dorsal pancreatic bud; *I,* urinary bladder of tetrapods; *J,* cloaca separated from the proctodeum by a cloacal membrane. The stomodeum is separated from the pharynx by a thin oral plate. The differentiated brain has five major subdivisions: telencephalon and diencephalon (forebrain), mesencephalon (midbrain), and metencephalon and myelencephalon (hindbrain).

jaws for acquiring, retaining, and, in some species, macerating food; and, in fishes, gills for respiration. These structures are correlated with a brain that is at least large enough to receive and process incoming information and to provide stimuli to the body musculature. Expansion of the brain over hundreds of millions of years has resulted in larger braincases that are increasingly movable independently of the trunk. **Cephalization** has developed to a greater degree in vertebrates than in any other group of animals.

The **trunk** contains the body cavity, or **coelom** (Figs. 1-1 and 1-2, *C*). Surrounding the coelom is the **body wall,** covered by skin and consisting chiefly of muscle, vertebral column, and ribs. The body wall must be opened to expose the viscera. The **neck** is a narrow extension of the trunk of reptiles, birds, and mammals, and lacks a coelom. It consists primarily of vertebrae, muscles, spinal cord, nerves, and elongated tubes—esophagus, blood vessels, lymphatics, trachea—that connect structures of the head with those of the trunk.

The **tail** commences at the anus or vent, hence is postanal. It consists almost exclusively of a caudal continuation of body wall muscles, axial skeleton, nerves, and blood vessels. Some adult vertebrates lack a postanal tail, although it is present in all embryos. The swimming larvae of frogs, toads, and wormlike amphibians (caecilians) have a tail, but it is resorbed at metamorphosis. Modern birds have reduced the tail to a nubbin, but the first

FIGURE 1-2

Typical vertebrate body in cross section. *A*, Dorsal aorta, giving off renal artery to kidney; *C*, coelom; *D*, kidney duct; *E*, epaxial muscle; *G*, future gonad (gonadal ridge); *H*, hypaxial muscle in body wall; *K*, kidney; *R*, rib projecting into a horizontal skeletogenous septum from the transverse process of a vertebra *(black)*. *1*, Dorsal root of spinal nerve; *2*, ventral root; *3*, dorsal ramus of spinal nerve; *4*, ventral ramus; *5*, parietal peritoneum; *6*, visceral peritoneum; *7*, ventral mesentery. A remnant of the notochord *(dark red)* lies within the centrum of a vertebra. The spinal cord *(light red)* lies above the centrum surrounded by a neural arch *(black)*.

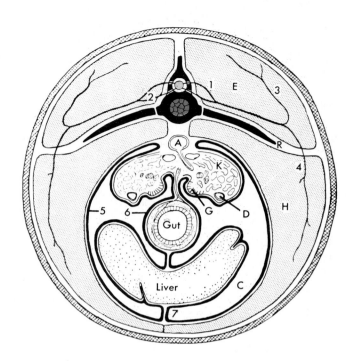

birds had long tails (Figs. 3-25 and 3-26). Human beings have a postanal tail early in embryonic life (Fig. 1-11).

Two pairs of locomotor **appendages**—pectoral and pelvic, supported by an internal skeleton and operated by contributions from the trunk musculature—are characteristic of vertebrates; but again they are sometimes vestigial or have been completely lost. The earliest known vertebrates lacked appendages such as those just described, and so do living jawless vertebrates (agnathans). The fins of fishes and the limbs of tetrapods are adaptations for locomotion in a specific environment.

Bilateral Symmetry and Anatomical Planes

Vertebrates have three principal body axes: a longitudinal (anteroposterior) axis, a dorsoventral axis, and a left-right axis. With reference to the first two, structures at one end of the axis are different from those at the other end. The left-right axis terminates in identical structures at each end. Thus the head differs from the tail and the dorsum differs from the venter, but right and left sides are mirror images of each other. An animal with this arrangement of body parts exhibits **bilateral symmetry.**

It is sometimes convenient to discuss parts of the vertebrate body with reference to three **principal anatomical planes.** Two axes define a plane. The transverse plane is established by the left-right and the dorsoventral axes. A cut in this plane is a cross section (Fig. 1-3). The frontal plane is established by the left-right and longitudinal axes. A cut in this plane is a frontal section. The sagittal plane is established by the longitudinal and dorsoventral axes. A cut in this plane is a sagittal section. Sections parallel to the sagittal plane are parasagittal. Acquainting oneself with these concepts is a simple exercise in anatomy and logic.

Metamerism

Vertebrates exhibit **metamerism,** the serial repetition of body structures in the longitudinal axis. It is clearly manifested in vertebrate embryos (Fig. 15-6) and is retained in many adult systems. No external evidence is seen because the skin is not metameric. If, however, the integument is stripped from the body of fishes, amphibians other than frogs and toads, and some reptiles, one sees a series of muscle segments that are reflections of the

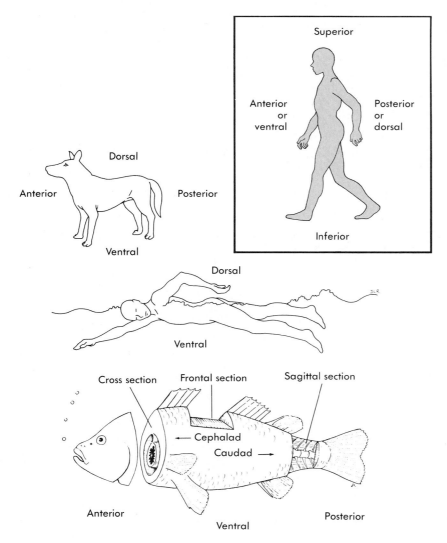

FIGURE 1-3

Terms of direction and position and planes of sectioning of the vertebrate body. Terms employed in human anatomy are shown in the box.

embryonic metamerism and are used chiefly for locomotion by lateral undulation of the vertebral column (Fig. 10-4). In addition, the serial arrangement of vertebrae, ribs, spinal nerves, embryonic kidney tubules, and segmental arteries and veins is a further expression of the fundamental metameric nature of vertebrates.

VERTEBRATE CHARACTERISTICS: THE BIG FOUR

Vertebrates exhibit a unique combination of four fundamental morphological features: (1) *a notochord in the embryo stage at least; (2) a pharynx with pouches or slits in its lateral walls in the embryo stage at least; (3) a dorsal, hollow, central nervous system; and (4) a vertebral column.* These are the "big four" vertebrate characteristics. The first three are chordate features. The fourth is unique to vertebrates. If a bizarre organism were to be discovered in the abyssal depths of the oceans, these four features alone, in combination, would admit this creature to the vertebrate category.

Other morphological details are unique among vertebrates—a large number, in fact—in every organ system of the vertebrate body. They vary from prominent (such as a skeletal braincase) to obscure (such as the presence of iodine-binding cells in the pharyngeal floor rather than elsewhere as in segmented worms, insects, or molluscs).

We will now examine the "big four" vertebrate characteristics. The notochord and vertebral column will be discussed together, since the latter occupies the same site as the former. Thereafter we will briefly consider a few of the many satellite features.

FIGURE 1-4

Transverse section of the vertebral column of a very young trout.

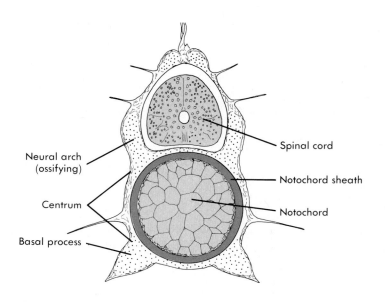

Neural arch (ossifying)

Centrum

Basal process

Spinal cord

Notochord sheath

Notochord

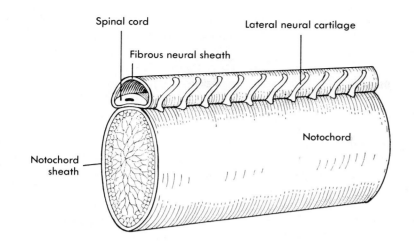

Spinal cord

Lateral neural cartilage

Fibrous neural sheath

Notochord

Notochord
sheath

FIGURE 1-5

Lateral neural cartilages of a
lamprey.

Notochord and Vertebral Column

The **notochord** is the first skeletal structure to appear in vertebrate
embryos. At its peak of embryonic development it is a rod of living cells
located immediately ventral to the central nervous system and dorsal to the
alimentary canal extending from the midbrain to the tip of the tail (Fig. 1-1).
The part of the notochord in the head becomes incorporated in the floor of
the skull, and, except in agnathans, the part in the trunk and tail becomes
surrounded by cartilaginous or bony **vertebrae.** These provide more rigid
support for the body than does a notochord alone. A typical vertebra consists of a **centrum** that is deposited around the notochord, a **neural arch** that
forms over the spinal cord, and various **processes** (Fig. 1-4). In the tail a
hemal arch may surround the caudal artery and vein (Fig. 7-1, *B, C,* and *E*).

The fate of the notochord in adult vertebrates is variable. In almost all
fishes it persists the length of the trunk and tail, although usually constricted within each centrum (Fig. 7-2). The same is true in many urodeles
and some primitive lizards (Fig. 7-3). However, in modern reptiles, birds,
and mammals the notochord is almost obliterated during development. A
vestige remains in mammals within the intervertebral discs that separate
successive centra. The vestige consists of a soft spherical mass of connective
tissue called the **pulpy nucleus** (Fig. 7-5, *D*). Modern reptiles and birds lack
even this vestige.

In agnathans the notochord grows along with the animal, and paired
lateral neural cartilages become perched on the notochord lateral to the
spinal cord (Fig. 1-5). These cartilages are reminiscent of neural arches, but
whether they are primitive vertebrae, vestigial vertebrae from an ancestor
that had a typical vertebral column, or entirely different structures is not
known.* When a notochord persists as an important part of the adult axial

---

*The backboneless, appendageless, jawless hagfishes and lampreys (agnathans) have been
admitted to the vertebrate hierarchy as a concession to logic. They are discussed in Chapter
3.

skeleton, it develops a strong outer elastic and inner fibrous **notochord sheath** (Fig. 1-4).

The notochord has been disappearing as an adult structure, but development of a notochord in every vertebrate embryo is a reminder that all vertebrates conform to a basic architectural pattern.

Pharynx

The **pharynx** is the region of the alimentary canal exhibiting pharyngeal pouches in the embryo (Fig. 1-6). The pouches may rupture to the exterior to form pharyngeal slits. These slits may remain throughout life, or they may be temporary. If they remain throughout life, the adult pharynx is the part of the alimentary canal having slits. If the slits are temporary, the adult pharynx is the part of the alimentary canal connecting the oral cavity and esophagus.

PHARYNGEAL POUCHES AND SLITS. The basic pattern of the vertebrate pharynx is expressed in all vertebrate embryos. A series of paired **pharyngeal pouches** arises as diverticula of the pharyngeal endoderm (Figs. 1-6 to 1-8). The pouches invade the pharyngeal wall and grow toward the surface of the animal. Simultaneously, an **ectodermal groove** grows toward each pharyngeal pouch (Figs. 1-7 and 1-8). Soon only a thin **branchial plate** separates the groove from the pouch. When the branchial plate ruptures, as it usually does, a passageway is formed between the pharyngeal lumen and the exterior. This embryonic passageway is a **pharyngeal slit.** The slits may be permanent or temporary.

FIGURE 1-6

Basic pharyngeal architecture as exemplified by a composite vertebrate embryo. A series of pharyngeal pouches has evaginated from the lateral walls of the digestive tract. Six aortic arches *(red)* connect the heart and ventral aorta with the dorsal aorta. (Typically, the first aortic arch disappears before the sixth one is formed.) Although a lung does not form in all vertebrates, it is an ancient structure and is represented by a swim bladder in fishes. The anterior end of the dorsal, hollow nervous system is enlarging to form the brain. The notochord commences at the level of the midbrain.

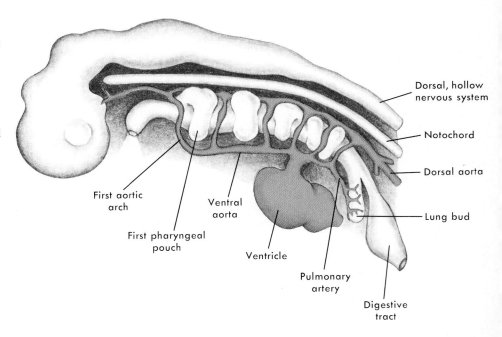

Dorsal, hollow nervous system

Notochord

Dorsal aorta

Lung bud

First aortic arch

Ventral aorta

First pharyngeal pouch

Ventricle

Pulmonary artery

Digestive tract

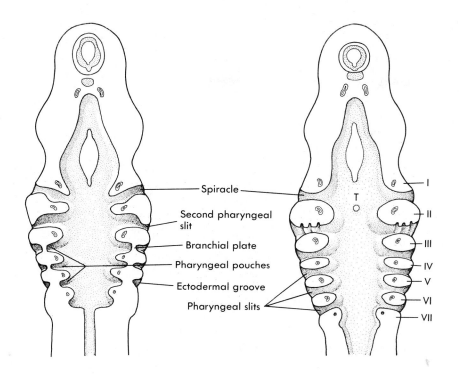

FIGURE 1-7

Pharyngeal arches (*I* to *VII*) and slits in embryonic shark, frontal section, looking down onto floor of pharynx. Early stage, *left;* later stage, *right.* *T,* Thyroid evagination.

Spiracle

Second pharyngeal slit

Branchial plate

Pharyngeal pouches

Ectodermal groove

Pharyngeal slits

I
II
III
IV
V
VI
VII

T

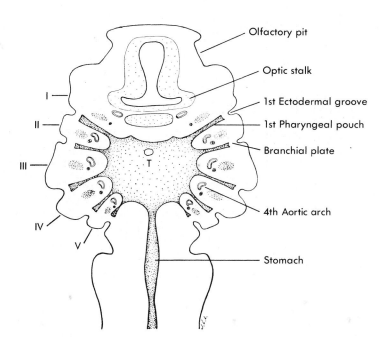

FIGURE 1-8

Frontal section of embryonic frog pharynx. *I* to *V,* Pharyngeal arches; *T,* thyroid evagination.

Olfactory pit

Optic stalk

1st Ectodermal groove

1st Pharyngeal pouch

Branchial plate

4th Aortic arch

Stomach

I
II
III
IV
V

T

FIGURE 1-9

Persistent pharyngeal slits in selected agnathans *(top)*, sharks *(center)*, and tailed amphibians *(bottom)*.

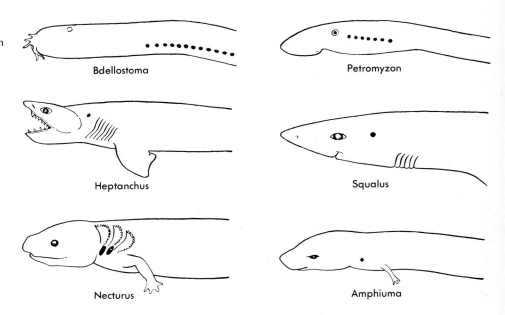

Pharyngeal slits are permanent in adults that live in water and breathe by gills (Fig. 1-9). Eight is the largest number of pouches and slits that forms in any jawed vertebrate, and this many are found only in primitive sharks. Agnathans have as many as 15 pouches and slits. Some urodeles retain one to three slits throughout life (Fig. 1-9, *Necturus* and *Amphiuma*, and Table 3-1).

Pharyngeal slits are temporary if the animal is going to live on land. Of the six pharyngeal pouches that form in frog embryos, four give rise to gill slits in tadpoles. These slits close when the tadpole metamorphoses into a frog. In reptiles, birds, and mammals no gills develop in the pouches, and the slits are transitory. Of the five pouches that develop in chicks, the first three rupture to the exterior and close again. Only one or two of the more anterior pouches of mammals may rupture. Cervical fistulas occasionally seen in human beings are usually the result of the failure of the cervical sinus (third and fourth slits) to close (Figs. 1-10 and 17-19).

Although the pharyngeal pouches of tetrapods rarely give rise to permanent slits, the first one becomes the auditory tube and middle ear cavity of tetrapods, and the second persists as the pouch of the palatine tonsil of mammals. The walls of several pouches give rise to endocrine tissue in all vertebrates.

FIGURE 1-10

Cervical fistula resulting from persistent pharyngeal slit.

PHARYNGEAL ARCHES. Each pharyngeal pouch or slit is separated from the next by a column of tissue known as a **pharyngeal arch** (Figs. 1-7, 1-8, and 1-11). Each arch, whether in an adult fish or a human embryo, typically contains four basic components or the blastemas from which these components develop. They are (1) *a supportive skeletal element*, illustrated in an adult shark in Fig. 8-1; (2) *striated muscles* that operate the arch (Fig. 10-22,

A); (3) *branches of the fifth, seventh, ninth, or tenth cranial nerves,* which innervate the muscles and provide sensory input to the central nervous system;
and (4) *an aortic arch* that connects the ventral and dorsal aortae (Fig. 1-6).
These basic components are found also in front of the first pouch and, with
some omissions, directly behind the last. Therefore a pharyngeal arch is a
column of tissue located between two successive pharyngeal pouches or
slits, in front of the first pouch or slit, and immediately behind the last. It is
covered by the integument and lined internally by endoderm. The skeleton
of the pharyngeal arches collectively constitutes a **visceral skeleton.** The
musculature is referred to as **branchiomeric muscle** because of its primitive
role as the musculature of the gills.

The upper and lower jaws and associated muscles, nerves, and vessels
constitute the first or **mandibular arch.** The second or **hyoid arch** is behind
the first pouch or slit. The remaining arches are referred to by number,
starting with the third. Arches that support gills are **branchial arches.**

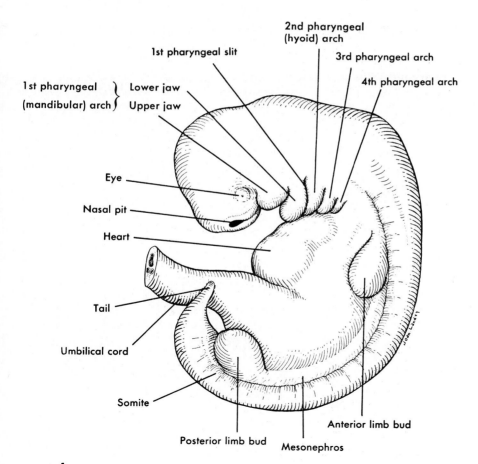

FIGURE 1-11

Human embryo approxi-
mately 4½ weeks after fertilization (5-mm stage).

1st pharyngeal slit

2nd pharyngeal
(hyoid) arch

3rd pharyngeal arch

4th pharyngeal arch

1st pharyngeal
(mandibular) arch

Lower jaw

Upper jaw

Eye

Nasal pit

Heart

Tail

Umbilical cord

Somite

Posterior limb bud

Mesonephros

Anterior limb bud

Actual crown-rump length

The boundaries of a pharyngeal arch can be determined from the exterior when there are ectodermal grooves or pharyngeal slits to serve as landmarks, and only then. If the grooves disappear or the slits close, the boundaries of the arches are lost and the components become reoriented. In most tetrapods, therefore, pharyngeal arches are anatomical entities in embryos only.

The primitive vertebrate pharynx was evidently a device for filtering food out of a respiratory water stream. The modifications that occurred in the pharynx of vertebrates who shifted from branchial to pulmonary respiration constitute one of the more fascinating chapters of vertebrate history.

Dorsal Hollow Central Nervous System

The central nervous system consists of brain and spinal cord and contains a cavity, the **neurocoel.** The dorsal location and the cavity result from the fact that the central nervous system typically arises as a longitudinal **neural groove** in what will become the dorsal surface of the embryo (Figs. 1-12 and 4-8). The groove closes over or rolls up and sinks beneath the surface to become a hollow **neural tube** located dorsal to the notochord (Figs. 1-1 and 1-6). The tube is wider anteriorly, and this part becomes the brain with its ventricles.

In cyclostomes, teleosts, and ganoid fishes the basic pattern of neurocoel formation has become slightly modified. Instead of forming a groove, the surface ectoderm dorsal to the notochord proliferates a wedge-shaped **neural keel** (Fig. 4-8, *E*). Eventually the keel separates from the surface and a cavity forms within it by rearrangement of the cells in its interior. The result is a typical dorsal hollow central nervous system.

FIGURE 1-12

Cross section of 24-hour chick embryo showing neural groove, notochord, and mesodermal somites.

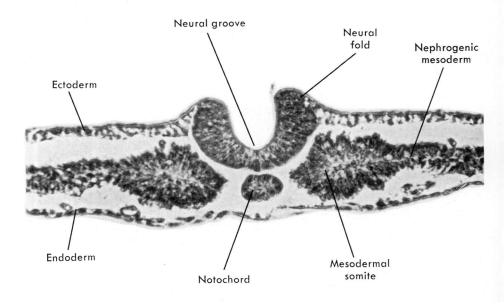

Cranial and spinal nerves connect the central nervous system with the various organs of the body. The nerves, along with associated ganglia and plexuses, constitute the **peripheral nervous system.** The spinal nerves of most vertebrates are metameric (Fig. 15-9), arising at the level of each embryonic body segment and passing to the skin and muscles derived from that segment, and to the viscera. Ten cranial nerves arise from the brain of fishes and amphibians and twelve in reptiles, birds, and mammals. The extra two are spinal nerves that have become "trapped" within the skull.

VERTEBRATE CHARACTERISTICS: SOME SATELLITE FEATURES

Vertebrates have inherited from earlier chordates many features other than the notochord, dorsal hollow nervous system, and pharyngeal pouches, including a significant number of features that are shared with some of the triploblastic (three germ-layered), deuterostomous invertebrates. (A deuterostome is a triploblastic organism in which the intitial opening from the exterior to the gut becomes the adult anus rather than the mouth.) *Vertebrates are* **triploblastic,** *and they are* **deuterostomous.** Among their heritage also were bilateral symmetry, metamerism, and a postanal tail, which have already been mentioned. Some of the other features that are accompaniments of vertebrae, although not universally, can be found in the thumbnail sketches of the coelom and organ systems that follow. Subsequent chapters provide the details.

Integument

The skin of vertebrates is unique in that it consists of **two layers,** an epidermis of ectodermal origin and an underlying dermis of mesodermal origin (Fig. 5-2). Many types of defensive, lubricatory, nutritive, pheromonal, and homeostasis-maintaining glands develop from the skin, and the skin is modified locally to form membranes such as the transparent conjunctiva of the eye and the mucous membranes of the lips. It also forms a variety of integumentary appendages, such as cornified scales, nails, feathers, and hair. The epidermis of vertebrates that live in water is different from that of vertebrates exposed to air. The epidermis of the latter protects against dehydration.

Respiratory Mechanisms

Most vertebrates carry on external respiration (exchange of respiratory gases between the animal and environment) by means of highly vascularized membranes located on the pharyngeal arches (gills), or derived from the pharyngeal floor (Fig. 1-6, lungs). In some species respiration takes place through the skin, the buccopharyngeal lining, and, in embryos, through special extraembryonic membranes that lie just inside the eggshell or in contact with the lining of the mother's uterus.

FIGURE 1-13

Chief subdivisions of the
coelom of vertebrates.

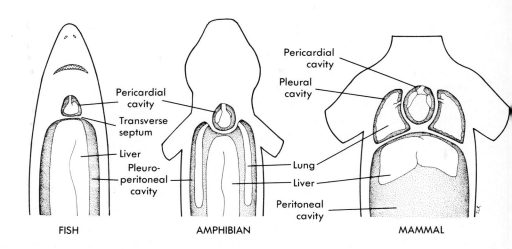

FISH   AMPHIBIAN   MAMMAL

Coelom

Like many invertebrates, vertebrates are built like a "tube within a tube," having in the trunk a body cavity, or coelom, between the body wall (the outer tube) and the digestive tube (the inner tube). The coelom is partitioned in fishes, amphibians, and some reptiles into a **pericardial cavity** that houses the heart and a **pleuroperitoneal cavity** that houses most of the other viscera, including the lungs (Fig. 1-13). The pericardial and pleuroperitoneal cavities in these vertebrates are separated by a fibrous **transverse septum.** In other reptiles, as well as in birds and mammals, each lung occupies a separate **pleural cavity** that is partitioned off from the rest of the coelom. The partitions—a tendinous oblique septum in reptiles and birds, a tendinomuscular diaphragm in mammals—arise as membranous folds of the dorsal and lateral parietal peritoneum (see below) and, in developing, incorporate the septum transversum in their structure. In many male mammals caudal outpocketings of the coelom house the testes, and these **scrotal cavities** are a fourth subdivision of the coelom.

The coelom is totally enclosed by a **peritoneal membrane.** The part of the membrane lining the body wall is the **parietal peritoneum** (Fig. 1-2, 5). The part lying on the viscera is the **visceral peritoneum** (Fig. 1-2, 6). The parietal and visceral peritonea are continuous via the dorsal mesenteries and, when present, via the ventral mesenteries. The few visceral organs that do not develop mesenteries—kidneys, for example—lie against the dorsal body wall just external to the peritoneum and are therefore retroperitoneal.

Digestive Organs

The digestive tract has specialized regions for the acquisition, processing, temporary storage, digestion, and absorption of food and for the elimination of the unabsorbed residue. Typical are the oral cavity, pharynx, esophagus (which is as long as the neck), stomach, and intestine. The last is often coiled, which increases the absorptive area without increasing the body length. The tract usually has a number of **ceca,** or diverticula, including a **liver** and **pancreas.**

The digestive tract in all but a few vertebrates terminates in a **cloaca,** a common chamber that also receives the urinary and reproductive ducts and opens to the exterior via the **vent.** All vertebrate embryos develop a cloaca, but in modern fishes it becomes so shallow as to be nonexistent, and in mammals other than monotremes it becomes subdivided into two or three passageways, each with its own exit to the outside (Fig. 14-42, *B* and *C*). The exit from the intestine is then the **anus.**

Urogenital Organs

Kidneys and gonads arise close together in the roof of the coelom (Fig. 1-2, *G* and *K*), and the two systems share certain passageways. Kidneys (nephroi) are the chief organs for elimination of water in those species in which this is necessary. (It is not necessary in marine or desert animals.) They also assist in maintaining an appropriate electrolyte balance. In the most primitive vertebrates, fluid and certain wastes were removed from the coelom by microscopic kidney tubules resembling somewhat the nephridia of earthworms. In most vertebrates, however, the substances to be excreted are collected directly from the blood by kidney tubules. The tubules transmit the substances to a pair of longitudinal ducts that empty into the cloaca, the urinary bladder, or to the exterior.

Reproductive organs include gonads, ducts, accessory glands, storage chambers, and copulatory mechanisms. Early in development all vertebrate embryos are bisexual, having gonadal and duct primordia for both sexes. If the animal is genetically constituted to become a female, the gonad primordia develop into ovaries, and only the female ducts differentiate. If the animal is to become a male, the gonad primordia become testes, and the male ducts differentiate. The duct system associated with the opposite sex largely disappears. Cyclostomes lack reproductive ducts. Their sperm and eggs are shed into the coelom and exit via a urogenital papilla located immediately behind the anus.

Circulatory System

The circulatory system pervades every cubic millimeter of the body of the smallest and largest vertebrate, a condition necessitated by logistics. Whole blood, consisting of plasma and formed elements—chiefly red and white blood cells and platelets—is confined to arteries, veins, capillaries, and sinusoids. Sinusoids are broad channels rather than tubes. They are a prominent feature of many invertebrate circulatory systems and are more common in fishes than in tetrapods. Vertebrates also have a system of lymph vessels that collect some of the interstitial tissue fluids and conduct them to large veins (Fig. 13-39).

The heart, which forms immediately ventral to the embryonic pharynx, remains in that location, close to the gills, in fishes. In tetrapods it is displaced somewhat caudad during subsequent development. The heart pumps blood forward into a **ventral aorta,** which is foreshortened in tetrapods, through **aortic arches,** and into the **dorsal aorta,** in which the blood courses caudad. Branches from the anteriormost aortic arches carry blood to the head (Fig. 1-6).

Fishes have a **single circuit heart.** Blood passes from the heart to the gills, to the body tissues, and then back to the heart to be recirculated. The advent of lungs resulted, after transitional stages, in **two-circuit hearts,** in which oxygenated blood is pumped to all parts of the body, returns to the heart, and is then pumped to the lungs, from which it again returns to the heart. Oxygen-rich blood is kept separate from oxygen-depleted blood within the heart by partitions in the heart chambers. Intermediate stages in the separation of oxygenated and unoxygenated blood are documented in Chapter 13.

Skeleton

Cartilage, bones, and ligaments constitute a jointed framework of rigid components that gives the body its shape, protects vital organs, and provides a site for attachment of locomotor and other muscles (Figs. 7-23 and 9-33). It consists of a longitudinal **axial skeleton,** principally cranium and vertebral column; a **pharyngeal skeleton,** best developed in fishes, in which it supports the gills; and an **appendicular skeleton.** As in invertebrates, mineralization of ordinary connective tissues gives skeletal components their rigidity.

Muscles

The axial skeleton is moved principally by trunk and tail muscles that are primitively metameric (Fig. 10-4). The appendicular skeleton is operated principally by buds from the trunk muscles (Fig. 10-16). The pharyngeal skeleton is operated by muscles of the pharyngeal arches (branchiomeric muscles). All of the foregoing muscles appear **striated** when viewed by light microscopy (Fig. 10-2). **Cardiac muscle**—the musculature of the heart—is a special variety of striated muscle tissue. Nonstriated or **smooth muscles** are found principally in the walls of hollow viscera other than the heart, and in the walls of tubes and vessels. Striated muscles other than cardiac can be operated voluntarily to the extent that higher brain centers are available for providing stimuli. Cardiac and smooth muscles are under reflex control exclusively.

Sense Organs

Vertebrates have a wider variety of sense organs, or receptors, than any other animals. These monitor the constantly changing external (ambient) and internal environments. Sense organs that monitor the external environment include **mechano-, chemo-, electro-,** and **thermoreceptors,** and **receptors for radiation** in the visible and infrared spectra. With the exception of those for radiation, these receptors are distributed widely on the head, trunk, and tail of some fishes, mechanoreceptors being ubiquitous. In tetrapods other than aquatic amphibians and snakes, receptors other than those for touch, pain, heat, and cold are confined to the head. **Proprioceptors** monitor the activity of the muscles, joints, and tendons, and **visceral receptors** monitor the rest of the internal environment. By "keeping in touch," the central nervous system is able to regulate skeletal muscle activity and visceral functions in a manner conducive to survival.

CONCEPTS RELEVANT TO MODERN VERTEBRATE MORPHOLOGY

As stated earlier, one aspect of the discipline of comparative vertebrate anatomy is the study of the structure of living and extinct species. The former reveals similarities and differences among the animals of today. The latter reveals what vertebrates have been like in the past. Assembling the data in a geological time frame reveals the changes that have taken place from the distant past to the immediate present, thereby providing the human species—the only species competent to comprehend—with a panorama of the events in the biotic world that preceded us. That panorama has led to the formulation of certain concepts, which we will now examine briefly. Their full import may not be apparent this early in the course; insight will grow as the terms are used in relevant contexts. Some, if not all, may be recalled from previous studies in the biological sciences.

Phylogeny and Ontogeny

Phylogeny is the evolutionary history of a taxon. It relates the taxon to ancestral taxa in a continuous evolutionary line. **Phylogenesis** is the process that produces the lineage. The two terms are sometimes used as synonyms. The operant in phylogenesis is a speciation, the formation of new species from existing ones.

Ontogeny is the history of an individual. It commences with the initiation of embryonic development, includes postembryonic changes associated with aging, and terminates at death. **Ontogenesis** is the process of development (embryogenesis) and of subsequent changes. The primary operants in ontogenesis are the genes. Phylogenesis requires tens of thousands of years; ontogenesis occupies a single lifetime. Change is the universal phenomenon that characterizes both processes.

In the chapters to come, the morphology—shape, form, structure—of organisms will be correlated with ontogeny and, to the extent this is enabled by current knowledge, with phylogeny.

VON BAER'S LAW. All vertebrates adhere to a similar architectural pattern both of structure and of development. This observation is incorporated in a generalization formulated in 1828 by the embryologist Karl Ernst von Baer (1792-1876), which states that *features common to all members of a major group* of animals—the subphylum Vertebrata, for example—*develop earlier in ontogeny than do the special features that distinguish the subdivisions of the* *group* (classes, orders, genera, species). This generalization has become known as **von Baer's law.** We have seen an example in the very early development of the notochord, the dorsal hollow nervous system, pharyngeal pouches, and aortic arches in all members of the subphylum. These early features become gradually modified in the direction of the species as development progresses.

What amounts to a corollary of von Baer's law has evolved from the ideas of von Baer and later biologists following the increasing acceptance of the concept of organic evolution. As it stands today, the corollary is, *features that develop earliest in ontogeny are the oldest phylogenetically,* having been inherited from very early common ancestors; *and features that develop later in ontogeny are of more recent phylogenetic origin.* This concept is sometimes referred to as the **biogenetic law.**

Taxonomy and Systematics

Taxonomy is the study of the principles and procedures that are employed in placing organisms in appropriate taxa in accordance with one or more predetermined criteria, *and of the techniques of ordering the taxa into hierarchies that reflect the criteria in a meaningful way.* These criteria principally are phylogenetic, which takes into account the genealogy of the group, and morphological, which takes into account the similarities and differences in phenotypes (morphological types resulting from the visible expression of specific gene complexes). The contribution of the taxonomist is, therefore, a **classification,** *an orderly arrangement of taxa* that, hopefully, reveals the intended interrelationships in the simplest and most informative manner possible, and with the greatest possible accuracy. Taxonomy and the study of the evolutionary processes that give rise to new taxa is known as **systematics.**

Conventional classification schemes currently in use are predicated on a mixture of similarity of structure, on evolutionary history, and on intuition, the latter being the last resort after all other data bases have been exhausted. An example of a scheme arrived at by this taxonomic procedure is the classification of animals with backbones that commences on p. 632. Taxonomists who utilize both morphology and genealogy, plus relevant scientific data from other disciplines (e.g., molecular biology, comparative embryology) after evaluating it intuitively—are **evolutionary** or **natural order taxonomists.** They represent one of three groups of present-day systematists.

A second group of taxonomists disregards genealogy and considers phenotypes (morphology) alone. Organismal groups are classified in accordance with the total number of shared phenotypic traits. The larger the number of shared traits, the more closely related any two categories are considered to be. This is considered a reasonable criterion for classification by proponents of the procedure, who are referred to as **numerical taxonomists** or **pheneticists,** the word being derived from "phenotype." A phenetic classification attempts to mirror only the degree of morphological similarity of the taxons.

A third group of taxonomists are known as **phylogenetic taxonomists** or

cladists. Cladistic systematics is based solely on genealogical relationships, and does not take into consideration the diverging morphology that accompanies the evolution of new taxonomic groups. Such a classification might place in the same category organisms with widely divergent morphological features. For example, cladistic systematists would remove birds from the class Aves to a class Archosauria along with crocodilians (currently in the reptilian order Archosauria) in recognition of the generally accepted conclusion that birds and crocodilians came from a common archosaurian ancestor. Such a classification would clearly reveal the genealogy of the two groups.

The conventional classification commencing on p. 632 is an attempt to relate the hierarchies phylogenetically and morphologically. It has been recognized for years by taxonomists that natural order classifications are only partially successful, and that the procedure is not unbiased because the final phase of construction of natural classifications involves intuitive decision making, and for a good reason. Once the taxonomist has employed all available scientific data and still cannot assign an organism or a group of organisms to one specific category in one specific hierarchical arrangement with the evidence at hand—which frequently has been the case—the final recourse is, of necessity, intuition. Natural order classification is a compromise between phenetics and cladistics, although it is the oldest of the three. Proponents of natural order systematics maintain that intuitive judgment cannot be wholly dispensed with in the foreseeable future, even in cladistics or phenetics, due to lack of sufficient detailed data for arriving at the equivalent of totally computerized taxonomic decisions. Cladists and pheneticists dispute this. The recently evolved struggle for accuracy in classification is evidence of a new vitality in the discipline of taxonomy. The result will be a purer science of systematics.

FAMILY TREES. Genealogical or "family" trees are regularly employed to present visually the ancestry of a taxon or a group of taxa. These visual aids provide certain data, but not others, depending on the pedagogical objective of the illustration. The family trees in Chapter 3 are designed to indicate major or otherwise significant taxa and their postulated ancestry. They do not show their distribution in time (hence which ones were contemporaries), the size of the population, or the relative number of species. Illustrations similar to Fig. 3-2 in this book show the distribution and relative abundance of the selected taxa in time, hence which of the major taxa represented were contemporaries. Space limitations restrict the extent to which they can indicate ancestry. In using such study aids, students should be aware of the purpose and limitations of the presentation. For a discussion of the topology of family trees the work of H.M. Savage (1983) may be consulted.

Homology

The concept of homology evolved slowly but persistently in anatomical circles during several centuries before evolutionary theory began to crystallize. As formulated during the seventeenth century, a homologue is *"the same organ in different animals under every variety of form and function"* (Boyden, 1943). Thus the stapes in the middle ear of mammals and the hyomandibular cartilage that suspends the lower jaw from the braincase of sharks are **homologous structures (homologues).** They are the same structure in different animals, despite the fact that they are not only dissimilar in appearance, but they also have different functions. The precaval vein of cats and the right common cardinal vein of lower vertebrates are also homologous, as are the intermaxillary bone of the human embryo (not an independent bone in adults) and the premaxillary bone of adult apes. The humerus of a crocodile, bird, and mammal are homologues with the same appearance, in the same location, and with the same function.

Since homologues in diverging organisms may diverge structurally and functionally as speciation (formation of new species) proceeds, a time may come when they no longer resemble one another in appearance or in their biological roles. How, then, can one be sure two organs in two different species are the same organ? If we are confronted with this problem with reference to structures in living species, comparative embryology can usually provide the answer. *Structures in two different species are homologous if they come from the same embryonic precursor* (anlage, blastema, fundament).* (This is equivalent to saying they are homologous if they have a common ancestry.) However, if confronted with an extinct and a living form, one has no access to embryological tissues from the extinct form. In such a case, the ideal procedure would be to follow the structure through the phylogenetic lines that link the species, if such is possible. If the structure is skeletal, and there are fossils of intermediate species, this may be fruitful; if it is soft tissue, it is unlikely to have been preserved in the fossil. In the latter situation, one must turn to whatever information is available from relicts, if any (last survivors of a related species facing extinction), and from impressions left in fossilized tissues, such as grooves or other impressions in bones, the location of foramina, and other tell-tale markings in the fossil, while recognizing the shortcomings of the process. The instances of homology cited earlier, wherein the structures had changed in appearance, location, and function, were not difficult to identify once embryos of suitable age were available. The evidence in those instances is indisputable.

The problem of deciding *which structures, though similar, are not the same phylogenetic structures* (that is, are **homoplasous** rather than homologous) sometimes presents itself, particularly in instances of evolutionary convergence and parallelism, to be discussed shortly. As a result, numerous alternative definitions of homology that attempt to better define the parameters of the concept have been debated, but none has yet been universally adopted. The concept itself, however, is not in dispute. De Beer (1971) presented an easily read analysis of the problems from the viewpoint of an embryologist. More recently, Wiley (1981) has written a treatise on the subject, and

---

*When this approach is inconclusive, as sometimes happens, other sources of data are useful (Chapter 10, Names and Homologies of Skeletal Muscles).

Northcutt (1984) has briefly reviewed some of the proposed definitions. Recognizing homologous structures, particularly those that have acquired new functions with time, adds to our understanding and appreciation of the awesome phenomenon of organic change that has been taking place on planet earth.

SERIAL HOMOLOGY. Since metamerism is characterized by repetition of paired structures in the long axis of the body, it should be pointed out that serial structures—spinal nerves, vertebrae, muscle segments, for instance—are homologous *as a series* with the same series in another species, but the *individual components of the series are not homologues.* For example, a spinal nerve that supplies a muscle of the anterior fin is not homologous with a spinal nerve that supplies a muscle of the posterior fin—not even in the same animal. They do not come from the same embryonic metamere. The term **serial homology,** which is used occasionally in referring to metameric structures of vertebrates and invertebrates, uses the term homology in a different sense than it has been used here. It has little relevance to phylogenetic studies.

A related problem arises in comparing the digits of two different tetrapods. *Collectively*, the digits of the manus (hand) of a frog, bird, and horse are homologous, but the first, second, or last digit of one species, if there are more than one digit, may not be homologous with the first, second, or last digit of the other. The homology of specific digits of different taxa of tetrapods has been studied extensively, and with increasing success. The single digit of the anterior limb of modern horses, for example, appears to be the middle digit of generalized vertebrates and of ancestral horses. The question ''Which finger is missing?'' is posed with reference to a tailed amphibian in Fig. 9-25.

ANALOGY. The term **analogy,** which refers to coincidental or chance resemblances, is probably familiar since it has application in many disciplines. In zoological usage, *two structures that have the same function but are not homologous are analogous.* Thus the jaw teeth of bony fishes and the horny ectodermal ''teeth'' of the jawless lamprey eels are analogous, as are the horns of cattle and those of rhinoceros. So, too, are the thymus gland and the bursa of Fabricius of birds (p. 626). The terms, although informative, are employed infrequently in current evolutionary literature.

Convergent and Parallel Evolution

The term **convergent evolution** *is applied when two or more unrelated species occupying the same kind of environment, concurrently or millions of years apart, develop a similar adaptive morphological trait.* The development of a streamlined sharklike body with dorsal fins, a bilobed tail, and paddlelike anterior limbs by ichthyosaurs, which are aquatic reptiles (Fig. 9-30), and dolphins, which are aquatic mammals (Fig. 3-39) are instances of convergence. So, too, are the development of webbed feet in frogs, ducks, and the platypus, and the development of wings in pterosaurs (Fig. 3-21), birds, and bats. In the latter example the *appendages* are homologous, their *functions* are analogous, and their *similar adaptations* are a result of convergent evolution.

Parallel evolution *denotes the development of similar structures in recently related but isolated taxa whose common ancestor lacked such traits.* Australian marsupials resemble placental mammals such as wolves, foxes, bears, and cats elsewhere in the world, often to the extent of being hardly distinguishable; yet these phenotypes in the two groups are not inherited from a common ancestor because the traits evidently appeared independently in the two groups after their isolation. This is an example of parallel evolution. The cause is conjectural.

For evolution to be parallel, there must be no opportunity for new mutations, as they arise, to be transmitted from one group to the other. The groups must be isolated in such a manner as to preclude interbreeding. Australian marsupials became geographically isolated from placental mammals when Australia separated from Gondwanaland (p. 91). It was this specific instance that inspired the term "parallel" evolution.

Note that in parallel evolution the taxa diverged from a recent common ancestor and then independently evolved certain similar traits that were not present in the ancestor; and that in convergent evolution the taxa are unrelated, or only remotely related, and approach one another in superficial resemblances as a result of clearly adaptive mutations.

Adaptation

The infinitive "to adapt," like the word analogy, should be familiar, since it is a common concept. Adaptations, whether a modification of a man-made instrument that enables its use in a new, specific context, or a modification having the same effect on a living organism, are analogous phenomena. *A biological adaptation is a hereditary modification of a phenotype that increases the probability of survival.* Since the phenotype is an expression of the genotype (the array of genes), adaptations are the result of genetic mutations. The streamlined fishlike body of an ichthyosaur (Fig. 9-30) assisted in adapting the ichthyosaur to life in the sea. It was, therefore, an adaptation. The thick, water-impervious skin of reptiles and mammals is an adaptation that inhibits water loss (dehydration) in animals on land. The vertebrate body, like the body politic, is a complex of adaptations—old and, sometimes, expendable, or new and vital. They enable the individual, hence the species, to survive in its new environment—always, however, at risk. The risk factors are part of the environment.

THE ROLE OF NATURAL SELECTION. Natural selection plays a significant role—although perhaps not the sole role (this is a matter of dispute)—in sorting out and perpetuating phenotypes, hence in propagating adaptations. Expressed in simple terms, the modern concept of natural selection is as follows: chance mutations and genetic recombinations result in variations in phenotypes, some of which may render a population of organisms better able than other populations to cope with the environment of which the populations themselves are a part. Therefore phenotypes with a larger number of elements of adaptiveness are more likely than others to be successful (that is, to maintain or increase their rate of reproduction generation after generation) in that specific environment. The result is a differential in survival rates of those phenotypic traits that confer an advantage. This is natural selection. It leads to speciation whenever the gene pool containing the information for the adapting trait is isolated from other gene pools so there is no further interbreeding. A modern biologist has defined natural selection as *the nonrandom differential reproduction of genotypes as a result of interaction of phenotypes with selective forces in the external environment.*

There is no scientific evidence that phenotypic changes take place because of need. This concept is known as **teleology**—*the use of design, purpose, or utility to explain a natural phenomenon.* It necessitates an **entelechy** (loosely, a supermechanical agency or "intelligence") embodied in, or guiding, natural phenomena. Populations that are locked into a changing environment and need some genetic mutation in order to remain successful and do not acquire it become extinct. Science attributes adaptations to fortuitous genetic changes that are induced by the natural laws of physics and chemistry.

Speciation

The formation of new species is preceded in almost every instance, if not invariably, by geographical isolation of a population from other populations of the same species. This could occur, for instance, in a migrant species in which some of the migrants become isolated from the others for one reason or another. This appears to be what happened to the landlocked freshwater lamprey with a marine name, *Petromyzon marinus dorsatus,* which lives in the Great Lakes of North America. Their very close relatives, *Petromyzon marinus marinus,* are marine lampreys that migrate into freshwater streams to spawn, then return to the sea until sexual maturity. There are several possible explanations—all conjectural—for the isolation of these two groups, but these are irrelevant in the present discussion. The important point is that further exchange of genes between the two groups has become impossible in the present circumstances. Consequently chance mutations, recombinations of genes, and the random genetic drift (irregular change in gene frequencies) that characterize all genetic complexes will result ultimately in genetic incompatibility between the two. Thereafter, if brought together again they would no longer be able to produce viable offspring. *Geographical isolation coupled with genetic change is the matrix out of which new species evolve.* Geographical isolation provides the time necessary for the species to drift apart genetically, since speciation within an interbreeding population appears not to be possible. *Natural selection channelizes the evolving product in the direction of increasing adaptiveness.* The length of time required for speciation depends on many factors, but the eventual emergence of new species, given the foregoing conditions, seems inevitable. The result of continuous speciation is phylogenesis, the formation of new taxa. Phylogenesis is the manifest effect of evolution.

Organic Evolution

The concept of organic evolution is a single, simple, easy-to-understand, unequivocal, yet widely misconstrued premise that is: *The plants and animals on earth have been changing,* and *the ones around us today are descendants of those that were here earlier.*

The conclusion that the animals and plants have been changing is based in part on geological evidence. It indicates that, if a human being from today could be transported backward in time several hundred million years, the plants and animals he would see would be exotic, alien, and unfamiliar. Five hundred million years ago he would see mostly water, and on the land there would be no land plants. If he happened to bring a fishing pole (!) he would probably catch no fish because the fish would be ostracoderms, and they were filter feeders. Four hundred million years ago fishing would have improved, but there would be no trout, perch, or salmon. The land would be higher and drier, there would be mosses and other simple land plants, and labyrinthodonts would be lumbering in and out of swampy waters, but there would be no frogs or toads. Three hundred million years ago the land would be still higher, and the swamps would be forested with seed ferns and conifers but no flowering plants. Cotylosaurs would be basking in the sun, but there would be no lizards or snakes. One hundred fifty million years ago the birds would have teeth, dinosaurs would be large and specialized, and hairy little animals with pouches to carry their babies in would be scurrying around in the forest, staying out of the way, but there would be no cats, rats, or monkeys. The traveler might be happy to return to the age of the mammals, if only because he is home, and home is familiar. It is in these geological findings that the theory of organic evolution has its roots, but by no means all of them.

The second part of the premise, that the animals and plants around us today are offspring of the animals of yesterday, is axiomatic. Life comes from preexisting life.

This is the theory of organic evolution, stripped of all satellite theories. It doesn't say that multicellular organisms came from protozoa; it doesn't say that man came from a monkey. It doesn't say *where* man came from—that is another theory, or several other theories. Any one of them could be called *the theory of where man came from,* but it is not the theory of organic evolution. Neither does the theory say what caused, or what causes, species to change. That, too, is another theory or several theories. The theory does not state how life began or how the universe began. There are theories about beginnings, but they are not the theory of organic evolution. These satellite or ancillary theories belong to scientific disciplines beyond the intended scope of this book. Finally, it is not a theory about a Supreme Intelligence. Science can neither affirm nor deny the existence of a Supreme Intelligence because it lacks the tools needed to collect the data. And without data science cannot conclude.

The theory of organic evolution might also be called the **theory of the mutability of species.** There is only one alternative, the **theory of the immutability of species.** Proponents of the latter theory must insist that every species on earth today is precisely like it was when it first appeared, and that the first member of the species appeared de novo, coming from no previous living organism. To consider the possibility of even a single change ("fishes appeared and from these came all kinds of fishes") is to abandon the theory of the immutability of species in favor of organic evolution. If it is accepted that *one* change can occur, it must be accepted that *two* changes can occur. And once this step in logic has been taken, there is no limit to the number of changes that can occur.

The concept of evolution appeared in early Greek writings, including those of Aristotle. However, under the restraints of scholasticism, the philosophy and theology of the Middle Ages of Western civilization, the concept was not nurtured, and it remained dormant until the Renaissance when inquiry into natural phenomenon revived. **Jean Baptiste de Lamarck** (1744-1829), a French naturalist, was the first of the Renaissance naturalists to publicly espouse the idea that species are not immutable and that complex species evolved from simpler ones. In an attempt to account for evolution, he propounded the **Doctrine of Acquired Characteristics** as the operative force. It postulated that characteristics that had been acquired through the use or disuse of morphological structures were inherited, thus becoming part of the legacy of succeeding generations. Thus when a part is employed in successive generations, it becomes stronger and better adapted for a specific role. Conversely, when a part is neglected in successive generations, it tends to become vestigial. This is how Lamarckism would account for the fact that the olfactory nerves of many aquatic mammals such as whales commence development in embryonic life, regress, and become vestigial. A lung-breathing animal could not inhale under water without drowning. Therefore, according to the theory, the lack of use of the olfactory apparatus resulted in its disappearance. Vampire bats have an esophagus so small that they cannot swallow anything other than liquids (Chapter 3). According to the Doctrine of Acquired Characteristics, the esophagus would have become too small to swallow solids because the only nourishment that passed through it for generations was mammalian blood. The doctrine as proposed is not acceptable at present because the current state of our knowledge provides no explanation as to how use or disuse of a part in individuals can be translated into alterations of the hereditary code stored in the sperm and eggs, which are set aside as germ plasm for the next generation early in embryonic life. The advent of modern genetics made this aspect of Lamarckism, at least for the present, untenable.

Charles Darwin (1809-1882), born 65 years after Lamarck, had a different explanation to account for evolutionary change. Darwin observed that variation in individual organisms is universal; no two individuals are identical. This is the starting point for Darwin's **Theory of Natural Selection** propounded in 1859. Individual variations, Darwin believed, result in varying degrees of success in competition between individuals and with the physical conditions of life. The resultant "preservation of favorable variations and the rejection of injurious variations I call Natural Selection," said Darwin, who paraphrased the concept as the "struggle for existence" in which the fittest survive. This is another way of stating that those best adapted are most likely to pass their characteristics on to another generation. *The concept of natural selection is Darwin's most important contribution to evolutionary theory.* It has, of course, been refined by subsequent studies. Darwin would be excited by the modern definition of natural selection given earlier.

The first paragraph of this section cited the *premise* of evolutionary theory. With the advent of modern genetics the *mechanics* of the process of evolution gradually unfolded, enabling speciation and evolution to be explained in a manner consistent with this new knowledge. In 1942 Julian Huxley coined the term "The Modern Synthesis" to refer to the body of knowledge that now exists in evolutionary theory. It is a synthesis of the basic premise set forth in the first paragraph of this section, natural selection, current genetic insight, and the findings of the disciplines of the life sciences—comparative morphology of plants and animals, paleontology, systematics, biochemistry, molecular biology, and others. It is a *natural,* rather than a supernatural, *explanation* for the variety and number of species on this planet, an estimated minimum of five million (Myers, 1985), just as the Law of Gravity is a natural explanation for the attraction between the earth and the moon. It does not answer the question of ultimate causes.

Paedogenesis, Neoteny, Paedomorphosis, and Heterochrony

In Chapter 2 the term **paedogenesis** will be employed to designate a phenomenon in larvaceans, a group of protochordates, in whom *the gonads become sexually mature and reproduction takes place in what is very clearly a larval organism.* Presumably, some time back in phylogeny the developmental rate of the gonads became accelerated as contrasted with the maturation rate of the other organs, as a result of which larvacean species reproduce and end their life cycle before they have metamorphosed. Alternatively, the rate of development of the other organs fell behind that of the gonads, resulting in reproduction taking place during the larval state. Reproduction in what are obviously larval species was given the name paedogenesis by von Baer in 1866, and the word became part of literature in the context described.

In Chapter 3 we look at an analogous condition in some tailed amphibians. These urodeles, of which *Necturus* is the best known example, are aquatic vertebrates that retain several larval gills and gill slits throughout life, employing them as primary or accessory respiratory organs. In these perennibranchiate tetrapods there has been a retardation of resorption of the gills; and it looks like an adaptation to permanently aquatic life. That it is *retardation of a somatic feature,* and *not precocious development of the gonads* is evidenced by the fact that some perennibranchiate amphibians that reproduce without resorbing their gills can be made to do so as well as to close their gill slits by the administration of thyroid hormone. These species, then, should not be considered sexually precocious, a term sometimes applied to them. The term **neoteny,** introduced in 1882, *refers to a condition in which adults retain certain larval traits.*

Paedogenesis and neoteny have tended to become synonyms, and this is reflected in some dictionaries. The term **paedomorphosis** has been preferred by some zoologists to include both conditions. Paedomorphosis may be defined as *the phenomenon in which larval or immature features of ancestors become adult characteristics of descendants.* All three terms—paedogenesis, neoteny, and paedomorphosis—refer to manifestations of a single phenomenon known as **heterochrony:** *the change in relative rates of development of characters during phylogeny.* Heterochrony is one way of bringing about evolutionary changes. The phenomenon has been the basis for a minority theory of the phylogenetic origin of cephalochordates. It will be discussed in Chapter 2.

ANATOMY FROM GALEN TO DARWIN

Cavemen had some knowledge of the internal organs of mammals, as did the Babylonians and ancient Egyptians, who practiced surgery and embalming. Embalming, which involved removal of most of the internal organs, was an advanced art. Aside from a small number of Egyptian medical papyri dating to about 3000 BC, the oldest anatomical works were written during the last 400 years BC by Greek philosophers and physicians. These works were incomplete, mostly superficial, and often imaginative. Anatomy of the classical era culminated in the works of Galen, a Greek philosopher-physician who practiced in Rome between AD 165 and 200. He assembled all available Greek anatomical writings, supplemented them with his own dissections of apes from the Barbary Coast (human dissection was at that time prevented by public opinion and superstition), and, in addition, wrote more than 100 treatises on medicine and human anatomy. Shortly thereafter, scholasticism took over, and during the next 1300 years Galen's descriptions were considered infallible and dissent was punishable. Subservience to authority became so accepted that (as has been said, probably partly in jest) if a scholar wanted to know how many teeth horses have, he saddled up and rode 100 miles, if necessary, to the nearest library to see what Galen said. (No doubt the peasants looked into the horse's mouth—an application of the scientific method.)

It was not until the fifteenth century that Leonardo da Vinci (1452-1519) and other Italian artists began to make anatomical observations of their own (at the peril of excommunication from the church), and a renaissance in human and animal anatomy began. In 1533 a young Flemish medical student named Vesalius at the University of Paris attended classes where Galen's works were read (by a "reader") while the professor tried, often with embarrassing lack of success, to harmonize Galen's descriptions with the dissection. (Recall that many of Galen's descriptions of man were written from dissections of apes.) After 3 years Vesalius quit Paris, earned a degree at Padua, stayed on to teach, and recorded his own anatomical observations. In 1543 he published *De humani corporis fabrica* (On the Structure of the Human Body), ushering in a renewed interest in anatomy of animals in general. His *Fabrica,* however, was influenced to a large degree by Galen, and many of the errors of Galen were perpetuated. Vesalius's figure of the hyoid and kidney were actually taken from a dog, which was his favorite subject of dissection. However, he was not interested in mammals other than man beyond the extent to which they might contribute to his knowledge of major organs of the human body. His interest was in descriptive human anatomy, not in species differences.

Eight years after publication of Vesalius's anatomy, Pierre Belon (1517-1564), a French naturalist and physician, published an important work on the dissection of cetaceans (dolphins, porpoises, and whales) and other marine animals, invertebrate and vertebrate, including fishes. He was also interested in birds, saying he never missed an opportunity to dissect one and had examined the internal parts of 200 different species. In 1555 he published what has become a classic illustration of a human and bird skeleton side by side, showing that their skeletons correspond almost bone for bone (see frontispiece). The bones were identified with names from human anatomy. He published the illustration because he had made the observations and thought the general public would be enlightened. The similarity of the structures in the two species, like the convergent evolutionary similarities that he observed between cetaceans and fishes, he attributed to the manifestation of a basic architectural plan, or *architype,* in the mind of the Creator. Because cetaceans lived in the sea, they were, to Belon, *fishes,* despite the fact that his studies demonstrated they were mammalian in all their basic features. Like other public figures of his day he was a teleologist. As employed today, the term *basic architectural plan* refers to the similarity of basic vertebrate features as a result of common ancestry.

The distinction of being considered the founder of comparative anatomy (and also of paleontology) should probably be awarded to Georges, Léopold, Chrétien, Frédéric, Dagobert Cuvier, Baron de Montbéliard (1769-1832), a French naturalist who studied at Stuttgart. A prolific author in the field of natural history, his works include, among many others, nine volumes entitled *Leçons d'anatomie comparée* (Paris, 1801-1805), four volumes of paleontological observations on fossil skeletons of quadripeds, a geological description of the environs of Paris, and a work that is now a classic, a four-volume 1816 abridgement of his course in comparative anatomy at the École Centrale du Pantheon in Paris, which he titled *Le règne animal distribué d'après son organisation pour servir de base à l'histoire naturelle des animaux et d'introduction à l'anatomie comparée* ("The animal kingdom arranged in accordance with its structure for the purpose of serving as a basis for the natural history of animals and as an introduction to comparative anatomy"). In this volume he summed up his lifetime of research on the morphology of fossil and living animals, which he divided, in the role of taxonomist, into four taxa (radiate, articulate, mollusc, and vertebrate) based on differences in structure of the skeleton and organs. His inspiration throughout life came from a copy of the first edition of Linnaeus's *Systema naturae* (1735), which in Cuvier's early years was his entire scientific library.* His final contribution was a twenty-two volume masterpiece, *Historie naturelle des poissons*

---

*For more about Linnaeus, see p. 57.

("Natural history of fishes"), published over a four-year span ending in 1832. He died in 1832 in Paris at the age of 56 after an energetic life dedicated to natural history and teaching, including the Chancellorship at the University of Paris. Throughout life he steadfastly maintained the absolute immutability of species, objecting vehemently to the mutability ideas of Lamarck (a contemporary), of Geoffroy Saint-Hillarie (1772-1844), with whom he had collaborated in several published works, and of others who began to envision the possibilities of changes in species with time.

Cuvier would have been a leading Darwin adversary in the great uproar in scientific circles, in academia, among theologians, and, consequently, in the press, that was evoked by the publication of Charles Darwin's masterpiece in 1859, *On the origin of species by means of natural selection, or the preservation of favoured races in the struggle for life.* The 1,250 copies sold out on the day of issue. The first four chapters explain the artificial selection of domesticated animals and plants by man, comparing this with natural selection as a consequence of the struggle of undomesticated species for existence. The fifth chapter discusses variations and causes of modification other than natural selection. The next five chapters examine the difficulties that stand in the way of a belief in hereditary changes (evolution) in species generally, as well as in natural selection as an operating factor. The last three chapters present the evidence for evolution from the field of paleontology, geographical distribution of species, comparative anatomy, embryology, and vestigial organs. The uproar and almost universal condemnation of Darwin was comparable to that resulting from Copernicus's treatise (1543) that the earth revolves around the sun. (Copernicus's treatise was probably completed by 1530, but was circumspectly withheld by him until he was on his death bed.) Neither Copernicus's nor Darwin's theory conformed to the accounts in Genesis, and both were therefore considered a denial of the authority of God. Darwin's theory was also seen as a denial of the story of Adam and Eve and of creation of the earth in seven days. As pointed out earlier, however, it in no way denies the existence of a Creator.

With the advent of the study of genetics the mechanism whereby change takes place was elucidated, and mankind was given a natural explanation for the diversity of species on the planet earth.

WORDS TO PONDER

Words are a necessity for conceptual thought. The more words one can command, the greater the variety of ideas one may entertain. The following words are sometimes employed in discussions of phylogeny. You will read them, hear them, and use some of them. There will be differences of opinion with reference to the connotations of most of them. However, if calling attention to these abstract terms stimulates thoughtful discussion, inclusion of them will have been justified.

Primitive is a relative term. It refers to a beginning or origin. A primitive trait is one that appears in a stem ancestor from which arose an array of subsequent species, some of which may retain the trait. The notochord is primitive, since it occurred in the first chordates. The placoderms were primitive fishes in that they gave rise to an array of later fishes. Ancient insectivorous mammals were primitive *placentals* because they gave rise to an array of later placentals. However, they were not primitive *vertebrates*. Somewhere in phylogeny there was a primitive primate and a primitive species of man. However, one cannot always be certain that a given structure is primitive. For example, the lateral neural cartilages of lampreys are primitive only if they reveal an original condition from which typical vertebrae later evolved.

Generalized refers to structural complexes that, at least in some of the descendants, have undergone subsequent adaptation to a variety of conditions. The hand of an insectivore was, and remains, a generalized mammalian hand. It was competent to evolve into the wing of a bat, the hoof of a horse, the flipper of a seal, and the hand of a primate. A generalized group of animals has demonstrated that it was genetically suitable for divergent evolution, that is, evolution in many directions. Labyrinthodonts were generalized tetrapods. The terms "generalized" and "primitive" come into contrast in that generalized connotes a state of potential adaptability and primitive connotes a state of being ancestral.

A **specialized** condition is one that represents an adaptive modification. Vertebrate wings are specializations of anterior limbs, and beaks are specialized upper and lower jaws. Beaks (Fig. 1-14) may be needlelike for extracting nectar from flowers (hummingbirds), chisel-like for drilling holes (woodpeckers), hooked for piercing and tearing captured prey (raptorial birds, such as hawks and eagles), long and pointed for capturing moving fish, lizards, and other prey (herons), or recurved for extracting grubs from burrows (the female huia of New Zealand). Increased specialization connotes increased adaptation. The greater the specialization, the less *may* be the potential for further adaptive changes.

Derived, or **modified,** connotes any state of change from a previous condition; a mutated state. If the presence of bone is a primitive trait, wholly cartilaginous skeletons are modifications of the condition. The modification (loss of the potential to form bone) was a specialization if it adaptively modified the animal. Perhaps not all modifications are adaptive (students of speciation disagree strongly about this); but if they are not, they may portend the demise of the species, since any change is statistically more likely to make the animal less competitive.

The terms **higher** and **lower** are used to express the relative position of major taxa on a conventional phylogenetic scale. They are not informative when applied below the class level. Birds and mammals evolved from cotylosaurs and hence are said to be higher than cotylosaurs. In this context the words have some meaning. Sometimes the terms are used to express relative mutational distance of a given taxon from a common ancestor when compared with some other taxon—but are mammals to be considered higher than birds? In this context the term may be misleading. The terms may also be meaningless when used to compare a genus within one taxon with a genus within another taxon, as when comparing a modern frog with a perch, or a human being with a hummingbird.

Simple is a relative term connoting a lack of complexity of component parts. A simple state is not necessarily a primitive one. The skull of a human being is simple compared to that of a teleost, but it is not primitive. The primitive may also be far from simple.

The term **advanced** should connote a modification in the direction of further adaptation. Unfortunately the word has overtones connoting progress and hence is subjective and misleading. It is a matter of opinion of one species—man—whether or not a modification in another species represented or represents progress. The phrase "more recent" or "more specialized" may be more informative than the term "advanced."

Degenerate is another value-judgment word. For example, it is sometimes applied to cyclostomes by those who think cyclostomes have lost jaw skeletons, paired appendages, bone in the skin, and other characteristics of typical vertebrates. However, the condition of the cyclostomes represents an adaptation to a semiparasitic state and as such may better be characterized as "specialized." These agnathans may have specialized themselves right into a state of neosimplicity! To call them "degenerate" would seem to discount the value of adaptive modification. Degenerate seems to be a term that should be avoided.

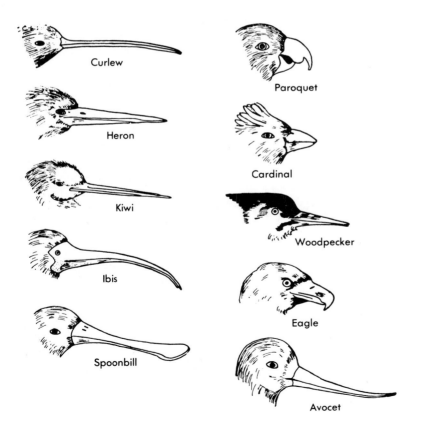

FIGURE 1-14

Adaptations of the beak for feeding in selected birds.

Curlew

Heron

Kiwi

Ibis

Spoonbill

Paroquet

Cardinal

Woodpecker

Eagle

Avocet

The words **vestigial** and **rudimentary** require explanation. A phylogenetic remnant that was better developed in an ancestor is vestigial. The pelvic girdle of whales is said to be vestigial, since ancestors of whales were tetrapods with functional tetrapod appendages. The yolk sac of the mammalian embryo is vestigial. The term "rudimentary" is used in two different senses, phylogenetic and ontogenetic. In the phylogenetic sense, structures that became more fully exploited in descendants are said to be rudimentary in the phylogenetic precursor. For example, the lagena of the inner ear of fish is sometimes referred to as a rudimentary cochlea, since it evolved into a cochlea in later forms. In the ontogenetic sense, a structure that is undeveloped or not fully developed is said to be rudimentary. The muellerian duct may be considered rudimentary in most male vertebrates. It is not always possible to be certain whether a structure should be called rudimentary or vestigial. The pseudobranch of the shark *Squalus acanthias* is vestigial if it represents a gill that, in ancestral sharks, was a full-fledged functional gill surface. However, if the pseudobranch is a potential future gill surface, then it is rudimentary. The majority opinion is that it is a vestigial gill with a new function.

If the foregoing thoughts trigger discussion, no matter how dissonant, the space employed in presenting them will have been well utilized. It must never be forgotten that words are sounds made by man to connote a concept. Otherwise, they may be nothing but noise. Or, do you wish to object, on a semantical basis, to the word "noise"?

Chapter Summary

Definitions briefly paraphrase those in the text.

1. Chordates are animals with a notochord and pharyngeal pouches in the embryo at least, and a dorsal hollow central nervous system. Vertebrates are chordates with vertebrae.

2. Vertebrates are bilaterally symmetrical, metameric, triploblastic, deuterostomous, chordates with marked cephalization, a postanal tail, two pairs of locomotor appendages, a two-layered skin, two to four coelomic chambers, a closed circulatory system with a single or double circuit heart, that respire by gills, lungs, skin, the buccopharyngeal lining, and, in embryos, by extraembryonic membranes.

3. The notochord is the first skeletal structure to appear in chordate embryos. In agnathans it is surmounted by lateral neural cartilages. In jawed vertebrates it becomes surrounded by cartilaginous or bony vertebrae, after which the notochord may disappear except for remnants.

4. Pharyngeal pouches tend to rupture to the exterior to form temporary or permanent pharyngeal slits. They persist as gill slits in fishes and some amphibians, and close permanently in other vertebrates.

5. A pharyngeal arch lies between each pouch or slit, and in front of the first pouch and behind the last. They support gills when the latter are present and are not definitive after the slits close. The first two are mandibular and hyoid arches.

6. Each pharyngeal arch contains a visceral skeletal element, branchiomeric muscles, a branch of a cranial nerve, and an aortic arch.

7. The central nervous system arises as a neural groove that sinks into the body wall to become a neural tube and, subsequently, brain and spinal cord. In a few fishes the neural tube forms from a neural keel. The peripheral nervous system consists of nerves, ganglia, and plexuses.

8. Phylogeny is the evolutionary history of a taxon. Phylogenesis is the process that produces evolutionary lines.

9. Ontogeny is the history of an individual. Ontogenesis is the process of embryonic development and subsequent aging.

10. Convergent evolution is the acquisition of similar structures in unrelated taxa as a result of mutations that are adaptive to similar environments.

11. Parallel evolution is the acquisition of similar structures in divergent but recently related groups the immediate ancestors of which lacked the trait.

12. Taxonomy is the classification of organisms and a study of the techniques of establishing taxonomic hierarchies. Classifications may be natural, phenetic, or cladistic.

13. Systematics is the study of taxonomy and speciation.

14. Homologous structures are the same organs in different animals under every variety of form and function. They arise from the same embryonic precursors.

15. Analogous structures are homoplasous structures having the same function.

16. An adaptation is a hereditary modification of a phenotype that increases the probability of survival.

17. Natural selection is the nonrandom differential reproduction of genotypes as a result of interaction of phenotypes with selective forces in the external environment.

18. Teleology ascribes natural phenomena to purposeful design rather than to natural laws.

19. Speciation, or formation of new species, is the result of genetic changes in geographically isolated fractions of a preexisting species.

20. Paedogenesis is reproduction by larvae with sexually mature gonads.

21. Neoteny is the retention of larval traits in adults.

22. Paedomorphosis is the phenomenon in which larval or juvenile features of ancestors become adult characteristics of descendants.

23. Heterochrony is a change in relative rates of development of characters during phylogeny.

24. Organic evolution is the concept that species have been changing and existing species are divergent descendants of earlier ones.

25. Darwin's major contribution to the life sciences was the Theory of Natural Selection. It was an alternative to Lamarck's Theory of Acquired Characteristics.

26. Cuvier should probably be considered the father of comparative anatomy.

SELECTED READINGS

Evolutionary Theory and Systematics

Alberch, P.: Ontogenesis and morphological diversification, American Zoologist **20:**653, 1980. *A discussion of the role of epigenesis in evolution.*

Alberch, P., and others: Size and shape in ontogeny and phylogeny, Paleobiology **5:**296, 1979.

Ayala, F.J.: Teleological explanations in evolutionary biology, Philosophical Society (London) **37:**1, 1970.

Bock, W.J.: The definition and recognition of biological adaptation, American Zoologist **20:**217, 1980.

Boyden, A.: Homology and analogy: a century after the definitions of "homologue" and "analogue" of Richard Owen, The Quarterly Review of Biology **18:**228, 1943.

Cole, J.F.: A history of comparative anatomy, London, 1944, The Macmillan Co., Ltd.

Craycraft, J.: The use of functional and adaptive criteria in phylogenetic systematics, American Zoologist **21:**21, 1981.

de Beer, G.R.: Homology, an unsolved problem, London, 1971, Oxford University Press.

Gould, S.J.: Ever since Darwin, New York, 1977, Norton.

Gould, S.J.: Ontogeny and phylogeny, Cambridge, Massachusetts, 1985, The Belknap Press of Harvard University.

Grene, M., editor: Dimensions of Darwinism: themes and counterthemes in twentieth-century evolutionary theory, New York, 1983, Cambridge University Press.

Hull, D.L.: Evolution and circularity in evolutionary taxonomy, Evolution **20:**174, 1976.

Huxley, J.: Evolution, the modern synthesis, London, 1942, Allen & Unwin.

Lauder, G.V.: Form and function: structural analysis in evolutionary morphology. Paleobiology **7:**430, 1981.

Lewontin, R.C.: Adaptation, Scientific American **239**(3):212, 1978.

Mayo, O.: Natural selection and its constraints, New York, 1983, Academic Press.

Mayr, E.: The growth of biological thought, Cambridge, Massachusetts, 1985, Harvard University Press. *A sweeping perspective of past and current ideas in the interrelated realms of systematics, evolution, and genetics.*

Minkoff, E.C.: Evolutionary biology, Reading, Massachusetts, 1983, Addison-Wesley Publishing Co. *A textbook of evolutionary theory.*

Moore, J.A.: Science as a way of knowing: evolutionary biology, American Zoologist **24:**467, 1984.

Myers, N.: The end of the lines, Natural History **94**(2):2, 1985. *A discussion of duration of species and extinction rates.*

Northcutt, R.G., and Gans, C.: The genesis of neural crest and epidermal placodes: a reinterpretation of vertebrate origins, Quarterly Review of Biology **58:**1, 1983.

Pollard, J.W., editor: Evolutionary theory: paths into the future, New York, 1984, John Wiley & Sons.

Savage, H.M.: The shape of evolution: systematic tree topology, Biological Journal of the Linnean Society, **20**(3): 1983.

Steele, E.J.: Somatic selection and adaptive evolution. On the inheritance of acquired characters, ed. 2, Chicago, 1981, University of Chicago Press.

Wake, M.H.: Evolution: the biology of whole organisms, American Zoologist **24:**443, 1984.

Wiley, E.O.: Phylogenetics: the theory and practice of phylogenetic systematics, New York, 1981, John Wiley & Sons.

Symposia in American Zoologist

Functional-adaptive analysis in systematics, **21:**3, 1981.

Models and mechanisms of morphological change in evolution, **15:**294, 1975.

PROTOCHORDATES AND THE ORIGIN OF VERTEBRATES

*In this chapter we will speculate concerning the invertebrate origins of verte-
brate animals. For background, we will gain some knowledge concerning protochordates—small aquatic animals with a notochord but lacking a vertebral
column. We will then hypothesize concerning their affinities with vertebrates. At the end of the chapter we will examine a larval lamprey, for the
reason that it bears a remarkable resemblance to protochordates, has been
mistaken for a protochordate, yet grows to become a lamprey eel.*

No study of vertebrate anatomy would be complete without providing some knowledge about the closest relatives of vertebrates, the protochordates. A sessile sea squirt squirting water out of its excurrent syphon seems a far cry from even a simple amphioxus, let alone the lowest jawed fishes. Yet, sea squirts, amphioxus, fishes, and tetrapods exhibit three fundamental structural features in common, a notochord, dorsal hollow central nervous system, and pharyngeal slits (Fig. 2-1). Notwithstanding the fact that the notochord becomes encased within vertebrae in vertebrates, and pharyngeal slits close sooner or later in tetrapods, the manifestation of these fundamental structures in all of them indicates that vertebrates are more closely related genetically to protochordates than to any other animals. Therefore, protochordates should provide some clues to the invertebrate origins of animals with backbones.

PROTOCHORDATES

In recognition of the postulate that protochordates and vertebrates are closely related, Ernest Haeckel in 1874 proposed the establishment of the **Phylum Chordata** incorporating three subphyla, **Urochordata** (sea squirts and certain other invertebrates), **Cephalochordata** (the amphioxuses), and **Vertebrata.** Members of the two lower subphyla have come to be known as protochordates, a convenient term, although without taxonomic status. All are marine organisms. Our interest in them stems from the realization that they and the vertebrates may have had common ancestors. Therefore, we will examine them briefly, emphasizing the morphological features that prompt this hypothesis. Their place in the chordate hierarchy as proposed by Haeckel and generally accepted today is as follows:

Phylum Chordata
  Subphylum Urochordata. (Tunicata)
    Class Ascidiacea. Sea squirts
    Class Larvacea. Larvaceans
    Class Thaliacea. Thaliaceans
  Subphylum Cephalochordata. Lancelets (amphioxuses)
  Subphylum Vertebrata. Chordates with vertebrae

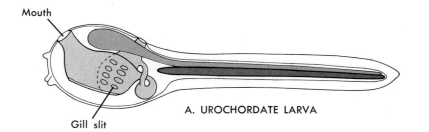

A. UROCHORDATE LARVA

B. AMPHIOXUS LARVA

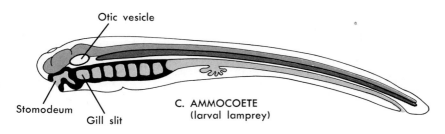

C. AMMOCOETE
(larval lamprey)

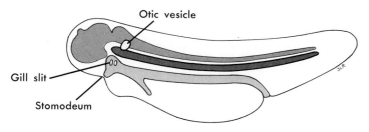

D. LARVAL AMPHIBIAN

FIGURE 2-1

Chordate larvae showing basic architectural pattern. **A** and **B** are protochordates, **C** and **D** are vertebrates. *Dark red,* notochord; *medium red,* dorsal nervous system; *light red,* alimentary canal. In **A,** a larval sea squirt, part of the left wall of the pharynx has been removed to show the gill slits in the right wall leading to the atrium.

Urochordates

The name urochordate signifies that the notochord is confined to the tail. Best known are the **ascidians,** or sea squirts (Fig. 2-2). Unlike other urochordates, sea squirts undergo complete metamorphosis, resorbing the notochord along with the tail to become sessile adults (Fig. 2-3). There are solitary and colonial species. **Larvaceans** (Fig. 2-4), also called appendicularians, fail to metamorphose and remain tiny, planktonic, free-swimming reproductive larvae. This is an instance of paedogenesis. The notochord remains throughout life, strengthening the tail, which is the locomotor organ. **Thaliaceans** have no tail, hence no notochord. We will not concern ourselves further with larvaceans or thaliaceans. All urochordates are enclosed within a tough cellulose-like tunic that is often beautifully colored and usually transparent. It inspired the name, tunicates.

LARVAL SEA SQUIRTS. The larvae of solitary sea squirts are tiny (0.5 mm to 11.0 mm), have a fleeting existence—as little as a few minutes, not longer than a few days—and they do not feed, subsisting on stored nutrients. The body consists of a trunk containing immature viscera still in the process of differentiation, and a muscular tail the sole function of which is locomotion during the few hours of their free-swimming existence. The notochord is a stiff rod composed of a small number of epithelial cells arranged in a single layer around a central core of matrix, and a filamentous sheath. Alongside of the notochord in solitary species there are about 36 uninucleate striated muscle fibers. (Striated muscle fibers of vertebrates are multinucleate.) Other species have as many as 1600. There are no tendons. The nervous system consists of a dorsal hollow nerve cord, several ganglia, and nerves. A **sensory vesicle** associated with the brain houses an **otolith** that stimulates nerve endings for statoreception, and a light-sensitive **ocellus** (a pigment-protected receptor cell, Fig. 2-7) is usually present in its wall. In some species neither definitive blood cells nor a functional heart differentiates until metamorphosis.

Respiratory water enters the branchial basket (pharynx) via the mouth, passes over gills lining the pharyngeal slits, and enters a chamber, the **atrium,** surrounding the pharynx. Oxygen-depleted water is ejected via an **atriopore.** At metamorphosis three **adhesive papillae** with sticky secretions attach the larvae to a permanent substrate. Larvae serve for dispersal of the species.

ADULT SEA SQUIRTS. At metamorphosis the notochord is resorbed, the nervous system is altered in location and structure, and a rearrangement of the viscera takes place (Fig. 2-3). The larval mouth becomes an **incurrent syphon** and the atriopore becomes an **excurrent** one. The forceful discharge of water from the latter whenever the polyp is irritated inspired the name sea squirt. Food in adults is now filtered out of the incoming water stream and trapped in mucus secreted by the **endostyle,** a glandular evagination of the pharyngeal floor. Food particles are moved by esophageal papillae and ciliary action into the stomach, while water passes over the gills and into the atrium. The polyp is now a filter feeder. Arising from each end of the

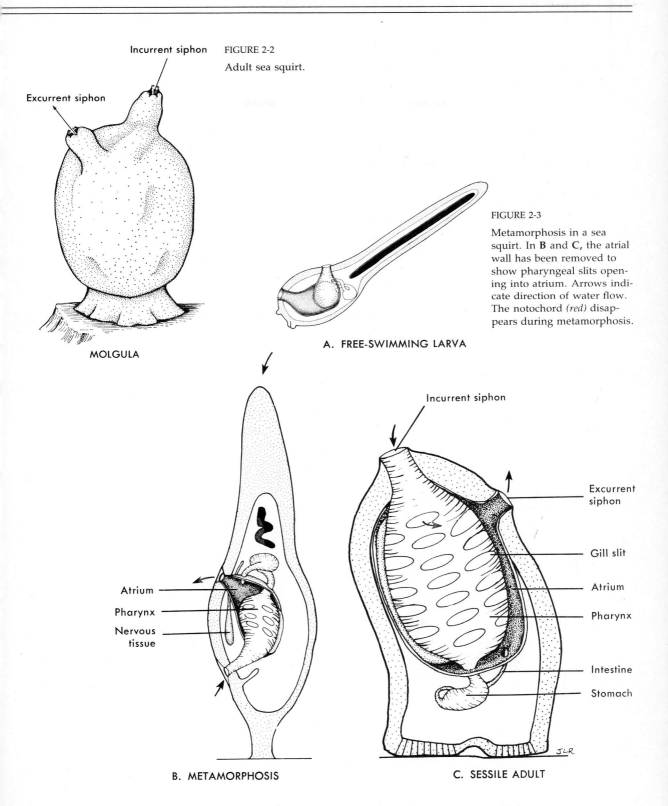

FIGURE 2-2
Adult sea squirt.

Incurrent siphon

Excurrent siphon

MOLGULA

FIGURE 2-3

Metamorphosis in a sea squirt. In **B** and **C**, the atrial wall has been removed to show pharyngeal slits opening into atrium. Arrows indicate direction of water flow. The notochord *(red)* disappears during metamorphosis.

A. FREE-SWIMMING LARVA

Atrium
Pharynx
Nervous tissue

B. METAMORPHOSIS

Incurrent siphon

Excurrent siphon

Gill slit

Atrium

Pharynx

Intestine

Stomach

C. SESSILE ADULT

FIGURE 2-4

A larvacean.

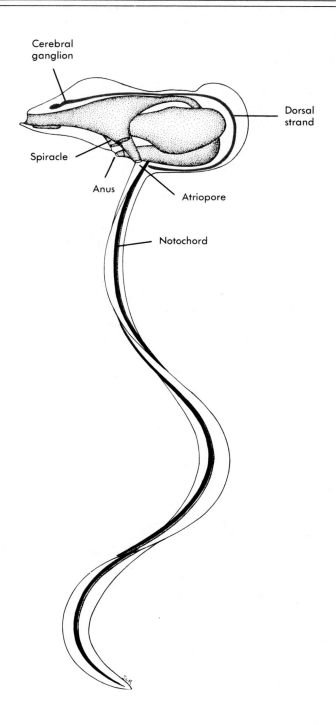

now functioning heart is a blood vessel. Blood is pumped alternately forward for a few beats, then backward, the heart pausing before reversing the flow.

It wasn't until the middle of the nineteenth century that it was discovered that these polyplike, sometimes colonial, marine animals have a free-living larval stage with a notochord and a dorsal nervous system. From the viewpoint of vertebrate kinships there is more to be learned from the larvae than from adult sea squirts.

Cephalochordates

Amphioxus means "sharp at both ends." Any member of the subphylum Cephalochordata may be called an amphioxus, or lancelet (little spear), but the correct generic name for the lancelet commonly studied in the laboratory is *Branchiostoma* (Fig. 2-5). *Asymmetron* is the only other genus in the subphylum.

Lancelets are found a short distance out from sandy beaches throughout most of the globe. They quickly burrow into sand with eel-like movements, make a U turn, and then emerge until only the oral hood area is protruding for filter feeding. Adults vary from less than 2 cm to more than 8 cm in length, the largest being *Branchiostoma californiense*. Off the coast of China, lancelets are collected in quantity and sold as a table delicacy.

An amphioxus is semitransparent but becomes opaque when immersed in preserving fluids. The body is practically all trunk and tail. A pair of longitudinal ridges of unknown function, the **metapleural folds,** hang along each side of the midventral line beneath the pharynx (Fig. 2-8, C).

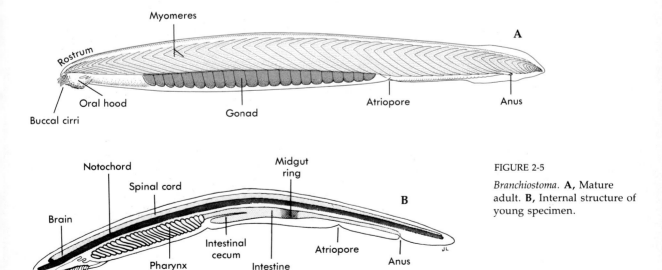

FIGURE 2-5

Branchiostoma. **A,** Mature adult. **B,** Internal structure of young specimen.

The notochord extends from the tip of the rostrum to the tip of the tail (Fig. 2-5, *B*). It consists of muscular discs arranged like a long column of coins separated by fluid-filled spaces. The muscle fibers in each disc run transversely and have dorsal extensions that end near nerve terminals from which they evidently receive their motor innervation. The muscle resembles that of invertebrates in that the protein is paramyosin. Contraction of the muscles increases the stiffness of the notochord, which may aid in swimming. Continuation of the notochord to the very tip of the rostrum—unlike in any other chordate—may be an adaptation for burrowing in sand. Surrounding the notochord is a thick collagenous connective tissue sheath. The only other skeleton is fibrous rods that support the gill bars, buccal cirri, and fins.

An amphioxus, like a vertebrate, has a **dorsal hollow central nervous system** consisting of a hollow brain and a dorsal nerve cord containing a central canal, both lined with a nonnervous supportive membrane, or **ependyma;** and, as in vertebrates, the caudal end of the cord consists of ependymal cells alone, all nervous elements having disappeared. A single delicate connective tissue membrane, the **leptomeninx,** surrounds the cord and brain.

Only two **brain** subdivisions can be identified, an anterior cilia-lined prosencephalon, and a posterior deuteroencephalon. Attempts at homologizing the parts of the brain with those of vertebrates have not been entirely successful. In an amphioxus the notochord extends anterior to the brain; in vertebrates it stops at the midbrain. Does this indicate the absence of a forebrain? Whether or not the nerves that supply the gills should be considered cranial nerves complicates the problem. If the branchial nerves are omitted, there are seven **cranial nerves,** including the (apical) terminal nerve. If the branchial and oral nerves are included, there are 39. Vertebrates have 10 or 12 cranial nerves. The absence of semicircular canals, eyes, lateral line system, and foramen magnum (there being no skull)

FIGURE 2-6

Cephalic end of an amphioxus shown in sagittal section. *1,* Vestibule bounded by oral hood; *2,* part of the wheel organ projecting into vestibule; *3,* velar tentacle; *4,* velum.

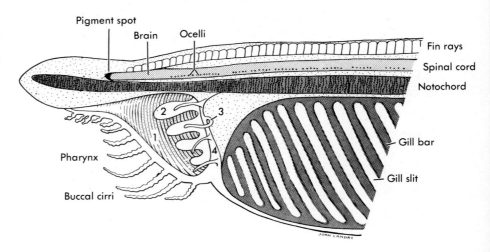

deprives us of landmarks that would be helpful. Because of these difficulties, it is not possible to decide where the brain ends and the cord begins.

Spinal nerves emerge from the cord metamerically. They consist of dorsal roots that contain sensory and motor fibers, the latter supplying only visceral organs. The "ventral roots" are not nerve roots, but tubes that conduct extensions of the body wall muscle cells into the spinal cord where they receive their innervation. Since the somites of the two sides are not exactly opposite one another, the roots of the left and right sides do not arise directly opposite one another.

The relatively small size of the brain is correlated with the paucity of **special sense organs.** There are no retinas, semicircular canals, or lateral line organs. It is doubtful whether an olfactory epithelium is present, but other chemoreceptors are especially abundant on the buccal cirri and velar tentacles, where they monitor the incurrent water stream. They are also scattered on other surfaces of the body, the tail being more sensitive than the trunk. Tactile receptors, which elicit withdrawal, are present over the entire body surface.

The most prominent sense organs are the light-sensitive, pigmented **ocelli** embedded within the ventrolateral walls of the spinal cord throughout its length (Figs. 2-6 and 2-7). Each ocellus consists of a receptor cell and a caplike melanocyte. The melanocyte lies between the receptor cell and the incoming light rays and is packed with large melanin granules. A conducting process extends from the base of the receptor cell to the central nervous system. Ocelli probably assist in orienting the animal as it burrows in the sand.

Cephalochordates, like sea squirts, are filter feeders. A collecting chamber, the **vestibule,** is bounded laterally by an **oral hood** and caudally by a perpendicular membranous **velum** perforated by the mouth, which opens into the pharynx (Fig. 2-6). The vestibule is broadly open to the sea ventrally. Cilia on the pharyngeal surface of the gill bars create a steady flow of water through the mouth and into the pharynx. A set of stubby projections in the vestibule, the **wheel organ,** is covered with sticky mucus that retrieves some of the heavier food particles that miss the mouth, and it directs these through the mouth and into the pharynx along with the water stream. **Buccal cirri** partially strain the water as it enters the vestibule, and monitor it chemically.

Once food is in the pharynx it is processed as follows: In the pharyngeal floor there is a **hypobranchial groove,** or **endostyle** (Fig. 2-8, C), and in the roof there is an **epibranchial groove.** On the gill bars, ciliated **peripharyngeal bands** connect the two grooves. The cells of these bands and grooves secrete mucus. Food particles trapped in the mucus are incorporated into a stringy food cord that is propelled by cilia dorsally into the epibranchial groove and then caudad into the midgut behind the pharynx. Here it is temporarily arrested by the **midgut ring** (Fig. 2-8, B) and mixed with digestive enzymes. Some of the digesting foodstuffs then pass beyond the ring into the hindgut, and some are propelled forward by cilia into the **intestinal**

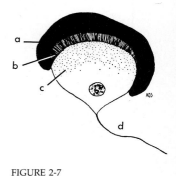

FIGURE 2-7

Ocellus (light receptor) from spinal cord of an amphioxus. *a,* Melanocyte; *b,* apical border of receptor cell; *c,* receptor cell; *d,* process for conduction of impulse.

FIGURE 2-8

Cross sections of an amphi-oxus. **A,** Anterior to mouth. **B,** Posterior to atriopore. **C,** Level of pharynx.

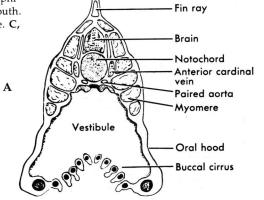

A

- Fin ray
- Brain
- Notochord
- Anterior cardinal vein
- Paired aorta
- Myomere

Vestibule

- Oral hood
- Buccal cirrus

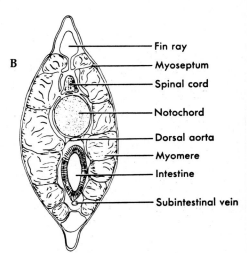

B

- Fin ray
- Myoseptum
- Spinal cord
- Notochord
- Dorsal aorta
- Myomere
- Intestine
- Subintestinal vein

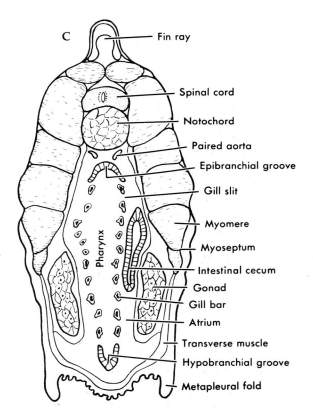

C

- Fin ray
- Spinal cord
- Notochord
- Paired aorta
- Epibranchial groove
- Gill slit
- Myomere
- Myoseptum
- Intestinal cecum
- Gonad
- Gill bar
- Atrium
- Transverse muscle
- Hypobranchial groove
- Metapleural fold

Pharynx

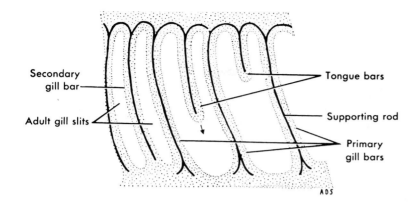

FIGURE 2-9

Tongue bars in the pharyngeal wall growing ventrad to
form secondary gill bars.

cecum. This evagination arises in the same manner as the vertebrate liver, but the two differ in function. The cecum secretes digestive enzymes for extracellular digestion within the cecum and intestine, and its lining cells phagocytose the smallest food particles and digest them by intracellular digestion. The intestine terminates at the anus.

Pharyngeal water relieved of food particles passes between the gill bars into the atrium and then to the exterior via the atriopore. A pterygial muscle innervated by dorsal roots in the cranial region alters the volume of the atrial chamber.

The number of **gill slits** varies but exceeds 60 in adults. During metamorphosis each larval slit is divided in two by the downgrowth of a **tongue bar** (Fig. 2-9). These become **secondary gill bars** (Fig. 2-11). Much respiration takes place through the skin.

The **skin** consists of a single layer of epidermal cells and a thin dermis (Fig. 5-1). Larval skin is ciliated for early locomotion, but the cilia disappear when the epidermis secretes a cuticle. Immediately internal to the dermis is the body wall musculature.

The **body wall musculature** is locomotor. It consists of an uninterrupted series of > -shaped muscle segments **(myomeres)** extending from the rostrum to the tip of the tail. Each myomere is separated from the next by a **myoseptum,** a connective tissue partition to which the longitudinal muscle bundles attach. Since myomeres are > -shaped, cross sections of the body include several successive myomeres.

Arterial blood courses forward beneath the pharynx in the muscular, contractile ventral aorta that commences in the sinus venosus. From the ventral aorta, afferent branchial arteries (aortic arches) pass up the gill bars. Before joining the dorsal aorta, these arteries contribute to vascular channels supplying protonephridia (excretory organs). The blood then enters the paired dorsal aortas. These pass caudad above the pharynx and unite just behind it to form an unpaired aorta. This distributes blood by paired vessels to the body wall and by median vessels to the visceral organs. The dorsal aorta continues into the tail as the caudal artery.

The **venous channels** (Fig. 2-10) are similar to the embryonic venous channels of vertebrates. From the capillaries of the tail a single caudal vein courses forward and divides into right and left posterior cardinal veins. These pass forward in the lateral body wall to a point just behind the pharynx. Here the posterior cardinal veins meet the anterior cardinal veins from the rostrum and pharyngeal wall. The blood then enters a common cardinal vein leading to the sinus venosus. Two parietal veins drain the dorsolateral body wall of the trunk. These also terminate in the sinus venosus.

Drainage from the visceral organs is via a median subintestinal vein arising from the caudal vein. The subintestinal vein passes cephalad along the ventral surface of the intestine. There it breaks up into smaller channels, receives tributaries from the intestine, and reconvenes to continue forward as a portal vein ending in the capillaries of the cecum. From the cecum the contractile cecal vein pumps blood to the sinus venosus.

Removal of coelomic wastes is accomplished by **protonephridia** lying beside the secondary gill bars (Fig. 2-11). Each protonephridium consists of clusters of **solenocytes** that project into the coelom, and of a chamber. The flagella cause a current of coelomic fluid to enter a solenocyte and flow down the stalk to the chamber, which opens into the atrium via a small

FIGURE 2-10

Basic venous channels and ventral aorta of an amphioxus, dorsal view. The cecum has been rotated 90 degrees around the long axis. The cecal portal vein in life is ventral to the cecum, and the cecal vein is dorsal. The conventional term "hepatic portal" for the cecal portal vein is not appropriate.

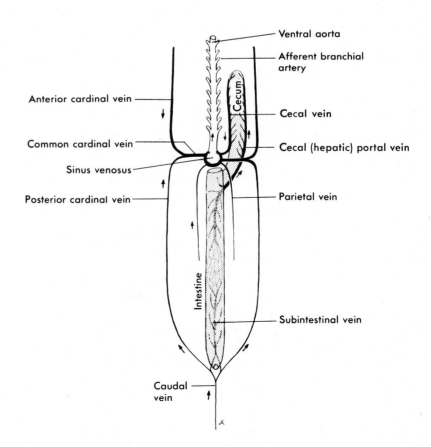

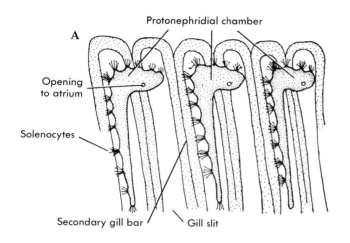

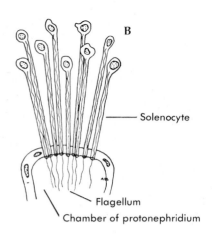

FIGURE 2-11

Excretory organs of an amphioxus. **A,** Three protonephridia. **B,** A cluster of solenocytes. They project into
the coelom at their free end
and empty into the protone-
phridial chamber at their
base.

pore. The excretory mechanism resembles the nephridia of marine annelids
and the flame cells of invertebrates.

Mature **gonads** are visible through the body wall and bulge into the
water-filled atrium, into which sperm or eggs are shed. They are then
flushed into the sea via the atriopore. The amphioxus is **dioecious,** that is,
ovaries and testes do not develop in the same individual.

AMPHIOXUS AND THE VERTEBRATES CONTRASTED. Although an
amphioxus resembles a vertebrate in many respects, differences are evident. An amphioxus has almost no cephalization and no paired sense
organs; it has a notochord but no vertebral column; it has pharyngeal (gill)
slits but in large numbers, emptying into an atrium; it has a dorsal, hollow
central nervous system, but the brain lacks the major vertebrate subdivisions; it has a segmented musculature, but the segments extend to the
anterior tip of the head; it has median fins but no paired ones; it has a
two-layered skin, but the outer layer is only one cell thick; it has arterial and
venous channels similar to the basic channels of vertebrates but no muscular heart; it is coelomate, but the coelom is greatly restricted; liquid wastes
are removed from coelomic fluid as in lower vertebrates, but the excretory
protonephridia resemble those of nonchordates. A hypothetical relationship of the amphioxus to other protochordates and to vertebrates is diagrammed in Fig. 2-13.

Hemichordates: Incertae Sedis

The term incertae sedis means "of uncertain status." It indicates that the
taxonomic status of a group of organisms is unsettled. Despite the advent
of modern techniques not all animals have been assigned to taxonomic
categories without dissent. Among these are acorn tongue worms—marine
organisms that live in the mud under shallow waters and, although fragile,
have been known to reach 1.5 meters in length (Fig. 2-12, *A*).

FIGURE 2-12

An acorn worm, *Saccoglossus.*
A, Entire worm. **B,** Head,
sagittal section (except proboscis).

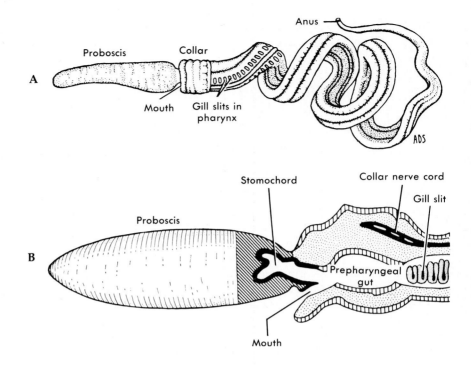

In 1870 Karl Gegenbaur established a new taxon, **Enteropneusta,** to
accommodate acorn worms. Four years later, Haeckel established the **phylum Chordata.** And after ten more years William Bateson in 1884 added
enteropneusts to the phylum Chordata as the lowest subphylum, Hemi-
chorda. His decision was based on the following observations.

1. Enteropneusts have a **dorsal strand of nerve cells** and fibers lying
within the epidermis in a longitudinal groove on the dorsal surface extending from the midcollar region to the end of the trunk. In the collar region
the groove sinks beneath the surface to become a **collar nerve cord** with a
continuous or discontinuous lumen. This seemed to Bateson to be a manifestation of the neural groove of embryonic chordates (Fig. 2-12, *B*).

2. Enteropneusts have **gill slits** in the lateral walls of the foregut leading
directly to the exterior. Pharyngeal slits are primary structures in chordates.

3. Enteropneusts have a short diverticulum of the foregut, the **stomochord,** extending forward into the proboscis. Bateson considered this to be
homologous with the notochord of chordates.

Bateson discounted certain facts that indicate a close kinship with echinoderms. Enteropneusts have bilaterally symmetrical larvae (tornaria) so
similar to those of some echinoderms that nineteenth-century biologists
frequently confused the two. This suggests an affinity of echinoderms and
enteropneusts. But even if this is an erroneous inference, enteropneust
larvae do not resemble protochordate larvae and could not be mistaken for

the latter. Second, similarities between enteropneusts and echinoderms in certain developmental processes, in muscle proteins, and in other traits that have been revealed by twentieth-century techniques seem to link enteropneusts closer to echinoderms than to chordates. Third, a detailed examination of the nervous systems of enteropneusts reveals a system quite unlike the dorsal, hollow, central nervous system of larval sea squirts and cephalochordates. In addition to the dorsal nerve strand, a second strand lies at the base of the ventral mesentery in the coelomic floor. Dorsal and ventral strands are united by a circumintestinal nerve ring at the caudal boundary of the collar. (There is also a visceral network of nerve cells and fibers in the lining of parts of the digestive tube, as in all coelomate animals.) But nowhere, collar or trunk, is there a region that can be regarded as a central nervous system, that is, a center that has been shown to integrate stimuli and responses. The collar nerve cord is not a brain in either the invertebrate or chordate sense, having no neurohistological specialization, and neither receiving or giving off nerves (i.e., *bundles* of nerve fibers).

The similarities between enteropneusts and echinoderms cited above coupled with the doubtful homology of stomochord and notochord, and the enigmatic nature of the enteropneust nervous system prompt the current majority view that enteropneusts should be classified, at least for the present, close to echinoderms, close to protochordates, but in a taxon independent of both. The invertebrate phylum **Hemichordata** usually includes, in addition to enteropneusts, two other classes, Pterobranchia and Planctosphaeroidea, that have little resemblance to one another or to enteropneusts.

There are competent zoologists who include Hemichordata in the phylum Chordata. Either classification is intuitive, and neither alters the fact that enteropneusts have much to contribute to the solution of the intriguing problem of the kinships and ancestry of vertebrates.

THE ORIGIN OF VERTEBRATES

The oldest vertebrates we know of are the ostracoderms (Figs. 2-13 and 3-4). These strange little fishes, 2 to 30 cm long and of diverse appearances, had no jaws, no paired fins, and many of the earliest ones were filter feeders. Broad bony plates in the skin formed a protective shield over the head and much of the trunk, inspiring the nickname, "armored fishes." Where there were no bony shields smaller bony scales fitted tightly together over the body like tiles. Ostracoderms are discussed further in Chapter 3.

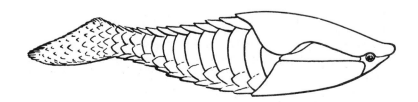

FIGURE 2-13

A heterostracan ostracoderm.

The broad outlines of vertebrate evolution after the ostracoderms have been remarkably well ascertained. The jawless ostracoderms were succeeded by jawed fishes and, eventually, by tetrapods. The perplexing problem is "Who were the ancestors of ostracoderms?"

To answer the question one can only speculate. There are no Ordovician fossils that might help. However, the absence of fossilization of transitional species suggests a paucity of mineralized tissues in those species, in which case the bone in the skin of ostracoderms may have appeared with comparative suddenness in some soft-bodied invertebrate that already had a notochord, dorsal nervous system, and pharyngeal slits. It is not necessary to design such an organism out of pure imagination: they already exist. They are the protochordates.

What clues to vertebrate ancestry might be found in a cephalochordate? On p. 47 an amphioxus was contrasted with vertebrates with respect to their morphological characteristics. We should now emphasize the similarities.

Cephalochordates have a notochord, pharyngeal slits that are employed in respiration, a dorsal hollow central nervous system with brain and cord, segmented body wall musculature that is used for locomotion, a two-layered skin, arterial and venous channels similar to the basic channels of fishes; they are deuterostomous, coelomate, and coelomic fluid is the collecting site for metabolic wastes. Not only do these features describe cephalochordates; they also describe lower fishes. Finally, cephalochordates are filter feeders, and so were many early ostracoderms. These similarities bespeak a close genetic relationship between protochordates and vertebrates. Some protochordate-like organism may have been the stock from which vertebrates arose.

We haven't reached an impasse in our search for vertebrate kinships when we arrive at this conclusion. There are additional pertinent facts. Echinoderms, like vertebrates, have mineralized tissue in their mesoderm (not in their ectoderm, as in molluscs, for example); like cephalochordates, they form their mesoderm and coelom as outpocketings of their early gut cavity, or archenteron (Fig. 4-3); and echinoderms, enteropneusts, urochordates, cephalochordates, and vertebrates are all deuterostomes, which means that they convert their blastopore, the first embryonic opening into their archenteron, into an anus and develop a new mouth, a trait found only in one other invertebrate taxon (Chaetognatha). On the basis of these and other morphological and embryological facts it is rational to hypothesize that vertebrates and protochordates are derived from a common prechordate ancestor.

A hypothetical scheme of phylogenetic relationships among the organisms we have been discussing is diagrammed in Fig. 2-14. It derives vertebrates from a hypothetical sessile or semisessile, bilaterally symmetrical, filter-feeding, deuterostomous stem chordate having a dorsal hollow nerve cord, pharyngeal gill slits, and a notochord confined to larvae, where it stiffened the larval tail for locomotion. The presence of a notochord

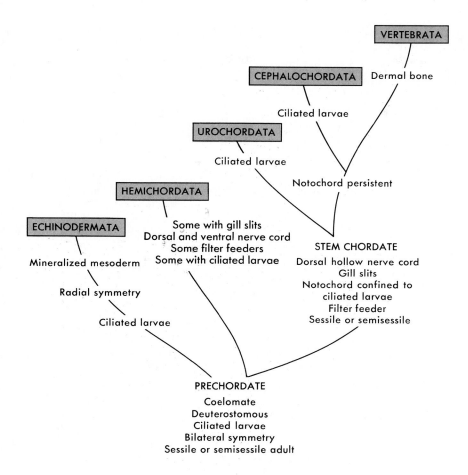

FIGURE 2-14

One of several postulated relationships between echinoderms, hemichordates, protochordates, and vertebrates.

throughout life in any chordate would then be an instance of neoteny. A cautionary note must be interjected at this point: the morphological similarities between protochordates and vertebrates that have been cited in the previous discussion as evidence of common ancestry may, in one or all instances, be attributable instead to evolutionary convergence.

One question about the origin of vertebrates is whether they originated in fresh or salt water. The question is far from settled. Paleontologists who think they originated in fresh water point to the absence of vertebrate fossils in marine Ordovician rocks. This is a formidable argument. Others insist that such fossils may well exist but have not yet been found because at that time vertebrates were few in number, and because disruption of sedimentary layers by slight elevations often expose accumulated sediments to erosion. They point out that negative evidence cannot be conclusive. Scientists agree. Arguments supporting both sides have been reviewed by Stahl.

THE AMMOCOETE: A VERTEBRATE LARVA

In addition to evidence concerning the origin of vertebrates provided in the preceding discussions, what has been interpreted as corroborating evidence is found in the free-swimming larvae of lampreys (Fig. 2-15). *Ammocoetes* was a genus at one time when these larvae were erroneously believed to be adult protochordates. An ammocoete, which lives from two to six years as a burrowing freshwater larva, is a filter feeder with all the basic chordate features, and lacks a vertebral column. The existence of such a larva supports the hypothesis that vertebrates share a common ancestry with protochordates. Therefore we will examine it at this time.

The dorsal hollow **central nervous system** consists of a brain with the three vertebrate subdivisions—forebrain, midbrain, and hindbrain—and spinal cord. The notochord underlies the hindbrain and extends almost to the tip of the tail. The pharynx has seven pairs of gill pouches opening directly to the exterior via gill slits. A complicated endostyle (subpharyngeal gland) lies ventral to the pharynx and opens into it by a short duct (Fig. 17-14). The gland secretes mucus that entraps food particles, chiefly diatoms, which enter the pharynx with the respiratory water stream. Much of the gland degenerates when a larva metamorphoses, since adult lampreys are not filter feeders. The cells that remain organize thyroid follicles like those of higher vertebrates.

The **body wall musculature** is arranged as myomeres that provide locomotion. The **epidermis** contains many unicellular mucous glands but, unlike that of cephalochordates, is multilayered. The **dermis** is thin and fibrous.

Sense organs are more numerous than in cephalochordates. From a median naris on the dorsum of the head a nasal canal slants downward and backward to open into a median olfactory sac, the receptor for smell. The sac lies just anterior to the brain, to which it is connected by a pair of olfactory nerves. Two middorsal eyes are located in tandem just under the skin of the head, attached to the forebrain by stalks containing nerve fibers. These are the pineal and parapineal organs (Fig. 16-19). Each consists of a simple lens and receptor cells. Paired lateral eyes lie deep under the skin and are vestigial. A pair of otic vesicles is located beside the hindbrain (Fig. 2-1,*C*). They arise as fluid-filled invaginations of the ectoderm and develop into inner ears. Other fluid-filled receptors, the neuromast organs (Fig. 16-2), lie in the skin of the head and open onto the surface via pores. Their function in bony fishes is discussed on p. 568.

The simple **digestive tract** is straight from mouth to anus. The most cephalic part of the early larval tract is the stomodeum (Fig. 2-1, *C*), an invagination of the embryonic ectoderm that grows toward, and finally achieves an opening into, the foregut. In older larvae the stomodeum becomes a buccal funnel and a buccal cavity that terminates at a velum. Beyond the latter is the pharynx. Water is drawn into the mouth and pharynx by pumping action of the pharyngeal muscles. The velum acts as a valve that prevents water from returning to the mouth during the compression phase of the pumping cycle. The water is forced through the gill slits while food is ensnared in mucus, as in cephalochordates. There is no stomach. Beyond the pharynx is a solid liver diverticulum and embedded in it is a gall bladder that disappears at metamorphosis when the food changes from diatoms to tissue juices of the prey.

The **circulatory system** is typically vertebrate. The heart consists of a sinus venosus, atrium, and ventricle. Blood is pumped by the ventricle forward into the ventral aorta, to the gills, then to the dorsal aorta, which courses caudad. Venous blood returns from the head via a pair of anterior cardinal veins and from most of the trunk and tail via posterior cardinals (Fig. 2-15). These all converge on paired common cardinals that empty into the sinus venosus. Blood returns from the intestine via a ventral intestinal vein (hepatic portal system) that ends in the capillaries of the liver. From the liver blood continues to the heart via a hepatic vein that empties directly into the sinus venosus.

Because the ammocoete lives in fresh water it is faced with the necessity of eliminating excess water from its tissues. This is accomplished by **kidneys** located anteriorly in the coelom (Fig. 2-15). The kidney of an early larva consists of three to six tubules with funnellike nephrostomes that collect coelomic fluid secreted by vascular tufts, the glomeruli (Fig. 14-2, *A*). Some of the constituents of the fluid are resorbed by the lining of the tubules. What remains passes caudad via longitudinal ducts (Fig. 2-15, pronephric duct) that empty to the exterior via a median papilla just behind the anus. Between the kidneys is a rudimentary **gonad** that is paired in early larvae but unpaired in later ones.

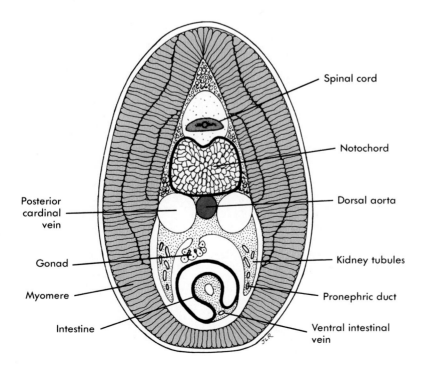

Spinal cord

Notochord

Dorsal aorta

Kidney tubules

Pronephric duct

Ventral intestinal vein

Posterior cardinal vein

Gonad

Myomere

Intestine

FIGURE 2-15

Relatively old ammocoete in cross section just behind liver. Several myomeres are cut in a single section because they curve caudad in passing inward from the skin.

AN ALTERNATE THEORY OF AMMOCOETE-AMPHIOXUS RELATIONSHIPS

The phenomenon of heterochrony, discussed in Chapter 1, has been used as a basis for a minority theory of the origin of cephalochordates that is directly opposite that accepted by most zoologists and presented in Fig. 2-14. If you will reexamine Fig. 2-1, you will see four chordate larvae. Two are protochordates, two are vertebrates. If, in the case of agnathans, a genetic differential in the rate of development of the gonads had become established such that the larval gonads were to mature and reproduction were to occur in the larval state, the evolutionary result would be a protochordate species remarkably similar to an amphioxus. And, in fact, the ammocoete larvae, which live from 2 to 6 years before metamorphosing, were at one time mistaken for protochordates and given the generic name *Ammocoetes*.

The minority theory of the origin of cephalochordates, or even of protochordates in general, derives them from an agnathan-like ancestor the larvae of which, through heterochrony, became sexually mature, giving rise to new species. This, according to the theory, explains the phenotypic similarities between an amphioxus and an ammocoete larva. You may wish to look back once more at the similarities and contrasts between amphioxuses and vertebrates as recounted earlier in the chapter. Recall, however, that the resemblances could also be explained on the basis of parallel evolution or convergence. Objections to the heterochrony theory of origin of protochordates include, among others, the lack in agnathan or other vertebrate larvae of an atrial chamber alongside the gills.

Chapter Summary

1. Protochordates are bilaterally symmetrical, filter-feeding marine chordates with notochord, gill slits, a dorsal hollow central nervous system, and no vertebral column. They constitute two subphyla, Urochordata (Tunicata) and Cephalochordata, in the phylum Chordata.

2. Urochordates include sea squirts, larvaceans, and thaliaceans. The notochord is confined to the tail of larvae. Thaliaceans have no tail and no notochord.

3. Cephalochordates are the lancelets (amphioxuses). The notochord extends the length of the body and remains throughout life.

4. Hemichordates are invertebrates of uncertain taxonomic status. They have gill slits, an invertebrate type of ventral nerve cord, a grooved dorsal nerve strand, a stomochord of uncertain homology, and no recognizable notochord.

5. Ammocoetes are free-swimming, filter-feeding protochordatelike larvae of lamprey eels. They have a notochord, gill slits, a dorsal hollow central nervous system, and no vertebral column.

6. Phenotypic similarity suggests a close phylogenetic relationship between protochordates and vertebrates. The latter may be derived from, or share common invertebrate ancestors with, protochordates.

7. Available evidence has been generally interpreted as indicating a freshwater origin of vertebrates, but the hypothesis is debatable.

SELECTED READINGS

Barrington, E.J.W.: The biology of Hemichordata and Protochordata, San Francisco, 1965, W.H. Freeman and Co. Publishers.

Barrington, E.J.W., and Jefferies, R.P.S., editors: Protochordates, Symposium of the Zoological Society of London, no. 36, New York, 1975, Academic Press.

Bullock, T.H., and Horridge, G.A.: Structure and function in the nervous systems of invertebrates, vol. 2, San Francisco, 1965, W.H. Freeman and Co. Publishers.

Cloney, R.A.: Ascidian larvae and the events of metamorphosis, American Zoologist 22:817, 1982.

Conklin, E.G.: The embryology of amphioxus, Journal of Morphology 54:69, 1932.

Flood, P.R.: A peculiar mode of muscular innervation in amphioxus: light and electron microscopic studies of the so-called ventral roots, Journal of Comparative Neurology 126:181, 1966.

Flood, P.R.: Fine structure of the notochord of amphioxus, Symposium of the Zoological Society of London 36:81, 1975.

Mallatt, J.: The suspension feeding mechanisms of the larval lamprey, *Petromyzon marinus*, Journal of Zoology, London 194:103, 1981.

Stahl, B.J.: Vertebrate history: problems in evolution, New York, 1974, McGraw-Hill Book Co.

Strathmann, R.R., and Bonar, D.: Ciliary feeding of tornaria larvae of *Ptychodera flava* (Hemichordata: Enteropneusta), Marine Biologist 34:317, 1976.

3

PARADE OF THE VERTEBRATES IN TIME AND TAXA

The purpose of this chapter is to introduce those vertebrates that will be referred to in subsequent chapters, relating briefly their natural history and phylogeny, so they will not be total strangers when you meet them again. In fact, it should prove rewarding to refresh your memory by consulting the chapter again at that time. The "cast of characters" for this vertebrate parade is the Abridged Classification, Appendix I, at the end of the book.

There are probably 50,000 different known species of animals with vertebral columns. Fortunately for Noah, more than half of these are fishes. However, many amphibians and some reptiles and mammals share with fishes the freshwater ponds and streams as their permanent abode, and a much smaller number share the seas. Amphibians that can tolerate salt water are rare. There are some. They don't live in the sea, but they forage there. Among living vertebrates a few snakes, turtles, iguanid lizards that live on marine algae, and crocodilians are predominantly or permanently marine, except that oviparous females must emerge to deposit their eggs. Among mammals, cetaceans and sirenians never leave the sea, and marine carnivores leave the sea only to breed. Although no bird lives in the sea, many are entirely dependent on marine organisms for food. All the foregoing animals other than fishes, however, are descendants of terrestrial ancestors and have returned to water. In some of the pages that follow we will examine the adaptations that make these different life styles possible.

VERTEBRATE TAXA

The classification scheme employed in this chapter is a conventional "natural classification" as defined in the discussion of taxonomy and systematics in Chapter 1. The chief vertebrate taxa are classes, subclasses, superorders, orders, suborders, families, genera, and species. Fig. 3-1 lists the vertebrate classes that are most widely recognized. However, all classifications, whether phenetic, cladistic, or natural, are tentative, being subject to revision with new findings. The lower four taxa in Fig. 3-1 are not assigned class status in all schemes. Agnatha, for example, are sometimes given superclass status, or even listed as a subphylum; and Chondrichthyes (cartilaginous fishes) are sometimes placed in a subclass along with Placodermi (ancient bony fishes). These differences mirror tendencies to

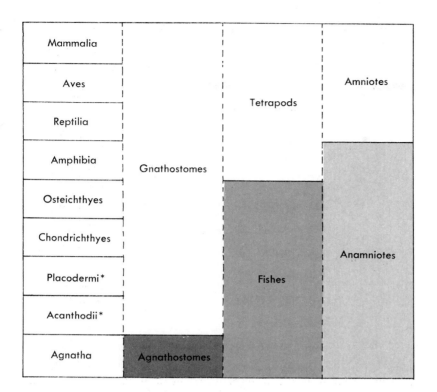

FIGURE 3-1

Major categories of vertebrates. Classes are listed in the first column. Asterisks indicate extinct groups.

incorporate aspects of phenetics or cladistics in a natural classification scheme. There will be no "correct" classification of vertebrates until the entire community of natural scientists agree on one. This seems far into the future! Meanwhile, the one used here has received wide acceptance.

Agnatha lack jaws, as the name implies, hence they are **agnathostomes.** All other vertebrates are **gnathostomes.** (Don't forget to take advantage of Appendix II throughout the course.) Commencing with amphibians, vertebrates are **tetrapods.** Commencing with reptiles, they have a unique extraembryonomic membrane, the amnion, hence they are **amniotes.** Fishes and amphibians lack this membrane, hence they are **anamniotes.**

Since the time of Carl von Linné (Latin, Linnaeus), a Swedish naturalist, taxonomic nomenclature has been latinized by agreement among zoologists of the world. This enables zoologists of all languages to understand one another without translation when the name of any animal is mentioned. For instance, not every zoologist knows what "un chat," "die Katze," or "el gato" is. All these words refer to the common domestic cat, but *Felis cattus* is the taxonomic name for this species. The generic name *Felis* separates certain cats—mountain lion, European wildcat, domestic cat, and others—from lions, tigers, leopards, and so forth, which have been placed in the genus *Panthera. Felis cattus* separates domestic cats from mountain lions *(F. concolor)* and from European wildcats *(F. sylvestrus).* The binomial designation for a species was introduced by Linnaeus in the tenth edition (1758) of his classic book *Systema Naturae.*

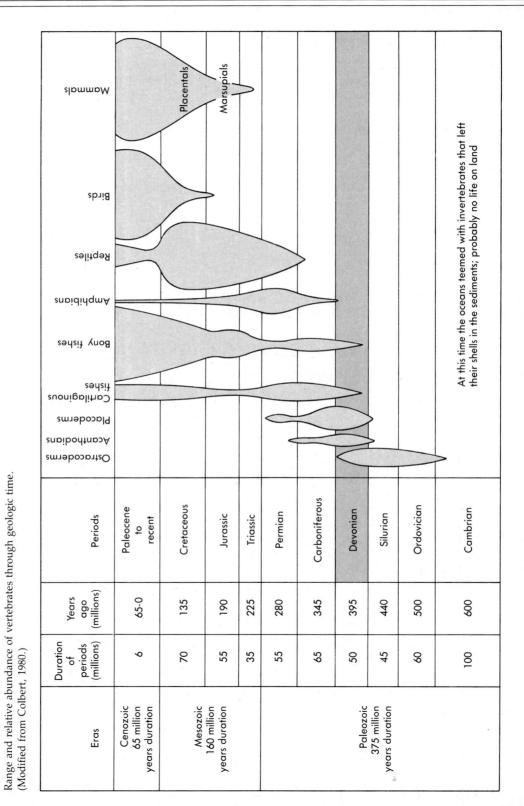

FIGURE 3-2

Range and relative abundance of vertebrates through geologic time. (Modified from Colbert, 1980.)

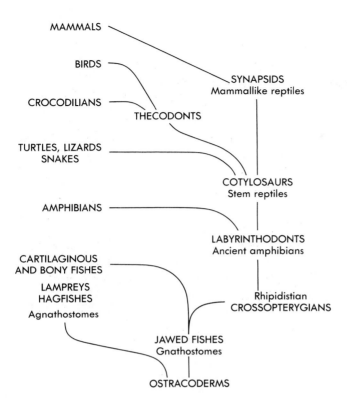

FIGURE 3-3

Mainstreams of vertebrate evolution. The left column represents extant taxa.

Fig. 3-2 shows the range and relative abundance of the major categories of vertebrates in geologic time. Fig. 3-3 summarizes the phylogeny of vertebrates that are alive today.

AGNATHA

The class Agnatha includes two groups of jawless fishes, the ancient bony **ostracoderms,** and the extant eellike boneless lampreys and hagfishes, commonly known as **cyclostomes.** They exhibit both contrasts and similarities.

Ostracoderms

Ostracoderms (Fig. 3-4) are the oldest known vertebrates, constituting a diverse assemblage that dates to the Ordovician. The entire body was covered with bony dermal armor consisting of broad plates and smaller tilelike scales. The plates were largest on the head, where they formed a bony shield. Ostracoderms lacked jaws and paired fins, although some had bizarre projections from the body wall that may have served as stabilizers. Many ostracoderms were only 2 or 3 cm long; a few attained a length of up to 30 cm.

The oldest known ostracoderms lived largely, if not exclusively, in fresh or brackish water, but by the late Silurian they were also marine. The term "ostracoderm" is a utilitarian name having no taxonomic status. They have been placed in four orders (Appendix I), of which **Osteostraci** are the most likely candidates for ancestors of jawed fishes.

Knowledge of the anatomy of Osteostraci, especially of structures beneath the fossilized head shields, is due largely to the painstaking research early in the twentieth century of E.A. Stensiö, a paleontologist at the University of Stockholm. His studies revealed that the head skeleton was a more or less flattened spicule-covered bony shield with four dorsal apertures. Two of these accommodated a pair of upward-staring eyes, a

FIGURE 3-4

A, A fossil ostracoderm of the order Heterostraci. **B,** The bony denticulated head shield of *Cephalaspis,* an ostracoderm of the order Osteostraci. *d* and *l,* Dorsal and lateral fields that may have been the sites of exteroceptors and electric organs, respectively. (**A,** Courtesy of the Palaeontological Association, London. From Special Papers in Palaeontology Number 18, Dineley, D.L., and Loeffler, E.J.: Ostracoderm faunas of the Delorme and associated Siluro-Devonian formations, North West Territories, Canada. Published by the Palaeontological Society, London, 1976. **B,** Reconstruction based on Dineley and Loeffler.)

A

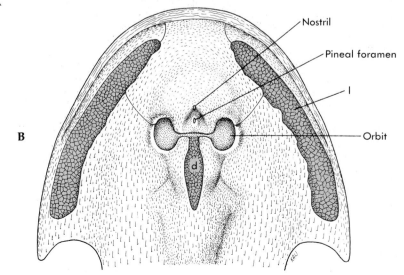

B

third accommodated the median, or pineal, eye, and a small, anterior opening was a single naris from which a nasohypophyseal duct led to an olfactory sac, and beyond. The shield turned under along its lateral edges, where it was then replaced beneath the gills by tilelike scales. Between the anterior edge of the shield and the scales a very small mouth opened into the oropharyngeal chamber, which was lined by gills. A curved row of external gill slits extended from the corners of the mouth to the caudal margins of the head shield. In addition to the shield, the head contained an endoskeleton of endochondral bone and considerable cartilage. Little is known of the postcranial skeleton except that the tail was heterocercal (Fig. 9-18, *A*).

The origin of ostracoderms is unknown. Possible phylogenetic ties with protochordates were discussed in the preceding chapter. Their numbers dwindled and they disappeared at the end of the Devonian, being replaced by jawed fishes.

Cyclostomes

Cyclostomes (Fig. 3-5) became separated from ostracoderms some 400 million years ago and, once established, seem to have undergone little additional morphological change. The fossil lamprey pictured in Fig. 3-6 is remarkably similar to modern lampreys. Extinct and living cyclostomes have been assigned to the orders **Petromyzontiformes** (lampreys) and **Myxiniformes** (hagfishes).

Many cyclostome features are primitive. For example, as in Osteostraci, median nostrils are connected with a single olfactory sac, there are no paired fins, and there are no skeletal elements homologizable with vertebrate jaws. Provided the fins and jaws have not been lost in the remote past, these traits in lampreys are primitive. Also, there are only two pairs of semicircular ducts instead of the three found in higher vertebrates. In other respects cyclostomes appear to be highly specialized. They have lost all bone everywhere in the body—no dermal armor or scales, no bony teeth, no bony skeleton—and they exhibit adaptations for parasitism such as a buccal funnel and the associated rasping tongue. Most interesting is the absence of recognizable vertebrae. As was mentioned in Chapter 1, no centra surround the prominent notochord, which remains the chief skele-

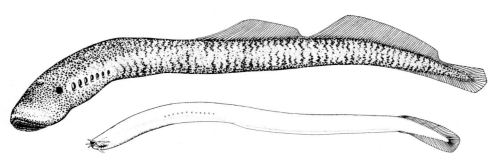

FIGURE 3-5

Lamprey *Petromyzon* above; hagfish *Bdellostoma* below.

FIGURE 3-6

Fossil lamprey *Mayomyzon* from the Carboniferous period. (From Bardack and Zangerl.)

ton of the trunk and tail throughout life. Whether this is primitive or highly specialized cannot be known, since little is known of the postcranial skeleton of their ostracoderm ancestors. A discussion of the terms primitive and specialized is found in Chapter 1.

LAMPREYS. The ammocoete larvae of lampreys have been described in Chapter 2. Many larval traits are retained by adults. A large buccal funnel lined with horny denticles helps keep the parasitic adult lamprey attached to the host while a tonguelike cartilaginous rod covered with horny teeth rasps the flesh of the victim, leaving only skin and skeleton. A single nostril is located dorsally on the head, and a nasohypophyseal canal leads from the nostril to the olfactory sac and then terminates blindly in a nasohypophyseal sac (Fig. 12-2). Seven pairs of gill pouches open separately via gill slits that conduct water into as well as out of the pouches, thus freeing the buccal funnel for feeding on the host.

Petromyzon marinus marinus is a species of **anadromous** lampreys. That is, members live in the sea but migrate upstream to lay their eggs. In 20 to 21 days the small, nonparasitic larvae emerge. After several years upstream they metamorphose into immature adults and migrate to the sea. There they attain sexual maturity and become physiologically adapted for the journey back to the spawning place. *Petromyzon marinus dorsatus* is a landlocked population inhabiting the freshwater Great Lakes between Canada and the United States. They, too, enter rivers to spawn. Most *Lampetra* are freshwater lampreys that do not migrate. They spawn on reaching sexual maturity and promptly die without eating!

HAGFISHES. Hagfishes, also known as slime eels, are marine cyclostomes with a shallow buccal funnel lacking denticles. They are chiefly bottom-feeding scavengers whose diet includes a variety of small invertebrates; but, like lampreys, they also parasitize slow-moving fishes. A fringe of stubby, fingerlike papillae surrounds the buccal funnel, and a single nostril is located just above the funnel. A canal leads from the nostril to the olfactory sac and then on to the pharyngeal cavity, carrying respiratory water (Fig. 12-1). The eyes are vestigial.

Myxine glutinosa, the Atlantic hagfish, has six pairs of gill pouches (occasionally five or seven) that open into a common efferent duct (Fig. 12-1). *Bdellostoma stouti,* common off the coast of California, has 10 to 15 pairs of gill pouches opening directly to the exterior. Hagfishes are not anadromous, and the larvae stay within the egg membranes until metamorphosis. This may be an adaptation to the saltwater environment in which the eggs are laid.

ACANTHODII AND PLACODERMI

Acanthodians and placoderms are two very different unrelated groups of early jawed fishes whose phylogenetic relationship with other early fishes is not clear. **Acanthodians** date to the Silurian and may be the oldest jawed fishes (Fig. 9-14, acanthodian). Like ostracoderms, their head and body were protected by a dermal armor of bony plates and scales. Most of them were only a few inches in length. Although sometimes called "spiny sharks," they were not sharks. Their skeleton consisted of bone and cartilage, and they had a large operculum. Their paired fins, supplemented by as many as five accessory pairs, were supported by hollow spines.

Paleontologists are unsure of the relationship of acanthodians to other Devonian fishes. The closest scrutiny of the structure of their jaws and their paired appendages has not enabled investigators to connect them with any of the known jawless forms that preceded them, nor to relate them beyond reasonable doubt to any of the cartilaginous or bony fishes that succeeded them. There are several points of view on the subject, all of them involving much conjecture. The result is, some systematists classify them as a separate *class* **Acanthodii** (the implication being that they were probably not ancestral to later Devonian fishes), whereas others classify them as a *subclass* in the class Osteichthyes. Stahl has reviewed the arguments that support these divergent opinions. This argument exemplifies the extent to which intuition ("educated guesses") dominate taxonomic classification. Nevertheless, intuition is an essential tool, and has proved to be a remarkably reliable tool, in systematics.

Placoderms were bizarre armored fishes that appeared a little later than ostracoderms and acanthodians and became abundant in the fresh waters

FIGURE 3-7

Two Devonian placoderms, each about one-third natural size. *Coccosteus* is an arthrodire; *Bothriolepis* is an antiarch. (From Colbert.)

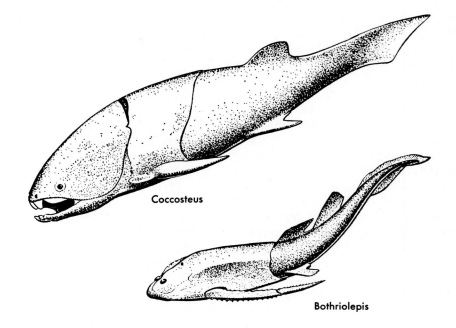

Coccosteus

Bothriolepis

of the Devonian while ostracoderms were disappearing. They had paired fins unlike those of any later fishes, and were swift predators. The best known placoderms are the **arthrodires** (Fig. 3-7, *Coccosteus*). A heavy bony dermal shield covered the head and gill region and another covered much of the trunk, the two shields meeting in a movable joint. The remainder of the body was covered by smaller bony scales or, in late species, was naked. **Antiarchs** (Fig. 3-7, *Bothrioplepis*) were small placoderms with atypical pectoral fins, dorsal eyes, and a flattened ventral surface that suggests they were bottom feeders. Placoderms are either lumped with cartilaginous fishes in a single class **Elasmobranchiomorphi** on the theory that placoderms and sharks, who were contemporaries in the Devonian, are closely related (cladistic taxonomy), or they are accorded class status, **Placodermi.** The fact is, we simply cannot say with a high degree of confidence who came from whom during the early periods of the Paleozoic.

CHONDRICHTHYES

Living Chondrichthyes, the cartilaginous fishes, are usually placed in two subclasses, **Elasmobranchii,** which includes sharks, rays, skates, and sawfishes, and **Holocephali,** the chimaeras, or ratfishes (Figs. 3-8 and 16-2, *B*). They have no bone in their body other than in their scales and teeth. The mouth is on the ventral surface rather than terminal, except in Paleozoic sharks such as *Cladoselache* (Fig. 3-8). Their unique scales, called placoid, are composed of a basal plate and a bony spine of dentin that protrudes through the epidermis, giving the skin of sharks the texture of sandpaper.

The skeleton of the pelvic fins of males is modified to form claspers used in transferring sperm to the female reproductive tract, fertilization being internal. The eggs are macrolecithal, and in oviparous species they are often encased in a horny or leathery shell with tendrils that entwine around vegetation as holdfasts.

Elasmobranchs

The subclass Elasmobranchii consists of two orders of Paleozoic sharks, **Cladoselachii** and **Pleuracanthodii,** an order of later extinct and modern sharks, **Squaliformes,** and one order of skates, rays, and sawfishes, **Rajiformes.** The gill slits are exposed (naked) rather than covered by a fleshy or bony operculum, and there are usually five pairs, although the sharks *Hexanchus* and *Haptanchus* have six and seven pairs, respectively. Seven slits may have been the primitive number in sharks, since that was the number found in *Cladoselache*. It is the largest number in any jawed fish. Anterior to the first gill slit there is usually a spiracle that bears in its walls a miniature gill-like surface, or pseudobranch. The caudal fin of Paleozoic sharks was heterocercal, the predominant type in the Paleozoic, and that of today's sharks, skates, and rays has not changed (Fig. 9-18, *A*).

Dissection of a shark is a valuable point of departure for study of comparative vertebrate morphology because of its generalized anatomy. If one were seeking a living blueprint of a generalized vertebrate—an architectural pattern that could provide preliminary insight into the systems of the bodies of all vertebrates—it would be difficult to find a better example than a shark. Most frequently studied is the viviparous spiny dogfish of the Atlantic, *Squalus acanthias,* named for the prominent horny spine associated with each dorsal fin. *Squalus suckleyi* is the Pacific spiny dogfish. *Mustelus,* the smooth dogfish shark, lacks dorsal spines.

Rays, skates, and sawfishes are compressed dorsoventrally, and the anterior fins are attached all along the sides of the head and trunk, forming broad, undulating, winglike locomotor organs (Fig. 3-8). The five gill slits

FIGURE 3-8

Representative Chondrichthyes. *Cladoselache* is primitive and extinct.

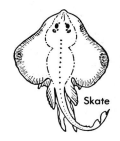
Skate

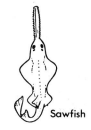
Sawfish

Chimaera

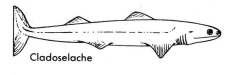
Cladoselache

Hammerhead shark

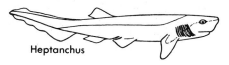
Heptanchus

are ventral (Fig. 12-8), but the spiracle, which is the chief incurrent route for respiratory water, remains on the dorsum. Most rays and skates subsist on molluscs scooped up from muddy bottoms, and the location of the spiracle removes it from the mud and debris that is stirred up by bottom feeding. Manta rays, on the other hand, are filter feeders. The tail of Rajiformes has changed from a muscular locomotor organ to a lean, often whiplike, organ of defense and, in some species, of offense. In electric rays (torpedos) it houses an electric organ capable of a high-voltage discharge that immobilizes prey. Stingrays have rows of spines along their tails that inflict wounds and often contain a poison. Sawfishes, not to be confused with sawfish sharks, have the rajiform shape to a lesser degree. The sword is used for impaling small fishes and disturbing the sea bottom in a search for burrowing animals.

Holocephalans

Chimaeras lack scales on most of the surface, have a fleshy operculum that hides their gill slits, and the spiracle is closed. The upper jaw, unlike that of elasmobranchs, is solidly fused with the cartilaginous braincase; and instead of teeth, hard flat bony plates on the jaws crush molluscan shells, the diet of chimaeras being similar to that of rays.

OSTEICHTHYES

The bony fishes can be traced to the Devonian. They have a skeleton composed partly or chiefly of bone, the gill slits are covered by a bony operculum that grows from the second visceral arch, and the mouth is usually terminal rather than on the ventral surface of the head as in cartilaginous fishes. Although one would seldom fail to recognize the members of the group as fishes—sea horses might pose a problem—there are extreme variations in structure among the taxa. Most bony fishes have a gas-filled swim bladder which, in air breathers, is connected to the pharynx by an air duct. The adult cloaca usually is so shallow as to be virtually nonexistent, except in dipnoans. There are two subclasses of bony fishes, the now abundant and ubiquitous ray-finned **Actinopterygii,** and a once prominent subclass that became a minority during the Mesozoic, the lobe-finned **Sarcopterygii.**

Ray-Finned Fishes

Actinopterygii are bony fishes in which slender fin rays are virtually the sole support for their otherwise membranous fins (Fig. 9-16). They also differ from other bony fishes in lacking internal nares. During the Paleozoic, their bony dermal armor and scales were covered with an enameloid called **ganoin** (Fig. 5-30, *A* to *C*) and their caudal fins, like those of Paleozoic sharks, were heterocercal. Both of these traits have all but disappeared from modern actinopterygians; traces of ganoin remain in the scales of a

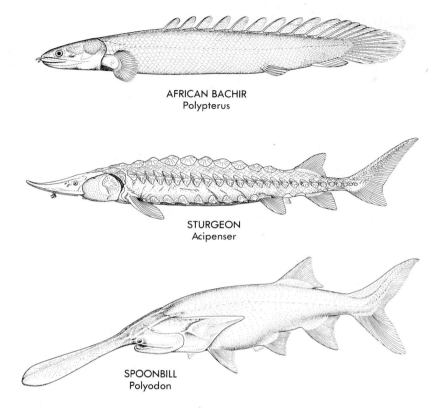

AFRICAN BACHIR
Polypterus

STURGEON
Acipenser

SPOONBILL
Polyodon

FIGURE 3-9

Three surviving chondrosteans.

few relics, the ganoid fishes. These relics have been placed in two super-orders, **Chondrostei,** which includes Paleozoic actinopterygians (paleoniscoids), and **Holostei.** Because of the unavoidable vagaries of taxonomy, these taxa include some older bony fishes that no longer qualify as ganoids because the ganoin has been lost.

The third superorder of ray-finned fishes, **Teleostei,** replaced ganoid fishes both in species numbers and actual numbers as a result of a burst of mutational activity that began about the time reptiles were undergoing radiation on land. The activity brought teleosts to a position of numerical dominance before the end of the Mesozoic.

CHONDROSTEANS. Living chondrosteans consist of sturgeons, spoonbills (paddlefishes), and two freshwater African genera, *Polypterus* and *Calamoichthyes* (Fig. 3-9). The African genera show a greater resemblance to their Paleozoic ancestors, having ganoid scales and a well-ossified endoskeleton. They therefore are true ganoids. The endoskeleton of sturgeons and spoonbills, on the other hand, remains largely cartilaginous, and their bony dermal plates or scales lack ganoin. Sturgeons (*Acipenser* and others) and spoonbills (*Polyodon* from the U.S.A. and *Psephurus* of China) are in a separate order, **Acipenseriformes.**

HOLOSTEANS. Only two genera of holosteans have survived, *Lepisosteus,* the garfishes (several species), and *Amia calva,* the bowfin, the latter being the sole surviving species of the genus (Fig. 3-10). Both are freshwater fishes. Gars are covered with lepisosteoid ganoid scales (Fig. 5-31). *Amia* is more modern. The head of both is covered with bony platelike scales, but these lack ganoin; and the trunk and tail of *Amia* have modern fish scales. Although much of the endoskeleton is ossified in both genera, the skull remains largely cartilaginous throughout life (Fig. 8-7).

TELEOSTS. Teleosts (Fig. 3-11) are the most recently evolved bony fishes, hence they are referred to as "modern." From the perspective of geologic time they are modern indeed, although some of them have been around for 65 million years, more or less. The tail is no longer heterocercal, the scales are no longer heavily bony and inflexible, the dermal bones of their skulls are much thinner, the jaws and palate have become more independently maneuverable, the pelvic fins are often far forward, and the body in general has become altered in innumerable other ways, resulting in organisms that can occupy all accessible aquatic niches of the planet. There are more than 20,000 species and they constitute 95 percent of all living fishes. There are long, slim ones who have lost paired appendages; short, fat ones with sails; transparent ones; fishes that stand on their tails; fishes with both eyes on the same side of the head; fishes that carry lanterns, that climb trees, that carry their unhatched eggs in their mouth, that appear to be smoking pipes, that possess periscopic eyes; and hundreds of other bizarre genera. They inhabit the abyssal depths far out from the continental shelves, they cavort in modest brooks, and some make nightly sorties onto land. They display a gamut of coloration, although a relatively small number of pigments assisted by myriads of light-dispersing crystals are responsible for all hues. With the exception of lobe-finned fishes, chondrosteans, and holosteans, any jawed fish caught on a hook, viewed in an aquarium, or sold in the marketplace is a teleost.

FIGURE 3-10

The two surviving genera of holosteans.

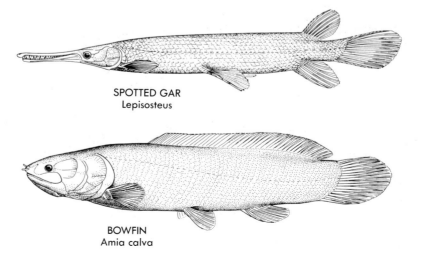

SPOTTED GAR
Lepisosteus

BOWFIN
Amia calva

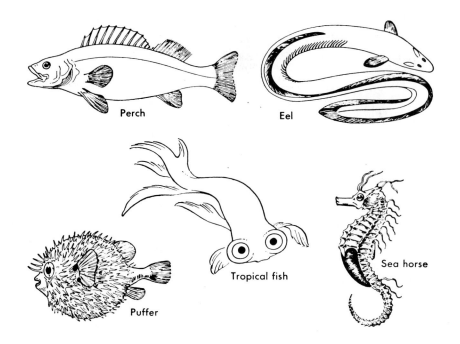

FIGURE 3-11
Group of teleosts.

Lobe-Finned Fishes

Sarcopterygii have a prominent fleshy lobe at the base of their paired fins (Fig. 3-12). The lobe contains part of the fin skeleton (Fig. 9-43; compare Fig. 9-16). Most lobe-fins also have internal nares that open into the oropharyngeal cavity. There are two orders, **Crossopterygii,** which includes the ancestors of amphibians, and **Dipnoi,** the true lungfishes. The two lines were differentiated by the start of the Devonian, but speciation has been a much slower process among lobe-fins than among ray-fins.

CROSSOPTERYGIANS. Crossopterygians were the most common fishes of the Devonian, but they are now extinct except for a single known relict, *Latimeria*. The skeletal elements within the fin lobe correspond closely to the proximal skeletal elements of early tetrapod limbs (Fig. 9-43), the skull was similar to that of the first amphibians (Fig. 8-8), they had swim bladders that were probably used by some species for lungs, and most had internal nares, although they probably were not used for breathing. They have been placed in two orders, the freshwater **Rhipidistia,** and a less typical side branch that lacks internal nares, **Coelacanthiformes,** of which *Latimeria* is a member. The freshwater rhipidistians are thought to have been the stem from which amphibians evolved.

DIPNOANS. These are the true lungfishes. There are three living genera, *Protopterus* from Africa, *Neoceratodus* from Australia, and *Lepidosiren* from Brazil. *Protopterus* and *Lepidosiren* have inefficient gills and suffocate if held under water, but *Neoceratodus* relies on gills for oxygen except when the oxygen content of the water is low. During the wet season these fishes

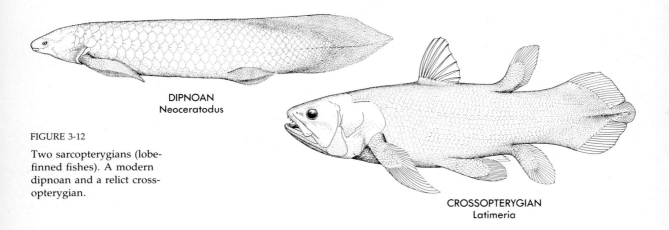

DIPNOAN
Neoceratodus

FIGURE 3-12

Two sarcopterygians (lobe-finned fishes). A modern dipnoan and a relict crossopterygian.

CROSSOPTERYGIAN
Latimeria

FIGURE 3-13

Major groups of fishes and their postulated ancestry. Asterisked groups are extinct.

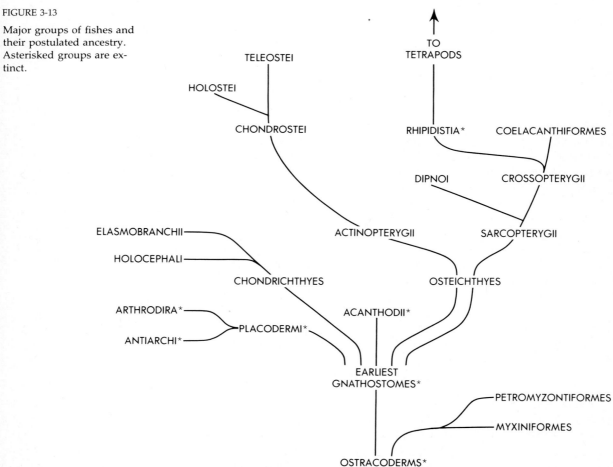

inhabit streams and swamps, but when the rains cease and the tropical sun dries up their environment, the African and Brazilian species dig deep burrows in the mud and spend the dry hot season in a state of aestivation. Their lowered metabolism minimizes water loss and reduces the need for nutrients and oxygen.

Dipnoans resemble amphibians in many respects, and the two groups may have had a common crossopterygian ancestor. In both, the swim bladder is supplied by a branch from the sixth aortic arch instead of from the dorsal aorta as in crossopterygians, the atrium of the heart is partially divided into two chambers, both usually have a larval stage with external gills, both have internal nares, and in both the swim bladders or lungs have ducts leading to the pharynx. The postulated phylogenetic lines of the major groups of fishes is given in Fig. 3-13.

AMPHIBIA

Amphibians have been placed in three subclasses, **Labyrinthodontia,** the first vertebrates to walk on land, **Lepospondyli,** a diverse group that was at their height in the upper Carboniferous, and **Lissamphibia,** which includes Triassic and modern amphibians. After amphibians appeared, they and the fishes evolved simultaneously, the former on land, the latter in the water. Successive mutations increasingly adapted labyrinthodont descendants for terrestrial life.

Labyrinthodonts

The oldest known amphibians were the swamp-dwelling labyrinthodonts (Fig. 3-14), so named because the dentin of their teeth was infolded in such a manner as to resemble a labyrinth when viewed in cross section. The subclass **Labyrinthodontia** consists of three orders, **Ichthyostegalia,** the oldest, who made their appearance in the late Devonian, **Temnospondyli,** a group that was common in the Permian, and **Anthracosauria,** a relatively small group some of whom were probably ancestral to reptiles.

Labyrinthodonts were a very diverse assemblage, and the question of their precise kinships is still unsettled. The structure of the vertebrae has played a prominent role in past attempts at a solution (Fig. 7-11). Particularly problematic are the kinships of lepospondyls, as will be seen shortly. One proposed set of hypothetical relationships is diagrammed in Fig. 3-15. It is not the only one, and there is at this time no "correct" one.

Labyrinthodonts had many features seldom seen in modern tailed amphibians, including minute bony scales in the dermis of the skin, a fish-

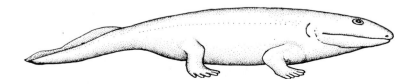

FIGURE 3-14

Reconstruction of an early labyrinthodont, *Ichthyostega*, from the Devonian. The skeleton is seen on p. 196.

FIGURE 3-15

A postulated family tree of amphibians in accordance with the monophyletic theory of amphibian origins. Any family tree of amphibians is tentative.

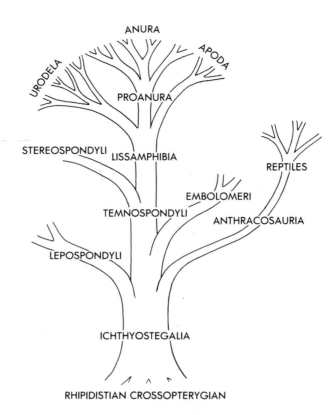

like tail containing dermal fin rays, and skulls so similar to those of rhipidistian crossopterygians that the two are easily confused unless there is an operculum to confirm that it is a fish (Fig. 8-8). Grooves in the skull bones just under the skin indicate that labyrinthodonts, like fishes, had a sensory canal system of neuromast organs that monitored the aquatic environment. Today's aquatic amphibians still have this system, but terrestrial species lose it at metamorphosis. Labyrinthodonts were as small as today's newts, and as large as the largest crocodiles. That they gave rise to the first reptiles is beyond question.

Lepospondyls

These small salamanderlike forms became extinct before the last of the labyrinthodonts. There are several orders, each as diverse as the three orders of modern amphibians. Among them were species that had lost their limbs by the beginning of the Carboniferous, species resembling modern urodeles, and species with large bizarre triangular skulls. The skull morphology of many lepospondyls suggests that those specific ones no doubt came from labyrinthodont stock, but it has been theorized by some paleontologists that at least some lepospondyls may have arisen independently of labyrinthodonts from rhipidistian crossopterygians (Diphyletic Theory of the Origin of Amphibians). Resolution of the problem has been handicapped by lack of critical fossils. It has also been proposed by competent paleontologists that urodeles were derived from lepospondyls. The current consensus is that lepospondyls were a very early side branch from the evolutionary line leading from labyrinthodonts to modern tetrapods, and that they left no descendants (Monophyletic Theory).

Lissamphibians

The subclass Lissamphibia includes three orders of modern amphibians, **Urodela** (Caudata), the modern tailed amphibians, **Anura,** which are the frogs and toads, and **Apoda** (Gymnophiona), the legless burrowing species. The problem of the origin of urodeles has already been mentioned. The ancestry of apodans is particularly conjectural because only a single fossil skeleton has been found, and it consists only of Cenozoic vertebrae. The skeletons of Triassic and Jurassic anurans, although primitive, resemble those of modern species to the extent that they provide no hint of who their immediate amphibian forebears might have been.

Amphibian eggs lack the membranes that enable reptilian embryos to develop in terrestrial habitats. Consequently, most amphibians must return to water to lay their fishlike eggs. There is usually an aquatic larval stage with external gills and neuromast organs that supplement other sense organs in informing them about their aqueous environment. At metamorphosis the gills are usually lost, even in aquatic species, although the latter usually retain neuromast organs. Amphibians for the first time develop a middle ear cavity, eardrum, and an ear ossicle, the columella. This sensory complex, not fully developed in all amphibians, replaces the piscine neuromast system by detecting sound waves transmitted in air and transmitting them to the membranous labyrinth inherited from fishes. The hind limbs are braced against the vertebral column by a single trunk vertebra, which provides leverage for amphibian-style locomotion, but they have little ability to keep the belly from sagging onto the ground. The skull articulates with the vertebral column by two occipital condyles instead of one as in their Paleozoic ancestors, and the joint is now movable. The skin has lost all traces of ancestral bony scales except in apodans, and a layer of keratinized water-impervious cells that retard desiccation in air tends to develop on the surface. These and other amphibian features are discussed in later chapters.

URODELES. Urodeles tend to be perennibranchiate, as shown in Table 3-1, which lists the eight extant families. Representative genera are illustrated in Fig. 3-16 and characterized briefly below.

Necturus is the only genus of Proteidae in the United States and Canada, where there are six species and subspecies. Its European counterpart, *Proteus*, is a blind cave dweller. A 2 cm larva has three pairs of external gills and slits. Adults retain the external gills and two pairs of slits and attain sexual maturity at five years of age. They are therefore neotenous. Unlike neotenous populations of *Ambystoma* and *Siren*, *Necturus* cannot be induced to discard its gills by administration of thyroid hormone.

Amphiuma is a North American eellike urodele up to 1 m long with tiny appendages. *Amphiuma pholeter* has one digit, *A. means means* has two, and *A. means tridactylum* has three.

Hynobius is a primitive terrestrial Asiatic genus. *Cryptobranchus* is Asiatic and North American. It looks ferocious because of its broad flat head and wrinkled skin, which often conceals the single gill slit.

Salamandra and *Notophthalmus* are Eurasian and Western Hemisphere salamanders. *Notophthalmus viridescens* (Fig. 3-17) after several months of an aquatic larval existence loses its gills and gill slits, sprouts legs, and, now an **eft,** emerges from the pond to commence a term of residence on land. The skin develops a thick cornified layer that blocks the openings of the sensory canal system and skin glands, the body assumes a bright orange-red color, and a series of dorsolateral black-bordered red spots appear. The red eft stage lasts one to three years, depending on the locality. At the approach of sexual maturity the eft, stimulated by the hormone prolactin, joins a mass migration to freshwater ponds where mating occurs and eggs are deposited. The tail commences to change from round to laterally compressed with dorsal and ventral keels, the thick cornified layer of epidermis is shed, exposing the openings to mucous glands and the sensory canal system, and the body gradually becomes olive green above and light yellow below, a protective coloration appropriate for life in a freshwater pond. The animal is now a **newt.** In some localities the larvae remain in the pond and retain vestigial gills throughout life. *Salamandra atra* is a viviparous species.

Ambystoma is terrestrial except in localities where, as in the higher altitudes of Mexico, S.A., the water is low in iodides needed for synthesis of thyroid hormone which induces metamorphosis. *Ambystoma mexicanum*, the mexican axolotl,* can be caused to discard its gills by administering

---

*Xolotl was an Aztec god of twins and monsters.

TABLE 3-1 Distribution of Gills, Pharyngeal Slits, and Lungs Among Adult Urodeles

Family	Representative Genera	Number of Pairs		Lungs
		Gills	Slits	
Proteidae	*Necturus*	3	2	Yes
Amphiumidae	*Amphiuma*	0*	1	Yes
Hynobiidae	*Hynobius*	0*	0	Occasionally
Cryptobranchidae	*Cryptobranchus*	0*	1	Yes
Salamandridae	*Notophthalmus*	0*	0	Yes
Ambystomatidae	*Ambystoma*	0*	0	Yes
Plethodontidae	*Plethodon*	0*	0	No
Sirenidae	*Siren*	3	3 to 1	Yes

*Some individuals are occasionally perennibranchiate.

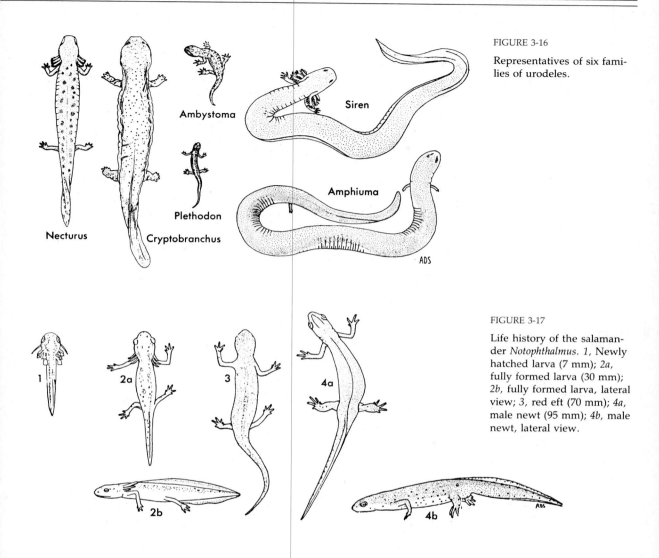

FIGURE 3-16

Representatives of six families of urodeles.

Ambystoma

Siren

Plethodon

Amphiuma

Necturus   Cryptobranchus

ADS

FIGURE 3-17

Life history of the salamander *Notophthalmus*. *1*, Newly hatched larva (7 mm); *2a*, fully formed larva (30 mm); *2b*, fully formed larva, lateral view; *3*, red eft (70 mm); *4a*, male newt (95 mm); *4b*, male newt, lateral view.

either iodine or the hormone. Even among neotenic populations some individuals metamorphose spontaneously, perhaps because of a more efficient iodine-concentrating mechanism in their thyroid follicles.

Plethodon belongs to a family most of the individuals of which lose their gills at metamorphosis, and none of whom develop lungs. They live somewhat remote from ponds and, excepting tropical forms, lay their eggs in moist sites such as under a log or in a cave. The larvae hatch with legs already formed and may never enter water. The larval gills are discarded a few days after hatching.

Siren lives in muck or swamplands. They are perennibranchiate and lack hind limbs. *Pseudobranchus*, another member of the family, comes closer to complete metamorphosis. There are three pairs of gills, but only the slit behind the first pair remains open. In one species the gills and slit, although present, are completely covered by skin.

FIGURE 3-18

An apodan. The annular structure is a specialization that probably aids in burrowing.

ANURANS. Anurans are tailless amphibians in which several caudal vertebrae are fused into one elongated urostyle (Fig. 7-13). No single morphological trait distinguishes frogs from toads. The most representative frogs are the long-legged, slender-bodied members of the family **Ranidae.** The most representative toads are in the family **Bufonidae.** Toads are more terrestrial and some have prominent parotoid glands (Fig. 5-10). Tree toads, also called tree frogs, are in the family **Hylidae.** Anurans breathe with lungs and skin and most are amphibious, entering only fresh water. A few, such as *Rana cancrivora*, the crab-eating frog of Thailand, can tolerate salt water.

A few frogs do not deposit their eggs in water-filled ditches or ponds. Robber frogs, for instance, like most anurans, breed only during and after rains, but the eggs are laid in rain-filled crevices, in a pile of leaf mold, or at the base of a grass hummock away from the water's edge and above the flood level. The larval stage is passed within the jelly envelopes, and the interval between egg-laying and the completion of metamorphosis is much shorter than in other anurans.

Some species of tree toads also do not lay their eggs in water. They carry developing eggs in a brood pouch under the skin of the back, and fully metamorphosed frogs emerge through an opening in the skin. A similar condition is found in the Surinam frog, *Pipa pipa*. The East African toad *Nectophrynoides vivipara* is viviparous. As many as 100 young develop in the uterus.

The earliest fossil anuran is *Triadobatrachus* (= *Protobatrachus*), a proanuran. The skull was quite similar to that of today's anurans, and the body was shortened as a result of reduction in the number of trunk vertebrae, approaching the condition in modern anurans. It had a tail with independent caudal vertebrae, the ribs were longer than they are today, the tibia and fibula were not fused as they are in today's anurans, and the abdomen was covered with bony scales.

APODANS. Apodans, or caecilians, are circumtropical limbless amphibians that, except for aquatic species, live in burrows in swampy places (Fig. 3-18). Their eyes are small and, in some of the 50 or more species, buried beneath the bones of the skull. Some species have minute scales in their skin. Most caecilians are about 30 cm in length, but some attain a length of over a meter and have as many as 250 vertebrae. The vent is almost at the end of the body, so the tail is very short. Terrestrial species lay large yolky eggs, and the larval stage is passed in the egg envelopes. Several aquatic genera are viviparous.

REPTILIA

From some group of early labyrinthodonts, perhaps anthracosaurs, there arose during the mid-Carboniferous a group of tetrapods destined to be named, 300 million years later, **cotylosaurs** (Fig. 3-19, *A*). These were the stem reptiles. From them evolved a varied and abundant assemblage of descendants who established themselves securely on land, conquered the air, and reinvaded the seas, flourishing as a tetrapod majority for 260 million years. Remaining are a relatively few descendants, the **turtles,** an ancient group that thus far has doggedly persevered in the struggle for existence; *Sphenodon,* a lizardlike relict; **modern lizards,** recent additions to reptilian society; **snakes and amphisbaenians,** which are lizards deprived of their appendages; and **crocodilians.** These survivors mirror the effects of mutations and the ravages of a changing environment on cotylosaurs and their descendants. Birds and mammals represent further mutations of reptilian chromosomes.

Reptiles experienced changes that made them better adapted for a permanent terrestrial existence than were their amphibious ancestors. Perhaps the major one was acquisition of three extraembryonic membranes, the **amnion, chorion,** and **allantois,** which emancipated them from the necessity of returning to water to lay their eggs. The amnion is a fluid-filled membranous sac in which the embryo develops. The fluid is secreted by the lining of the amnion. To state the situation figuratively, instead of the mother having to go to water to deposit her eggs, the embryo is provided with its own private pond, and the egg can be laid on land. Thus, reptiles became the first amniotes. The chorion and allantois constitute a vascular-

FIGURE 3-19

A, Stem reptile. B, "Living fossil" reptile. C, Modern iguanid lizard.

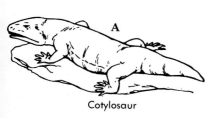

A

Cotylosaur

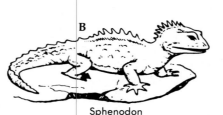

B

Sphenodon

C

Lizard

FIGURE 3-20

Anolis in process of hatching. (Courtesy Carolina Biological Supply Co., Burlington, N.C.)

ized membrane that lies against the porous eggshell, another reptilian innovation, taking the place of larval gills for respiration. The role of these membranes in viviparous amniotes is discussed in Chapter 3.

Because of these three extraembryonic membranes, oviparous reptiles were able and, indeed, forced, to lay their eggs on land. The young hatch fully formed, without a larval stage, ready to seek their own food (Fig. 3-20). Not only were reptiles liberated from returning to water to lay their eggs; aquatic oviparous species must go onto land to do so, since porous eggs would become water-logged in an aqueous environment.

Reptiles exhibit other adaptations to terrestrial life. The body surface is covered with a thick layer of cornified epidermal cells that is organized into plaques, shields, or surface scales unlike the bony dermal scales of fishes and early amphibians. The scales are impervious to water, which results in water conservation, a necessity for animals living in air and often remote from water. Reptiles have developed a neck by specialization of several postcranial vertebrae. This modification, combined with their single occipital condyle, enables reptiles to scan the horizon. The pelvic girdle now articulates with two sacral vertebrae, providing a stouter brace for more powerful hind limbs. The digits are supplied with claws, and a new kidney, the metanephros, has come into existence as a modification of the amphibian kidney. The heart is partially or completely divided into right and left chambers, thereby separating the systemic and pulmonary circulations to an extent not achieved in anamniotes.

Living reptiles, like fishes and amphibians, are **ectotherms,** that is, they cannot maintain a more or less constant body temperature in the face of variations in the ambient temperature. *Sphenodon* and some lizards have retained a parapineal organ that is exposed to the environment and aids in

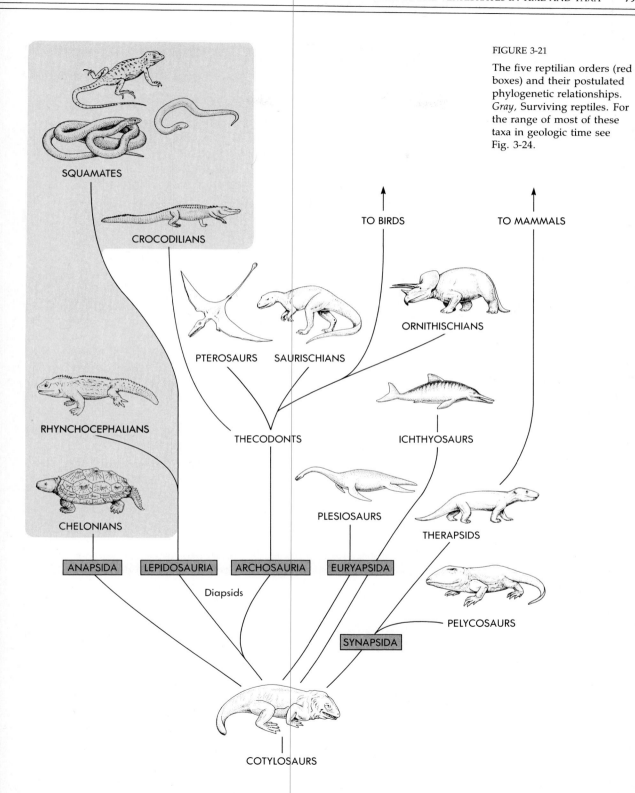

FIGURE 3-21

The five reptilian orders (red boxes) and their postulated phylogenetic relationships. *Gray*, Surviving reptiles. For the range of most of these taxa in geologic time see Fig. 3-24.

thermoregulation by monitoring the duration of exposure of the animal to the sun's rays (Fig. 16-18).

These then are the reptiles: scaly, clawed, mostly terrestrial tetrapods lacking feathers and hair, which, excepting viviparous species, lay macrolecithal, shell-covered (cleidoic) eggs on land, the embryos of which develop within an amnion, and the young of which are hatched fully formed. We will look briefly at the five subclasses, **Anapsida, Lepidosauria, Archosauria, Euryapsida,** and **Synapsida,** which have been erected according to the number of temporal arches and fossae in the skull (Fig. 8-15). Their postulated phylogenetic relationships are diagrammed in Figure 3-21.

Anapsids

The subclass Anapsida includes cotylosaurs and turtles. The name "anapsid" refers to the absence of a bony temporal arch. **Turtles (**order **Testudinata,** or **Chelonia)** are ancient reptiles that have probably remained unchanged for 175 million years. They are characterized by their shell of bony dermal plates to which ribs and trunk vertebrae are solidly fused. A loss of most trunk musculature occurred, but this could not have been detrimental since the rigid shell rendered them useless. Turtles have lost their teeth, but there are still carnivorous, herbivorous, and omnivorous species. One of the earliest turtles from the Triassic is illustrated in Fig. 3-21.

Lepidosaurs

There are two orders of living lepidosaurs, **Rhynchocephalia,** with a single survivor, *Sphenodon,* the tuatara, and **Squamata,** the lizards, snakes, and amphisbaenians. Primitive lepidosaurs had two temporal fossae, hence diapsid skulls. *Sphenodon* still has a diapsid skull, but squamates have only remnants of one or both arches. The loss is correlated with the expansion of increasingly powerful jaw muscles.

RHYNCHOCEPHALIANS. The **tuatara** (Fig. 3-19, *B*) is a primitive generalized lizardlike reptile, but with quite different scales, teeth, and internal morphology than lizards. At one time they inhabited New Zealand, but they are now confined to the islands of Tasmania and New Guinea, all three of which were once connected. The tuatara, now government protected, is becoming extinct. They feed on small vertebrates and insects, mature in 20 years, and their normal life span may exceed a century. They attain a length of 0.75 m (2.5 ft).

SQUAMATES. These, the scaly reptiles, include **lizards** (suborder **Lacertilia**), **snakes** (suborder **Serpentes**), and the snakelike, burrowing **amphisbaenians** (suborder **Amphisbaenia**). Lizards are the most versatile reptiles. Because of well differentiated appendicular muscles and a suitably structured skeleton some lizards run agilely on their hind limbs alone; some are amazing broad jumpers; some, the nocturnal arboreal geckos, have on their toes suction discs that adhere to smooth surfaces when climbing; and some, like *Draco*, the flying dragon, glide through the air on rib-supported extensions of the lateral body wall. A few, the skinks, are limbless. Some lizards are blind, and some have transparent eyelids **(spectacles).** There is a third eyelid, the **nictitating membrane,** and the teeth are in sockets. Iguanid lizards (family Iguanidae) are more generalized. *Iguana* is an arboreal, strictly vegetarian species that enters sea water to forage on sea weed. The largest lizard, the Komodo dragon of Indonesia, reaches 2.75 m in length and 115 kg in weight. The horned toad of the North American desert, *Phrynosoma*, is a lizard.

Snakes evolved from lizards. Lacking limbs, they have achieved other modes of locomotion, including a unique set of muscles that connects the ribs with the large ventral scales, or scutes. Despite the absence of limbs snakes inhabit mountains, deserts, trees, fresh water, and the sea. Most sea snakes are viviparous, thereby eliminating the necessity to return to land to lay their eggs.

Amphisbaenians are subterranean lizards 30 to 70 cm long, mostly limbless, with annulated bodies like those of the burrowing apodans. The eardrum and eyes are covered with opaque skin. There are about 120 species.

Archosaurs

Archosaurs were the dominant land vertebrates during the Mesozoic. The subclass includes **Thecodontia,** the stem archosaurs, **Pterosauria,** flying reptiles, **Saurischia,** dinosaurs with a reptilian pelvis, **Ornithischia,** dinosaurs with a birdlike pelvis, and **Crocodilia,** the latter being the sole surviving archosaurs (Fig. 3-22). All have diapsid skulls.

Thecodonts are so named because their teeth are set in deep sockets, unlike those of other early reptiles. Pterosaurs, like birds, had pneumatic bones, but their wings were more batlike, being supported by an elongated fourth finger (Fig. 9-29, *B*). They had a long neck and a tail. The largest had a wingspread of 10 m (35 ft).

Saurischians and ornithischians are differentiated on the basis of their pelvic girdles. That of a saurischian dinosaur, shown in Fig. 9-12, *C*, is quite different from that of an ornithischian, shown in Fig. 9-10 along with that of a goose. Dinosaurs came in many sizes and body builds, many were biped-

FIGURE 3-22

Representative archosaurs.

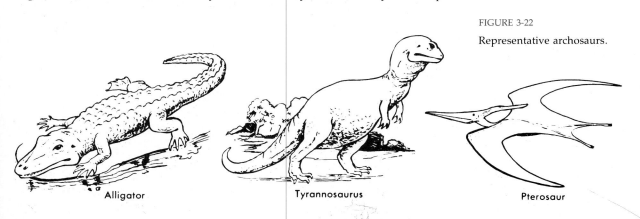

Alligator   Tyrannosaurus   Pterosaur

al, and massive ones like *Tyrannosaurus,* a saurischian, do not represent the majority. Many saurischians were swift predatory carnivores, whereas ornithischians were herbivores lacking front teeth, although some had horny beaks. Some small unidentified bipedal ornithischian is thought to have given rise to birds.

Crocodilians include the subtropical and tropical alligators, crocodiles, caimans, and a single species of gavials. Crocodiles can be distinguished from alligators by the shape of the snout, which is more slender and triangular in crocodiles, broad and rounded in alligators; and by the enlarged fourth tooth of the lower jaw, which in crocodilians fits into a notch on the upper jaw and can be seen when the mouth is closed, whereas in alligators it fits into a pit medial to the upper jaw tooth line and cannot be seen. Caimans and alligators look very much alike and belong to the same family (Alligatoridae). Caimans are South American species; gavials are from northern India. The gavial snout is long and very slender because the left and right halves of the mandible are united by a long symphysis that extends all the way to the fifteenth tooth.

Euryapsids

Euryapsida, also known as **Parapsida** and **Synaptosauria,** is a varied group that includes not only extinct marine reptiles, the best known of which are the ichthyosaurs and plesiosaurs, but also some primitive lizard-like terrestrial species. The taxon was established to accommodate reptiles with a single dorsal temporal fossa on each side, but its validity is questionable because there is doubt that the fossae are homologous throughout the group.

Ichthyosaurs were fishlike in outward appearance with no visible neck (Fig. 9-30). The largest were about 3 m long. **Plesiosaurs,** which reached 12 m in length, had a tiny head, long neck, short flattened trunk, short tail, and paddlelike limbs (Fig. 3-21).

Synapsids

Synapsids diverged early from other reptilian lines and are the reptiles from which mammals emerged. They are distinguished by a single lateral temporal fossa (Fig. 8-15, *B*). **Pelycosaurs,** the earliest synapsids, were recovered from the late Carboniferous. They had primitive reptilian fea-

FIGURE 3-23

Therapsid reptile about the size of a large dog, from the Triassic. (From Colbert.)

Cynognathus

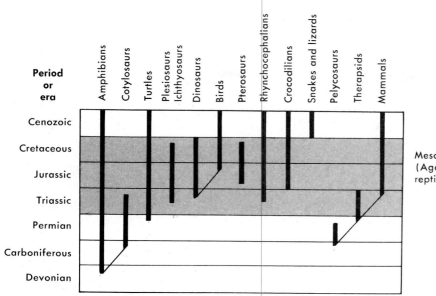

FIGURE 3-24

Range of selected reptiles through time. Connecting lines indicate probable origins of cotylosaurs, birds, and mammals.

tures including a parietal foramen, which indicates the presence of a median eye. Mammal-like synapsids, the **Therapsida** from the middle Permian to the Jurassic, are thought to be ancestral to mammals (Fig. 3-23). They had two occipital condyles, a secondary palate, and heterodont dentition with incisors, canines, and grinding molars. The dentary was the largest bone of the lower jaw, presaging the condition in modern mammals (Fig. 8-36, C). Like other reptiles, therapsids had a tiny braincase and a single bone in the middle ear cavity. Nothing is known of the skin which, of course, bears hair in mammals. The distribution of selected reptiles in geologic time is shown in Fig. 3-24.

AVES

Birds are endothermic vertebrates with feathers. Endothermy is the ability to maintain a relatively stable body temperature despite fluctuations in the ambient temperature. A less precise term is "warm blooded." Bipedal Ornithischian dinosaurs, from which birds probably arose, could stand and run on their hind limbs, which freed the forelimbs for other functions including, ultimately, flight in birds. Birds have retained reptilian scales on their beak, legs, and feet, a single occipital condyle, and a diapsid skull which, however, has been modified by partial or total loss of one of the temporal arches.

Feathers are exquisitely structured keratinized appendages of the epidermis that have replaced reptilian scales on all surfaces except the beak, legs, and feet. Feathers make bird flight possible. Whereas pterosaurs and bats have a wing membrane stretched between the forelimbs and trunk, or even the tail, that serves as an airfoil, feathers attached to the bones of the forearm and hand are the only airfoil birds have. They also insulate against seasonal heat and high-altitude cold. (When birds produce excessive heat during flight, it is eliminated in expired air.) Feather pigments facilitate recognition by other members of the species and often provide protective coloration. Seasonal plumage, the result of molting, signals the sex and reproductive state of the individual.

In addition to feathers and wings birds have other adaptations for flight. Body weight has been reduced in several ways. Long bones have become slender, and most bones, including vertebrae, have lost the central marrow, which leaves cavities that contain air-filled extensions of air sacs from the lungs. The skull has been lightened by a thinning of the compact layers of the membrane bones, but it remains durable because sutures have been eliminated and the spongy bone that remains is architecturally strong. All teeth have been lost, having been replaced functionally by a lighter-weight horny beak that is used also for getting food into the mouth. The bones of the wrist, palm, and digits have been reduced in number and size (Fig. 9-27); consequently the volume of muscle *within* the forelimb has been greatly reduced. The urinary bladder has been lost, and the large intestine has been shortened. All these modifications have reduced the cost of flight in terms of energy expended.

Modifications of the skeleton include a large sternal keel, or **carina,** for attachment of the massive flight muscles (Fig. 7-21). It is absent in most birds that do not fly. The entire skeleton of the trunk except the ribs has become a rigid unit by ankylosis of trunk vertebrae and their fusion with the pelvic girdle to form a **synsacrum** (Fig. 7-20). The ankle is rigid, and the angle of approach of the femur to the ornithischian-type girdle has been altered. Uncinate processes of ribs strengthen the thoracic wall and provide additional surfaces for attachment of some of the flight muscles of the scapula (Fig. 7-21).

Many soft parts have been modified. The esophagus often has a crop for storage of seeds and grain, and the stomach has become a grinding gizzard like that of crocodilians. The size and shape of the eyes have been modified, which result in increased acuity of vision from high in the air, and the orbits have become enlarged. The cerebral hemispheres have a thick new stratum of nerve cells, the hyperstriatum (Fig. 15-24), which is responsible for stereotyped avian behavior patterns such as nest building. The cerebellum is enlarged as a consequence of increased neural input from wing and hind limb muscles. The modified lungs and air sacs are discussed in Chapter 12.

Archaeornithes

The oldest birds known until 1986 were those in the genus *Archaeopteryx*. Two fossils about the size of a crow were found during the nineteenth century in Late Jurassic limestone in Bavaria, West Germany. The first specimen was complete except for a few cervical vertebrae, the right foot, and lower jaw. The second specimen, illustrated in Fig. 3-25, was even more complete; and three additional specimens, not exactly alike, have been recovered elsewhere. *Archaeopteryx* had a long reptilian tail, thecodont

FIGURE 3-25

Fossil *Archaeopteryx* embedded in limestone. Wings and their skeleton are near bottom of photograph, the long feathered tail is at top. (Photograph courtesy The Museum für Naturkunde an der Humboldt-Universität, Berlin.)

FIGURE 3-26

A, *Archaeopteryx,* from the Jurassic period. **B,** A carinate bird (pigeon) for comparison. How many skeletal changes can you find? (From Colbert.)

teeth on both jaws, and feathers on the wings and tail that were no different from today's. The skull was more reptilian than avian, the nostrils were far forward, there was no beak, and the braincase had not expanded to accommodate an enlarged brain. The cervical vertebrae were not saddle shaped at the ends as in today's birds, trunk vertebrae were not rigidly fused, and the synsacrum was not well developed. The sternum was small, indicating weak flight muscles. These birds may have soared more than they flew. Fig. 3-26 contrasts some of the more obvious features of the skeleton with that of a pigeon. *Archaeopteryx* has been placed in the **subclass Archaeornithes** along with **Protoavis** (next paragraph). All other birds are currently in the **subclass Neornithes.**

In 1986, fossil remains of two crowlike Triassic birds 75 million years older than *Archaeopteryx* were discovered in a mudstone quarry in the state of Texas. These specimens, which have been placed in the genus *Protoavis,* are more dinosaur-like than *Archaeopteryx,* and had smaller wings. The paleontologists who discovered them think *Protoavis* may have been closer to the evolutionary line leading to modern birds, and that *Archaeopteryx* may have become extinct without leaving descendants.

Neornithes

The **subclass** Neornithes includes all extinct and living birds other than *Archaeopteryx*. It is separated into two groups, **Odontognathae** (Neornithes with teeth) and **Neognathae** (toothless birds). The only odontognath known for sure is *Hesperornis* (Fig. 3-27) from marine deposits in Kansas. The wings were vestigial, so it did not fly; but it swam and dove for fish, catching them with sharp, conical teeth.

Most neornithes fly, but some, the **ratites,** cannot. They have small incompetent wings, but they have powerful leg muscles that enable them to run well. It is likely that their ancestors could fly. Many are known only as fossils, and the survivors are in danger of extinction by humans. Ostriches, emus, cassowaries, rheas, kiwis, and penguins are the living ratites. The extinct moas were nearly 4 m (13 ft) tall and laid eggs more than 30 cm (1 foot) in diameter.

Neornithes that fly are **carinates;** that is, they have a large carina. Penguins have a large carina but they cannot fly, partly because their forelimbs are powerful flippers that make them good swimmers but can't serve as wings. The largest living carinate is the Andean condor, with a wingspread of 3 m (10 ft) and weighing 15 kg (35 lbs). But the Giant Teratorn *(Teratornis),* which lived in Argentina 5 million years ago and is thought to have been a flier, had a wingspread of 7.5 m (25 ft), weighed 360 kg (160 lbs), and measured 3.5 m (11 ft) from beak to tip of tail. Some pterosaurs were larger.

Many carinates are annual migrants. The Arctic tern spends several months above the Arctic Circle and the remaining months in the Antarctic,

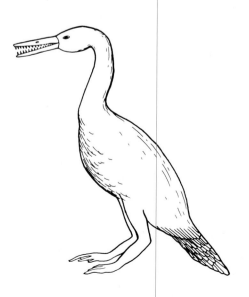

FIGURE 3-27

A reconstruction of *Hesperornis,* a flightless toothed bird from the Cretaceous seas of Kansas in the United States.

traveling 22,000 miles round trip each year! During migration, birds move in mass flights, often at night and at an elevation of approximately 2,000 feet. Those destined for the same geographical location may pass over approximately the same flyways year after year. A great flyway is located directly over the Gulf of Mexico between the Yucatan peninsula and the Gulf Coast of the United States. Other flyways are located overland between Central America and the United States, and from the West Indies via the Florida peninsula. Migration is associated in part with mating and is triggered by a specific spectrum of hormones. How birds navigate has long challenged ornithologists. The sun, stars, barometric pressure, polarized light, and low-frequency sound all seem to provide input during flight at one time or another.

There are fewer species of birds today than at the start of the Cenozoic. Climate changes resulting from glaciation of the northern hemispheres probably accounted for much of the loss, but increasing populations of mammals that competed with birds for food, especially insectivorous and carnivorous birds, may have played a role. The spread of human civilization has been the most recent threat to avian species.

MAMMALIA

Mammals succeeded therapsid reptiles at the end of the Triassic. They are amniotes with a synapsid skull, hair and, except monotremes, mammary glands and nipples. Further distinguishing modern mammals from other vertebrates are the single dentary bone on each side of the lower jaw articulating with the squamosal bone; three bones in the middle ear cavity; a muscular diaphragm separating thoracic and abdominal cavities; sweat glands (in most mammals); absence of an adult cloaca in all but the lowest order; heterodont dentition (except in toothed whales); only two sets of teeth (milk teeth and a permanent set) instead of continual replacement as in lower vertebrates; biconcave, enucleate red blood cells that are circular (except in camels and llamas); loss of the right fourth aortic arch; a pinna, or sound-collecting lobe, accessory to the outer ear; a more specialized larynx; and extensive development of the cerebral cortex.

Because of many variations in limb structure, mammals have been able to achieve a greater diversity of habitats than any other vertebrates except perhaps reptiles. They burrow in the ground, hop, lumber, or gambol over the plains, scramble over mountain crags, swing through trees, propel themselves through the air in true flight, and inhabit the oceans, each lifestyle made possible by modifications of body structure.

Mammals are divided into three subclasses, **Prototheria,** which lay eggs and have a cloaca throughout life, **Metatheria,** viviparous mammals that use the yolk sac for a placenta, and **Eutheria,** which have a chorioallantoic placenta. There is only one order of prototherians, **Monotremata,** and one order of metatherians, **Marsupialia.** Unless prototherians emerged independently from reptilian ancestors, they diverged very early from other mammals (Fig. 3-28).

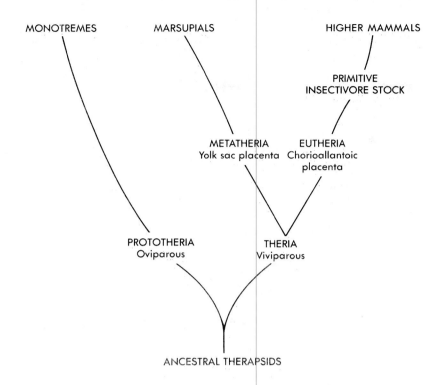

FIGURE 3-28

Major categories of living mammals and their postulated origins. Whether prototherians arose independently of therians is unsettled.

Monotremata

The platypus, or duckbill, *Ornithorhynchus*, and two genera of spiny anteaters, or echidnas, all from Australia or nearby Tasmania and New Guinea, are the sole surviving monotremes (Fig. 3-29). Like reptiles, they lay heavily yolked eggs and retain a cloaca throughout life. The single cloacal aperture inspired the name of the order (Fig. 14-42, *A*). Like reptiles, too, they have a ventral mesentery extending the length of the abdominal cavity, the testes are within the abdomen, the outer ear has no pinna, and the brain lacks the great transverse fiber tract, corpus callosum, that connects the two cerebral hemispheres in other mammals. The malleus and incus are larger than in other mammals, resembling the articular and quadrate bones of therapsids. They have no nipples, but a milky fluid exudes from modified sweat glands onto tufts of hairs in shallow pits on the abdomen, from which it is licked up by the young. Monotremes are endothermic, but their body temperature is less stable than that of higher mammals, fluctuating as much as 13° C.

Duckbills live in pairs in burrows up to 180 m (60 ft) long, in the banks of creeks and rivers, and spend most of their life in the water. They have webbed feet that enable them to walk on muddy river bottoms, and a soft, rubbery, sensitive, beaklike muzzle with which they detect invertebrates, especially molluscus, in the mud and dredge them. The food is then stored in a cheek pouch until it is crushed later with horny teeth that replace a

PLATYPUS

ECHIDNA

FIGURE 3-29

The two surviving mono-
tremes, the platypus *Ornithorhynchus*, and one of the
spiny anteaters.

prenatal set of regular ones. During breeding seasons the female moves out
of the burrow and constructs a new one just above the water line, where
she lays one to three eggs in a nest she has constructed. The eggs are nearly
round, about 2 cm in diameter, and covered with a pliable white shell. She
incubates the eggs for about 2 weeks, after which they hatch.

Echidnas are terrestrial and have a long sticky tongue that is used to
gather insects. Except on the abdomen they are armed with sharp quills
interspersed among coarse hairs, and they roll into a ball for protection. A
single egg, about 4 mm in diameter, is incubated in a temporary pouch that
develops as a thin flap of abdominal skin in females. The milk-secreting
glands are on the abdominal wall in the pouch, and the young hatch and
are carried in the pouch for several weeks until they can forage for food.

Marsupialia

Marsupials (Fig. 3-30) are primitive mammals in which the fetal yolk sac
(in contact with the chorion) serves as a placenta. The young are born in
almost a larval state and are transported, incubated, and nursed after birth
in a maternal abdominal pouch (**marsupium**) of muscle and skin until they
are old enough to become independent. The walls of the pouch are supported by two slender marsupial (epipubic) bones that project forward from
the pelvic girdle. In several South American genera the pouch is incomplete
or absent. Newborn young make their way to the pouch by squirming and
wriggling and by using the claws on their forelimbs, which are considerably

larger than the hind limbs at birth. The lips are sealed at the angles of the mouth, which therefore consists of only a small circular opening. Once the young has taken a nipple into its mouth the tip of the nipple swells, and the young cannot easily drop off.

The only mammals indigenous to Australia other than bats are monotremes and marsupials. Bats were found there by the earliest explorers of record. The apparent trail of marsupials to Australia from a source in the northwestern part of North America via what were to become the future continents of South America and Antarctica has been revealed by fossils of increasingly later ages that have been found along the southbound route

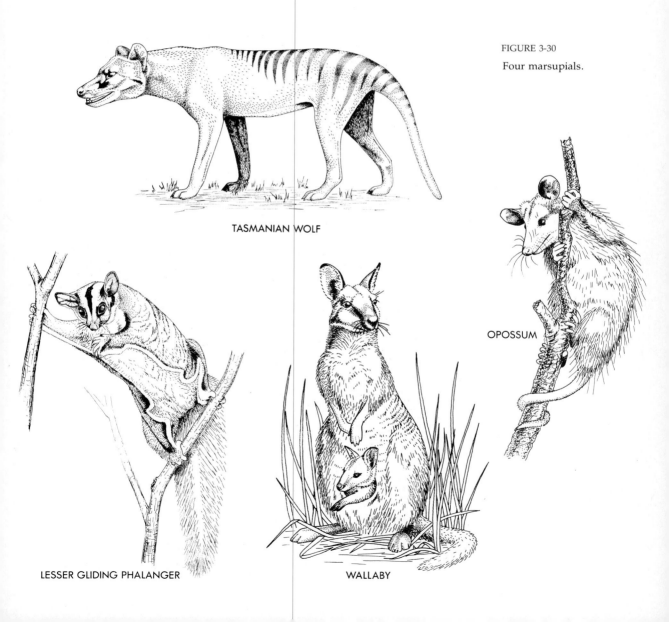

FIGURE 3-30

Four marsupials.

TASMANIAN WOLF

OPOSSUM

LESSER GLIDING PHALANGER

WALLABY

(Zinsmeister, 1986). The breakup of Gandwana into present day South America, Africa, Antarctica, Australia, and India, which commenced in the late Mesozoic, was completed to its present state when Australia separated from Antarctica 60 million years ago and commenced to slowly drift several thousand miles into the warmer regions of the South Pacific. Marsupials had arrived in Australia before its isolation. During the continental drift, Australian marsupials were undergoing an extensive adaptive radiation while those in the rest of the world were facing extinction. With the exception of New World opossums and an obscure family of South American marsupials (Caenolestidae), living native marsupials are now found only in Australia, Tasmania, New Guinea, and a few adjacent islands. The absence in Australia of indigenous placental mammals and fossils of placentals indicates that placentals had not yet reached Australia when that land mass separated from Antarctica.

Marsupial fossils found in Europe attest to their presence there until about 15 million years ago. No marsupials or marsupial fossils have been found in Alaska, Asia, or New Zealand. The New Zealand islands broke away from Australia and began their drift before Australia separated from Antarctica, and marsupials had no opportunity to colonize these islands. Bats, however, found their way there just as they did to Australia. Rats probably arrived much later, debarking no doubt from early wooden galleys.

Among Australian marsupials are kangaroos, wallabys, Tasmanian wolves (doglike marsupial carnivores), bandicoots (rabbitlike marsupial rats), wombats, Australian anteaters, and phalangers. Australian marsupials resemble many eutherian mammals—wolves, foxes, bears, rabbits, mice, and cats—in surprising detail. Some Australian phalangers resemble flying squirrels, and there are marsupial moles. The close resemblances between Australian marsupials and placental mammals elsewhere are instances of parallel evolution.

Insectivora

Insectivores (Fig. 3-31) are generalized mammals closely related to the primitive stock from which higher mammals arose. Although at one time abundant, they are represented today chiefly by small, retiring, often subterranean survivors: moles, shrews other than tree shrews, tenrecs, hedgehogs, solenodons, and "flying" lemurs that do not fly, but glide, and are not lemurs. Most insectivores subsist on a diet of insects, worms, and other small invertebrates.

Among primitive characteristics are a flat-footed (plantigrade) gait; five toes; smooth cerebral hemispheres; small, sharp pointed teeth with incisors, canines, and premolars poorly differentiated; a large embryonic allantois and yolk sac; and, in some species, a shallow cloaca. The testes are retained in the abdominal cavity in some genera (a primitive trait), and they never descend fully into scrotal sacs in any.

Moles have short, stout anterior limbs, with forefeet that are broad and more than twice as large as the hind feet, an adaptation for digging (Fig.

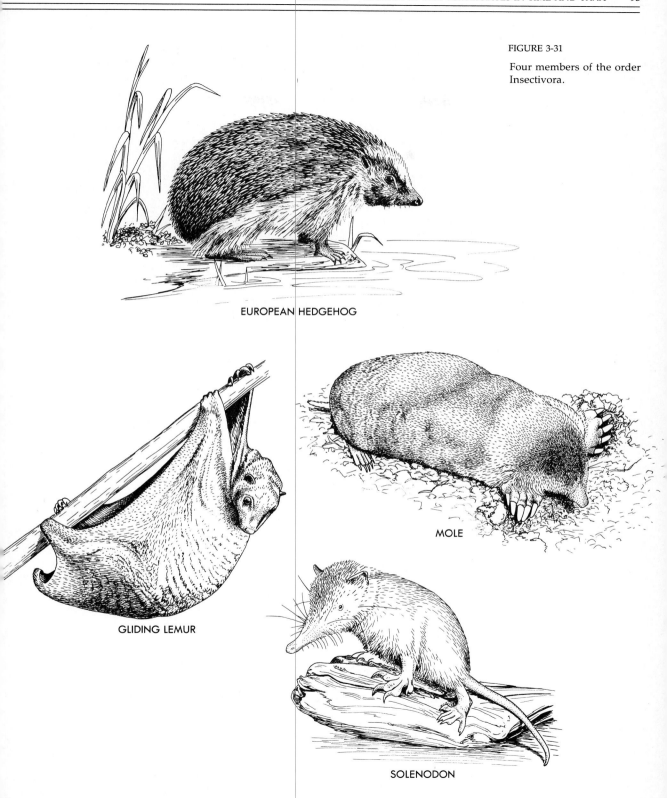

FIGURE 3-31

Four members of the order Insectivora.

EUROPEAN HEDGEHOG

GLIDING LEMUR

MOLE

SOLENODON

FIGURE 3-32

Tree shrew. (Courtesy Delta
Regional Primate Research
Center, Covington, La.)

FIGURE 3-32

Tree shrew. (Courtesy Delta Regional Primate Research Center, Covington, La.)

9-23). The neck is short, and the shoulder muscles are so powerful that head and trunk seem to merge. Their tiny eyes are practically useless, but an acute sense of smell locates distant food, and the sensitive tip of the elongated snout tells them when they have encountered it. Shrews superficially resemble mice. They are shy, busy little fighters with a keen sense of hearing. They have an elongated, bewiskered, sensitive snout, and their incisor teeth are long and curved. The pigmy shrew, weighing about 3 g, is the smallest mammal on earth.

Tree shrews (Fig. 3-32) are classified by some authorities in the order Insectivora, by others as the most primitive primate. They have traits that bridge the gap between the two orders.

Chiroptera

Bats (Fig. 3-33) comprise a large mammalian order that is probably derived from a primitive insectivore. They, pterosaurs, and birds are the only known vertebrates to have achieved true flight. The wing, or **patagium,** is a double membrane of skin stretched along the length of the body between the trunk, forelimbs, and hind limbs, and extending from there to the tail. It incorporates four greatly elongated, clawless fingers. The first digits, or thumbs, project from the anterior margins of the wing membranes and bear claws that aid in clambering about on rough vertical surfaces. All five digits of the hind limbs bear claws, and these are used for hanging upside down, wings folded, from crevices in caves, from small branches, or within hollow trees. The pectoral muscles are strong, and the sternum is keeled, although not as much as in carinate birds. All the bones are slender, but not pneumatic. Nipples, usually a single pair, are limited to the thoracic wall. Bats have large external ear auricles (pinnas), and their facial glands are prominent, giving the head a bizarre appearance. The development of a wing in pterosaurs, birds, and bats are instances of convergent evolution.

Bats are insectivorous, frugivorous (fruiteaters), or sanguinivorous (subsisting on the blood of other mammals). Vampire bats have received attention because of sanguinivorous habits. Incisor teeth occur on the upper jaw only, and there is only one pair. They are razor sharp and point toward one another so they slit the skin of prey. As blood oozes from the wound, the bat licks it up without awakening the sleeping victim, which is often a domestic animal. Associated with the vampire habit of taking only fluid nourishment is the very small lumen of the esophagus, through which no solid food could possibly pass.

The adaptations of the teeth and esophagus of vampire bats and the fluid diet inspire the speculative question of whether a small esophageal lumen forced these bats to adopt a diet of mammalian blood (the only available fluid containing all the nourishment needed by most mammals) or whether the fluid diet during many generations was, as Lamarck would have said, a cause of the decrease in the size of the lumen. If the former was the case, it is fortunate that vampire bats happened upon blood as a source of nourishment; otherwise they could not have survived the change. If, as theorized by Lamarck, the change in the esophageal lumen resulted from the disuse of this part for solid foods (Doctrine of Acquired Characteristics, Chapter 1), modern genetics has thus far been unable to fathom the mechanism whereby disuse might be translated into a change in the genotype.

FIGURE 3-33

A chiropteran.

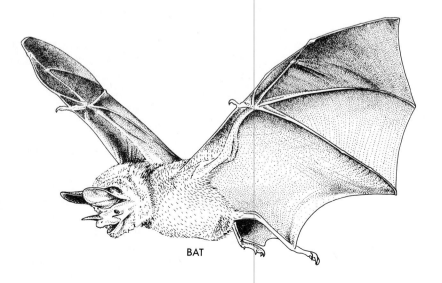

BAT

Primates

Primates (prī-mā'-tēz) are primarily arboreal mammals that arose as an offshoot of Cretaceous insectivore stock. One of many classification schemes divides them into four suborders, **Lemuroidea,** lemurs and lorises; **Tarsioidea,** tarsiers and their relatives; **Platyrrhini,** South American monkeys and small squirrel-like marmosets; and **Catarrhini,** Old World monkeys, apes, and man. Platyrrhines and catarrhines are often referred to as **anthropoids.**

Among primate specializations is the grasping hand so built that the thumb can be made to touch the ends of the other four fingers of the same hand. The big toe is also opposable in most primates. At least some of the digits are provided with nails instead of claws. Often there is a prehensile tail, which supplements the hands for grasping during arboreal locomotion. The cerebral hemispheres of the brain are larger than those of any other mammal. The snout has been shortened, with the result that both eyes can look forward. There is frequently only one pair of nipples, and these are on the thorax. Among primitive features are a flat-footed gait, five digits, a large clavicle, a central wrist carpal in many primates, and generalized dentition.

LEMUROIDS. Lemurs, the largest of which is the size of a domestic cat, receive their ghostly name from the habit of swinging silently through the trees while most higher primates are asleep. The long axis of the head is in line with the long axis of the body, as in most other mammals. The long tail is not prehensile. The second finger and toe have a claw instead of a nail.

FIGURE 3-34

Tarsius, a lower primate.

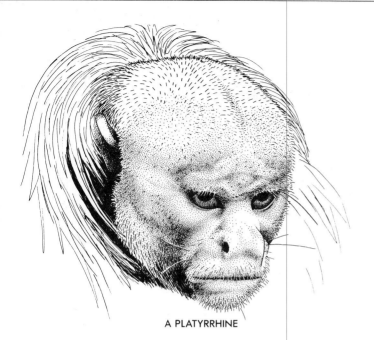

A PLATYRRHINE

A CATARRHINE

The uterus is duplex. The placenta is nondeciduate; that is, the fetal part of the placenta does not become rooted into the uterine lining of the mother, and so there is no trauma of the uterus at birth. These are primitive traits.

Lorises are slow-moving nocturnal primates of India, Sri Lanka, and S.E. Asia. They have no tail and the index finger is vestigial. In the same family are the pottos and bush babies.

TARSIOIDS. Tarsiers (Fig. 3-34) resemble higher primates more than lemurs do. Their eyes are close together and directed forward so there is an overlap in the left and right fields of vision; the head is more nearly balanced at right angles to the vertebral column; all five fingers have nails and so do all toes except the second and third; and the placenta is deciduate.

HIGHER PRIMATES. The higher primates fall into two natural groups, platyrrhines and catarrhines. These two groups can be differentiated on the basis of the direction in which the nostrils open (Fig. 3-35). Those of the New World group are separated by a broad internarial septum and open to the side. They are the **platyrrhines.** The nostrils of the Old World group lie close together and open downward. They are the **catarrhines.** The human being is a catarrhine. In both groups the head is at an angle to the long axis of the vertebral column, the eyes are directed forward and are close together, the cerebral hemispheres are maximally developed, there are 32 teeth in the permanent set, the placenta is deciduate, and only one baby is normally born at a time. Despite these specializations, the primates, including humans, are much less specialized than are the members of several other mammalian orders—whales and Perissodactyls, for example.

FIGURE 3-35

Two anthropoids. The platyrrhine is an ouakari, a small S. American monkey with a nonprehensile tail. The catarrhine is an orangutan.

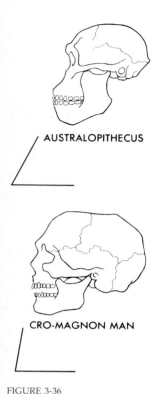

AUSTRALOPITHECUS

CRO-MAGNON MAN

FIGURE 3-36

Facial angles of two hominids. The angles are shown beneath the skulls.

Platyrrhines include *Cebus,* the capuchin, *Ateles,* the spider monkey, and *Alouatta,* the howler monkey. The latter are named for the screeching cries made with the enlarged hyoid bone and larynx (Fig. 12-13). Catarrhines are in two superfamilies, **Cercopithicoidea** and **Hominoidea.** Among the cercopithicoids are baboons, mandrills, and the macaque, or rhesus monkey, from which name the symbol "Rh" for designating human blood types was derived. Hominoids include chimpanzees, gibbons, gorillas, orangutans, and man, all of whom no longer have a tail except as an anomaly.

Since the discovery in 1856 of Neanderthal man, who lived in Europe 100,000 years ago, fossil parts of prehistoric man have been found in considerable numbers. The latest fossils indicate that mankind lived in Africa at least 4 million years ago. With each discovery the problem arises whether to include the new member in existing species, or to erect new species, or even genera. For example, Neanderthal man is sometimes classified as a subspecies of modern man (*Homo sapiens neanderthalensis*) and sometimes placed in a separate species (*Homo neanderthalensis*). The rules for classifying other vertebrates are applied to the taxonomy of man. The oldest known prehuman remains are assigned to the genus *Australopithecus.* They walked upright, were about 5 feet tall, they had a humanlike body but the facial contours of an ape, and they had an ape-sized brain. They used simple bone tools. *Homo erectus* is a more recent hominid. He fashioned tools and used fire. Modern man, *Homo sapiens sapiens,* who added nuclear energy to his array of tools, occupied Europe with Neanderthal man and replaced him either by competition or absorption.

In the emergence of man from an earlier anthropoid many changes occurred. An S-shaped curve in the vertebral column permitted an erect posture; the facial angle became less acute (Fig. 3-36); the teeth, especially the canines, became smaller; the frontal lobes of the cerebral hemispheres enlarged, resulting in a larger braincase and a more prominent forehead; the eyebrow ridges became reduced; the nose became more prominent; the tail became confined to embryonic stages; the arms became shorter; a metatarsal arch developed in the otherwise flat feet; the big toe moved in line with the other toes and ceased to be opposable; articulate speech appeared; and many other changes occurred.

It is to the massive development of the frontal lobes of the cerebral hemispheres, to the opposable thumb, and to articulate speech that man owes his present dominance in the animal kingdom. With his fingers he can construct instruments of offense and defense and machines to lighten his burden, and with them he can scrawl symbols that convey experiences and techniques to the corners of the earth and to generations unborn. With his voice he can communicate with contemporary fellow creatures and exchange ideas with delicate shades of meaning. With his brain he can associate sensory stimuli that are currently received with those recalled from earlier experiences, and after meditation elect a mode of action that he calls "intelligent." With his brain, too, he can enjoy the esthetic beauty of the imponderable universe, search for ultimate truth, and dream of Utopia, which, but for his residual animal nature, might be his heritage.

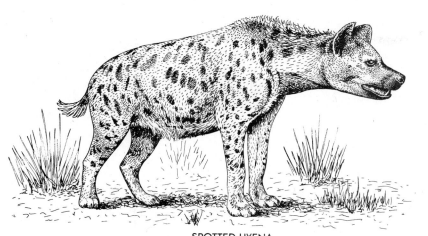

SPOTTED HYENA

FIGURE 3-37

Representatives of four families of terrestrial Carnivores. Pandas belong to the racoon family.

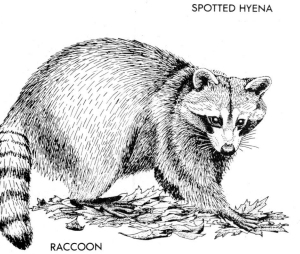

RACCOON

PANDA

AFRICAN CIVET

RIVER OTTER

FIGURE 3-38

Sea lions (marine carni-
vores). (Courtesy The American Museum of Natural History, New York.)

Carnivora

Carnivores are a large and diverse group of flesheaters that includes terrestrial and marine species. They have powerful jaws and, except in water forms, long sharp canine teeth capable of spearing and tearing flesh. The cerebral cortex is convoluted and the animals are capable of considerable learning.

Terrestrial carnivores, suborder **Fissipedia** (Fig. 3-37), include seven families (domestic cats, panthers, lynx, others), the dog family, bears, pandas and racoons, civets, hyenas, and mustelids (mink, otters, skunks, ferrets, badgers, and others). Mustelids have strong scent glands near the anus. Most terrestrial carnivores have five toes, a few have four, with sharp retractible claws.

Marine carnivores are placed either in a suborder, **Pinnipedia,** or in a separate order with the same name. The taxon includes sea lions, seals, and walruses (Fig. 3-38). Among adaptations for life in the water are their webbed paddlelike limbs that are more or less included within the body wall and which, unlike those of their land-dwelling close relatives, usually lack claws. The hind limbs of "wriggling" seals are permanently bound to the tail and movements on land consist of pulling themselves awkwardly along or flopping about. Despite their adaptations for aquatic life young marine carnivores are born on land and are unable to swim.

Cetacea

Whales, dolphins, and porpoises are permanently aquatic marine mammals (Fig. 3-39). The weight of the largest whale is 23 million times that of the smallest shrew. The cetacean tail has a horizontal terminal fin filled with fibrous tissue but without fin rays, and is the means of propulsion. A dorsal fin is sometimes present. The forelimbs are paddlelike balancers

FIGURE 3-39

Representative cetaceans. Baleen is seen suspended from the palate of the blue whale.

BOTTLE-NOSE DOLPHIN

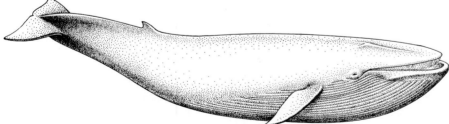

BLUE WHALE

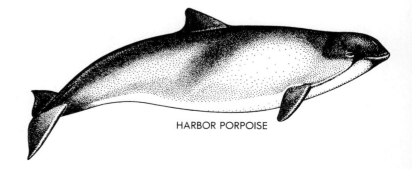

HARBOR PORPOISE

with the finger bones embedded at the end (Fig. 9-32). Hind limbs and girdles are either mere vestiges embedded in the trunk or, as in some species, they are entirely missing. The frontal and nasal bones are short, with the result that the nostrils are on top of the head and frequently united to form a single large blowhole. The nostrils have valves that close underwater. The thoracic diaphragm is unusually muscular and the waterspout, composed mostly of vapor that forms when expired air is expelled, may last from 3 to 5 minutes. A heavy layer of blubber, or subcutaneous fat, conserves body heat.

Most cetaceans have teeth, but whalebone whales have, instead, frayed horny sheets of whalebone, or baleen, hanging from the palate. These strain several tons of small fish, crustaceans, or plankton, depending on the species, out of the sea each day. A few hairs occur on the muzzle of cetaceans, but elsewhere the skin is naked. Babies are born in the sea. The mother or an assistant bites the umbilical cord to separate the neonate from the afterbirth, and the mother nudges the baby to the surface for the first inhalation of air. Thereafter, the baby hangs onto one of the four inguinal teats as the mother swims about.

Cetaceans have good eyesight, but they also scan their environment by echolocation, which can tell them even the texture of another object. They have an excellent sense of taste but no sense of smell, since their olfactory nerves do not develop. They would be useless, anyway, since, if the animal were to inhale under water, it would drown. Cetaceans communicate with one another by a descriptive language that is best developed in dolphins.

FIGURE 3-40

Two orders of higher insectivorous mammals with unusual dentition.

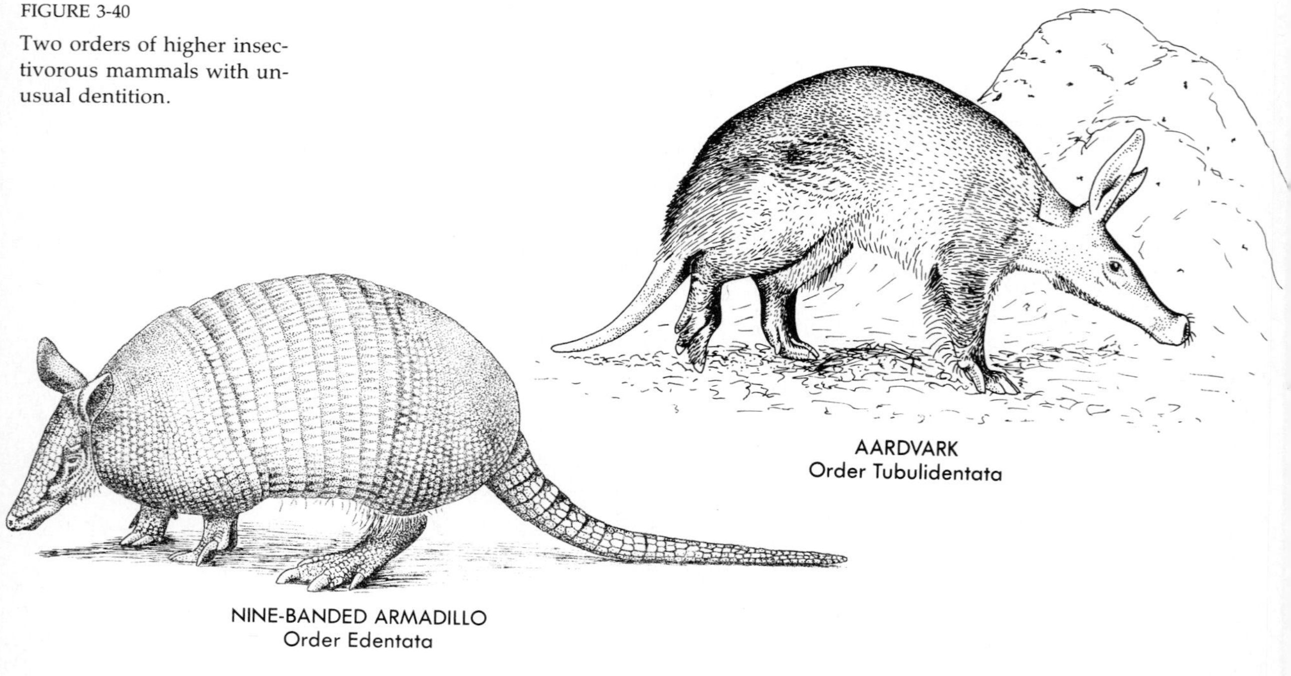

AARDVARK
Order Tubulidentata

NINE-BANDED ARMADILLO
Order Edentata

FIGURE 3-41

A pangolin. The surface is covered with epidermal scales and hairs. (From Flower and Lydekker: Introduction to the study of mammals, London, A. & C. Black, Ltd.)

Edentata

Edentates are chiefly South American mammals that have diverged from insectivores. They include tree sloths, South American anteaters, and armored edentates, best known of which is the armadillo (Fig. 3-40). Several giant armored fossil edentates are known. All teeth are absent in South American anteaters (hence the name "edentate"), but many other species have cheek teeth that lack enamel. Histological evidence of enamel organs has been found, however. Armadillos always give birth to identical quadruplets from a single fertilized egg.

Tubulidentata

Tubulidentates are another order with few teeth. There is only one living tubulidentate, the insectivorous aardvark from South Africa (Fig. 3-40). It is a burrowing anteater about 1.5 m (5 ft) long with coarse, sparse fur, an elongated piglike snout, and a long sticky tongue that is used for capturing termites.

Pholidota

There is only one genus of pholidotans, the toothless pangolins (*Manis*), known as scaly anteaters (Fig. 3-41). There are seven species. The scales constitute a horny armor. Hairs grow between the scales. Their relationship to other mammals seems remote, although they are presumably descendants of ancestral insectivores.

FIGURE 3-42

The three suborders of rodents. The cavy is a feral guinea pig.

BEAVER
Sciuromorpha

CAVY
Caviomorpha

PORCUPINE
Caviomorpha

MEADOW VOLE
Myomorpha

Rodentia

Rodents comprise the largest mammalian order and are distributed worldwide. They have a single pair of long, curved, upper and lower incisor teeth that are used for knawing. These teeth have enamel on their outer surface only, which provides a chisel-like edge as the softer dentin wears away. The teeth grow throughout life. Canine teeth are absent, so there is a diastema, or a stretch of toothless jaw, between the incisors and the first grinding teeth. Rodents are cellulose eaters, and at the beginning of the large intestine there is a long, coiled cecum that houses commensal cellulose-digesting microorganisms.

There are three suborders. **Sciuromorpha** includes squirrels, chipmonks, woodchucks, gophers, beaver, and others. **Myomorpha** includes micelike rodents such as rats, voles, hamsters, and lemmings. **Caviomorpha (Hystricomorpha)** includes porcupines, cavys, nutria (coypus), chinchillas, and others (Fig. 3-42).

Lagomorpha

There are only two families of lagomorphs, **Ochotonidae,** the pikas (Fig. 3-43), and **Leporidae,** the hares and rabbits. Lagomorphs differ from rodents in having *two* pairs of incisors on the upper jaw, a small pair lying immediately behind, not alongside of, a much larger pair. The front pair are rodentlike and grow throughout life. The smaller pair lack cutting edges.

Rabbits differ from hares in being born in a fur-lined nest, blind, virtually hairless, and helpless. Hares are born without benefit of a nest, with eyes open, and able to run around a few minutes after being delivered. Some hares of the genus *Lepus* acquire white fur in winter; no rabbits do. Jack rabbits of western North America and snowshoe rabbits of the arctic are hares, but the so-called Belgian hare is a rabbit! Rabbits and hares

PIKA

FIGURE 3-43

A lagomorph.

have a split upper lip, which inspired the term "harelip" for the human anomaly.

Pikas live above the timber line in Western North America and Northern Asia. They differ from rabbits and hares in having a smaller body, shorter ears, and fore and hind limbs of about equal length.

Ungulates and Subungulates

Bringing up the rear of this vertebrate "parade" are five orders of mammals that are either ungulates, or their closest relatives. These are the orders **Perissodactyla** and **Artiodactyla,** which are ungulates, **Proboscidea** and **Hyracoidea,** which are subungulates, and **Sirenia,** which are thought to be aberrant descendants of primitive ungulate stock.

Ungulates walk on the tip of their toes, which are protected by hoofs. They have no more than four toes on each foot and some, such as horses,

FIGURE 3-44

Two perissodactyls.

INDIAN RHINOCEROS

BRAZILIAN TAPIR

FIGURE 3-45
An artiodactyl.

PECCARY

have only one. Ancestral ungulates probably had five, the generalized tetrapod condition. Reduction in the number of toes is documented by studies of the evolution of horses, the small Eocene ancestors of which had four toes in front and three behind. Ungulates have no clavicle, which may facilitate grazing, and they are the only mammals with horns.

All ungulates, subungulates, and sirenians are herbivores, and their molar teeth are high crowned and deeply grooved for grinding vegetation. There is little morphological difference between premolars and molars.

Perissodactyla

There are three families of perissodactyls, **horses and horselike mammals, tapirs,** and **rhinoceros** (Fig. 3-44). They walk on the hoofed tips of one, three, or occasionally four toes and are distinguished by the fact that most of the body weight is borne on a single digit (Figs. 9-35 and 9-36). This is a **mesaxonic foot.** Perissodactyls are usually called odd-toed ungulates, but tapirs and some rhinos have four toes on the forefeet.

Artiodactyla

Artiodactyls are ungulates in which the weight of the body is borne on two toes (Fig. 3-45). This is a **paraxonic foot** (Figs. 9-35 and 9-36). Living artiodactyls have an even number of toes, but at least one extinct artiodactyl had five toes on the forelimb. Artiodactyls include pigs, hippopotamuses, peccaries, cattle, camels, llamas, deer, antelope, and giraffes. With the exception of pigs, hippopotamuses, and peccaries, they have stomachs divided into not fewer than three chambers, sometimes four. They bolt their food, which then passes to the rumen, the first segment of the stomach. Such animals are **ruminants.** At their leisure, they force undigested food balls back up the esophagus and masticate this cud more thoroughly.

Proboscidea

Proboscidea is an order of subungulates that includes elephants, mastodons, and their relatives. They have a proboscis composed of a greatly drawn out upper lip, which is accompanied in its overgrowth by the nostrils (Fig. 11-14, mastodon). They have scanty hair and thick, wrinkled skin. The incisor teeth of one or both jaws are elongated to form tusks, canine teeth are absent, and the molars are large grinders, as in ungulates. Proboscideans are bulky animals, and the limbs are almost vertical pillars of bone and muscle. They have five toes that end in thick, hooflike nails. An elastic pad on the back of each toe bears much of the body weight. The pad is not present in ungulates.

Hyracoidea

Hyracoidea is an order of subungulates containing two genera of hyraxes (Fig. 3-46). They have short ears, they hunch the body when at rest, the upper lip is split (harelip), and the incisor teeth grow continuously. Although they are plantigrade, all digits except one end in small, flat hooves, fingers have been reduced to four and toes to three, and they have ungulate-like molar teeth.

Sirenia

Sirenians, or sea cows, are the exclusively vegetarian manatees and dugongs. They are stout, clumsy-appearing, freshwater or marine mammals with an overgrown, wrinkled, almost pathetic-looking snout covered with scattered, coarse bristles (Fig. 3-47). The rest of the body is naked except for a few scattered hairs. The forelimbs are paddles, but within them the tetrapod complement of bones is intact. Hind limbs are absent, but there are internal skeletal vestiges of them and of the pelvic girdle. The tail

FIGURE 3-46

A rock hyrax, order Hyracoidea.

HYRAX

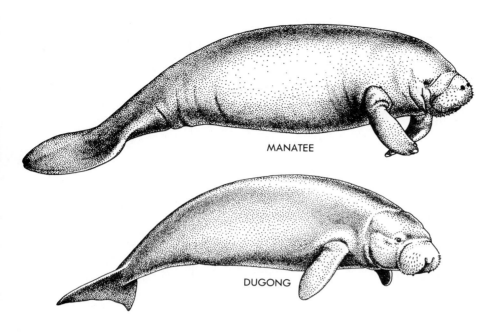

FIGURE 3-47
Representatives of the two families of the order Sirenia. Notice the notch in the fluke of the dugong, and the vestigial nails on the flippers of the manatee.

MANATEE

DUGONG

in some is flattened horizontally like that of whales. They are thought to be aberrant descendants of primitive ungulate stock. Sirenians and cetaceans are the only completely adapted aquatic mammals. All others must return to rookeries to breed.

VARIATION AMONG INDIVIDUALS

In this chapter we have had a glimpse of the variety of vertebrate life. But variation is not restricted to differences between species. Variation occurs within populations, and even among littermates. Indeed, no two products of sexual reproduction are identical. The number of kinds of gametes or of homozygous genotypes possible with n pairs of genes is 2^n. The number of different kinds of genotypes possible is 3^n. Even if there were only 100 pairs of genes, 48 digits would be needed for writing the number of possible genotypes. And there are thousands and thousands of pairs of genes. In addition, there may be as many as 400 mutational sites on a single gene. Therefore the number of possible combinations of hereditary traits staggers the imagination. For example, the gastrohepatic artery in a population of *Felis cattus* is sometimes 20 mm long, sometimes 1 mm long, or the gastric and hepatic arteries may arise independently and there is no gastrohepatic artery. The number of variations that actually occurs in arterial channels alone in any interbreeding population is such that one could identify a single shark, or a single human being, from his arterial channels alone. In practice, each of us does better than that! We identify a single human being simply by looking at the face or at a print made from a fingertip. We could, of course, recognize an individual organism by looking at the sequence of nucleotides in its DNA. Genetic variation, coupled with geographical isolation, seems to be the matrix out of which new species evolve.

Chapter Summary

1. Ostracoderms were the first vertebrates. They had dermal armor, no jaws, and no paired appendages. The order Osteostraci may have been ancestral to jawed fishes.

2. Lampreys and hagfishes are agnathans that lack jaws, paired appendages, a vertebral column, bone, and scales.

3. Acanthodians were the earliest jawed fishes. They were followed shortly by placoderms. Both were heavily armored and had paired fins.

4. Chondrichthyes, the cartilaginous fishes, include elasmobranchs and holocephalans. The former have a spiracle, naked gill slits, and placoid scales that become teeth on the jaws. Holocephalans have a fleshy operculum, nearly naked skin, and bony plates instead of teeth.

5. Osteichthyes are bony fishes. There are two groups, ray-finned and lobe-finned, with ganoid, ctenoid, or cycloid scales and a bony operculum. Speciation among ray-fins has been active since the early Devonian, with dermal armor gradually replaced by flexible scales. Chondrostei and Holostei are relict ray-fins, teleosts are recent.

6. There are two phylogenetic lines of lobe-finned fishes, crossopterygians and dipnoans. The former gave rise to amphibians, the latter are the true lungfishes. Most lobe-fins have internal nares. Speciation among lobe-fins has been conservative.

7. Labyrinthodonts were the first amphibians. Lissamphibia are recent and include anurans, urodeles, and apodans. Typical modern amphibians are scaleless, have a glandular skin, deposit their eggs in water, and have an aquatic larval stage. A few are viviparous. Some urodeles are perennibranchiate.

8. Reptiles, birds, and mammals are amniotes. Embryos develop within an amniotic sac in association with chorionic and allantoic membranes.

9. Reptiles are ectotherms with cornified epidermal scales, plaques, or scutes, and claws. Most of them lay large yolky eggs surrounded by a porous shell. Some are viviparous.

10. Cotylosaurs, the stem reptiles, arose from labyrinthodonts. Subsequent speciation produced five subclasses, Anapsida, Lepidosauria, Archosauria, Euryapsida, and Synapsida, established on the basis of the number and location of temporal fossae.

11. Birds are feathered endothermic descendants of bipedal archosaurs. Reptilian characters include a modified diapsid skull, reptilian scales on the legs and feet, claws, and cleidoic eggs. The forelimbs are modified for flight, but the wings of most ratites are vestigial. Archaeornithes, the oldest birds, had teeth and long tails.

12. Mammals are descendants of therapsid reptiles. Prototheria (monotremes) have a cloaca and lay reptilian eggs. They have primitive mammary secretions and no teats. Metatheria (marsupials) are viviparous and employ the yolk sac as a placenta. Eutheria employ the chorioallantoic membrane as a placenta.

13. Genetic variation coupled with geographical isolation seems to account for most new species.

SELECTED READINGS

See also Comprehensive References.

Aidley, D.J., editor: Animal migration. Society for Experimental Biology Seminar, Series 13, Cambridge, 1981, Cambridge University Press.

Bond, C.E.: Biology of fishes, Philadelphia, 1979, W.B. Saunders Co.

Brodal, A., and Fänge, R., editors: The biology of Myxine, Oslo, 1963, Universitetsforlaget (Norway).

Buffetaut, E.: Evolution of the crocodilians, Scientific American **241**(4):16, 1979.

Carrol, R.L.: The origin of lizards. In Andrews, S.M., Miles, R.S., and Walker, A.D., editors: Problems in vertebrate evolution, New York, 1977, Academic Press.

Charig, A.J.: A new look at dinosaurs, New York, 1983, Facts on File, Inc.

Colbert, E.H.: Dinosaurs, an illustrated history, Maplewood, New Jersey, 1983, Hammond, Inc.

Colbert, E.H.: Evolution of the vertebrates: A history of the backboned animals through time, ed. 3, New York, 1980, John Wiley and Sons.

Eisenberg, J.F.: The mammalian radiations, Chicago, 1981, University of Chicago Press.

Gingerich, P.I., Wells, N.A., and Shah, M.I.: Origin of whales in epicontinental remnant seas: New evidence from the early Eocene of Pakistan, Science **220**:403, 1983.

Griffiths, M.: The biology of monotremes, New York, 1978, Academic Press.

Hardesty, M.W.: Biology of the cyclostomes, London, 1979, Chapman & Hall, Ltd.

Hill, J.E., and Smith, J.D.: Bats: a natural history, Austin, Texas, 1984, University of Texas Press.

Hoage, R.J., editor: Animal extinctions. What everyone should know, Baltimore, 1985, Smithsonian Institution Press.

Hotton, N. III, MacLean, P.D., Roth, J.J., and Roth, E.C., editors: The ecology and biology of mammal-like reptiles, Baltimore, 1986, Smithsonian Institution Press.

Hunsaker, D., editor: The biology of marsupials, New York, 1977, Academic Press.

Kemp, T.S.: Mammal-like reptiles and the origin of mammals, New York, 1982, Academic Press.

Nickel, R., Schummer, A., and Seiferle, E.: Anatomy of the domestic birds, New York, 1977, Springer-Verlag.

Panchen, A.L., editor: The terrestrial environment and the origin of vertebrates, New York, 1980, Academic Press.

Pough, F.H.: Feeding mechanisms, body size, and the ecology and evolution of snakes, American Zoologist **23**:339, 1983.

Ridgway, S.H., and Harrison, R., editors: Handbook of marine mammals, vol. 3, The sirenians and baleen whales, New York, 1985, Academic Press.

Schmalhausen, I.I.: The origin of terrestrial vertebrates (translated from the Russian by Leon Kelso), New York, 1968, Academic Press.

Stahl, B.J.: Vertebrate history: problems in evolution, New York, 1974, McGraw-Hill Book Co.

Sturkie, P.D., editor: Avian physiology, ed. 3, New York, 1976, Springer-Verlag.

Tarling, D.H., and Tarling, M.P.: Continental drift: a study of the earth's moving surface, ed. 2, London, 1977, Bell.

Taylor, E.H.: The caecilians of the world, Lawrence, Kansas, 1968, The University Press of Kansas.

Thompson, K.S.: The biology of the lobe-finned fishes, Biological Reviews **44**:91, 1969.

Vial, J.L., editor: Evolutionary biology of the anurans, Columbia, Missouri, 1973, University of Missouri Press.

Zinsmeister, S.J.: Fossil windfall at Antarctica's edge, Natural History **95**(5):61, 1986.

Symposia in American Zoologist

Behavioral and reproductive biology of sea turtles **20**:1980.

Evolutionary morphology of the actinopterygian fishes **22**:237, 1982.

EARLY VERTEBRATE MORPHOGENESIS

In this chapter we will learn that all vertebrates pass through similar, though not identical, early formative stages. We will discover the steps whereby a hollow ball of totipotent embryonic cells is converted into an early embryo with three indispensable germ layers, and the contribution of each layer to the adult body. At the end, we will examine the extraembryonic membranes and their roles prior to hatching or birth.

Specimens studied in comparative anatomy or vertebrate morphology laboratories are viewed in a single time frame, the immediate present. A bit of knowledge from two additional time frames will enrich our understanding of morphology: the specimen's remote past, which requires knowledge of its ancestors; and its recent past, which includes its early ontogeny. Chapter 3 introduced the ancestors. This chapter recounts the highlights of early embryogenesis. The total picture—phylogeny, ontogeny, and morphology—is more meaningful than is morphology alone, if only because the first two account for the last.

VERTEBRATE EGGS
Egg Types

Eggs of vertebrates vary in the amount of yolk they contain and in the distribution of yolk within the egg. Eggs with very little yolk, such as those of the amphioxus and placental mammals, are **microlecithal.** Eggs with moderate amounts of yolk, such as those of freshwater lampreys, ganoid fishes, lungfishes, and amphibians, are **mesolecithal.** Eggs with massive amounts of yolk, such as those of marine lampreys, elasmobranchs, teleosts, reptiles, birds, and monotremes, are **macrolecithal.**

The yolk in microlecithal eggs is evenly distributed throughout the cytoplasm as fat droplets and small yolk globules. Eggs with an even distribution of yolk are said to be **isolecithal.** In mesolecithal and macrolecithal eggs the large yolk mass tends to be concentrated at one end, the **vegetal pole** (Fig. 4-1). The opposite pole, containing the nucleus and relatively yolk-free cytoplasm, is the **animal pole.** Eggs in which cytoplasm and yolk tend to accumulate at opposite poles are **telolecithal.**

Oviparity and Viviparity

Animals that spawn (lay their eggs) are said to be **oviparous.** Eggs of oviparous species contain sufficient nourishment in the form of yolk, and

FIGURE 4-1

Cleavage and the blastula of amphibian egg. *Gray,* animal pole, the cells of which contain little yolk; *red,* vegetal pole containing much yolk. In **I,** gastrulation by epiboly is under way. (After Eycleshymer.)

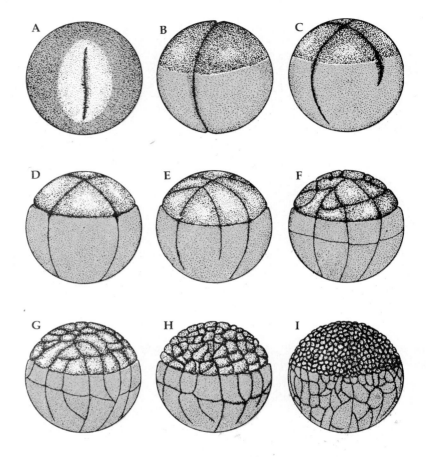

sometimes albumen, to support development into a free-living organism that is able to take food orally. If the yolk is massive the young may hatch fully formed, as in birds. If there is less yolk, the young hatch in a larval state, as in frogs. When there is very little yolk, as in the egg of the amphioxus, the free-living, self-nourishing state must be achieved very quickly after the egg is deposited. Accordingly, the amphioxus hatches into an externally ciliated, free-swimming embryo 8 to 15 hours after fertilization, at which time the notochord is a mere ridge in the roof of the primitive gut and there are no gill slits.

In many vertebrates the egg is retained within the mother's body during embryonic development, and living young are delivered. Such species are said to be **viviparous.** The relationship between mother and embryos varies from one in which the mother provides only protection and oxygen, nourishment having been stored in the egg, as in viviparous sharks, to one in which the embryos are dependent on maternal tissues for all nourishment, for oxygen, and for carrying away waste products of metabolism, as in placental mammals. The term **ovoviviparity** has been coined to designate the former condition. **Euviviparity** designates the condition in which the embryo cannot develop without nourishment being constantly supplied by the mother from maternal tissues. There are many intermediate conditions, and reptilian embryos that are initially ovoviviparous may become euviviparous before pregnancy has terminated. A reptilian egg in a viviparous species has a shell, but it is an uncalcified semipermeable membrane.

Viviparity in one degree or another has evolved in every class of vertebrates except agnathans and birds. It has developed independently more than a dozen times in teleost fishes, at least 10 times in lizards, and at least six times in snakes. In fact, 20% of all squamates are viviparous. There are no viviparous crocodilians or turtles. Viviparity occurs in 40 families of sharks and rays, and there is fossil evidence for viviparity in holocephalans and chondrosteans. It is thought to have occurred in the extinct Mesozoic marine reptile *Ichthyosaurus*. There are viviparous urodeles and anurans that give birth to fully metamorphosed young in terrestrial environments, and there are viviparous aquatic apodans.

The dogfish shark, *Squalus acanthias*, is an example of an ovoviviparous organism. The unborn pups are nourished entirely by the yolk from their own yolk sac (Fig. 4-10), but they receive oxygen from greatly enlarged and highly tortuous blood vessels in the uterine lining of the mother. The pups may be removed from the mother 2 to 3 months before birth (the gestation period is 20 to 22 months) and will complete their development, utilizing the nourishment from their yolk sac, as long as they are confined in finger bowls or plastic tubes containing oxygenated seawater.

In viviparous teleosts the egg may be fertilized and the young may develop in the ovarian follicle, there having been no ovulation. The young of *Gambusia* develop in this way. Or the embryos may develop in the ovarian cavity (Fig. 14-33). Among adaptations of the embryo for development in these locations are enlargement of the embryo's pericardial sac, which lies in contact with the maternal tissues of the ovary and absorbs necessary substances; enlargement of the embryonic gut, which lies in contact with maternal tissues; and long villuslike projections (**trophotaeniae**) of the embryonic gut that protrude through the vent of the embryo into the surrounding nutrient-rich medium. In some teleosts the young develop for a while in the follicular chamber and exhibit an enlarged pericardial sac, then pass into the ovarian cavity and develop absorptive villi. In some species developing eggs or larvae are ingested by other larvae.

Maternal tissues, under the influence of hormones, exhibit diverse adaptations for viviparity. The wall of the ovarian follicle occupied by teleost embryos may develop vascular folds or villi that secrete nutrients, and these may even project into the mouth or opercular cavity of the embryos. In *Dasyatis americana*, a euviviparous sting ray, the gravid uterus is lined with villi 2 to 3 cm long, which produce a copious secretion that nourishes the embryo. Secretions of uterine glands provide nourishment for unimplanted blastocysts of mammals, and perhaps for implanted blastocysts throughout pregnancy in perissodactyls and artiodactyls. **Histotrophic (embryotrophic) nutrition** is the term applied to nutrition by *glandular secretions* from maternal tissues, as contrasted with nutrition by substances exchanged via a placenta.

Internal and External Fertilization

In viviparous vertebrates, fertilization takes place within the body of the female. Fertilization is also internal whenever eggs are covered by an impenetrable shell before being extruded.

In oviparous fishes, frogs, and toads external fertilization is the rule. This is possible only because mating takes place in water and very large numbers of sperm and eggs are shed. In apodans and urodeles, however, fertilization is usually internal even though the eggs are subsequently laid. Some male urodeles deposit a sac of sperm embedded in jelly (**spermatophore**) in the immediate vicinity of a female during the mating ritual. The spermatophore is seized by the cloaca of the female or is placed in the cloaca by the male. The sperm then escape from the jelly and migrate up the female reproductive tract to the eggs. Packaging sperm in jelly makes it possible for those few urodeles that live far from water to convey viable sperm to a female despite lack of a copulatory organ.

EARLY DEVELOPMENT OF REPRESENTATIVE CHORDATES
Cleavage and the Blastula

Early cell divisions of the zygote are initiated by fertilization of the egg, and are referred to as **segmentation,** or **cleavage.** As a result, the zygote becomes subdivided into smaller and smaller cells that form (usually) a hollow sphere, or **blastula.** Each cell of the blastula is a **blastomere** and the cavity is the **blastocoel** (Fig. 4-2).

FIGURE 4-2

Blastula and gastrulation by involution in an amphioxus, sagittal section. Cells in *black* surround the blastopore (entrance to archenteron) and are the most active, mitotically. *Red* indicates presumptive endoderm in the blastula, definitive endoderm in the gastrula. The notochord establishes the dorsum and the long axis of the body.

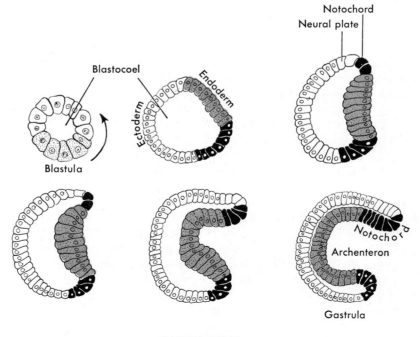

GASTRULATION
Amphioxus

When there is very little yolk in an egg, the blastomeres are approximately of equal size (Figs. 4-2 and 4-7, blastulae). When cell division is impeded by a moderate amount of yolk, as in amphibian eggs, the cells at the vegetal pole divide more slowly and are larger (Fig. 4-1). When there is a massive yolk, as in the eggs of reptiles and birds, segmentation is confined to the animal pole and results in a **blastoderm** perched like a skullcap on the massive yolk (Fig. 4-6). The embryo develops from this blastoderm.

The eggs of placental mammals exhibit an animal-vegetal polarity, just as do the eggs of reptiles, even though they have almost no yolk. The first cleavage division divides the egg into two blastomeres, one representing the animal pole, the other the vegetal pole (Fig. 4-7, cleavage). Further cell division results in a blastula. The descendants of the vegetal pole rapidly become a nutritional membrane, the **trophoblast,** for absorbing nourishment from the uterine fluids. The animal pole cells become the blastoderm, better known in mammals as the **inner cell mass** and, later, the **embryonic disc.** Because of the cystlike nature of the mammalian blastula, it is called a **blastocyst.**

As a result of cleavage, vertebrates typically pass through a hollow, spherical stage of development known as a blastula. During this stage, areas of distinctive developmental potential—the rudiments of future ectoderm, endoderm, mesoderm, and notochord—are established.

Gastrulation: Formation of the Germ Layers

Gastrulation is a dynamic process of cellular movements whereby presumptive endoderm, mesoderm, and notochord cells of the blastula migrate to the interior, thereby generating the primary germ layers from which the organs will form. In the process, bilateral symmetry is established. These cellular migrations are referred to as **formative,** or **morphogenetic movements,** because in the process the cells that until then were totipotent (capable of becoming any kind of cell) become restricted with respect to the direction of further differentiation. The restriction is imposed in part by enzymes from nearby cells. Rapid cell proliferation maintains a continuing supply of additional cells for these formative movements. Because in an amphioxus gastrulation is unencumbered by yolk, we will examine the process in this protochordate first.

AMPHIOXUS. In an amphioxus, presumptive endoderm from the surface of the blastula invaginates into the blastocoel (Fig. 4-2, *red*). The process, **involution,** produces the earliest gut, or **archenteron,** and results in a gastrula (Fig. 4-2, gastrula). The entrance to the archenteron is the site of involution, the **blastopore.** Soon, presumptive notochord cells from the surface of the gastrula flow inward to become the beginning of the notochord, located temporarily in the roof of the archenteron (Fig. 4-2, gastrula). Then, presumptive mesoderm cells pass inward to form bands of undifferentiated mesoderm lateral to the notochord (Fig. 4-3, *A*). At this stage the amphioxus embryo consists of an outer tube of ectoderm and an inner tube chiefly

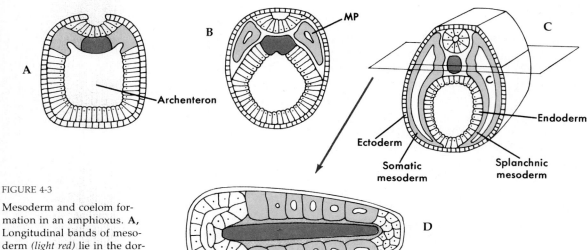

FIGURE 4-3

Mesoderm and coelom formation in an amphioxus. **A,** Longitudinal bands of mesoderm *(light red)* lie in the dorsolateral wall of the archenteron lateral to the notochord *(dark red).* **B,** Mesodermal pouches *(MP)* have formed. **C,** Pouches have grown ventrad between ectoderm and endoderm to form a coelom, C. **D,** Early larva in frontal section showing segmented coelom.

of endoderm surrounding the archenteron. The roof of the archenteron is the elongating notochord and the paired bands of mesoderm. Additional proliferation of cells from the rim of the blastopore results in elongation of the embryo.

Once mesoderm has been established as a pair of bands in the roof of the archenteron it continues to differentiate and, in the process, produces a coelom. The actively mitotic bands fold upward, pinch off, and form a series of hollow **mesodermal pouches (coelomic pouches)** commencing at the anteriormost end of the band (Fig. 4-3, *A* and *B*). About the time two pouches have formed, the embryo hatches into a free-swimming larva. As the larva elongates, additional pouches develop.

After the mesodermal pouches have been established, they grow ventrad, pushing between ectoderm and endoderm (Fig. 4-3, *C*). They finally meet underneath the gut, and fuse, forming thereby a temporary **ventral mesentery.** The outer wall of each pouch lies against the ectoderm and is **somatic mesoderm.** Together with the ectoderm it constitutes the **somatopleure,** or body wall. The inner wall of each pouch lies against the endoderm and is the **splanchnic mesoderm.** Together with the endoderm it constitutes the **splanchnopleure,** which becomes chiefly the digestive tract. The cavity between somatic and splanchnic mesoderm is the **coelom** (Fig. 4-3, *C*).

The coelom of an amphioxus is segmented for a while because of its origin from a series of pouches (Fig. 4-3, *D*). Later, the walls between the pouches disappear, establishing a single long coelom on each side. Still later, the ventral mesentery disappears, and the left and right coelomic cavities become confluent underneath the gut, as in vertebrates. The germ layers are now in a position to form the organs.

FROG. In vertebrates other than placental mammals gastrulation is complicated by the presence of moderate to large quantities of yolk. Amphibians manage to tuck the unwieldy yolk of their mesolecithal eggs inside the embryo. This is accomplished partly when the small cells of the animal pole grow downward over the larger cells of the vegetal pole, a process called **epiboly** (Figs. 4-1 and 4-4, *A, C,* and *E*). Yolk then constitutes the archenteric floor (Figs. 4-4, *D* and *F,* and 4-5). Presumptive endoderm, initially on the surface of the blastula, migrates to the interior via the blastopore in a sequence of invaginations and migrations that are more complicated than in the microlecithal eggs of an amphioxus, but the results are similar. Definitive endoderm comes to line the elongating archenteron, excepting the roof. The outer cellular layer of the embryo is then ectoderm.

Mesoderm is not formed in vertebrates as outpocketings of the roof of the archenteron. In frogs specifically, some presumptive mesoderm flows over the dorsal rim of the blastopore along with presumptive notochord

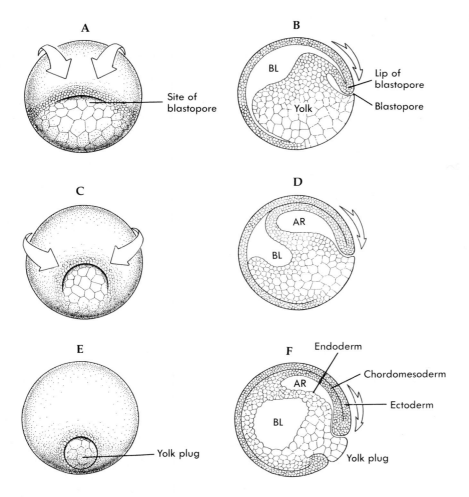

FIGURE 4-4

Epiboly and gastrulation in a frog. **A,** Blastula stage slightly older than that in Fig. 4-1, *I.* **B,** Hemisection of **A** in vertical plane. Involution commencing at blastopore. **C** and **D,** Later stages in epiboly, gastrulation continuing. **E** and **F,** Yolk plug stage. Compare **F** with Fig. 4-5. *AR,* Archenteron; *BL,* blastocoel. *Arrows* indicate downgrowth of animal pole cells over yolk cells.

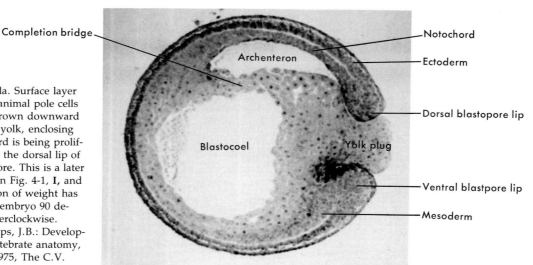

FIGURE 4-5

Frog gastrula. Surface layer consists of animal pole cells that have grown downward around the yolk, enclosing it. Notochord is being proliferated from the dorsal lip of the blastopore. This is a later stage than in Fig. 4-1, **I**, and redistribution of weight has rotated the embryo 90 degrees counterclockwise. (From Phillips, J.B.: Development of vertebrate anatomy, St. Louis, 1975, The C.V. Mosby Co.)

cells to occupy a temporary position in the roof of the archenteron. This migrating stream is **chordomesoderm** (Fig. 4-4, *F*). The notochord organizes from the midline portion of the chordomesoderm and is therefore temporarily in the roof of the archenteron (Fig. 4-5). Chordomesoderm immediately lateral to the notochord gives rise to a pair of longitudinal bands of mesoderm paralleling the notochord. These bands become the **dorsal mesoderm.** The dorsal mesoderm in the future trunk and tail region subsequently segments and reorganizes to form hollow **mesodermal somites** similar to the mesodermal pouches of an amphioxus (Fig. 4-8, *B*). The remaining presumptive mesoderm in frogs reaches the interior by flowing over the lateral and ventral rim of the blastopore. Once inside, it streams forward as an advancing sheet of rapidly expanding **lateral-plate mesoderm,** pushing its way cephalad between ectoderm and endoderm. The lateral-plate mesoderm remains unsegmented and soon splits into two sheets, an outer sheet, the **somatic mesoderm,** and an inner sheet, the **splanchnic mesoderm** (Fig. 4-8, *B*). The cavity between them is the coelom. A longitudinal strand of unsegmented **intermediate mesoderm (nephrogenic mesoderm)** connects the dorsal and lateral-plate mesoderm. It gives rise to the kidney tubules and their ducts.

CHICK. We will use the chick to illustrate gastrulation in amniotes with macrolecithal eggs, since gastrulation in reptiles and birds differs only in details. The cells of the animal pole do not *initially* grow downward around the massive yolk as they do in epiboly. Instead, the blastoderm organizes into an upper sheet of cells, the **epiblast,** and a lower sheet, the **hypoblast** (Fig. 4-6). This process, **delamination,** hastens the formation of the blood vessels necessary for transporting yolk to the developing embryo. Eventually, as if more leisurely, cells from the hypoblast grow outward over the surface of the yolk, then downward around it, to become the endodermal lining of a yolk sac. The epiblast is chiefly presumptive ectoderm and chordomesoderm. The hypoblast is chiefly presumptive endoderm. Delamination temporarily bypasses formation of a blastopore. If a homologue of the archenteron exists, it is not recognizable.

Gastrulation is accomplished by much streaming laterad and caudad of migrating epiblast and hypoblast cells above the huge yolk. The details require many pages of explanation and can be found in embryology textbooks. The culmination of this initial cell migration is the formation of a **primitive streak** and **Hensen's node,** the latter being homologous with the rim of the blastopore of lower vertebrates (Fig. 4-8, *C*). It defines the caudal end of the future embryo, which can then be constructed. From Hensen's node a notochordal process pushes forward beneath the epiblast, presumptive mesoderm cells stream forward alongside the notochord to establish the dorsal mesoderm, and others migrate beneath the epiblast forward, laterad, and downward, become continuous with the dorsal mesoderm, and form intermediate and lateral-plate mesoderm.

Lateral-plate mesoderm consists initially of a sheet of loosely aggregated mesenchyme. While the dorsal mesoderm of the trunk and stubby tail region is segmenting to form somites, the lateral-plate mesoderm is splitting into somatic and splanchnic mesoderm (Fig. 4-8, *D*). The resulting cavity between the somatic and splanchnic mesoderm is the coelom.

By the end of the second day of incubation the mesoderm of the splanchnopleure (splanchnic mesoderm plus endoderm) has given rise to the first blood cells and to a network of delicate vessels (**vitelline,** or **omphalomesenteric**) that collect yolk globules and transport them to a simple, twitching, S-shaped heart (Fig. 15-6). The heart propels this early stream of cells and nutrients to the rapidly growing tissues.

As a result of complex morphogenetic movements and accompanying cell proliferation, the chick embryo establishes ectoderm and endoderm, dorsal, intermediate and lateral-plate mesoderm, a coelom, and an extraembryonic yolk sac. The germ layers thereby become situated in locations from which they can proceed with organogenesis.

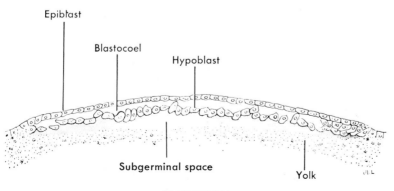

Epiblast

Blastocoel

Hypoblast

Subgerminal space

Yolk

BLASTODERM
Bird

FIGURE 4-6

Blastoderm at animal pole of avian egg, sagittal section. Head end is to left. Only a fraction of the massive uncleaved yolk is shown.

The processes outlined above for species with mesolecithal and metalecithal eggs differ in details among the vertebrate classes. The greatest variation is seen among fishes for the reason that they represent so many divergent lineages.

PLACENTAL MAMMALS. At the time gastrulation commences in mammals the blastocyst exhibits an **inner cell mass** derived from the animal pole of the blastula (Fig. 4-7, blastocysts). The wall of the blastocyst of mammals does not contribute to formation of the embryo, having no presumptive ectoderm, mesoderm, or endoderm. The embryo develops from the inner cell mass. The blastocyst wall, called a **trophoblast,** establishes contact with, or implants itself into, the lining of the maternal uterus, becoming the first membrane through which the conceptus (embryo plus extraembryonic membranes) receives nourishment and oxygen directly from maternal tissues.

The source of the first endoderm is not precisely the same in all mammalian orders. In some it arises from cells proliferated from the inner cell mass (Fig. 4-7). In others it is described as forming by a splitting of the inner cell mass into epiblast and hypoblast, the latter being considered presumptive endoderm, as in macrolecithal eggs. In any case, endoderm eventually spreads along the inner surface of the trophoblast, forming a yolk sac around a yolk that isn't there (Fig. 4-7, gastrula)! The trait bespeaks the reptilian ancestry of mammals.

FIGURE 4-7

Cleavage, blastula, and gastrula-like stage of a mammal. At cleavage (upper left) the red cell at the animal pole gives rise to the embryo. The other cell gives rise to the trophoblast. The first endoderm (hypoblast) comes from the inner cell mass. The yolk sac contains no yolk.

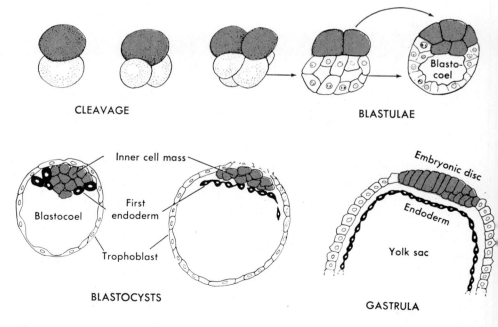

CLEAVAGE

BLASTULAE

Blastocoel

Inner cell mass

First endoderm

Blastocoel

Trophoblast

BLASTOCYSTS

Embryonic disc

Endoderm

Yolk sac

GASTRULA

Mammal

The inner cell mass eventually consists of an epiblast and a hypoblast in all placental mammals, and it is then called a **blastodisc (embryonic disc)**. The epiblast is a thick plate of columnar cells. The hypoblast, thickened by cell proliferation, is immediately beneath the epiblast and in contact with it, unlike in birds. Thereafter, gastrulation and formation of the germ layers resemble the processes described in vertebrates with macrolecithal eggs. There is much streaming of presumptive ectoderm, mesoderm, and endoderm from the blastodisc to definitive locations, the result of which, as in chicks, is establishment of the germ layers and, simultaneously, of the extraembryonic membranes vital to the euviviparous embryo.

Before concluding the discussion of gastrulation a final word must be added. Because of the diversity of the specific processes whereby the various vertebrates achieve a bilaterally symmetrical, three-germ-layered stage with notochord and neural tube rudiments, it is not possible to specify a common end point for the process of gastrulation that would be applicable to all species. Gastrulation places the germ layers in their respective positions vis-á-vis one another. When this has been accomplished, gastrulation is complete and organogenesis, the formation of organ systems, can commence.

Mesenchyme

Mesenchyme consists of aggregations of undifferentiated cells before they come under the influence of enzymes and any other factors that initiate differentiation toward a specific cell type. Although it is an embryonic tissue, aggregations of mesenchyme are ubiquitous in adults and are the source of new cells for growth and tissue repair throughout life. Most mesenchyme is mesodermal; some is ectodermal, or ectomesenchyme. If its source is the embryonic neural tube, ectodermal placodes, or neural crests, it is neurectoderm.

Neurulation

The central nervous system is established by a simple procedure that is essentially concurrent with gastrulation. Presumptive neural ectoderm is situated in a wide band on the dorsal surface of the embryo overlying the notochord and segmental mesoderm. It commences at or slightly behind the blastopore and extends into the future head region beyond the notochord, where it overlies the endoderm of the cephalic end of the foregut. Close to the time of commencement of gastrulation the band commences to differentiate, forming a thickened **neural plate,** the lateral margins of which become elevated to form a pair of **neural folds** bounding a **neural groove** (Fig. 4-8). The anterior end of the neural groove is widest, being the future brain. The rest is future spinal cord. The neural plate sinks into the dorsum of the embryo, and the neural folds grow toward one another above the groove, meet in the middorsal line, then fuse, converting the neural plate and groove into a **neural tube.** The process of closure commences near the

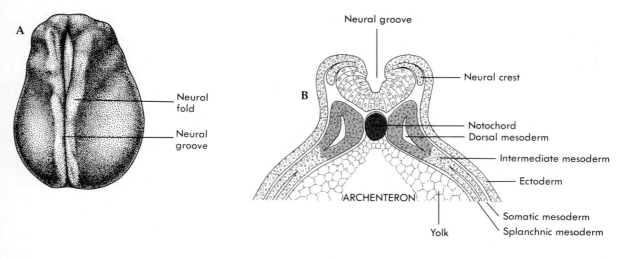

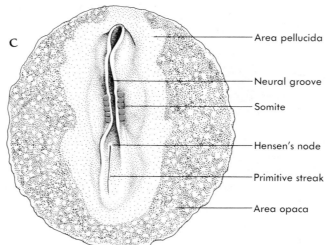

FIGURE 4-8

Neurulation in a frog, chick, and teleost. **A,** Frog neurula near end of gastrulation, removed from its jelly envelope. The nearly closed blastopore is at the caudal end of the neural groove. **B,** Cross section of **A** at level of future midgut. **C,** Chick embryo at about 26 hours of incubation. The amniotic folds have not yet appeared. The undersurface of the area opaca, but not of the area pellucida, had adhered to the yolk. The dark splotches in area opaca are blood islands.

caudal end of the brain and sweeps cephalad and caudad. Finally, only a small opening remains at each end of the tube **(anterior and posterior neuropores).** These later close. As a result of formation of a neural tube by closure of the neural groove, the central nervous system is ectodermal, dorsally located, and hollow. The cavity is the **neurocoel.** It remains as the ventricles of the brain and central canal of the spinal cord. The process that establishes the neural tube is **neurulation.** Further differentiation of the walls of the neural tube to form brain and cord is discussed in Chapter 15.

Neurulation differs among the vertebrate classes in minor details. The major variation is seen in cyclostomes and ray-finned fishes, in which neural folds do not form, and the neural plate does not roll into a tube. Instead, it becomes a wedge-shaped **neural keel** that invades the dorsal body wall

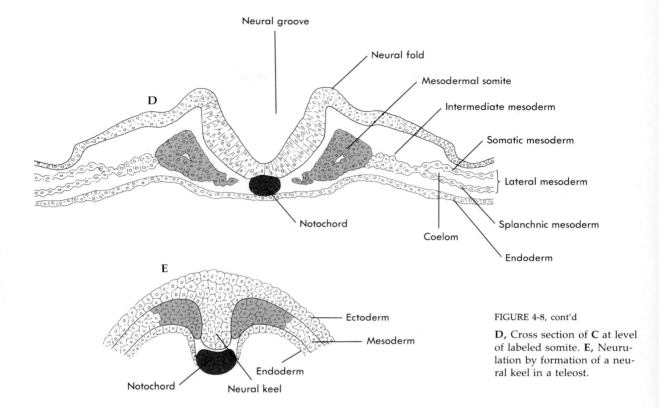

FIGURE 4-8, cont'd

D, Cross section of **C** at level of labeled somite. **E,** Neurulation by formation of a neural keel in a teleost.

(Fig. 4-8, *E*). Subsequently it becomes detached from the overlying surface ectoderm and organizes a neurocoel by rearrangement of the interior cells. The result is a typical dorsal hollow central nervous system.

FATE OF THE ECTODERM

Embryonic ectoderm gives rise to (1) the neural groove, (2) the epidermis and all of its derivatives, including all glands that lie within the dermis, (3) the lining of the stomodeum and proctodeum and their derivatives, and (4) a broad spectrum of other tissues as a result of differentiation of ectomesenchyme derived from neural crests and ectodermal placodes. Among ectomesenchymal derivatives are tissues that, except for their specific location, would have arisen from mesoderm.

Stomodeum and Proctodeum

Formation of the oral cavity is initiated by a midventral invagination of the embryonic head to form a **stomodeum** (Figs. 1-1 and 16-6, *A*). Rupture of the thin oral plate separating the stomodeum from the foregut establishes an anterior opening for the digestive tract. Only a small portion of the resulting oropharyngeal cavity is lined by ectoderm. The major portion is derived from the foregut and is lined by endoderm. Stomodeal ectoderm gives rise to ameloblasts, which produce the enamel of the teeth of mammals at least,* to the mucosa of the stomodeal portion of the oropharyngeal cavity (oral cavity of tetrapods, including approximately the anterior one third of the mucosa of the tongue of mammals) and to some of the more anterior glands of the oral cavity. (Once the oral plate has ruptured it becomes difficult to determine the germ layer origin of some of the epithelia immediately on either side of the rupture.) An evagination from the roof of the stomodeum, **Rathke's pouch,** becomes the adenohypophysis in all vertebrates other than hagfishes (Fig. 17-6).

An invagination similar to the stomodeum develops in relation to the embryonic hindgut, the two being separated temporarily by a cloacal plate. The invagination is the **proctodeum** (Fig. 1-1). It becomes the lining of the terminal portion of the cloaca. No discernible proctodeal remnant remains in adults in which the cloaca disappears as a discrete entity during embryogenesis.

Neural Crests

As the neural groove sinks into the dorsal body wall some of the ectoderm along its lateral edges becomes dissociated from both the surface ectoderm and the developing neural tube to form an independent strand of neurectoderm paralleling the neural tube on each side (Fig. 4-8, *B*, neural crest). Shortly thereafter the strand segments, and the cells aggregate to form several clusters alongside the midbrain and hindbrain, and a cluster at the site of each developing spinal nerve. These clusters are **neural crests,** named for their location near the crest of the neural tube.

Neural crest ectoderm gives rise to a broad spectrum of differentiated tissues, many of the cells having migrated far from their origin. Some become pigment cells, located not only in the skin and iris diaphragm but also within the walls of blood vessels of fishes and amphibians, between the striated muscle fibers of some mammals, in the meninges of the brain and spinal cord, and in other deep locations. Other migrating neural crest cells become chromaffin cells, some of which contribute to the adrenal complex and secrete amines such as epinephrine. Still other migrating neural crest cells become calcitonin cells of the thyroid, parathyroid, and ultimobranchial glands. Neural crest cells surround the spinal cord and brain to form the leptomeninx, and when present, the arachnoid meninx. Cells from neural crests along the length of the trunk stream ventrad around the notochord and dorsal aorta to form the autonomic ganglia of the trunk, and from neural crests in the head to form the parasympathetic ganglia of the head.

Cells from neural crests of the head region stream ventrad into the developing visceral arches, where they form mesenchymal blastemas for the pharyngeal skeleton while others are forming blastemas for the branchio-

---

*Enamel differs among the vertebrate classes. Enamel-like substances (enameloids) of some lower vertebrates are produced by odontoblasts. Whether the latter are of neural crest origin has not been ascertained.

meric musculature. Some contribute to the anteroventral portion of the neurocranium and, in chicks at least, to many of the membrane bones that invest the neurocranium. The lingual cartilages and branchial basket of lampreys are formed by the neural crest cells under the inductive influence of the branchial endoderm. These are some of the tissues that, were it not for their location in the head, would be of mesodermal origin. Experimental studies, past and continuing, are elucidating the role of nearby embryonic tissues in inducing the differentiation of ectomesenchyme in the direction of fibroblasts, scleroblasts, and myoblasts (Fig. 6-1).

We might suppose that, with all this emigration, no cells would be left in the neural crest! This is not the case, because mitosis continually replenishes the supply until organogenesis is completed. A significant number remain in the neural crest, being either neuroblasts or precursors of Schwann cells. These specific neuroblasts become the cell bodies of sensory neurons whose processes extend centrally into the brain or cord and peripherally to receptors. Consequently, neural crests emerge ultimately as sensory ganglia of the spinal nerves and, with exceptions noted in the next paragraph, of the cranial nerves. Schwann cells form the neurilemma, a membrane that ensheathes nerve fibers outside the central nervous system.

Ectodermal Placodes

Ectodermal placodes are paired localized thickenings of the embryonic ectoderm that, to one degree or another, sink beneath the surface and give rise to certain sensory neurons or to sense organs as follows:

1. A series of placodes on the side of the head of all vertebrates (called **epibranchial placodes** in fishes) sink into the head to a position just lateral to the hindbrain, where they are often mistaken for neural crests. They contribute many or all of the neurons in the sensory ganglia of cranial nerves V, VII, VIII, IX, and X, augmenting those from cranial neural crests to an extent depending on the species. In effect, therefore, they represent an alternate method of neural crest formation.

2. A pair of **olfactory placodes (nasal placodes)** form on either side of the anterior end of the head just above the stomodeum and anterior to the cephalic extremity of the forebrain. The placodes sink into the head to become nasal pits, but retain an opening to the exterior, the nostril, or "naris" (Fig. 12-8). Some of the cells of each placode become olfactory receptor cells whose processes grow back into the forebrain and collectively constitute one of the two olfactory nerves.*

3. A **linear series of placodes** extending the length of the trunk and tail of fishes and aquatic amphibians, and continuing onto the head in a variable pattern, sink into the dermis or beneath it to become mechanoreceptive neuromast organs of the cephalic and lateral line canals. Similar placodes on the head of fishes, not part of the linear series, become thermoreceptive and electroreceptive neuromast organs.

---

*Olfaction is not considered an exteroceptive sense because of the predominantly visceral connections of the olfactory nerve within the brain.

4. A pair of **otic placodes** located lateral to the hindbrain of all vertebrates sinks deep into the head to become the membranous labyrinths (inner ears).

5. A **lens placode** forms opposite each developing retina and sinks under the surface to become the lens of the eye. (The retina is ectodermal because it arises as an evagination from the forebrain.)

■ ■ ■

From the facts presented in reference to neurulation, neural crests, and ectodermal placodes, it is evident that the neural components of the entire nervous system, central and peripheral, many of their supportive and nutritive cells of the nervous system, and the sensory epithelia of all the special sense organs of the head other than taste buds are either of neural plate, neural crest, or ectodermal placode origin. It is also evident that the embryonic ectoderm as a whole is much more active in organogenesis than might be assumed from its position on the surface of vertebrate gastrulae.

FATE OF THE ENDODERM

Endoderm gives rise to the epithelium of the entire alimentary canal between the stomodeum and proctodeum. The taste buds and oral glands caudal to the stomodeum are endodermal. The epithelioid components of the parathyroid glands, ultimobranchial glands, thymus, auditory tube, and middle ear cavity are endodermal because of their origin from pharyngeal pouches. Midventral evaginations of the pharynx, including thyroid, lungs, and swim bladders and their ducts, if any, are lined with endoderm or have endodermal components. Caudal to the pharynx the endoderm evaginates to form crop sacs, liver, gallbladder, pancreas, and various gastric and intestinal ceca. Most urinary bladders and the urogenital sinus of mammals are lined with endoderm, because these are derived from the cloaca. Any other structure that arises as an evagination of the embryonic foregut, midgut, or hindgut has endodermal components.

FATE OF THE MESODERM
Dorsal Mesoderm (Epimere)

The somites constitute collectively the dorsal mesoderm. They are aligned beside the notochord and neural tube the entire length of the trunk and tail, and develop in the head in varying numbers (Figs. 4-8, C and D, and 15-6). A somite exhibits three regions, **myotome, sclerotome,** and **dermatome** (Fig. 4-9). Myotomes contribute mesenchyme that migrates to form the skeletal muscles other than those of the pharyngeal arches. Sclerotomes of trunk somites give rise to the vertebral column and proximal portion of the ribs; those of head somites give rise to the posteroventral components of the neurocranium, chiefly parachordal cartilages, and to one or another of the sense capsules, depending on the species. Dermatomes contribute mesenchyme that forms the dermis of the skin of the

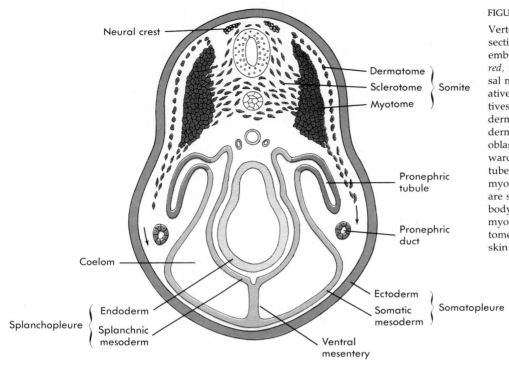

Neural crest

Dermatome
Sclerotome } Somite
Myotome

Pronephric tubule

Pronephric duct

Coelom

Ectoderm
Somatic mesoderm } Somatopleure

Splanchopleure {
Endoderm
Splanchnic mesoderm

Ventral mesentery

FIGURE 4-9

Vertebrate embryo in cross section showing fate of the embryonic mesoderm. *Dark red,* mesodermal somite (dorsal mesoderm) and its derivatives; *medium red,* derivatives of intermediate mesoderm; *light red,* lateral mesoderm. Sclerotome cells (scleroblasts) are streaming toward notochord and neural tube to form a vertebra; myotomal cells (myoblasts) are streaming into lateral body wall *(arrows)* to form myotomal muscle; dermatome will form dermis of skin of part of the back.

dorsalmost part of the back. (Most of the dermis arises from lateral-plate mesoderm, and in the head from neural crests.)

Lateral-Plate Mesoderm (Hypomere)

Lateral-plate mesoderm is confined to the trunk and consists of somatic and splanchnic mesoderm (Figs. 4-8, *B* and *D,* and 4-9). **Somatic mesoderm** gives rise chiefly to the connective tissues and blood vessels of the body wall and to the skeleton of the body wall, girdles, and limbs. The dermis of the body wall is its outermost product, and the parietal peritoneum is the innermost. **Splanchnic mesoderm** gives rise to smooth muscles and connective tissue of the digestive tract and its outpocketings, to the heart, and to the blood vessels of the viscera. The visceral peritoneum is its outermost derivative. Lateral-plate mesenchyme streams into the developing head in chicks and some other vertebrates; its contribution to the head is under investigation. Much of what was once thought to arise from these cells has been shown by more modern marking techniques to be of cranial neural crest origin.

Intermediate Mesoderm (Mesomere)

Intermediate, or **nephrogenic mesoderm,** consists of a pair of longitudinal ribbons of unsegmented mesoderm extending the length of the trunk just lateral to the somites (Fig. 4-8, *D*). It gives rise to kidney tubules and the longitudinal ducts that drain them.

EXTRAEMBRYONIC MEMBRANES

Most embryonic vertebrates are provided with special membranes that extend beyond the body. These extraembryonic membranes arise early in ontogeny and perform important services for the embryo until hatching or birth. The chief extraembryonic membranes are the **yolk sac, amnion, chorion,** and **allantois.** The last three are found only in reptiles, birds, and mammals. The yolk sac is the most primitive.

Yolk Sac

The yolk sac surrounds the yolk (Figs. 4-10 and 4-11). It empties into the midgut and is usually lined by endoderm, although not in bony fishes. The sac is highly vascular, and its vessels (**vitelline arteries** and **veins**) are confluent with circulatory channels within the embryo proper. Yolk particles in the sac are usually digested by enzymes secreted by the lining of the sac and then transported to the embryo by vitelline veins. (In sharks, yolk also enters the intestine directly from the sac, propelled by rapidly beating cilia that line the sac and stalk.) The yolk sac grows smaller as the embryo grows larger. As it shrivels, it slowly disappears into the ventral body wall. The intracoelomic remnant of the sac is finally incorporated into the wall of the midgut, or it may remain as a small diverticulum.

In embryonic sharks a large diverticulum of the yolk sac develops within the coelom close to where the sac opens into the duodenum. In pups ready to be born and with the yolk sac almost completely retracted, the diverticulum is easily demonstrated. It is distended with yolk, which has also spilled over into the spiral intestine. This remaining yolk serves to nourish the newborn pup for several days until it is able to obtain food from the environment.

Despite the fact that there is no yolk in typical mammalian eggs, embryonic mammals develop a yolk sac—a reminder of their genetic relationship with egg-laying reptiles. In humans a vestige of the yolk sac (Meckel's diverticulum) remains in about 2% of the adult population. Its average position is on the ileum, 30 cm above the ileocolic valve. Its average length is about 5 cm.

Because the yolk sac in viviparous anamniotes is highly vascularized and lies close to maternal tissues, it often serves as a membrane for absorbing oxygen from the parent. After the yolk in the sac is depleted, or if there is none, the sac may absorb nourishment from maternal tissues. When functioning in either capacity, it constitutes a **simple yolk sac placenta.**

Amnion and Chorion

The embryos of reptiles, birds, and mammals develop within two membranous sacs, the amnion and chorion (Figs. 4-11 and 4-12). These, along

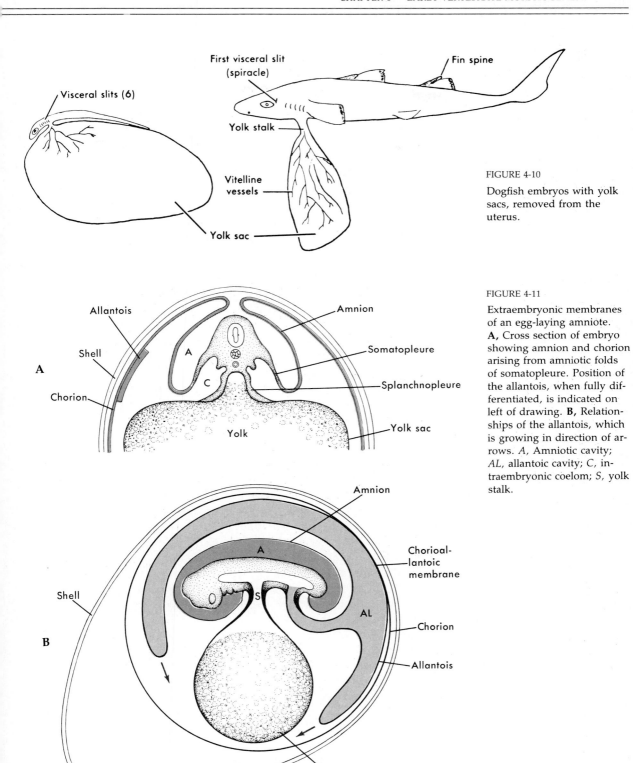

FIGURE 4-10

Dogfish embryos with yolk sacs, removed from the uterus.

FIGURE 4-11

Extraembryonic membranes of an egg-laying amniote. **A,** Cross section of embryo showing amnion and chorion arising from amniotic folds of somatopleure. Position of the allantois, when fully differentiated, is indicated on left of drawing. **B,** Relationships of the allantois, which is growing in direction of arrows. *A,* Amniotic cavity; *AL,* allantoic cavity; *C,* intraembryonic coelom; *S,* yolk stalk.

FIGURE 4-12

Gravid canine uterus, ventral view. One uterine horn has been opened to expose conceptuses, and the chorionic sacs have been partially cut away.

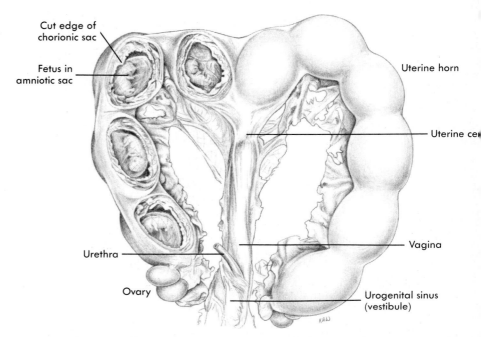

Labels on figure: Cut edge of chorionic sac; Fetus in amniotic sac; Urethra; Ovary; Uterine horn; Uterine ce[rvix]; Vagina; Urogenital sinus (vestibule)

with the semiporous eggshell, made it possible for ancestral reptiles and their descendants to lay their yolk-laden eggs on land. The membranes were subsequently inherited by placental mammals. The sacs are formed when upfoldings of the embryonic somatopleure meet above the embryo and fuse (Fig. 4-11, *A*).

The **amniotic sac** contains the embryo and is filled with a slightly salty amniotic fluid secreted by the lining of the sac. The source of the fluid is chiefly metabolic water produced by the embryonic tissues as a result of cellular respiration. Turtle eggs that are laid in moist locations absorb considerable water from the environment through the eggshell, but this is not a common source of amniotic fluid. When the embryonic kidneys commence to function, nitrogenous wastes are added to the fluid. **Amniotic fluid** buffers the fetus against mechanical injury and helps prevent dessication of embryos in oviparous species. In viviparous mammals the embryo is suspended in the amniotic fluid by the **umbilical cord,** which connects the fetus with the placenta (Fig. 13-36).

The **chorion** is a larger sac that surrounds the amniotic sac and lies in intimate relationship with either the lining of the eggshell or with the lining of the mother's uterus, depending on the species. Thus the amnion with its amniotic fluid protects the fetus, whereas the chorion keeps it in communication with its source of nutrients and oxygen.

Allantois

The **allantois** is an extraembryonic membrane sac that arises typically as a midventral evagination of the embryonic cloaca (Fig. 4-11, *B*). It ordinarily grows until it comes in contact with the inner surface of the chorion to form a **chorioallantoic membrane** (Fig. 4-11, *B*). The chorioallantoic membrane in reptiles, birds, and monotremes is in contact with the eggshell where, except in those viviparous squamates that depend on the mother late in pregnancy for at least some nourishment, it serves chiefly as a respiratory organ (Fig. 4-11, *A*). In eutherian mammals the chorioallantoic membrane comes in direct contact with the lining of the maternal uterus, where it constitutes the fetal part of a **chorioallantoic placenta.** Here it performs two additional functions, being also the site of transfer of nutrients from maternal tissues to the embryo and of transfer of metabolic wastes from embryo to mother.

In some eutherian mammals (cats, rabbits, humans, others) the allantois extends only part way out the umbilical cord before dwindling and terminating as a blind sac, but its vessels, the allantoic (umbilical) arteries and veins, continue toward, and vascularize, the chorion. In many marsupials, neither the allantois nor the allantoic vessels reach the chorion, and the yolk sac and its vitelline vessels take the place of the allantois as part of a **choriovitelline placenta.**

The base of the allantois—the part closest to the cloaca—becomes the **urinary bladder** of amniotes (Fig. 13-36). The part extending between the tip of the bladder and the umbilicus in mammals may remain after birth as a **middle umbilical ligament,** or **urachus** (Fig. 14-41, *F*). The portion within the umbilical cord is discarded at hatching or at birth. It is likely that the allantois is the old amphibian urinary bladder with added functions.

Placentae

The term *placenta* in its broadest sense refers to any region in a viviparous organism where maternal and embryonic tissues of any kind are closely apposed, and which serves as a site for exchanges between parent and embryo. In this sense, viviparous fishes that develop in an ovarian follicle sometimes exhibit a placenta. In a more restricted sense, a **placenta** is an organ composed of (1) the highly vascular region of an extraembryonic membrane (yolk sac, choriovitelline membrane, chorioallantoic membrane, or chorion alone), and (2) the associated highly vascular lining of the maternal uterus.

The yolk sac frequently serves as the fetal part of a placenta. In viviparous fishes and amphibians it may lie in immediate contact with the uterine lining **(simple yolk sac placenta).** In most marsupials the yolk sac, being "trapped" within the chorionic sac, lies against the chorion as part of a **choriovitelline placenta.** Some squamates and marsupials have both a choriovitelline and a chorioallantoic placenta. (In viviparous reptiles the thin egg shell lies between the fetal portion of the placenta and the uterine lining.) As stated earlier, mammals above marsupials have a **chorioallantoic placenta.**

The intimacy of the relationship between fetal and maternal tissues in mammals varies greatly. In marsupials and most ungulates the extraembryonic membranes lie in simple contact with the uterine lining **(endome-**

trium), and at birth the fetal membranes simply peel away without any shedding of the lining. This is a **contact,** or **nondeciduous, placenta.** In more intimate relationships, **chorionic villi,** which are fingerlike outgrowths of the chorionic sac, become rooted to one degree or another into the endometrium, or even dangle into uterine blood sinuses. When the fetal part of such a placenta disengages at parturition, the invaded portion of the uterine lining, the **decidua,** is shed, and some bleeding occurs. This is a **deciduous placenta.** In any case, the extraembryonic membranes and any sloughed uterine tissues are delivered as the afterbirth.

Chorionic villi are distributed on the surface of the mammalian chorionic sac (Fig. 4-13) in isolated patches **(cotyledonary placentae),** in a band encircling the sac **(zonary placentae),** in a single large discoidal area **(discoidal placentae),** or diffusely over the entire surface of the chorion **(diffuse placentae).** The mammalian placenta is one source of hormones essential for the maintenance of pregnancy.

FIGURE 4-13

Chorionic sacs of mammals showing placental areas. Pig embryo surrounded by the amnion can be seen through the exceptionally thin chorionic sac.

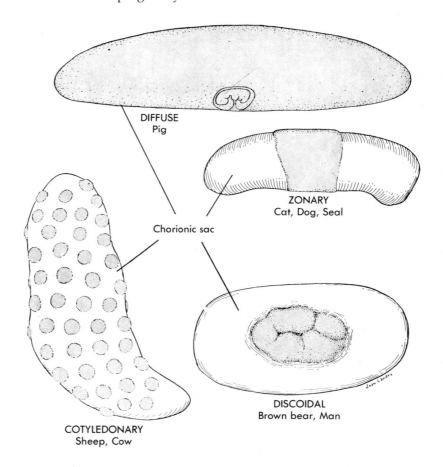

DIFFUSE
Pig

ZONARY
Cat, Dog, Seal

Chorionic sac

DISCOIDAL
Brown bear, Man

COTYLEDONARY
Sheep, Cow

Chapter Summary

1. Eggs are microlecithal, mesolecithal, or macrolecithal, according to the quantity of yolk. Eggs are isolecithal or telolecithal, depending on the distribution of yolk within the egg.

2. Oviparous vertebrates deposit eggs that contain enough nourishment to support development until birth or hatching.

3. Cephalochordates deposit microlecithal eggs with very little yolk that hatch quickly into immature, free-living, ciliated larvae with rudimentary organ systems.

4. Viviparous vertebrates develop in the body of the mother. In ovoviviparous species the eggs are laden with yolk, and the mother provides oxygen and protection. In euviviparous species the eggs are practically devoid of yolk, and all nourishment is supplied from the mother's tissues.

5. In viviparity highly vascularized embryonic tissues (e.g., pericardial sac, gills, villi, opercular lining, extraembryonic membranes) are sites for absorption of necessary substances.

6. Viviparity is found in every vertebrate class except agnathans and birds.

7. Fertilization is internal in viviparous species, in urodeles and apodans, and in species that cover the egg with a shell.

8. Cleavage follows fertilization. It produces a blastula with a blastocoel.

9. Gastrulation is characterized by notogenesis, neurulation, establishment of bilateral symmetry, and formation of the germ layers. The basis of the process is morphogenetic streaming of undifferentiated cells.

10. Gastrulation is initiated by involution in cephalochordates, epiboly in species with mesolecithal eggs, and delamination of a blastoderm in species with macrolecithal eggs.

11. In placental mammals an inner cell mass gives rise to the embryo. The process is similar to that in species with macrolecithal eggs.

12. Neurulation results in formation of a brain and spinal cord.

13. Ectoderm contributes epidermis, the nervous system and some of its membranes, several varieties of sense organs, the lens of the eye, the stomodeum and proctodeum, the pharyngeal skeleton and its musculature, some of the neurocranium, pigment cells, chromaffin cells, and additional miscellaneous tissues.

14. Endoderm gives rise to the epithelium of the digestive tract from stomodeum to proctodeum, and to the epithelioid components of all organs that arise as evaginations from the embryonic foregut, midgut, or hindgut.

15. Dorsal mesoderm forms mesodermal somites consisting of a sclerotome, myotome, and dermatome. These give rise, respectively, to vertebrae and the proximal part of the ribs, the dermis of the dorsum, and skeletal muscles other than those of the pharyngeal arches.

16. Lateral-plate mesoderm splits into somatic and splanchnic layers. The somatic layer, with ectoderm, becomes the somatopleure. The splanchnic layer, with endoderm, constitutes the splanchnopleure.

17. Intermediate mesoderm gives rise to kidney tubules and their ducts.

18. The coelom is the cavity between the somatic and splanchnic mesoderm. It is initially segmented in cephalochordates.

19. The chief extraembryonic membranes are yolk sac, amnion, chorion, and allantois. The trophoblast serves as a temporary nutritive membrane in eutherian embryos.

20. The yolk sac serves as a simple yolk sac placenta in viviparous anamniotes, and as part of a choriovitelline placenta in some viviparous amniotes.

21. Amniotic fluid bathes the amniote embryo, preventing desiccation and helping to prevent mechanical trauma.

22. The chorion is applied to the eggshell or to the lining of the maternal uterus. In the latter case it is part of the placenta.

23. The allantois is a midventral evagination of the cloaca of amniotes. It participates in formation of chorioallantoic placentae. The part proximal to the cloaca becomes the urinary bladder. The distal part becomes the urachus in some mammals.

24. The most common placentae are chorioallantoic, choriovitelline, and simple yolk sac placentae. They may be deciduous or nondeciduous. Villi may be arranged in zonary, cotyledonary, discoidal, or diffuse patterns.

SELECTED READINGS

Amoroso, E.C.: Viviparity in fishes, Symposium of the Zoological Society of London, **1**:153, 1960.

Arey, L.B.: Developmental anatomy, ed. 7 (revised), Philadelphia, 1974, W.B. Saunders Co.

Austin, C.R., and Short, R.V., editors: Embryonic and fetal development. Reproduction in mammals, Book 2, Cambridge, 1982, Cambridge University Press.

Balinsky, B.I.: An introduction to embryology, ed. 5, Philadelphia, 1981, W.B. Saunders Co.

Ballard, W.W.: Morphogenetic movements and fate maps of vertebrates, American Zoologist **21**:391, 1981.

Gans, C., editor: Biology of the reptiles, vols. 14 and 15, Development A and B. Vol. 14 focuses on turtles, crocodilians, and *Sphenodon*. Vol. 15 concentrates on lizards and snakes. New York, 1985, John Wiley and Sons.

Gorbman, A.: Early development of the hagfish pituitary gland: Evidence for the endodermal origin of the adenohypophysis, American Zoologist **23**:639, 1983.

Hopper, A.F., and Hart, N.H.: Foundations of animal development, ed. 2, New York, 1985, Oxford University Press.

Long, W.L.: Cell movements in teleost fish development, BioScience **34**:84, 1984.

Meier, S.: Development of the chick embryo mesoblast: morphogenesis of the prechordal plate and cranial segments, Developmental Biology **83**:49, 1981.

Nelsen, O.E.: Comparative embryology of the vertebrates, New York, 1953, The Blakiston Co., Inc. This work has not been equalled in breadth of coverage.

Northcutt, R.G., and Gans, C.: The genesis of neural crest and epidermal placodes: A reinterpretation of vertebrate origins, The Quarterly Review of Biology **58**(1):1, 1983.

Oppenheimer, S.B.: Introduction to embryonic development, Boston, 1980, Allyn & Bacon, Inc.

Patten, B.M.: Early embryology of the chick, ed. 5, New York, 1971, McGraw-Hill Book Co.

Patten, B.M., and Carlson, B.M.: Foundations of embryology, New York, 1974, McGraw-Hill Book. Co.

Slack, J.M.W.: From egg to embryo, New York, 1984, Cambridge University Press.

Tracy, R.C., and Snell, H.L.: Interrelations among water and energy relations of reptilian eggs, embryos, and hatchlings, American Zoologist **25**:999, 1985.

Trinkhaus, J.P.: Cells into organs: The forces that shape the embryo, ed. 2, Englewood Cliffs, New Jersey, 1984, Prentice-Hall, Inc.

Weston, J.A.: Motile and social behavior of neural crest cells. In Bellairs, R., Curtis, A., and Dunn, G., editors: Cell behavior, Great Britain, 1982, Cambridge University Press.

Wourms, J.P.: Viviparity: The maternal-fetal relationship in fishes, American Zoologist **21**:473, 1981.

Symposium in American Zoologist

Gastrulation, **24**:535, 1984.

5

INTEGUMENT

In this chapter we will study the basic structure of the integument. We will examine the epidermis of aquatic vertebrates, then see how it has been modified for life in air. We will also note some of the remarkable structures that form from epidermis. After that, we will examine the dermis and learn that bone is present in much modern skin and that its absence is a specialization. As a recapitulation, we will review the integument class by class. Finally, we will be reminded of some of the roles the integument plays in survival.

It would be difficult to imagine an organism the surface of which is not structured specifically for its role as an interface between organism and environment. Even amoebae have plasma membranes that wall off the protoplasm from the surrounding water. The integument of multicellular animals subserves many functions that contribute to the survival of the organism. It covers all external surfaces, including the exposed portion of the eyeball, where it is usually transparent and is called the **conjunctiva;** it covers the eardrum; and it is directly continuous with the mucous membranes that line all passageways opening onto the surface.

The integument of vertebrates from fishes to mammals is constructed in accordance with a single basic morphologic pattern, consisting of a multilayered **epidermis** derived from ectoderm and a **dermis** derived chiefly from mesoderm. Variations in the morphologic features of the epidermis and dermis of the several vertebrate classes involve (1) the relative number and complexity of skin glands, (2) the extent of differentiation and specialization of the most superficial layer of the epidermis, and (3) the extent to which bone develops in the dermis. The skin of an amphioxus exhibits epidermis and dermis, but the epidermis is only one cell thick (Fig. 5-1).

PREVIEW: SKIN OF THE EFT

As an introduction to the integument we will take a quick look at a skin that is unencumbered by scales, feathers, or hair—the skin of the red eft, the juvenile land stage of the aquatic urodele *Notophthalmus* (Fig. 5-2). It does not reflect the oldest feature of vertebrate skin, the presence of bony plates or scales in the dermis, but this single species illustrates the contrasting adaptations of vertebrate skin for life on land as opposed to life in water.

The epidermis of the eft is a multilayered glandular epithelium. The columnar cells in the basal, or **germinal,** layer are constantly undergoing mitosis, replacing those lost from the surface. Proliferation from the basal layer causes older cells to be pushed outward. As they approach the surface they synthesize keratin, a scleroprotein that is insoluble in water, and they become flattened **(squamous).** They are then said to be **keratinized,** or **cornified.** Keratinization causes the cells to die. In efts, these cornified cells do not remain for long on the surface. This is in contrast to the condition in permanently terrestrial vertebrates, in which the cornified layer becomes much thicker, constituting a definitive **stratum corneum.**

The glands of the skin develop from the epidermis and, in efts, are simple multicellular sacs that bulge into the dermis, where they are in the immediate vicinity of capillaries that supply nutrients and oxygen, and carry away metabolic wastes. They are able to synthesize mucus, but in the eft stage in the life of the organism they are quiescent, although not totally inactive.

The dermis of efts consists chiefly of connective tissue that supports the bases of the glands, blood vessels, lymphatics, small nerves, and pigment cells. It adheres closely to the underlying body wall muscle.

The life cycle of *Notophthalmus* dramatically illustrates the direct correlation between a terrestrial habitat, lack of surface mucus, and the phenomenon of keratinization. The larvae of this species live in water, have many active mucous glands, and no cornified epidermis. When they metamorphose and assume life on land most of the skin glands become quiescent, keratinization of the epidermis commences, and it persists as long as the eft lives on land. At the onset of sexual maturity the eft migrates back to water, the mucous glands become active, and the cornified cells are shed and do not reappear. The skin of larvae and adults resembles that of fishes, whereas the skin of efts is like that of terrestrial amphibians.

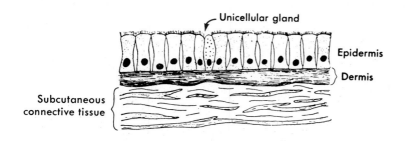

FIGURE 5-1

Skin of young amphioxus.

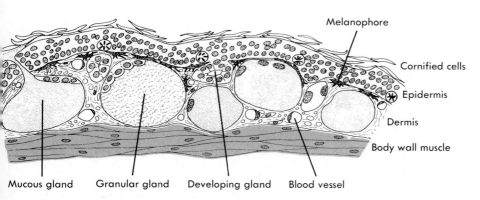

FIGURE 5-2

Skin of the red eft *Notophthalmus*. Note mitotic figures in germinal layer of epidermis.

FIGURE 5-3

Keratinized (cornified) cells of the stratum corneum on the surface of human skin. (Courtesy Johnson & Johnson Research, New Brunswick, N.J.)

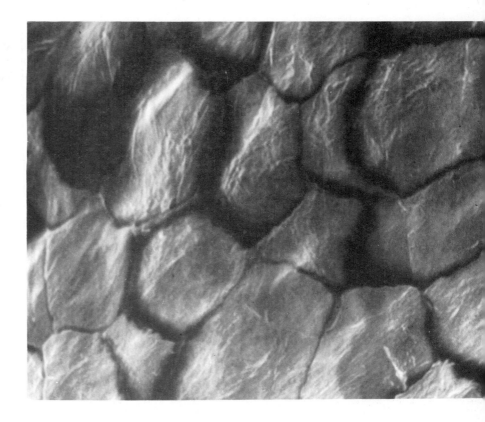

THE EPIDERMIS

Having completed our preview, we are better prepared to study the integument in detail. We will study the epidermis separately from the dermis because the two have different roles. The epidermis is the interface between organism and environment, and this is reflected in its structure. The dermis provides physiologic support for the interface.

Two kinds of nonliving coverings overlie the living epidermis of vertebrates. In fishes and most aquatic amphibians it is a **thin coat of mucus,** not identical with the mucus secreted by the multicellular glands. It appears to be secreted continually by every epidermal cell that reaches the surface. In terrestrial vertebrates the covering is a layer of **water-impervious cornified cells,** the stratum corneum (Fig. 5-3). The survival value of the stratum corneum is clear. It minimizes water loss through the skin of vertebrates that are exposed to the air. The survival value of the constant mucous coat is unknown, but it probably has a role as vital in fishes as the stratum corneum is in terrestrial vertebrates.

Embryonic skin has a temporary surface layer of flattened epidermal cells, the **periderm,** that protects the underlying actively proliferating germinal layer. In mammals it lies on the surface in association with the embryonic hair, or **lanugo,** and is therefore called the **epitrichium.** The periderm is lost at the time of hatching or birth, being replaced either by protective mucus or by a stratum of cornified cells.

The epidermis is glandular to some degree, and that of fishes is exceptionally so. Most of the epidermal glands of fishes and larval amphibians are unicellular, whereas those of adult amphibians and terrestrial vertebrates are predominantly multicellular. The latter are either straight, coiled, or branching **tubular glands,** or they are composed of one or more sacs, or alveoli (**simple, branched,** or **compound alveolar glands**). Although of epidermal origin, they invade the dermis, where they establish an intimate relationship with capillary beds.

Epidermis of Fishes and Aquatic Amphibians

MUCOUS GLANDS. Mucous glands abound in fishes and aquatic amphibians. In fishes most mucous glands are unicellular and of various shapes, including goblet cells (Figs. 5-4 and 5-5). A few, usually with an

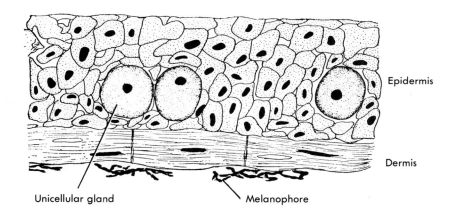

Epidermis

Dermis

Unicellular gland   Melanophore

FIGURE 5-4

Skin of a larval lamprey. The dermis contains very dense collagenous fibers.

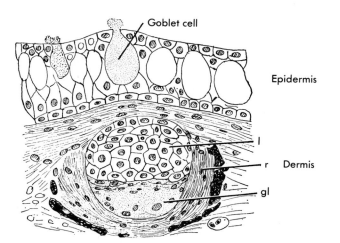

Goblet cell

Epidermis

l

r   Dermis

gl

FIGURE 5-5

Skin and light organ (photophore) of a luminous fish. *gl,* Luminous cells; *l,* lens cells; *r,* reflector cells (absent in some species) surrounded by pigment cells (*dark*). The goblet cells are unicellular mucous glands.

FIGURE 5-6

Glandular skin of the dipnoan *Protopterus*. Multicellular,
g, and unicellular, *u*, glands
in the epidermis. Pigment
cells are seen in the dermis.

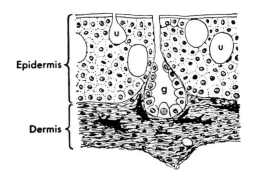

FIGURE 5-7

Undersurface of skin re-
flected from the underlying
muscle, tail of *Necturus*,
showing mucous glands *(arrow)* projecting into subcutaneous tissue just external to
the body wall musculature.

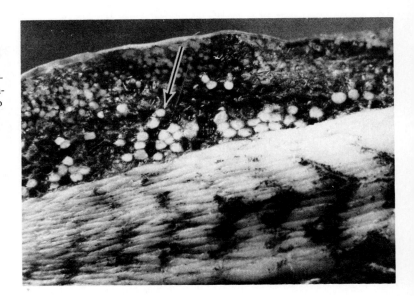

FIGURE 5-8

Frog skin. The epidermal
glands are mucous glands in
different stages of activity.

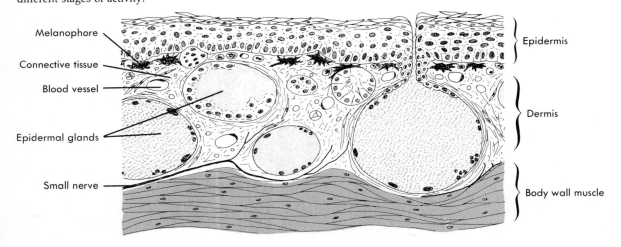

added ingredient that makes the mucus slimy, are multicellular (Fig. 5-6). In amphibians they expand until they become subcutaneous (Figs. 5-7 and 5-8). They release a secretion in special situations, including in response to stressful external stimuli. When threatened with capture they exude a slimy mucus in such abundance that they become very slippery. This is especially true of hagfishes, whose multicellular slime glands are surrounded by striated muscle fibers. The lungfish *Protopterus* surrounds itself with a slimy cocoon before aestivating in a burrow during the dry season. Some teleost mothers secrete nutritious mucus that is eaten by hatchlings. It can be assumed that mucus keeps the skin moist in species that make short excursions out of water and that this enables the skin to continue to perform its respiratory role until the mucus dries. The role of the mucus secreted by the ampullae of Lorenzini of sharks is purely speculative. Mucus in fishes and aquatic amphibians plays an important role related to maintenance of an appropriate internal environment and in meeting some of the challenges of the external environment.

GRANULAR (SEROUS) GLANDS. Granular glands, the predominant type of skin glands in terrestrial toads and reptiles, are present only in small numbers in fishes, and are rare in aquatic amphibians. They have a granular cytoplasm, and their secretions are basically protective, being mostly noxious or poisonous.

PHOTOPHORES. Some of the multicellular glands of fishes, especially deep-water teleosts, have become light-emitting organs, or photophores. Like other multicellular skin glands, photophores arise in the epidermis and invade the dermis. In one variety (Fig. 5-5), the upper part of the gland consists of mucous cells that serve as a magnifying lens. Surrounding the base of the photophore are a blood sinus and a concentration of pigment cells. The light emitted by photophores is not intense and is of many hues. The light serves for species and sex recognition, sometimes as a lure, and sometimes as an aid in concealment by countershading.

KERATINIZED APPENDAGES. Most fishes synthesize little or no keratin, but some aquatic amphibians have a thin stratum corneum, especially species that venture onto land. In cyclostomes conical horny epidermal spines and teeth develop in the buccal funnel and on the rasping tongue (Fig. 5-9),

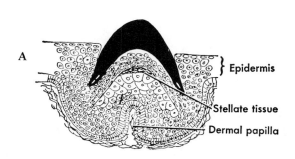

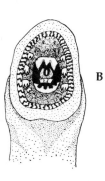

FIGURE 5-9

Lamprey teeth and buccal funnel. **A,** Cornified tooth *(black).* The stellate tissue may be the beginning of a replacement tooth. **B,** Buccal funnel. The horny teeth are shown as white against a dark background. The tip of the tooth-bearing lingual cartilage, sometimes called a tongue, occupies the rear of the cavity.

and tadpoles have horny teeth, lips, and jaws that enable them to feed by rasping vegetation during the herbivorous larval stage. Tadpoles shed these at metamorphosis. Calluslike caps develop on the toes of aquatic urodeles subjected to buffeting in mountain streams. In general, however, keratin is not a feature of the skin of fishes and aquatic amphibians.

Epidermal Glands of Terrestrial Vertebrates

Mucous glands have all but disappeared from the skin of terrestrial vertebrates. The synthesis of mucus could dehydrate a vertebrate that lacks a rather immediate source of either drinking water or available moisture that can be absorbed through the skin. Any mucous glands that remain in terrestrial vertebrates are multicellular, and mostly amphibian. It is presumed that unicellular glands lost survival value when they became covered with a continuous sheet of cornified cells. **Granular glands,** on the other hand, proliferated in reptiles to become a variety of specialized types the secretions of which are effective only in air. Most granular glands are **holocrine:** the cells constitute the secretion instead of simply manufacturing it, as in **merocrine** glands. **Apocrine** glands occupy an intermediate position. The apical portion of the cell ruptures, and part of the cell contents—both cytoplasm and secretion—are released. Endotherms have added **oil glands** and mammals have added **sweat glands** to the glandular repertoire of vertebrate skin.

AMPHIBIANS. The amphibians that come nearest to being terrestrial are apodans and toads with thick warty skin, but both require a moist habitat. Toads produce irritating or toxic alkaloids from granular glands associated with warty spots on the back and legs, and a prominent granular gland, the **paratoid,** is located just behind the eye in many toads (Fig. 5-10). The noxious secretions of these glands are protective. Granular glands are uncommon in other amphibians. The predominant skin glands of anurans are mucous glands.

REPTILES. Reptiles are the first class of truly terrestrial vertebrates, and their predominant epidermal glands are granular glands that secrete either noxious alkaloids or **pheromones.** Pheromones are substances that, when

FIGURE 5-10

Warty skin of toad *(Bufo)*. A granular gland, the parotoid, is seen behind the eye.

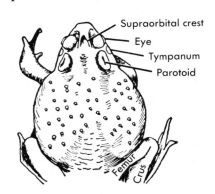

secreted into the environment, affect the behavior or physiology of other organisms. As many as four different sets of granular glands, including pheromonal glands, encircle the vent of lizards. When smeared on tree branches they attract insects, which are then eaten. Musk turtles exude a yellowish fluid from two glands on each side of the trunk just below the carapace, and a row of glands of unknown function, probably pheromonal, lies along the back in crocodilians. Femoral glands on the medial side of the hind limbs of male lizards secrete a substance that hardens to form temporary spines that restrain the female during copulation.

BIRDS. Only two integumentary glands have been described in birds, and both are oil glands. The **uropygial gland** is a prominent swelling at the rump. It is best developed in aquatic birds, but is readily demonstrable in domestic fowl. Its oily secretion is transferred to feathers during preening. Smaller oil glands line the outer ear canal and encircle the vent in some species.

MAMMALS. Integumentary glands reach a peak in complexity and specialization in mammalian skin. All appear to be variants of two basic types, sebaceous glands and sudoriferous glands. Mammary glands appear to be a variant of sebaceous glands.

Sebaceous glands are holocrine alveolar glands with an oily exudate. They are present wherever there are hairs, and the secretion, sebum, is usually exuded into a hair follicle (Fig. 5-11). Fur and human hair glisten after brushing because of the oil. In the outer ear canal ceruminous glands secrete cerumen, which in association with hairs traps insects that otherwise might go deeper into the canal and touch the painfully sensitive eardrum. Tarsal, or meibomian glands secrete onto the conjunctiva of the eye. These glands are embedded within a plate of dense connective tissue, the tarsus, in each eyelid. The ducts open in a single row on the inner surface of the lids just internal to the eyelashes. A duct occasionally becomes occluded, causing, in humans, a chalazion that can be seen as an inflamed swelling on the lid. Sebaceous glands open independent of hairs on the lips, glans penis, and labia minora.

Sudoriferous glands are long, slender, coiled tubular glands that extend deep into the dermis of mammals (Fig. 5-11, sweat glands). Their secretion oozes onto the surface via tortuous canals that open as pores. The cooling effect of the evaporation of sweat, their predominant secretion, is thermoregulatory. In furry mammals sweat glands are confined to the least furry regions, for example, the feet of cats and mice, the lips of rabbits, and the side of the head of bats. In hippopotami they are only on the ears, which are above the water line when a hippo is in its preferred habitat. Some mammals lack sweat glands. Among these are pangolins, whose skin is scaly, sirenians and cetaceans, whose environment precludes cooling by evaporation, and echidnas. Ciliary glands that open into the follicles of eyelashes are sudoriferous glands. Humans, with less hair than most mammals, have the largest number of sweat glands per centimeter of surface area.

FIGURE 5-11

Mammalian skin. Supporting the constituents of the dermis are bundles of collagenous connective tissue. Epidermal derivatives are red. The tubular sweat glands are tortuous, hence are seen only in cross section. They are longer than the dermis is deep.

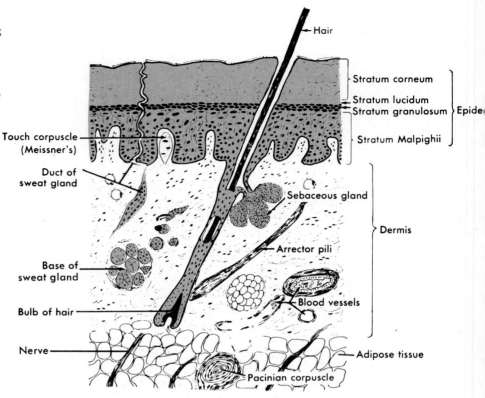

Sebaceous and sudoriferous glands produce a variety of scents, most of which are pheromonal or defensive. Glands on the feet of goats leave scent trails that other members of the species are able to recognize; anal glands of skunks drive away enemies; and those of male musk deer signal the sex. Kangaroo rats have sebaceous glands on the back, where they are exposed when the back is arched in defense. Male elephants have a temporal gland that swells during breeding seasons. Natives say that it signals danger. A gland above the eye of a peccary looks like a navel. Some male lemurs have a hardened patch of skin on the forearm, under which is a gland the size of an almond. Most of the odors at a well-maintained mammalian zoo are caused by scent glands, not by unhygienic conditions in the pens. One species, *Homo sapiens,* takes the pheromone from the anal gland of a musk deer, adds other odorants, and dabs it behind the ears. It is called perfume, but it performs the role of a pheromone.

Mammary glands develop in both sexes from a pair of elevated ribbons of ectoderm called **milk lines,** which extend along the ventrolateral body wall of the fetus from axilla to groin (Fig. 5-12, *A*). Patches of future mammary tissue develop at one or more sites along each milk line, depending on the species, and invade the dermis (Fig. 5-13). They later spread *beneath* the dermis, and a nipple forms above each patch. As females approach sexual maturity, rising titers of female hormones cause the juvenile duct system to spread and branch. Later, during pregnancy, a battery of hormones causes the formation of alveoli at the ends of the duct system. Like sebaceous glands, mammary glands are holocrine glands with a fatty secretion.

The number and location of mammary glands, and therefore of nipples, depend on the number of young that is typical of the species and on the survival value of one location over another. Cats, dogs, pigs, rodents, edentates, and many other mammals have axillary, thoracic, abdominal, and inguinal nipples that develop along most of the milk line. During pregnancy in these species adjacent glands may expand toward each other until they have formed two long continuous masses of considerable weight. Species with smaller litters have fewer nipples. Insectivores and some lemurs have one thoracic and one inguinal pair. Flying lemurs and marmosets have a single pair, in the armpits. Monkeys, apes, and humans have one pair located where the nursing baby can be protected in the mother's arms while she monitors the environment for enemies. Cetaceans have a single pair near the groin, where the baby porpoise or whale can hold on and nurse while the mother feeds, surfaces, and dives. Nutrias, which are at home in water, have four nipples on the back, and the babies ride along

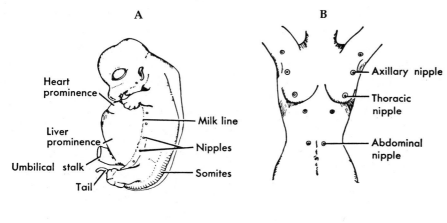

FIGURE 5-12

A, Milk line and nipples in a 20 mm pig embryo. **B,** Supernumerary nipples in a human female.

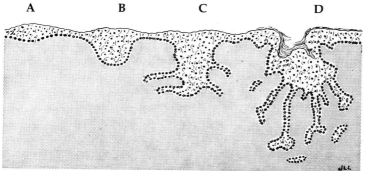

MAMMARY GLAND
Morphogenesis

FIGURE 5-13

Successive stages of mammary gland development. **A,** Equivalent to a human embryo at 6 weeks; **B,** equivalent to a human embryo at 9 weeks and to a mouse embryo at 16 days; **C,** intermediate stage; **D,** at birth. Gray area represents dermis.

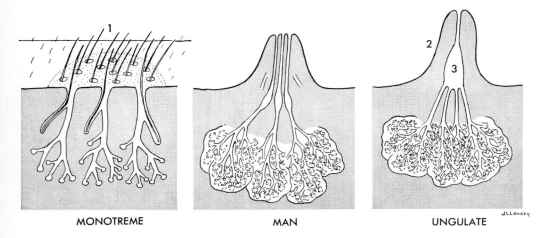

MONOTREME   MAN   UNGULATE

FIGURE 5-14

Mammary glands, ducts, and nipples. The monotreme lacks nipples, and the glands resemble modified sweat glands. *1*, Hairs. *2*, Nipple. *3*, Cistern.

above the water line while nursing. The nipples originate from a ventrolateral milk line but are displaced dorsad by differential growth rates on the two sides of the milk line. This causes the nipples to be slowly displaced dorsad during ontogeny. Supernumerary nipples may develop in any mammal (Fig. 5-12, *B*).

Monotremes do not develop typical mammary glands or nipples. Instead, in both sexes modified sweat glands produce a nutritious secretion that is lapped off a convenient tuft of hairs by the young (Fig. 5-14, monotreme). Nipples would probably be useless in the duckbill, because it is doubtful that the young, hindered by horny beaks and lacking muscular cheeks and lips, could nurse. Except during lactation, the nipples of the opossum are stored in depressions within the skin.

The Stratum Corneum

Keratinization provides protection against desiccation, a necessity for life on land. This is the primary role of the stratum corneum in amphibious vertebrates. With the acquisition of an amnion, which completed the liberation of vertebrates from the water, the stratum corneum became increasingly specialized in various regions of the body for protection against abrasion, for offense and defense, and finally, as an adjunct to thermoregulation. Early specializations—scales, claws, and horny protuberances—were followed by hair and feathers, the latter being the most remarkable of all. A mammalian stratum corneum is seen in Fig. 5-11.

EPIDERMAL SCALES. Epidermal scales are repetitious thickenings of the stratum corneum that are found only in amniotes. In **squamates**—lizards and snakes—the stratum corneum is disposed in overlapping folds (Figs. 5-15 and 5-16). The cells on the exposed surface of each fold are especially heavily cornified, forming an overlapping scale. Because of the continuity of the stratum corneum from fold to fold (Fig. 5-15, *B*), individual epidermal scales cannot be dissected out of the skin. Thinning of the stratum corneum

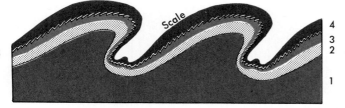

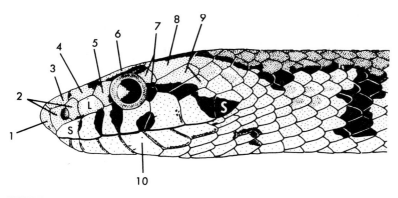

FIGURE 5-15

Squamate scales. **A,** Collared lizard. The entrance to the outer ear canal is seen as a dark crescent behind the corner of the mouth. **B,** Diagrammatic section of skin of a snake or lizard. *1,* Dermis; *2,* actively mitotic layer of epidermis; *3,* newly cornified layer of epidermis; *4,* older cornified layer to be shed at next molt. **C,** Scales from the banded water snake, *Nerodia fasciata.* (**A,** courtesy F.W. Schmidt, naturalist photographer, La Marque, Tex. **C** courtesy T.P. Forks.)

FIGURE 5-16

Epidermal scales on head of milk snake *Lampropeltis triangulum. 1,* Rostral; *2,* nasal; *3,* internasal; *4,* prefrontal; *5,* preocular; *6,* supraocular; *7,* post-oculars; *8,* parietal; *9,* anterior temporals; *10,* fifth of seven infralabials; *L,* loreal; *S,* first and last supralabials. (Courtesy Kenneth L. Williams.)

FIGURE 5-17

Warty and spiny skin of horned toad, a lizard. (Courtesy Carolina Biological Supply Co., Burlington, N.C.)

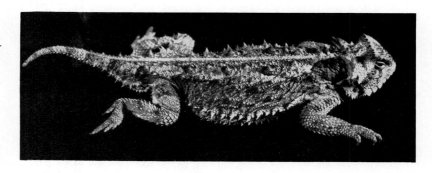

FIGURE 5-18

Epidermal scales of the turtle *Chrysemys*. **A,** Carapace. **B,** Plastron. On the carapace the scutes shown are, *c*, costals; *m*, marginals all around the periphery, including the nuchal *(nu)*, and pygals *(p)*; *n*, neurals. On the plastron the scutes shown are, *g*, gulars; *h*, humerals; *p*, pectorals; *a*, abdominals; *f*, femorals; *an*, anals.

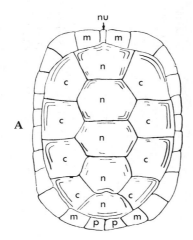
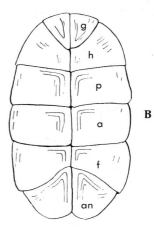

FIGURE 5-19

Armadillo skin showing epidermal scales and interspersed hairs.

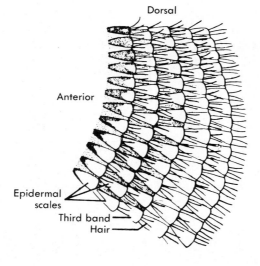

at scale joints permits mobility of the skin. The scales of lizards sometimes assume bizarre shapes (Figs. 5-17 and 16-14).

The continuity of the stratum corneum is readily demonstrable in **crocodilians,** which have small, heavily cornified, nonoverlapping scales on the limbs, and larger, more or less regular thickenings on the rest of the body (Fig. 5-36, *A*).

Large, thin, quadrilateral or polygonal scales are scutes. Snakes have them on the belly, where they are used for locomotion. **Turtles** have them on the plastron, which slides along the ground, thinner ones on the carapace (Fig. 5-18), and small scales elsewhere. Turtle scutes and scales do not overlap.

Epidermal scales in **birds** are found where there are no feathers: the facial area, legs, and feet. Armadillos have hair and scales interspersed over the entire body (Fig. 5-19), but the scales of most **mammals** are confined to the legs and tail, as in rats and beavers. The scales of pangolins appear to be agglutinated hairs, and are of recent origin rather than inherited from reptilian ancestors (Fig. 3-41).

Lizards and snakes have two distinct layers of stratum corneum, an inner layer in the process of being deposited and an outer layer that will be shed at the next molt (Fig. 5-14, *B*). In lizards the outer layer flakes off in large patches, but in snakes the outer layer of the entire body, including the spectacle, a thick lenslike conjunctiva, is shed in one piece. Crocodilians and turtles do not molt; their stratum corneum wears off gradually, like that of humans.

CLAWS, HOOFS, AND NAILS. Claws, hoofs, and nails are modifications of the stratum corneum at the ends of digits. Claws first appeared in reptiles and have persisted in birds and most mammals. (The "claws" of the African clawed toad are not comparable to reptilian claws.) Claws evolved into nails in primates, and into hoofs in ungulates. Claws, hoofs, and nails have the same basic structure. They consist of two curved parts, a horny dorsal plate, the **unguis,** and a softer ventral plate, the **subunguis** (Fig. 5-20). The two plates wrap partially around the terminal phalanx, which is usually pointed when associated with a claw, blunt when associated with a hoof or nail. A still softer calluslike, cornified pad, the **cuneus** (called the "frog" by horsemen), is frequently present in ungulates, partially surrounded by the subunguis. The thick, hard unguis of a hoof is U or V shaped, and, because it consists of dead cells a shoe can be nailed into it. The unguis of a nail has become flattened, and the subunguis is much reduced. As a result, nails cover only the dorsal surfaces of digits; but, if permitted to grow, nails become clawlike.

Although claws in birds are often thought of as associated only with the feet (Fig. 5-21), sharp claws are frequently borne on one or more digits of the wings (ostriches, geese, some swifts, and others). *Young* hoatzins use claws on the wings for climbing about on the bark of trees, but the claws cease growing and disappear at maturity. *Archaeopteryx* had three claws on each wing. Only squamate claws are shed. Claws, hoofs, and nails of other animals are worn down by friction.

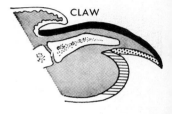

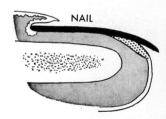

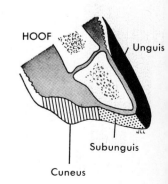

FIGURE 5-20

Claw, nail, and hoof with terminal phalanx diagrammed in sagittal section.

FIGURE 5-21

Claw on the middle toe of a great blue heron. (From Atwood, W.H.: Comparative anatomy, ed. 2, St. Louis, 1955, The C.V. Mosby Co.)

FEATHERS. Feathers are remarkably complicated cornified epidermal appendages of the integument. There are three morphologic varieties, contour feathers, down feathers, or plumules, and hairlike feathers, or filoplumes. The roles of feathers are described in Chapter 3.

Morphologic Varieties of Feathers. **Contour feathers** are the conspicuous feathers that give a bird its contour, or general shape (Fig. 5-22). A typical contour feather consists of a horny **shaft** and two flattened **vanes.** The base of the shaft is the **calamus,** or quill. The vane-bearing segment is the **rachis.** Each vane consists of a row of **barbs** that have **barbules** and **flanges** (Fig. 5-22, *B*). The barbules have **hooklets** that interlock with the flanges of adjacent barbs, stiffening the vane. When a contour feather is ruffled, the barbules have become unhooked. Preening rehooks the hooklets. During preening the oily secretion of the uropygial gland is applied to the barbs. On the smaller contour feathers of the wing (covert feathers) hooklets are missing from the lower barbules, so this region of the feather is fluffy. In ostriches and a few other birds all feathers are fluffy.

Arising from a notch, the **superior umbilicus,** at the base of the rachis of a contour feather is an **afterfeather.** It is usually shorter than the main feather, but in emus and cassowaries the two are the same length, resulting in "double feathers."

Although contour feathers cover most of the body, the follicles from which they grow are usually disposed in feather tracts, or **pterylae** (Fig. 5-23). A few birds, including ostriches and penguins, lack these.

Inserting on the walls of the feather follicles in the dermis are smooth erector muscles **(arrectores plumarum),** which, along with extrinsic integumentary muscles, enable a bird to fluff its feathers.

Down feathers are small, fluffy feathers lying underneath and between contour feathers. Very young birds lack contour feathers and are covered with down. Down feathers may be ancestral to contour feathers. They have a short calamus, with a crown of barbs arising from the free end, and no hooklets. Eiderdown, used in pillows, is the down feathers of the eider duck.

Filoplumes are hairlike feathers consisting chiefly of a threadlike shaft. They are scattered throughout the skin among the contour feathers. The very long colorful feathers of a peacock are filoplumes.

Development of a Feather. Feather formation is initiated by a mesodermal papilla that organizes beneath the epidermis, becomes vascularized, and induces mitotic activity in the basal layer of the overlying epidermis. This initial activity results in a minute pimplelike elevation on the surface of the skin, the feather primordium (Fig. 5-24). The primordium elongates, and a pit lined with epidermis, the **feather follicle,** develops around its base.

At the base of the feather follicle a mitotically active growth zone proliferates tall columns of epidermal cells that push toward the distal tip of the growing feather between the dermal papilla and the epidermis, now a **feather sheath.** These epidermal columns separate from one another, cornify, and develop into barbs. A growing feather still surrounded by its sheath is a **pinfeather.** When the feather sheath splits open, the fluffy barbs stretch out of their cramped quarters and the shaft elongates.

When the feather is full grown the dermal papilla in the shaft dies and becomes pulp. The living basal portion of the papilla withdraws from the base of the shaft, leaving an opening, the **inferior umbilicus.** New feathers develop from reactivated dermal papillae that have already given rise to

FIGURE 5-22

A, Contour feather from a grouse. **B,** Two successive barbs showing two barbules interlocked by hooklets with flanges. **C,** Cross section of a flange showing interlock.

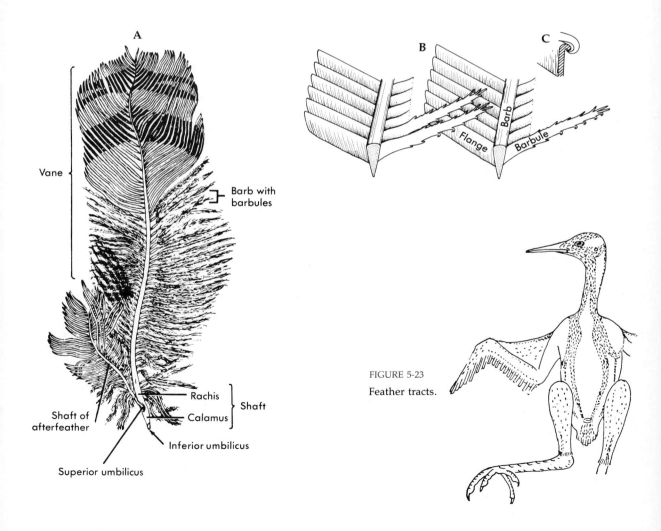

FIGURE 5-23

Feather tracts.

FIGURE 5-24

A feather primordium.

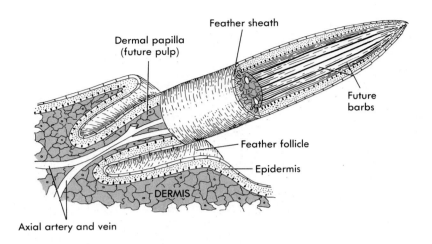

feathers. In many birds—perhaps in all—old feathers are passively pushed out of the follicles by incoming feathers.

Origin of Feathers. A question frequently asked is, "How did feathers originate?" One hypothetical answer is that they arose from reptilian scales, because early developmental stages are similar. Development of both is initiated by formation of a vascularized dermal papilla, after which the overlying epidermis thickens and then forms a scale or feather. However, subsequent development of the two is dissimilar, and the hypothesis is viewed with skepticism. The only other hypothesis that is justified at present is that, like hairs, they are new evolutionary structures lacking a reptilian precursor.

HAIR. Hairs, like feathers, are keratinized appendages of the skin. They may form a dense, furry covering over the entire body, or there may be only one or two bristles on the upper lip, as in some whales. Where fur is dense, there are usually short, fine hairs (underfur) as well as long, coarse ones.

Hairs have an insulating effect when dense enough, and are also sensitive tactile organs. The root of each is surrounded by a basketlike network of sensory nerve endings, and displacement of the root initiates a train of sensory impulses to the brain. Disturb a single hair on the back of your hand and note the sensation evoked. Vibrissae (stiff long whiskers on the face of many mammals) perform this role exclusively.

Hairs grow in isolated groups of two to a dozen or more, and linearly between scales when scales are present. Some monkeys have groups of three, apes have groups of five, and humans have groups of three to five. A glance at the side of your hand, near the base of the thumb, will reveal the linear arrangement of hairs in that location.

Morphology of a Hair. Hairs grow from hair follicles and elongate as a result of continual mitosis in the **bulb** (Fig. 5-11), within which a highly vascular dermal papilla provides nutrients and oxygen required for growth. The **root** is the part within the follicle where the hair cells are cornifying and dying but where the hair has not yet separated from the follicular wall. The remainder of the hair is the **shaft,** which commences just below the openings from the sebaceous glands. Within the follicle the shaft is surrounded by sebum.

A single hair consists of dense keratin from disintegrated or partially disintegrated cornified cells, trapped air vacuoles, and varying quantities of melanin granules that were released when the hair cells disintegrated. (The source of the granules and an explanation of hair color are discussed under "Dermal pigments".) Each hair is covered by a membranous **cuticle** composed of thin, transparent, cornified squamous cells arranged like shingles with free edges directed away from the root. Coarse or spiny hairs contain a medulla composed of irregular, shrunken cornified cells separated by large amounts of air and connected by intercellular bridges of keratin.

Inserting on the wall of each hair follicle is a tiny smooth muscle, the **arrector pili** (Fig. 5-11). When the arrectores pilorum contract, the hairs are elevated and the skin is pulled into tiny mounds resembling "gooseflesh." Elevation of the hairs gives carnivores a ferocious appearance. It is also a device for thermoregulation, increasing the insulating effect of the fur.

The scales of pangolins, and perhaps the horns of rhinoceros, may be evolutionary products of agglutinated hairs. Other modifications of hair include bristles, the spines of spiny anteaters, and porcupine quills.

Development of a Hair. The development of a hair is initiated by a cylindrical ingrowth of the epidermis into the dermis (Fig. 5-25, *A*). Beneath the ingrowth and indenting its base, a dermal papilla organizes. This contrasts with the initiation of development of a feather, which requires a dermal papilla at the outset, as do reptilian scales. With continued proliferation of epidermal cells, the hair primordium grows deeper and deeper into the

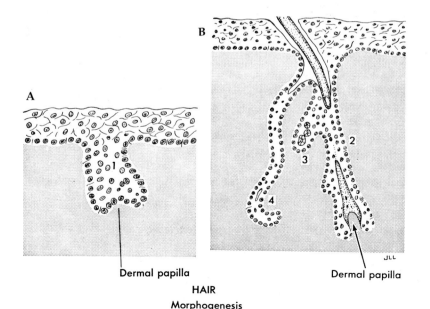

FIGURE 5-25

Successive stages in the development of hair and associated glands. *1,* Initial epidermal ingrowth into dermis; *2,* hair follicle; *3,* developing sebaceous gland; *4,* developing sweat gland. *Gray area,* dermis.

Dermal papilla

Dermal papilla

HAIR

Morphogenesis

dermis. When a bulb has differentiated sufficiently at the base of the primordium, cornified cells commence to appear, and a hair shaft rises out of the follicle.

Origin of Hair. The phylogenetic origin of hair is conjectural. A **protothrix hypothesis** proposed by Elias and Bortner suggests that hairs are modifications of sensory pits similar to those found in the integument of lizards (p. 594). These pits are distributed singly or in groups of up to seven between scales. An epidermal bristle **(protothrix)** emerges from each pit and transmits tactile stimuli to an innervated receptor cell in the pit. The relationship of sensory pits to reptilian scales is reminiscent of the interscale distribution of hairs in armadillos, which makes possible the receipt of tactile stimuli via a scaly mammalian skin. The protothrix theory has not received general support. That hairs are derivatives of reptilian scales would seem to be ruled out on the basis of differences in the ontogeny of the two structures.

HORNS AND ANTLERS. Ungulates are endowed with organs of offense and defense, the horns and antlers. The term "horn" implies that the surface is composed of keratin. Three varieties of mammalian horns meet this criterion: **bovine horns** and **hair horns** (Fig. 5-26), and the **horns of pronghorn antelopes.** Antlers and giraffe horns are not really horns.

Bovine Horns and Pronghorns. Members of the family Bovidae (oxen, sheep, goats, true antelopes, gnus, gazelles, and others) and pronghorn antelopes (family Antilocapridae) have true horns. They consist of a core of dermal bone covered by a sheath of horn. Bovine horns are usually curved or recurved, and are never shed. When the sheath is removed from the underlying bone by surgery the sheath is hollow (Fig. 5-26, true horn). Bovine horns are usually found in both sexes, but there are exceptions. Polled cattle have lost their horns by selective breeding. The chief difference between bovine horns and pronghorns is that pronghorns are branched, and the horny covering, but not the bony core, is shed annually.

Hair Horns. Rhinoceros have hair horns (Fig. 5-26). Hair horns differ from other horns in being composed of agglutinated keratinized hairlike epidermal fibers that form a solid horn perched on a roughened area of the nasal bone. Both sexes have them, and they are not shed. Some African rhinos have two horns, one behind the other.

Antlers and Giraffe Horns. Antlers are characteristic of the deer family. They are not cornified structures but branched dermal bone attached to the frontal bone. New growing antlers are said to be "in velvet" because they are covered with a soft vascular skin and velvety hair. They develop only in males, with the exception of caribou and reindeer. Typically at the approach of autumn, when antlers are full-grown, the blood supply to the velvet is occluded at the base of the antlers and the skin, unable to be maintained, rubs off, leaving bare bone. Later, at the end of the rutting season, when male territory no longer must be defended and testosterone concentration declines in the circulation, the antlers are shed with some

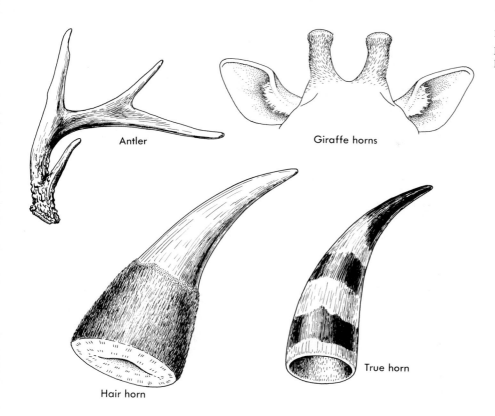

FIGURE 5-26

Mammalian horns and antlers.

Antler

Giraffe horns

Hair horn

True horn

help from the deer, who evidently find them annoying. They are replaced annually.

The "horns" of giraffes resemble stunted antlers. They are short bony projections of the frontal bones that remain in velvet throughout life.

BALEEN AND OTHER CORNIFIED STRUCTURES. Toothless whales have from 100 to 400 broad, thin, horny sheets of oral epithelium called baleen, or whalebone (Fig. 5-27), that hang into the oral cavity from the palate along its length. Each sheet is fringed along the edge, and the fringes act like combs or sieves that strain food out of the water passing between them. The sheets of the huge right whale exceed 3 m in height. Differences in the arrangement and configuration of the sheets are correlated with feeding habits. A blue whale swims up to a dense swarm of small fish or crustaceans, opens it mouth, and takes in water and food until a huge pouch under the tongue, pharynx, and chest wall is filled with about 70 tons of food and water. The mouth is then closed, and the pouch is emptied by forcing the water through the sieve and back into the sea. The collected food is swallowed leisurely through a very small throat. Other whales skim the surface for plankton with the mouth open or feed near the bottom. Baleen is continually being worn away and replaced, like hair or fingernails.

FIGURE 5-27

Sheets of baleen (whalebone) removed from the oral cavity of a whalebone whale. The sheets, which may be from 2 to 12 feet (3.5 m) long, are seen *in situ* in the mouth of the blue whale in Fig. 3-39. (Courtesy General Biological Supply House, Inc., Chicago.)

FIGURE 5-28

Rattles from a rattlesnake.

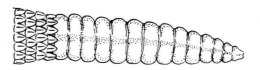

FIGURE 5-29

Tori and friction ridges.

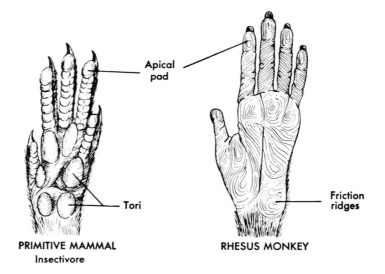

Rattlesnake **rattles** are rings of horny stratum corneum that remain attached to the tail after each molt (Fig. 5-28). **Beaks** are covered with a horny sheath, and roosters' **combs** are covered with a thick, warty stratum corneum. Monkeys and apes sit on thick **ischial callosities,** and camels kneel on **knee pads. Tori** are epidermal pads that most mammals other than ungulates walk on (Fig. 5-29). Cats "pussyfoot" by retracting their claws and walking stealthily on tori. At the ends of digits, tori are called **apical pads. Corns** and **calluses** are temporary thickenings of the stratum corneum that develop where the skin has been subjected to unusual friction.

THE DERMIS

The basic component of dermis, whether of fishes or human beings, is collagenous connective tissue. Other components include, universally, blood vessels, small nerves, and pigment cells; and in one species or another, lymphatics, naked and encapsulated exteroceptors, the bases of multicellular glands, and in endotherms, the bases of hairs or feathers and their erector muscles. In addition, *the dermis has an ancient and persistent potential to form bone,* and bone is a constituent of the dermis in some members of every vertebrate class except Aves. Early fishes had so much bone in the skin that they are called armored fishes. When bone is lacking in the dermis, it is because bone salts were not deposited on the collagenous matrix. Cowhide, from which leather is made, the dermis of leatherback turtles, the skin of cyclostomes—all are exceptionally tough because the collagen bundles are densely packed as though prepared for a shower of bone salts that, for lack of an enzyme or some other reason, never materialized.

The Bony Dermis of Fishes

The armor of ancient fishes varied in histologic details, but a generalized pattern consisted of lamellar bone, spongy (vascular) bone, dentin, and a surface layer of enamel or enameloid (Fig. 5-30, *A*). Frequently the surface was roughened by **denticles,** which were spiny or knobby enameloid-covered elevations of dentin. The armor was disposed as broad bony plates or

FIGURE 5-30

Dermal bone and dermal scales through the ages, diagrammatic. *1,* Lamellar bone; *2,* spongy bone; *3,* dentin; *4,* enameloid of one kind or another, including ganoin; *5,* fibrous plate characteristic of modern fish scales. The surface elevations in ancient armor are denticles consisting of layers *3* and *4.* The lamellar bone in modern fish scales is acellular. Cosmoid scales (not illustrated) resemble ancient armor.

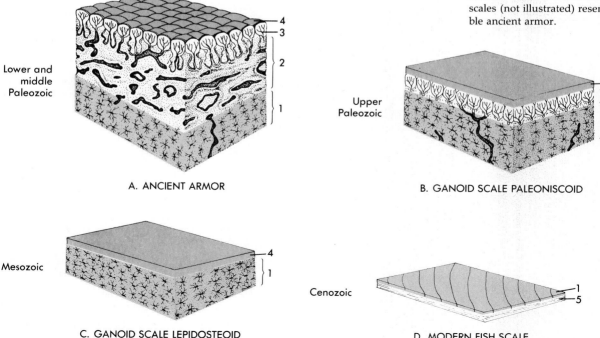

Lower and middle Paleozoic

A. ANCIENT ARMOR

Upper Paleozoic

B. GANOID SCALE PALEONISCOID

Mesozoic

C. GANOID SCALE LEPIDOSTEOID

Cenozoic

D. MODERN FISH SCALE

small bony scales that covered much of the body (Figs. 3-4 and 3-7). It was protective, but it may have also served as a storage site for calcium and phosphates. In time, the large plates on the trunk and tail gave way to smaller and thinner scales, those on the head contributed to the skull, and those in the shoulder region contributed dermal bone to the otherwise endochondral endoskeleton of the pectoral girdle.

Dermal scales and plates are classified as cosmoid, ganoid, placoid, and modern, the latter being ctenoid and cycloid. **Cosmoid scales and plates**

FIGURE 5-31

Ganoid scales. **A,** Scales on trunk of a garfish. The tail is at the right. **B,** Canaliculi and lacunae within a single scale (microscopic).

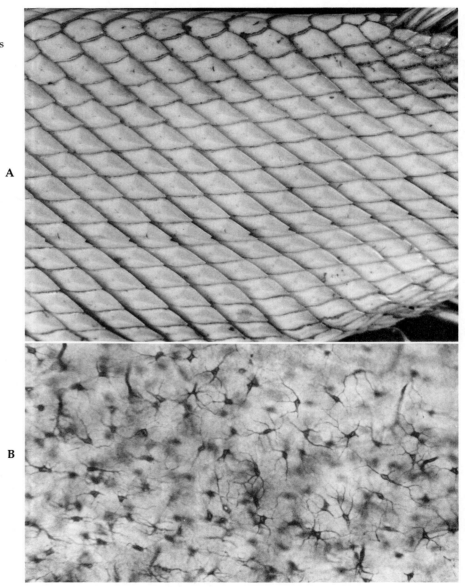

A

B

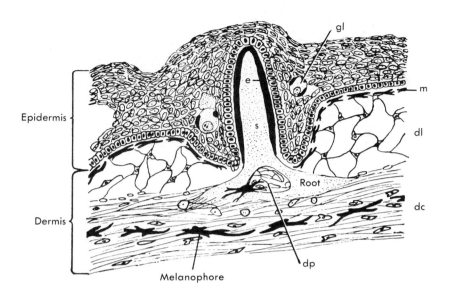

FIGURE 5-32

Developing placoid scale in skin of embryonic shark. Dermis consists of compact layer, *dc,* and loose layer, *dl; dp,* dermal pulp within root and spine; *e,* enamel; *gl,* unicellular epidermal gland; *m,* melanophore; *s,* spine, composed of dentin, enamel, and a core of pulp.

were present on early lobe-finned fishes and resembled the ancient dermal armor seen in Fig. 5-30, *A.* The dentin in those scales is sometimes called cosmine, which accounts for their name.

Ganoid scales and plates covered paleozoic fishes when they were at their zenith. The earliest ganoid scales, called paleoniscoid (Fig. 5-30, *B*), are found today only on *Polypterus, Calamoichthyes,* and *Latimeria,* although on the latter the dentin layer has been reduced. The scales of today's gars have lost the spongy bone and dentin (Fig. 5-30, *C*), but they still form a continuous shiny bony armor on the entire trunk and tail (Fig. 5-31). On the head they have become large or small bony plates (Fig. 8-11, *A*). Other surviving chondrosteans and holosteans have reduced the bony layer even more, and ganoin is lacking. In the bowfin *Amia,* scalelike bony plates lacking ganoin are confined to the head and modern flexible scales cover the rest of the body.

Placoid scales were present on paleozoic sharks, and are still present on today's sharks, skates, and rays. They have the same structure as paleoniscoid scales, consisting of lamellar bone, dentin, and an enameloid covering. But arising from the flat bony scale in the dermis (**basal plate,** or root) is a **spine** of enamel-covered dentin that erupts through the epidermis (Fig. 5-32). Thus, like the armor of old, that of sharks is still denticulate. Placoid scales become teeth at the edges of the jaws (Fig. 11-9).

Cycloid and ctenoid scales are found on teleosts and modern lobe-finned fishes. There is no difference between them except that the ctenoid scales have a comblike (ctenoid) free border (Fig. 5-33). These modern scales consist of a very thin layer of acellular lamellar bone underlain by a plate of dense collagen (Fig. 5-30, *D*). As a result they are flexible and translucent. The overlying epidermis is very thin.

The ability to form scales has been lost by cyclostomes, catfish, bony eels, and some other fishes; nevertheless, scale anlagen develop transitorily

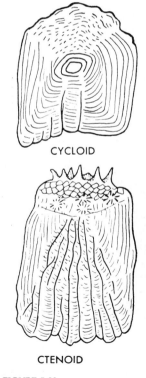

CYCLOID

CTENOID

FIGURE 5-33

Modern flexible fish scales. The upper edges are caudal free borders.

in the embryos. In contrast to scaleless teleosts, sea horses and pipefish have dense bony plates in their dermis over most of the body surface.

Dermal Ossification in Tetrapods

The outstanding difference between the dermis of one vertebrate class and another is the extent to which each contains bone. When the first tetrapods lumbered onto land they brought with them versions of the cosmoid scales of their lobe-finned ancestors. Some labyrinthodonts, and even some cotylosaurs, had large bony plates in the dermis; others had minute bony scales, referred to also as **osteoderms** in tetrapods. Osteoderms are still present in a few species of amphibians and in most reptiles. Bony dermal plates are also present in armadillos, but whether they are vestigial or newly acquired is not known.

Among amphibians, caecilians and some tropical toads have osteoderms. Those between the furrows in caecilians are microscopic, perpendicular to the surface, and lie in circumferential bands separated by glandular skin. Those within furrows are just large enough to be seen with the naked eye when loosened with a scalpel (Fig. 5-34).

Crocodilians have oval osteoderms, especially along the back, where they are often associated with cornified crests (Figs. 5-35 and 5-36). The young of some species of lizards have osteoderms under the epidermal scales of the head until some time after hatching or birth, after which they usually fuse with the underlying membrane bones of the skull. Turtles other than leatherbacks are truly armored, being boxed in by large bony plates that meet in immovable sutures (Fig. 5-37). The carapace and plastron are united by bony lateral bridges that must be sawed through to expose the viscera.

FIGURE 5-34

A single osteoderm from a caecilian.

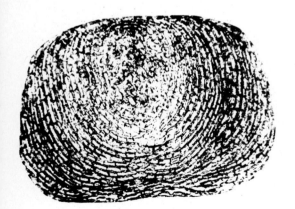

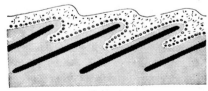

Teleost

FIGURE 5-35

Site of dermal scales (*black*) in a teleost and of osteoderms in a crocodilian.

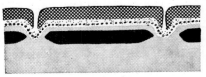

Alligator

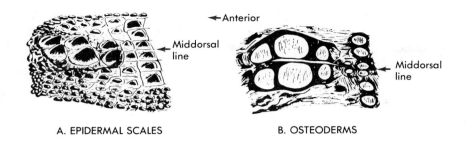

A. EPIDERMAL SCALES   B. OSTEODERMS

←Anterior

Middorsal line

Middorsal line

FIGURE 5-36

Alligator skin. **A,** From dorsum of neck, showing epidermal scales. **B,** Same section turned upside down to show osteoderms (dermal scales) embedded in skin beneath the tall crests. The middorsal line may be used as a reference point.

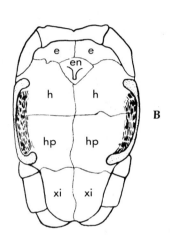

FIGURE 5-37

Dermal plates of the carapace, **A,** and plastron, **B,** in the turtle *Chrysemys,* from an internal view. On the carapace the plates shown are *c,* costals, united with ribs; *n,* nuchal; *p1* and *p2,* precaudals; *v,* vertebrals (six of the eight are labeled). Marginals encircle the periphery and include the pygal, *py.* On the plastron the plates shown are *e,* epiplastrons; *en,* entoplastrons; *h;* hyoplastrons; *hp,* hypoplastrons; *xl,* xiphiplastrons.

Among mammals armadillos alone have dermal armor as a normal condition. It consists of identical small polygonal bones immovably united and extending almost to the midventral line. The bone is covered by epidermal scales (Fig. 5-19). Bone forms in the skin of other mammals, including human beings, as a pathologic condition (dermostosis). An earlier observation is worth repeating: the dermis of vertebrates has an ancient and persistent potential to form bone.

Dermal Pigments

Chromatophores are cells that contain pigment granules. They are found in many locations of the body, but those in the skin are in the dermis, and they are responsible for skin coloration. All have processes that are permanent extensions from the cell body, and in some the pigment granules can be dispersed into the processes or aggregated in the vicinity of the nucleus by neurotransmitters and by hormones such as intermedin and melatonin. Chromatophores with dispersible granules are responsible for **physiologic color changes,** which are rapid changes such as are seen in chameleons.

Chromatophores are named for the color of their granules. **Melanophores** contain melanin granules **(melanosomes)**, which are varying shades of brown; **xanthophores** contain yellow granules; and **erythrophores** contain red granules. Xanthophores and erythrophores are sometimes called lipophores because the granules are soluble in lipid solvents. **Iridophores** contain a prismatic substance, guanine, which reflects and disperses light, producing silvery or iridescent skin.

The pigment granules in hair and feathers, and any in the epidermis, are not in the cells that produced them. Deep within the dermis the branched processes of dermal melanophores ramify among the epidermal cells of the growth zones of feather and hair follicles deep and actively *inject* pigment granules into cells that are being added to the growing hair or feather. Similarly, melanin granules in the mammalian epidermis have been injected into cells of the germinal layer.

Hairs receive only melanin granules, and hair color (black, brown, red, blonde) is attributable to the specific distribution and density of the granules and to the number of air vacuoles in the medulla of the hairs. Gray or white hair is the result of large numbers of air vacuoles and few melanin granules.

Feathers receive brown, yellow, and red pigments. Despite what appear to be blue feathers, vertebrates have no blue granules. When viewed under a light microscope by transmitted light, the "blue" feather is brown, the color of the melanin granules beneath the prismatic layer. The blue color observed in reflected light is a dispersion phenomenon, like the blue of the sky. A red feather is red even when viewed in transmitted light because of red pigment granules. The iridescence of feathers is also a dispersion phenomenon.

FIGURE 5-38

Dermal chromatophore unit responsible for physiological color changes in anurans. Adaptation to a dark background is illustrated, with processes of melanophores overlying iridophores. (Modified from Bagnara, J.T., Taylor, J.D., and Hadley, M.E.: The Journal of Cell Biology **38:**67, 1968.)

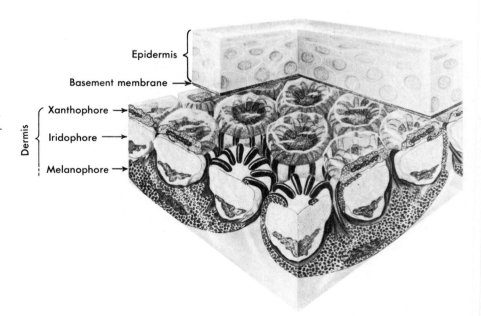

Physiologic color changes occur only in ectotherms. Dispersal of granules creates a pigmentary blanket that masks any underlying pigments; aggregation exposes the latter. Inasmuch as not all varieties of chromatophores respond similarly to the same stimuli, various color combinations result. A functional association of dermal chromatophores is illustrated in Fig. 5-38.

Most vertebrates cannot change color reflexly because the granules in their chromatophores cannot aggregate and disperse. These animals change color only as pigment granules are synthesized in response to long-term stimuli such as exposure to sunlight ("getting a tan," for example) or when hairs or feathers are shed and replaced with others. These are **morphologic color changes.** They are generated most commonly by seasonal biological rhythms that regulate hormone synthesis.

Pigment cells are not confined to the dermis. They are found in many deep locations, including the meninges and striated muscles. Pigment cells in these locations contain only melanin granules and are usually referred to as **melanocytes.** The granules are permanently dispersed. All pigment cells arise from neural crests.

DERMAL INDUCTION OF EPIDERMAL DIFFERENTIATION

Whether an epidermis produces feathers, scales, hair, or ciliated, striated, glandular, or simple squamous epithelium is determined by instructions in the form of specific amino acids exported during embryogenesis by the underlying dermis at that specific site. Feathers can be induced to grow on the feet of chicks if dermis from a potential feathered area is transplanted under the epidermis of the embryonic foot at the critical time during embryonic development, that is, when the controlling genes in the dermis are "turned on." Even the chorionic epithelium can be made to produce feathers or scales by carefully timed transplants of appropriate dermis.

THE INTEGUMENT FROM FISHES TO MAMMALS: A CLASS-BY-CLASS OVERVIEW

This section constitutes, in essence, a summary of the integument on a phylogenetic basis.

Agnatha (Figs. 5-4 and 5-9)

Living agnathans lack scales, but their multilayered **epidermis** is characterized by an abundance of unicellular mucous glands, which earned hagfishes the descriptive name "slime eels." The epidermis, although multilayered, is not stratified. All layers are mitotic, including the surface cells, which secrete a thin cuticlelike substance onto the surface. The only cornified structures are horny denticles in the buccal funnel. These are periodically shed and replaced.

The **dermis** is thinner than the epidermis, but it is exceptionally tough because of very compactly interwoven bundles of collagenous connective tissue. It contains many melanophores, and adheres tightly to the under-lying body wall musculature, especially at the myosepta.

Chondrichthyes (Fig. 5-32)

The **epidermis** of cartilaginous fishes consists of many more layers of cells than does that of cyclostomes, the cells are compactly disposed, and unicellular epidermal glands are much less abundant than in cyclostomes (Fig. 5-32). Multicellular epidermal glands are present but they are few in number and in restricted locations, such as at the base of the claspers of males; and photophores, which are light-emitting multicellular glands of epidermal origin, are found in the dermis in a number of species. Sting rays have aggregates of toxin-secreting goblet cells at the base of the sting. The epidermis of elasmobranchs is pierced by the enamel-covered spines of placoid scales located in the dermis.

The **dermis,** which is considerably thicker than the epidermis, is com-posed of two more or less well-defined layers and is unique in the presence of placoid scales. Just beneath the epidermis is a continuous sheet of mela-nophores, which is far more dense on the dorsum than on the venter, except in deep sea species. Other pigment cell types may also be present. The dermis of each fin contains long, flexible rods of fibrous connective tissue, the ceratotrichia, which stiffen the fin.

Holocephalans have very few scales and many more mucous glands than elasmobranchs. Consequently the skin is slippery rather than having the sandpaper texture of sharks and rays.

Osteichthyes (Figs. 5-5, 5-6, 5-30, 5-31, 5-33, and 5-35)

The skin of bony fishes is similar to that of Chondrichthyes, the chief differences being the larger number of mucous glands in bony fishes, espe-cially teleosts, and the nature of the scales.

Epidermal glands of bony fishes are almost universally unicellular mucous glands. Multicellular mucous glands secrete the mucoid cocoon that envelopes aestivating lungfishes in dry seasons, preventing desicca-tion, or they perform other specialized functions in one or another species. A few teleosts have toxin-secreting multicellular glands; and some teleosts that inhabit the dark depths of the oceans have variously disposed and sometimes bizarre luminescent organs that facilitate self-species recogni-tion or serve as lures.

The **dermis** of bony fishes is characterized by the presence of cosmoid, ganoid, cycloid, or ctenoic scales. Some of the scales of dipnoans are cos-moid-like, others are cycloid or ctenoid. Ganoid scales are found on *Lati-meria, Polypterus, Calamoichthyes,* and garfishes; bony plates or scales lack-ing ganoin are found on part or all of the body of other actinopterygians below teleosts, although those of spoonbills are very tiny and sparse. (*Amia* has bony plates on the head, modern scales on the trunk and tail.) Teleosts, except for a few naked ones, have ctenoid or cycloid scales, sometimes both on the same fish; and some scaleless teleosts develop abortive scales during embryonic life. Lepidotrichia, the flexible dermal fin rays of bony fishes, consist of tiny bony scales that are aligned end to end.

Amphibia (Figs. 5-2, 5-7, 5-8, 5-10, 5-34, and 5-38)

The skin of living amphibians differs from that of fishes in three major respects: (1) scales are absent except in occasional species; (2) epidermal glands are predominantly multicellular; and (3) the epidermis of terrestrial species exhibits an incipient stratum corneum.

In the **epidermis** of aquatic and semiaquatic amphibians **multicellular glands** abound. Mucous glands maintain the skin in a moist state, a vital role in any amphibian that breathes partly through the skin when on land. Granular glands, rare in amphibians other than toads, secrete only when the animal requires the protection of acrid, sometimes toxic, secretions. Although integumentary glands arise from the epidermis, most multicellular glands invade the dermis, pushing dermal elements aside. Amphibians are the lowest vertebrates in which multicellular glands are commonplace.

The epidermis of frogs and some urodeles exhibits a thin layer of **cornified cells,** toads have a heavy layer, and aquatic species have little or none. Efts have a thin covering of cornified cells until they return to water as sexually mature adults, at which time the cornified cells are shed, epidermal glands hypertrophy, and the skin becomes soft and slippery.

Cornified appendages of the integument of amphibians are rare. Callus-like caps develop on the ends of the digits of urodeles subjected to buffeting in mountain brooks, and tadpoles have horny teeth that rasp aquatic mosses and algae, which is their larval diet. At metamorphosis the horny teeth are shed and the adult becomes insectivorous.

The **dermis** performs the same roles as in other vertebrates. It is firmly attached to the underlying musculature in apodans and urodeles, but the relationship of the two in anurans is unique. Separating skin and muscle in anurans are voluminous broad lymph-filled spaces **(subcutaneous lymph sinuses),** the survival value of which is conjectural. The skin of anurans therefore is less intimately attached to the underlying tissues than in other vertebrates. The chromatophores in some species of amphibians are capable of physiologic color changes.

Tiny **bony scales** 1 to 2 mm in diameter are embedded in the dermis of some caecilians, often in bands alternating with bands of glandular epidermis, and a few tropical toads have bony scales in the dermis of the back. Unless these scales are of recent origin, they are the last remnants of ancient bony armor in amphibians.

Reptilia (Figs. 5-15 through 5-18, 5-28, and 5-35 through 5-37)

The skin of reptiles differs from that of fishes and amphibians in the following respects: (1) reptilian skin exhibits a relatively thick stratum corneum; (2) reptilian scales are derivatives of the epidermis rather than of the dermis, although dermal scales (osteoderms) are also present; (3) reptiles exhibit numerous cornified integumentary appendages including claws, horny spines, rattles, and, in turtles, beaks; (4) integumentary glands are not abundant, and those that are present are granular glands.

Epidermal scales, scutes, and **plaques of stratum corneum** are focal characteristics of the epidermis of reptiles. They are repetitious modifications of a continuous sheet of stratum corneum. The scales of squamates overlap because they form on the surface of overlapping folds of epidermis. Scutes on the ventral surface of snakes and on the plastron of turtles are smooth, quadrilateral or polyhedral sheets; those on the carapace of turtles are similar, but usually geometrically sculptured. The stratum corneum of crocodilians is warty, crested, or composed of nonoverlapping scales or thick plaques, depending on the region of the body. The epidermal scales of amphisbaenians are disposed in successive concentric rings that encircle the body, imparting an annulate appearance to the skin similar to that of caecilians.

The epidermal scutes on the carapace of turtles lie in three groups, neurals, costals, and marginals, the latter including nuchal and pygals. On the plastron are paired gular, humeral, pectoral, abdominal, femoral, and anal scutes. Small, nonoverlapping epidermal scales cover the limbs, tail, neck, and the membrane connecting the carapace and plastron.

Epidermal glands of reptiles are almost entirely granular and are restricted to localized regions of the body where their protective and pheromonal secretions will be most effective. Inasmuch as granular glands secrete only in response to specific discontinuous stimuli, the skin of reptiles is dry.

Periodically, snakes shed the entire outer layer of stratum corneum in one piece, or "cast," the process being **ecdysis.** Lizards slough it in patches. Other reptiles wear it off by abrasion.

The presence of **dermal bone** in at least some living representatives of all three reptilian subclasses makes reptilian dermis notable compared with that of modern amphibians. Most conspicuous is the shell of turtles, consisting of carapace, plastron, and lateral bridges that connect the two. Carapace and plastron are composed of large dermal plates arranged in a specific pattern, but one that does not coincide with that of the overlying scutes (compare Figs. 5-18 and 5-37). Soft-shelled and leatherback turtles have a flexible leathery skin because of the absence of bone in the dermis. Crocodilians have osteoderms in localized regions of the body. Along the dorsum of the trunk osteoderms develop prominent crests that give the back an awesome prehistoric appearance. A few lizards have small osteoderms underlying the epidermal scales.

Aves (Figs. 5-21 through 5-24)

The term "thin skinned," sometimes applied figuratively to humans, is literal when applied to birds. The epidermis and dermis constitute a delicate membrane that is loosely applied to the underlying muscle, hence is quite mobile. Only on the feet and head is the skin relatively thick and intimately attached. In these regions are found the horny beak, epidermal scales, spurs, and claws. Elsewhere the body is clothed in feathers. Skin glands are few, and there are no osteoderms.

Horny **scales,** similar to the small scales of turtles, are limited to the feet and the base of the beak. The horny sheath of the **beak** may be in one or several pieces, and it may extend dorsally over the nasal region to form a frontal shield. The segments of compound sheaths appear to be homologous with specific preorbital scales of reptiles. The beak sometimes bears horny toothlike structures that are correlated with specific feeding habits. One or two sharp **claws** are present on the digits of the wings of chimney swifts and numerous other species. A young hoatzin uses them for climbing about on branches. Bony dermal **spurs** on the legs of male gamecocks and some other species are covered with horns and are fused to the tarsometatarsal bone. In some other species, both sexes have a spur on the carpometacarpus.

Except for two varieties, **integumental glands** are lacking in birds. A prominent uropygial gland, best developed in water birds, is located at the base of the foreshortened tail. Its oily secretion is transferred to the feathers during preening. The only other integumentary glands are small oil glands that line the outer ear canal of some Galliformes, and sometimes encircle the vent.

The **dermis** supports the feather follicles and arrectores pilorum. Its exceptional thinness, and the mobility of the skin as a whole, may be correlated with the thermoregulatory role of feathers, which can be elevated to a greater degree than hairs.

Mammalia (Figs. 5-3, 5-11 through 5-14, 5-19, 5-20, 5-25 through 5-27, and 5-29)

The notable features of mammalian skin are (1) hair, (2) a greater functional variety of epidermal glands than in any other vertebrate class, (3) a highly stratified epithelium, and (4) a dermis that, for the first time, is many times thicker than the epidermis.

Hairs are cornified epidermal appendages that arise from follicles deep within the dermis. Hair color results from varying combinations of melanin granules and air vacuoles within the shaft.

Epidermal glands are of two types: sebaceous, which synthesize oily secretions, and sudoriferous, which are predominantly sweat glands. Mammary glands appear to be a variety of sebaceous glands. There are no integumentary mucous glands.

The epidermis exhibits a **stratum malpighi,** with an actively mitotic germinal layer, and a dense **stratum corneum** that is thickest on the palms and soles. Between these, in sites where the stratum corneum is exceptionally thick, are two additional layers where keratinization takes place, a **stratum granulosum** containing granules of keratohyaline, and a **stratum lucidum,** where the granules have dissolved and the nuclei have disappeared.

The epidermis gives rise to reptilian structures including, in one species or another, scales, claws, and horns. Epidermal **scales** are confined to armadillos, in which they are thick and horny, and to the tail and feet of opossums and many rodents. Pangolins develop nonreptilian scales that appear to be agglutinated hairs. Scales also develop on the skin of fetal bears and European hedgehogs, but these are shed before birth. **Claws** have become hoofs in ungulates and flat nails in primates and some other orders. **Horns** are either composed entirely of keratin, as are the unique hair horns of rhinoceros, or they have an outer horny sheath overlying a bony core, as in bovines and pronghorn antelopes. Antlers and giraffe horns lack a horny sheath, hence these are not true horns.

The exceptional thickness of mammalian **dermis** is a result of the presence of many hair follicles, erector muscles of the hairs, sweat and sebaceous glands, the requisite supportive connective tissue, and its exceptional vascularity. The rich blood supply is necessary to satisfy the metabolic demands of the other structures, and it also performs a unique function, thermoregulation, by regulating heat loss from the body surface. Encapsulated exteroceptors are abundant in dermal papillae that indent the underside of the epidermis, and one variety, pacinian corpuscles, lie deep in the dermis (Fig. 5-11). Dermal bone is present only in armadillos, where it forms a continuous sheet of thin scalelike bones beneath the horny epidermal scales.

Because the skin is hidden by fur in most mammals, dermal pigments other than those injected into the hairs play little role in animal coloration. The vividly colored ischial callosities of baboons are among the exceptions. In naked species, melanophores protect against the actinic rays of the sun in proportion to their density.

Mammalian skin is separated from the underlying muscle by a cushion of loose connective tissue, the **superficial fascia.** The accumulation of **contour-shaping fat,** of which whale blubber is an extreme example, is in this fascia.

FUNCTIONS OF THE INTEGUMENT

The vertebrate integument, like the liver or a gill, is an organ, and it plays many roles in the survival of a species. Some of these roles are primary, others are supplementary. Among the roles are protection, exteroception, excretion, respiration, thermoregulation, locomotion, maintenance of homeostasis (regulation of the ratio of water to salts in body fluids), nourishment of young, sexual and species signaling, and additional miscellaneous functions. Examples of some of these roles have been mentioned earlier.

The **protective value** of dermal armor seems obvious. When in the head it protects the brain and some of the sense organs from mechanical injury. Skin glands secrete noxious substances or keep the conjunctiva of the eye free of irritants. Pigments provide protective coloration and, in naked skin, protect underlying tissues from harmful actinic rays of the sun. The bristling coat of an angry or frightened mammal and the ruffled plumage of a bird give them an ominous appearance. Claws, nails, horns, spines, barbs, needles—all confer advantages in the struggle for existence.

Exteroception is subserved by naked or encapsulated nerve endings in the skin (Figs. 5-11 and 16-23). These endings are stimulated when the skin comes in contact with another object. They alert all organisms to the *fact* of contact, which in some instances can be a danger alert; and to successively higher vertebrates they provide increasingly complex information about the texture and temperature of an object. With respect to objects encountered by the primate hand, they provide information concerning shape (stereognosis). These and other cutaneous sensory phenomena are discussed in Chapter 16.

Excretion of carbon dioxide in some aquatic amphibians is entirely via the skin, and a small amount of nitrogenous wastes are excreted by sweat glands.

Respiration via the skin is the rule in many aquatic urodeles, which acquire as much as three fourths of their oxygen from the water via the skin.

Thermoregulation is in part a function of the skin in endotherms. Fur and feathers insulate against heat and cold, sweat cools by evaporation, and dilation of blood vessels within the dermis increases heat loss by radiation. When heat conservation is necessary, the vessels constrict.

Locomotion is subserved by adhesive pads and claws that assist in climbing, scutes that assist in slithering, and feathers that provide an airfoil for aerial locomotion. Webbed feet of ducks and frogs facilitate swimming.

Homeostasis is subserved in some fishes by the dermal scales, which serve as reservoirs for calcium and phosphate ions that can be drawn on as needed. Chloride-secreting glands excrete excess chloride ions acquired from a marine environment, and stratum corneum conserves water in terrestrial species. Under the influence of neurohypophyseal hormones, the skin of aestivating lungfishes and of toads *absorbs* water from moist surroundings, thereby preventing desiccation.

Some teleost hatchlings are **nourished** by feeding on mucous secreted by the mother's skin, and newborn mammals are nourished by maternal mammary glands.

Pheromonal secretions and sex-linked pigment patterns signal sex and identify other members of a species. Pheromones also serve as alarms. Vitamin D is synthesized from ergosterol in the skin of some higher animals, including humans. No wonder an earlier biologist characterized the integument as "a jack of all trades!"

Chapter Summary

1. The vertebrate integument consists of a nonvascular epidermis and a vascular dermis.

2. The epidermis of fishes has a constant thin mucous covering of unknown function that is produced by nonglandular surface cells. The epidermis of terrestrial vertebrates is covered by a layer of cornified cells that retard dehydration.

3. Unicellular and multicellular mucous glands are characteristic of fishes and aquatic amphibians. Granular glands are characteristic of toads and reptiles. Oil glands are found in birds and mammals. In mammals they are sebaceous glands. Mammals also have sudoriferous glands, which produce chiefly sweat.

4. Photophores are light-emitting multicellular epidermal glands of fishes.

5. Epidermal glands of one vertebrate class or another produce mucus, slime, noxious or toxic substances, lipids, sweat, nourishment for new offspring, and pheromones.

6. The epidermis produces a number of cornified appendages, chiefly in amniotes, including epidermal scales, scutes, claws, nails, hoofs, horns, baleen, rattles, horny teeth, beaks, feathers, and hair.

7. A typical contour feather consists of an afterfeather and a main feather, each with its own shaft and vane, the vane consisting of barbs, barbules, flanges, and usually hooklets. Down feathers, lacking hooklets, are fluffy. Filoplumes consist primarily of a threadlike shaft. Feathers are erected by arrectores plumarum.

8. Hairs consist of a bulb, root, and shaft. Sebaceous glands secrete into hair follicles, and arrectores pilorum insert on the walls of the follicles.

9. Modifications of hairs include spines, quills, bristles, vibrissae, and the scales of scaly anteaters. Rhinoceros horns may be agglutinated hairs.

10. The basic constituent of dermis is connective tissue, chiefly collagenous. It supports blood vessels, lymphatics, nerves, chromatophores, and, in one species or another, encapsulated receptors, the bases of multicellular glands, and feather or hair follicles and their erector muscles.

11. Bone in the form of scales or plates is a prominent feature of the dermis of several vertebrate classes. It is confined to a single mammal (armadillo), and is absent in birds.

12. The dermal scales of fishes are cosmoid, placoid, ganoid (paleoniscoid or lepidosteoid), ctenoid, or cycloid. The latter two are flexible modern scales with minimal bone.

13. Chromatophores are pigment cells with permanent branching processes. Physiologic color changes are reflexive, and result from movement of pigment granules into or out of the processes. Morphologic color changes are slow, depend on pigment synthesis, and are commonly generated by seasonal biological rhythms.

SELECTED READINGS

Appleby, L.G.: Snakes shedding skin, Natural History 89(2):64, 1980.

Bagnara, J.T., and Hadley, M.E.: Chromatophores and color change, Englewood Cliffs, New Jersey, 1973, Prentice-Hall, Inc.

Birch, M.C., editor: Pheromones, New York, 1974, Elsevier/Excerpta Medica/North-Holland.

Elias, H., and Bortner, S.: On the phylogeny of hair, American Museum Novitiates **1820**:1, 1957.

Goss, R.J., illustrated by Wendy Andrews: Deer antlers, regeneration, function, and evolution, New York, 1983, Academic Press.

Lucas, A.M., and Stettenheim, P.R.: Avian anatomy: integument. Parts I and II. Agriculture handbook 362, Washington, D.C., 1972, United States Government Printing Office.

Montagna, W., and Parakkal, P.F.: The structure and function of the skin, ed. 3, New York, 1974, Academic Press.

Nelson, D.O., Heath, J.E., and Prosser, C.L.: Evolution of temperature regulatory mechanisms, American Zoologist **24**:791, 1984.

Regal, P.J.: The evolutionary origin of feathers, The Quarterly Review of Biology **50**:35, 1975.

Sengel, P.: Morphogenesis of skin, Cambridge, England, 1976, Cambridge University Press.

Sokolov, V.E.: Mammalian skin, Berkeley, 1983, University of California Press.

Spearman, R.I.C.: The integument: A textbook of skin biology, Cambridge, England, 1973, Cambridge University Press.

Symposia in American Zoologist

The vertebrate integument, **12**:12, 1972.

Chromatophores and color changes, **23**:461, 1983.

6

MINERALIZED TISSUES: AN INTRODUCTION TO THE SKELETON

In this chapter we will be introduced to the various types of skeletal tissues, see how they are formed, and look briefly at their invertebrate origins. We will discover that mineralized tissues have a role in homeostasis, and we will learn why growing skeletons must undergo continual remodeling.

The skeleton is composed of mineralized connective tissue and of ligaments, tendons, and bursae. The mineralized tissue for the most part is bone, but there is also dentin (often considered a variety of bone), cartilage, and enamel or enameloid substances. Osteoblasts produce bone, odontoblasts produce dentin, chondroblasts produce cartilage, and ameloblasts produce most enamels. These specialized cells arise from less differentiated scleroblasts that arise from mesenchyme (Fig. 6-1).

A preliminary step in formation of skeletal tissues is the synthesis of collagen by fibroblasts. Collagen is a proteinaceous **fibril** demonstrable only by high-power electron microscopy. Fibrils aggregate to form **collagen fibers** that are visible with light microscopy. The fibers form dense **collagen bundles** that are woven into a compact network of dense connective tissue like that found in dermis, tendons, and ligaments (Fig. 6-2). It is on such a network that minerals are deposited to form cartilage and bone.

BONE

The gross structure of a long bone is shown in Fig. 6-3. Bone tissue is composed of a matrix of collagenous fibers, the spaces between which have been impregnated with **hydroxyapatite crystals** that are composed of calcium, phosphate, and hydroxyl ions [$3Ca_3 (PO_4)_2 \cdot Ca(OH)_2$]. The crystals are deposited under the influence of osteoblasts. A **cementing substance** composed of water and mucopolysaccharides binds the crystals to the collagenous matrix. In most bone the osteoblasts ultimately become trapped by the bone they have laid down around them, and, as osteocytes, occupy tiny fluid-filled pools, or **lacunae,** of interstitial fluid (Fig. 6-4). Interconnecting the lacunae are fluid-filled canals, **canaliculi,** that house protoplasmic processes extending from the osteocytes. The fluid in the lacunae and canaliculi contains calcium and phosphate ions, which are constantly being deposited on the collagenous matrix or withdrawn from it, depending on the level of serum calcium.

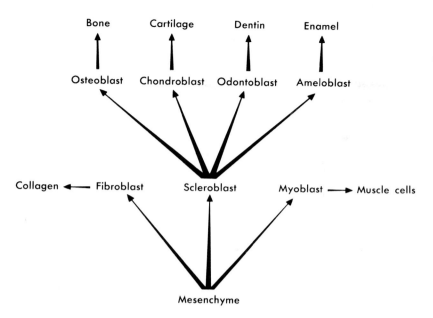

FIGURE 6-1

Some differentiated products of mesenchyme.

FIGURE 6-2

Collagen bundles in tendon. The individual collagen fibers in each bundle appear as dots in cross section. The cells are fibrocytes.

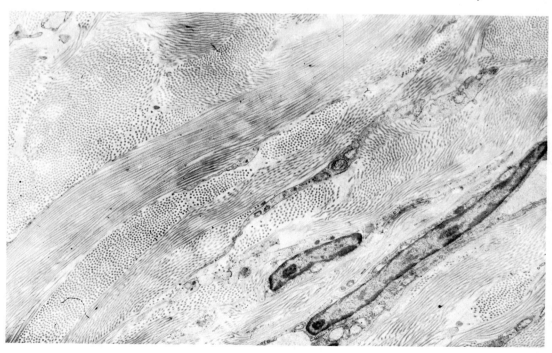

FIGURE 6-3

Metatarsal bone in longitudinal section.

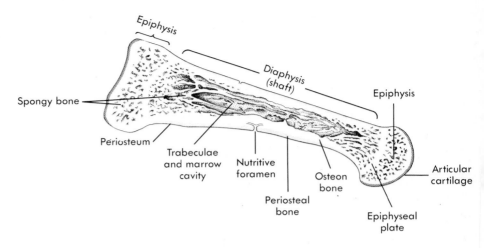

Compact Bone

Compact bone consists of lamellae of mineralized collagenous bundles arranged concentrically around a haversian canal (Figs. 6-4, *A*, and 6-5), as in the shaft of long bones, or in more or less flat sheets that lack canals, as in periosteal bone. **Haversian canals** contain an arteriole, a venule, a lymphatic, and nerve fibers. The canal and its surrounding lamellae constitute an **osteon,** or **haversian system.** The blood vessels are responsible for the configuration of the haversian systems, and because the vessels branch and rebranch, haversian systems do likewise. The vessels in the canals are continuous with those in the bone marrow. In the shaft of long bones the haversian systems more or less parallel the long axis of the bone (Fig. 6-4, *B*). The flat lamellae of periosteal bone are formed by osteoblasts on the inner surface of the **periosteum,** the dense fibrous membrane that encloses all bones except at their articular surfaces (Fig. 6-3). Haversian systems are characteristic of amniotes; most amphibians, a few reptiles, and some small insectivores and rodents lack them.

Spongy Bone

Spongy, or **cancellous, bone** and compact bone are alike in their essential elements but differ in appearance when observed grossly. Spongy bone consists of bony trabeculae and marrow (Fig. 6-3). **Trabeculae** are an assemblage of beams, bars, and rods that, like architectural trusses, form a rigid framework that provides maximum strength at areas of stress. They are composed of irregularly arranged lamellae without haversian systems. **Marrow** consists of a reticulum of connective tissue fibers that support blood vessels, nerve fibers, and adipose tissue (yellow marrow), and in some bones hemopoietic tissue, which is the source of red blood cells and some types of white blood cells (red marrow). A core of spongy bone and marrow sandwiched between two layers of compact surface bone characterize **flat bones,** such as ribs, the scapula, and the membrane bones of the skull.

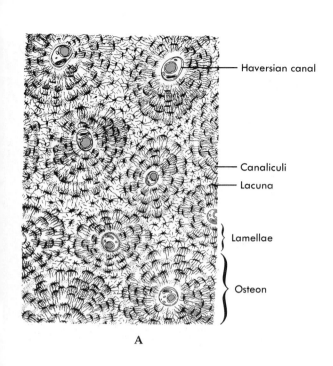

Haversian canal

Canaliculi

Lacuna

Lamellae

Osteon

A

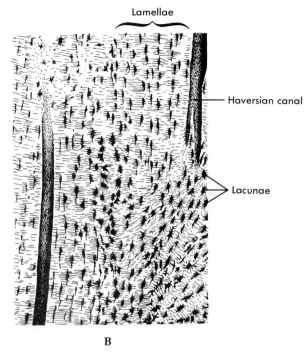

Lamellae

Haversian canal

Lacunae

B

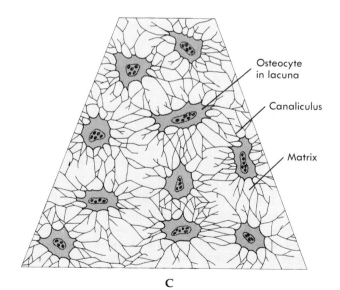

Osteocyte
in lacuna

Canaliculus

Matrix

C

FIGURE 6-4

A, Osteon bone in cross section. *Red,* arterioles in haversian canals. **B,** Osteon bone in longitudinal section. **C,** Nonlamellar membrane bone.

FIGURE 6-5

Cross section of one osteon (haversian system).

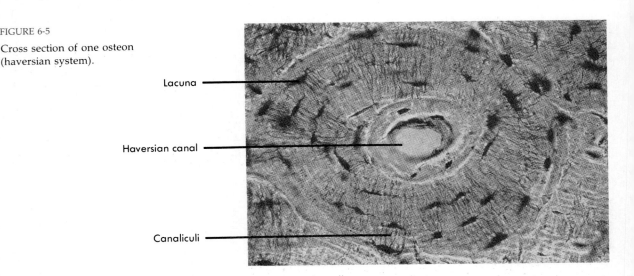

Lacuna

Haversian canal

Canaliculi

Dentin

Dentin has the same constituents as compact and spongy bone, but the odontoblasts are not trapped in lacunae during osteogenesis. They retreat as they deposit dentin, hence are always at its inner border (Fig. 6-6). However, they leave behind protoplasmic processes in canaliculi, which in dentin are called **dentinal tubules.** The tubules extend all the way to the surface of the dentin. Dentin forms only in the outer layer of the dermis, just beneath the epidermis, and is frequently coated on its surface by enamel or enameloid (Fig. 5-30, *A* and *B*). The enameloid of placoid scales is produced by odontoblasts, hence is a very hard dentin. Although a constant feature of the dermis of the earliest vertebrates, dentin is found today only in the scales of ganoid and elasmobranch fishes and in teeth.

Acellular Bone

There is one type of bone in which the osteoblasts not only retreat as they deposit bone but in addition leave behind no processes or canaliculi. It is the thin layer of **acellular bone (aspidin)** that constitutes the fibrous plates of the flexible scales of modern fishes.

Membrane Bone and Replacement Bone

Before bone or cartilage can be deposited, a preskeletal blastema must develop. A **blastema** is any aggregation of mesenchyme that, given the appropriate stimulus, differentiates into some tissue, such as muscle, cartilage, or bone. Once a preskeletal blastema has formed, some of the mesenchyme cells become fibroblasts and secrete collagen. Others become either osteoblasts or chondroblasts and secrete the enzymes essential for formation of bone or cartilage.

MEMBRANE BONE. Bone deposited directly within a membranous blastema without having been preceded by a cartilaginous model is **membrane bone.** Intramembranous ossification gives rise to certain bones of the lower jaw, skull, and pectoral girdles; to dentin and other bone that forms in the dermis of the skin; to vertebrae in teleosts, urodeles, apodans only; and to a few miscellaneous bones. Membrane bone may be compact or spongy, and lamellar or nonlamellar. Because of the arrangement of the blood vessels that participate in its deposition, it lacks haversian canals (Fig. 6-4, C). Periosteal bone is membrane bone.

Any bone derived ontogenetically or phylogenetically from the dermis of the skin may be called a **dermal bone.** The term specifies its history, not its histologic features. It is membrane bone. Many, though not all, membrane bones are of dermal origin, as for example, the bones that constitute the dermatocranium, the bones that invest Meckel's cartilage of the lower jaw, and the membrane bones of the pectoral girdles. These bones are phylogenetic derivatives of the integument. (For a possible explanation see page 221, "How it may have begun.") Not much bone remains in the dermis of recent vertebrates, but that which does—the bony plates and scales of lower vertebrates—arises in situ.

REPLACEMENT BONE. Replacement bone is deposited where hyaline cartilage already exists (Fig. 6-7). In the process the existing cartilage undergoes degenerative changes and disappears. The processes of endochondral and intramembranous ossification are the same in that both consist of impregnation of a collagenous matrix with hydroxyapatite crystals. The difference is, in endochondral ossification cartilage must be removed before bone can be deposited. In both cases the immediate result is formation of temporary spongy bone. The latter, in turn, is eroded and replaced by compact bone, spongy bone, or a marrow cavity, depending on its location.

In a typical long bone of a tetrapod appendage the process of endochondral ossification commences midway in the shaft and spreads toward the two epiphyses. Shortly thereafter, an endochondral ossification center appears in each epiphysis. Meanwhile, cartilage continues to be deposited ahead of all zones of ossification, so that the future bone continues to grow

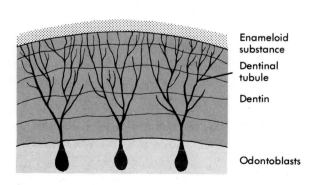

Enameloid
substance

Dentinal
tubule

Dentin

Odontoblasts

FIGURE 6-6

Dentin covered with enameloid. The canaliculi (dentinal tubules) contain processes of the odontoblasts. See also Fig. 5-31, *A* and *B*.

FIGURE 6-7

FIGURE 6-7

Endochondral ossification. *Red,* blood vessels of bone marrow invading hyaline cartilage, digesting old matrix in preparation for deposit of new matrix by osteoblasts.

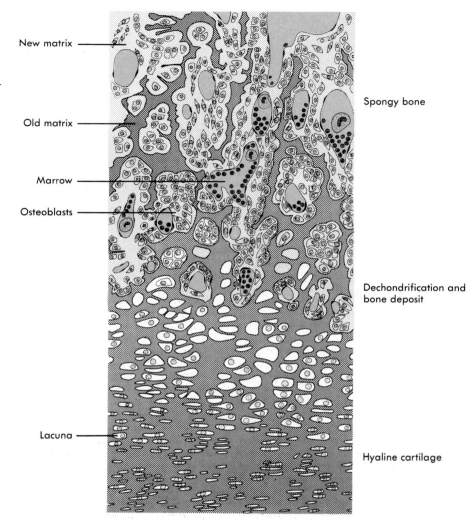

New matrix

Old matrix

Marrow

Osteoblasts

Lacuna

Spongy bone

Dechondrification and bone deposit

Hyaline cartilage

in diameter and length. Eventually, the epiphyses are separated from the diaphysis by only a narrowing **epiphyseal plate** of cartilage, where chondrogenesis continues for a time (Fig. 6-8). The bone continues to elongate so long as the epiphyseal plates are elaborating cartilage to replace that being ossified. In ectotherms this is throughout life. In birds and mammals the epiphyseal centers cease elaborating cartilage shortly after attainment of sexual maturity, the epiphyseal plates ossify, the epiphyses become sutured to the shaft, and the bone stops growing. Bones of other shapes exhibit other combinations of ossification centers (Fig. 8-5).

While endochondral ossification is going on within a bone, periosteal bone is being deposited at the surface. As a result of this activity combined with continual resorption and remodeling (p. 182), a bone reaches adult shape, size, and proportions. Details of the histogenesis of bone will be found in textbooks of histology.

Cartilaginous fishes either no longer have the genetic codes necessary for endochondral ossification or there is a genetic reason why they cannot give expression to them.

CARTILAGE

Cartilage resembles bone in that it is formed within a matrix of collagenous connective tissue and the cells lie in lacunae. The intercellular matrix, however, contains a sulfated mucopolysaccharide instead of hydroxyapatite crystals. Unlike bone, cartilage has no canaliculi demonstrable by light microscopy, and no blood vessels penetrate it except those en route to other organs. The lacunar cells **(chondrocytes)** are supplied with oxygen and nutrients by diffusion from the nearest capillaries. Cartilage formation is initiated by chondroblasts, which become chondrocytes after being trapped in lacunae.

Hyaline cartilage is the least differentiated variety. It is translucent because all components have the same refractive index to light. Little hyaline cartilage remains after the growth of replacement bones has ceased. Thereafter it is found chiefly on the articular surfaces of bones within joints (Fig. 6-8, *B*). Most any other hyaline cartilage that forms is likely to be transformed into fibrocartilage, elastic cartilage, or calcified cartilage.

Cartilage with exceptionally thick, dense collagenous bundles in the interstitial matrix is **fibrocartilage.** The intervertebral discs of mammals are fibrocartilage. **Elastic cartilage** contains, in addition to collagenous fibers, a network of elastic fibers. In mammals it is found in the pinna of the ear, in the walls of the outer ear canal, in the epiglottis, and elsewhere. **Calcified**

FIGURE 6-8

A, Proximal end of tibia of youth aged 16 years. The epiphyseal plate is unossified; hence the bone is still growing. **B,** Diagramatic section of a simple diarthrodial joint, ligaments removed. *Red,* synovial fluid.

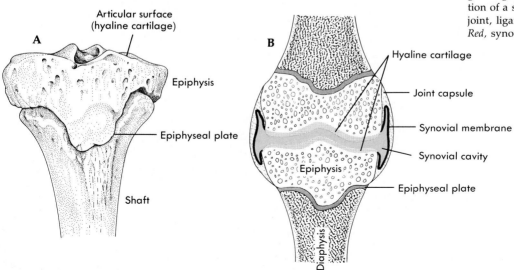

Articular surface
(hyaline cartilage)

A

Epiphysis

Epiphyseal plate

Shaft

B

Hyaline cartilage

Joint capsule

Synovial membrane

Synovial cavity

Epiphysis

Epiphyseal plate

Diaphysis

cartilage is formed when calcium salts are deposited within the interstitial substance of hyaline cartilage or fibrocartilage. It is often mistaken for bone. The jaws of sharks, no matter how large, are calcified cartilage. Calcification is sometimes seen at the site where hyaline cartilage is degenerating in the course of endochondral ossification.

SKELETAL REMODELING

Bones not only provide mechanical support for the vertebrate body but, with scales and teeth, constitute important storage places for calcium and other mineral salts. They therefore participate in maintaining homeostasis. Calcium is constantly being deposited and withdrawn in response to dietary intake and cellular demands. When serum calcium levels are rising, excess calcium that is not excreted is deposited along with inorganic phosphate as hydroxyapatite crystals. When serum calcium is falling, calcium is withdrawn from the skeleton and other depots, thereby maintaining a normal serum calcium level. Withdrawal is regulated by parathyroid hormone and calcitonin; deposit may not be under endocrine control.

In addition to bone resorption in response to lowered serum calcium levels, cartilage and bone are constantly being resorbed and replaced in another process, skeletal remodeling. For example, a skull the size of that of a newborn child could not accommodate a 21-year old brain without growing (Fig. 6-9). But growth by accretion would not provide a larger brain cavity; it would only provide a thicker skull. The entire skull must be con-

FIGURE 6-9

Comparative size of skull of newborn and 21-year-old human being. Continued remodeling enlarges the brain cavity.

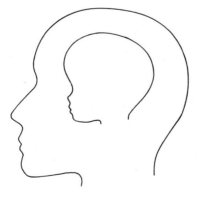

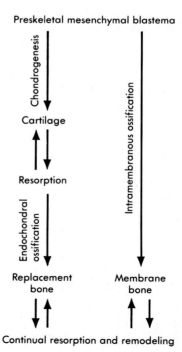

FIGURE 6-10

Steps in osteogenesis and remodeling.

tinuously remodeled by constant resorption of existing cartilage and bone and replacement of these in a new alignment to provide a wider, higher skull with a cavity large enough to accommodate the growing brain. This entails not only remodeling of the cranium but also of the facial bones and lower jaw. Cartilage and bones elsewhere in the body are also subject to remodeling. Although remodeling is a more active process in growing skeletons, it continues to some degree throughout life, accelerating during periods of skeletal repair (Fig. 6-10).

Remodeling of bones also occurs in response to mechanical stress resulting from continuing use of inserting muscles or from weight bearing. A bone becomes thicker at sites where stress is high, thin when stress is withheld. Roughened surface areas, bony ridges, and other prominences to which muscles attach enlarge with sustained muscle use. The arrangement of the latticelike components of spongy bone (Fig. 6-3) bears a striking resemblance to lines of force generated by the bone's natural load.

TENDONS, LIGAMENTS, AND JOINTS

Tendons and ligaments consist chiefly of closely packed bundles of collagen. **Tendons** connect muscles with bones, have a shiny white appearance, and the collagen bundles are so arranged as to offer maximal resistance to the tension created when a muscle contracts. Where the tendon is attached to muscle the collagenous bundles of the tendon are directly continuous with those of the fibrous muscle sheath (epimysium); and where it attaches to cartilage or bone they are continuous with the collagenous bundles of the perichondrium or periosteum. **Ligaments** connect bone to bone. The arrangement of the collagen bundles is less regular, but they are directly continuous with those of the periosteum. Tendons and ligaments that are flat and very wide are **aponeuroses.** The longest ligament in mammals is the nuchal ligament in the back of the neck of grazing herbivores. Extending between the occiput of the skull and the neural spines of some of the thoracic vertebrae, it provides a resilient mechanical support for the head. It is least developed in mammals with short necks.

The term ligament is sometimes applied to fibrous mesenteries or cords that support visceral organs or hold them in place, as for instance, the falciform ligament and the round ligament of the ovary. However, these are not skeletal structures.

In some species certain tendons and ligaments become mineralized as a normal phenomenon. Turkeys, for example, have ossified tendons in their legs, and ornithischian dinosaurs had them millions of years ago. **Sesamoid cartilages** or **bones** are mineralized nodules in tendons or ligaments. Best known is the patella, or kneecap. It is endochondral bone in some species, membrane bone in others.

A **joint** is the site where two bones or, occasionally, cartilages meet. If the joint is movable in one or more planes it is a **diarthrosis.** The articular surfaces of the bones in a diarthrosis are covered with hyaline cartilage, and the space between them is enclosed in a fibrous **joint capsule** that contains a lubricatory **synovial fluid** (Fig. 6-8, *B*). The fluid is secreted by the synovial membrane that lines the capsule. Ligaments hold the bones in their proper alignment in the joint.

In sutured joints, or **synarthroses,** no movement is possible. The joints in the calvarium (roof of the skull) are synarthroses (Fig. 8-25). If the suture between two bones has been obliterated during development, the condition is an **ankylosis.** The premaxilla and maxilla of the human embryo ankylose and as a result cannot be distinguished as two bones in an adult. In cats they are sutured but not ankylosed. A **symphysis** is a joint in the midline, in which two bones are separated by a pad of fibrocartilage and movement is severely restricted, if not impossible. The pubic symphysis of pregnant female mammals becomes movable by hormone-initiated dissolution of the fibrocartilage shortly before the onset of labor.

MINERALIZED TISSUES AND THE INVERTEBRATES

Vertebrate mineralized tissues are not unique. In fact, two thirds of the living species of animals that contain mineralized tissues are invertebrates. The matrix is collagen, and is as old as the sponges, but the inorganic crystals are more often calcium carbonate than calcium phosphate. Cartilage is found among invertebrates, including squids, some gastropods, and protochordates; and bone, dentin, cartilage, and enameloid were all present in Ordovician ostracoderms. For that reason it cannot be said that one of these is phylogenetically older than the other.

REGIONAL COMPONENTS OF THE SKELETON

The skeleton may be divided regionally into the following components, which include associated ligaments and tendons.

Axial skeleton
    Notochord and vertebral column
    Ribs and sternum
    Skull and visceral skeleton
Appendicular skeleton
    Pectoral and pelvic girdles
    Skeleton of paired fins and limbs
    Skeleton of median fins of fishes

HETEROTOPIC BONES

In addition to the skeletal components outlined above, miscellaneous **heterotopic bones** develop in aberrant locations in amniotes by endochondral or intramembranous ossification. They are usually missing from routinely prepared skeletons. Among heterotopic bones are an **os cordis** in the interventricular septum of the heart of deer and bovines, a **baculum** (os penis) in the septum between the spongy bodies of the penis of dogs, lower primates, and many other mammals (Fig. 6-11), and an **os clitoridis** in many female mammals. In walruses the baculum is nearly 60 cm long.

A heterotopic bone forms in the gizzard of some doves, in the tongue of at least one species of bats, in the gular pouch of a South American lizard, in the muscular diaphragm of camels, in the syrinx of some birds (pessulus, Fig. 12-15), and in the upper eyelid of crocodilians (adlacrimal, or palpebral, bone). A similar plate of fibrous tissue, the tarsus, develops in human beings. A rostral bone develops in the snout of swine, and is used in routing in soil, and a cloacal bone develops in the ventral wall of the cloaca of some lizards. All these are incidental bones lacking phylogenetic significance.

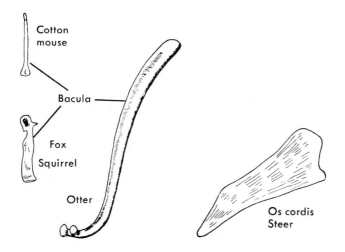

FIGURE 6-11

Some heterotopic bones.

Chapter Summary

1. The chief mineralized tissues are bone (from osteoblasts), dentin (from odontoblasts), cartilage (from chondroblasts), and enameloids (from ameloblasts).

2. Bone consists of collagen, inorganic salts, cementing substance, and osteocytes in lacunae usually connected by canaliculi.

3. Bone is compact or spongy (cancellous). In deep compact bone the lamellae are concentric around haversian canals. In periosteal bone lamellae are flat and lack the canals.

4. Dentin is a form of bone with peripheral odontoblasts, dentinal tubules, and no lacunae. It is found in the skin and teeth.

5. Mature acellular bone lacks cells. It is found in modern fish scales.

6. Ossification may be intramembranous, producing membrane bone, or endochondral, replacing preexisting cartilage and yielding replacement bone.

7. Dermal bones are membrane bones originating ontogenetically or phylogenetically from the dermis of the skin.

8. Cartilage consists of collagen, chondrocytes in lacunae, and canaliculi. It may be hyaline, fibrocartilage, or elastic, and is sometimes calcified.

9. Bone and cartilage are constantly being deposited and resorbed, thereby facilitating orderly growth, homeostasis, and tissue repair.

10. Tendons attach muscles to bone. Ligaments connect bone to bone. Both may be calcified or contain sesamoid bones.

11. Joints are diarthroses, synarthroses, or symphyses. Ankyloses are junctions between bones in which the sutures have been obliterated.

12. Cartilage, bone, dentin, and enameloid were present in ostracoderms, and it cannot be said which of these may be phylogenetically older.

13. Heterotopic bones form in miscellaneous locations by endochondral or intramembranous ossification.

SELECTED READINGS

Burt, W.H.: Bacula of North American mammals, University of Michigan Museum of Zoology Miscellaneous Publications **113**:1-76, 1970.

Currey, J.: Comparative mechanical properties and histology of bone, American Zoologist **24**:5, 1984.

Moss, M.L.: The origin of vertebrate calcified tissues. In Ørvig, T., editor: Current problems of lower vertebrate phylogeny, New York, 1968, Interscience-Wiley.

Moss, M.L.: Skeletal tissues in sharks, American Zoologist **17**:335, 1977.

Murray, P.D.F., with introductory essay by B.K. Hall: Bones, New York, 1985, Cambridge University Press.

Ørvig, T.: The dermal skeleton: general considerations. In Ørvig, T., editor: Current problems of lower vertebrate phylogeny, New York, 1968, Interscience-Wiley.

Patterson, C.: Cartilage bones, dermal bones and membrane bones, or the exoskeleton versus the endoskeleton. In Andrews, S.M., Miles, R.S., and Walker, A.D., editors: Problems in vertebrate evolution, New York, 1977, Academic Press.

Todd, J.T., and others: The perception of human growth, Scientific American **242**(2):132, 1980.

7

VERTEBRAE, RIBS, AND STERNA

In this chapter we will look at some typical and some not-so-typical vertebral columns of fishes and tetrapods, consider their ontogeny and phylogeny, and note the advantages conferred by specialization in tetrapods. Along the way we will of necessity consider ribs; and at the end we will examine the sternum in the four tetrapod classes.

The vertebral column, skull, ribs, and tetrapod sternum, along with their ligaments, constitute the major components of the **axial skeleton.** The vertebral column forms around and, sometimes, within the notochord during ontogeny; the ribs and sternum form in the lateral and ventral body wall.

VERTEBRAL COLUMN

The vertebral column is the keystone of the skeleton. It is a usually flexible arch to which the head is attached and from which all the rest of the body is suspended in fishes, and all of the trunk between appendages in most tetrapods. It provides a protective bony tunnel for the spinal cord and, in gnathostomes, it is an essential participating skeletal structure for locomotion. In no vertebrate is its locomotor importance more readily demonstrable than in fishes and limbless tetrapods.

Fishes are surrounded by an environment that offers considerable resistance to forward progress and a buoying effect on the body. During locomotion, resistance is overcome by pushing laterally against the water with rhythmical, undulating, side-to-side movements of the trunk and tail. These movements are brought about by segmental muscles attached to vertebrae and myosepta. The architecture of the vertebral column of most fishes permits only side-to-side flexibility of the column. When vertebrates ventured onto land they brought with them the fishlike method of locomotion, but it was an awkward way to move about. Unlike water, land is beneath the animal rather than around it, it is not level, and it is littered with obstacles that must be avoided or clambered over. Eventually, selective forces altered the vertebral column to provide dorsoventral flexibility, which is more suited to locomotion on land, while at the same time providing a strong skeletal arch for suspending the trunk above the ground between forelimbs and hind limbs. These changes were achieved at the expense of some side-to-side flexibility of the column, and they ultimately resulted in regional specialization of trunk vertebrae in tetrapods.

Centra, Arches, and Processes

Vertebrae typically consist of a **centrum, neural arch,** and one or more processes, or **apophyses,** that project from the arch or centrum (Fig. 7-1). Centra occupy the position occupied early in ontogeny by the notochord. Neural arches are perched on the centra, and the successive arches and their interconnecting **interarcual ligaments** enclose a long vertebral canal occupied by the spinal cord. In most vertebrates **hemal arches,** known also as **chevron bones** in amniotes, are inverted beneath the centra of the tail and house the caudal artery and vein (Fig. 7-1, *B, C,* and *E*).

Diapophyses (transverse processes) are the most common processes. They project laterad into the horizontal skeletogenous septum that separates the epaxial and hypaxial muscles (Figs. 1-2 and 7-6). There they serve as attachments for some of the muscles that extend or flex the vertebral column. **Zygapophyses** are paired processes at the cephalic end of trunk vertebrae **(prezygapophyses),** and at their caudal end **(postzygapophyses),** chiefly in tetrapods (Fig. 7-1, *D* to *F*). The articular facets of prezygapophyses articulate with facets of the postzygapophyses of the next vertebra in such a manner as to restrict dorsoventral flexion of the column in the trunk region. (The facets on the prezygapophyses face upward and articulate against the downward directed facets of the postzygapophyses.) Tetrapod tails are highly flexible because zygapophyses are absent. **Parapophyses** are lateral projections from some of the centra of *tetrapods,* and, when present, serve as the articulation site for the capitulum of a bicipital rib (Fig. 7-24, *B*). Hypapophyses are prominent midventral projections from the centra of some amniotes, especially snakes (Fig. 7-1, *D*). They serve as sites of attachment for certain tendons or muscles. Other processes develop in occasional species.

Morphogenesis of Vertebrae

A typical vertebra arises from mesenchyme cells that stream out of the sclerotomes of mesodermal somites, surround the notochord and neural tube, and produce the blastema for a future vertebra (Fig. 4-9). Chondroblasts that differentiate within the blastema subsequently deposit a cartilaginous centrum and neural arch, and, in the tail, a hemal arch. The result is a cartilaginous vertebra with the notochord constricted within each centrum. Later, except in cartilaginous fishes, the cartilage is removed and bone is deposited. Remnants of the notochord may remain in the centrum to varying degrees, depending on the species. In Chondrichthyes the cartilaginous vertebrae are never replaced by bone. In teleosts, urodeles, and apodans, membrane bone rather than cartilage is deposited in the perichordal blastema, but the arches arise from cartilage.

In many fishes, and in amphibians except anurans, not only is cartilage or membrane bone deposited *around* the notochord **(perichordal cartilage or bone),** but chondroblasts invade the notochord sheath and deposit cartilage *within* the sheath and notochordal tissue. Consequently, the centra of these vertebrates contain **chordal cartilage** that may calcify or ossify (Figs. 7-2 and 7-3). The result of these variations in the process of centrum for-

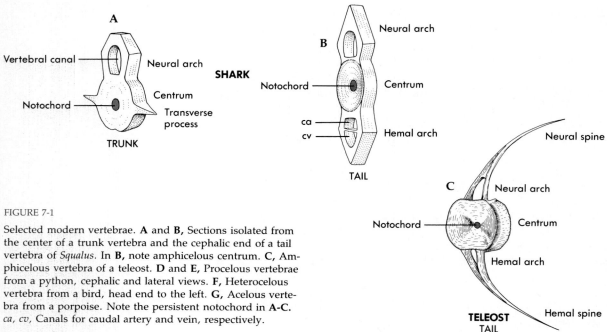

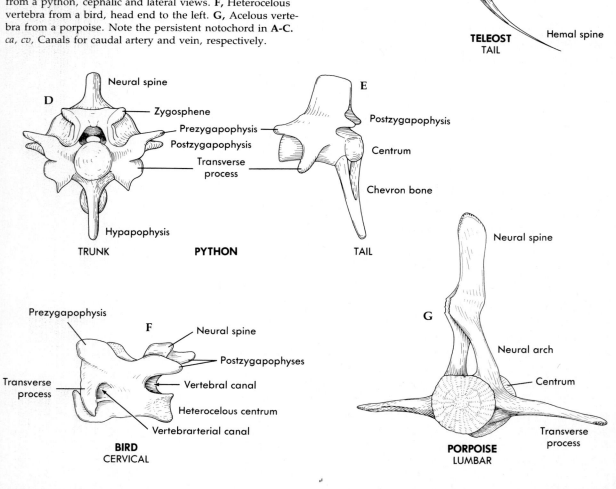

FIGURE 7-1

Selected modern vertebrae. **A** and **B,** Sections isolated from the center of a trunk vertebra and the cephalic end of a tail vertebra of *Squalus.* In **B,** note amphicelous centrum. **C,** Amphicelous vertebra of a teleost. **D** and **E,** Procelous vertebrae from a python, cephalic and lateral views. **F,** Heterocelous vertebra from a bird, head end to the left. **G,** Acelous vertebra from a porpoise. Note the persistent notochord in **A-C.** *ca, cv,* Canals for caudal artery and vein, respectively.

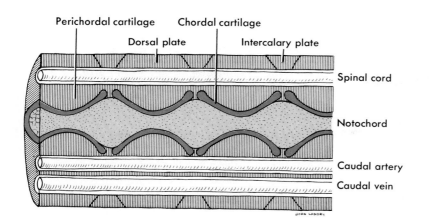

Perichordal cartilage   Chordal cartilage

Dorsal plate   Intercalary plate

Spinal cord

Notochord

Caudal artery

Caudal vein

FIGURE 7-2

Caudal vertebral column of *Squalus*, sagittal section. Calcified chordal cartilage has been deposited in the notochord sheath *(dark red)*. The intercalary plates of *Scyllium* are seen from lateral view in Fig. 7-6.

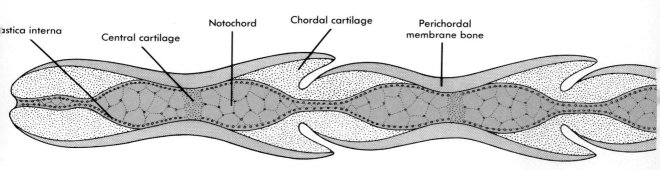

Elastica interna   Notochord   Chordal cartilage   Perichordal membrane bone

Central cartilage

FIGURE 7-3

Two trunk vertebrae from the plethodontid salamander *Gyrinophilus*, frontal section, head end to the left. Note opisthocelous condition. Chordal cartilage is the outer portion of the chondrified notochord sheath; the elastica interna is the innermost portion. (Modified from Wiedersheim.)

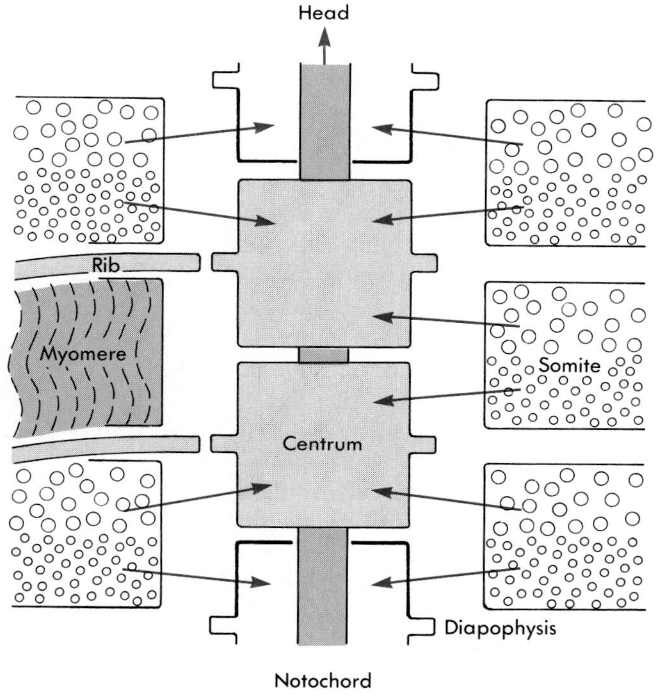

Head

Rib

Myomere

Centrum

Somite

Diapophysis

Notochord

FIGURE 7-4

Intersegmental (intersomitic) position of centrum and myoseptum or rib. Each centrum is a contribution from two successive somites *(arrows)*.

mation can be illustrated by contrasting the centra of urodeles with those of anurans. The centra of urodeles, and also of apodans, consist of perichordal membrane bone, chordal cartilage, and, at the center, unaltered notochordal tissue (Fig. 7-3). The centra of anurans, on the other hand, consist entirely of replacement bone deposited in a perichordal blastema, and the notochord disappears. The difference has been interpreted by some paleontologists as supporting evidence for a diphyletic origin of Lissamphibia.

In the formation of centra, scleroblasts from the caudal half of one somite and the cephalic half of the next somite stream to an intersegmental location around the notochord to establish the perichordal blastema. Consequently, vertebrae are intersegmental, whereas myomeres are segmental (Fig. 7-4).

Vertebral Column of Fishes

The vertebrae of fishes exhibit a broad spectrum of morphological diversity from species to species. This is attributable to the varying degrees to which scleroblasts invade the notochord sheath and the notochord proper, to the nature and extent of the contribution of the sheath and notochord to the adult centra, to the number of chondrification and ossification centers that develop in the perichordal blastemas, and to the extent to which separate foci of skeletogenic activity remain independent or unite during

FIGURE 7-5

Vertebral types based on articular surfaces. Midsagittal sections, head to the left. Amphicelous vertebrae are found in fishes, generalized urodeles, caecilians, and primitive lizards, opisthocoelous in garfishes and salamanders, procelous in anurans and modern reptiles, and acelous in mammals. A heterocelous vertebra is seen in Fig. 7-1, F. Color indicates notochordal tissue.

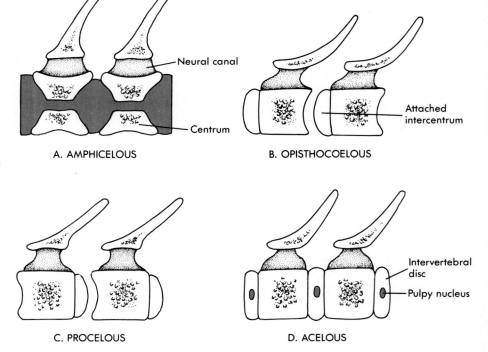

A. AMPHICELOUS — Neural canal, Centrum

B. OPISTHOCOELOUS — Attached intercentrum

C. PROCELOUS

D. ACELOUS — Intervertebral disc, Pulpy nucleus

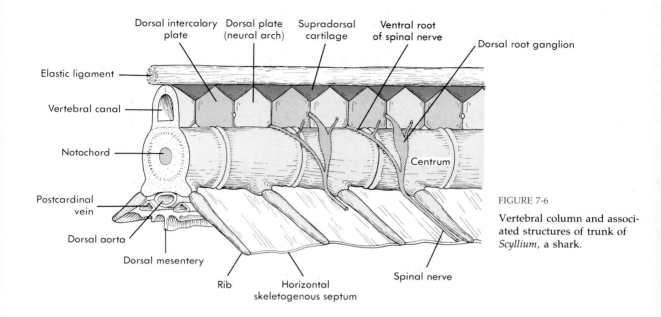

Dorsal intercalary plate
Dorsal plate (neural arch)
Supradorsal cartilage
Ventral root of spinal nerve
Dorsal root ganglion
Elastic ligament
Vertebral canal
Notochord
Centrum
Postcardinal vein
Dorsal aorta
Dorsal mesentery
Rib
Horizontal skeletogenous septum
Spinal nerve

FIGURE 7-6

Vertebral column and associated structures of trunk of *Scyllium*, a shark.

ontogeny. Although a fish has only two varieties of vertebrae, **trunk** (dorsal) and **caudal,** both may differ gradually along the length of the column from head to vent, and from vent to the tip of the tail.

Agnathans have a strange vertebral column if, indeed, they may be said to have one. The only skeletal elements are **lateral neural cartilages** (Fig. 1-5), one or two pairs per body segment, depending on the species. Caudally, the lateral neural cartilages fuse to form a single dorsolateral cartilaginous plate perforated by foramina for spinal nerves. In hagfishes, lateral neural cartilages are limited to the tail. These lateral neural cartilages may be vestigial vertebrae, primitive vertebrae, or they may bear no phylogenetic relationship to vertebrae. Whether you call them vertebrae depends on how you wish to define "vertebra." How would *you* define the term?

In sharks the notochord is present throughout the length of the vertebral column and is constricted within each centrum. As a result, the centra, composed of chordal and perichordal cartilage, are concave at each end. A centrum that is concave at each end is said to be **amphicelous** (Fig. 7-5). The spinal cord is enclosed in a continuous tunnel, the **vertebral (neural) canal,** consisting of paired **dorsal plates** that constitute a neural arch, paired **dorsal intercalary plates** between the arches, and, in some species, paired wedge-shaped **supradorsal cartilages** (Fig. 7-6). *Squalus acanthias* lacks supradorsals. The neural arches and intercalary plates of *Squalus* are perforated by foramina for the dorsal and ventral roots of spinal nerves and for blood vessels. In *Scyllium* they emerge between plates. Hemal arches consist of paired **ventral plates. Ventral intercalary plates** are common between hemal arches. A fibroelastic ligament, present in all fishes, overlies and connects the neural spines along the entire length of the vertebral column.

In holocephalans, chordal cartilage is deposited in the notochord sheath, which becomes greatly thickened. Thereafter, calcification may convert the sheath into calcified cartilaginous rings that are much more numerous than the body segments (Fig. 7-7). The rings constitute the major component of the column in holocephalans. Any minor components resulting from perichordal chondrification differ along the length of the column and from species to species.

Modern dipnoans, chondrosteans other than *Polypterus,* and the coelacanth *Latimeria* have no centra. The notochord is present and unconstricted throughout its length, and its thick fibrous sheath contains little chordal cartilage or bone (Fig. 7-8). Associated with the notochord in each body segment are paired **basidorsal, basiventral, interdorsal,** and **interventral** cartilages, which are products of isolated skeletogenic loci in the embryonic perichordal blastema. The chondrostean *Polypterus* and living holosteans have well-ossified complete centra.

Teleosts have well-ossified amphicelous vertebrae (Fig. 7-1, C). The core of each centrum is a dumbbell-shaped vacuole where the notochord previously existed, and the space between successive vertebrae is occupied by a porous cartilage-like material that probably includes notochordal tissue. The centra and neural arches of successive vertebrae are interconnected by a complex system of collagenous and elastic ligaments that facilitate lateral undulation of the body for locomotion. In the posterior trunk and tail these ligaments are frequently ossified to form a long, delicate, bony rod. The

FIGURE 7-7

Calcified vertebral rings from an extinct holocephalan. The rings consist of calcified cartilage deposited in the notochord sheath. They are more numerous than the somites. Neural arches have been removed.

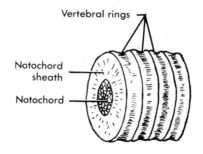

Vertebral rings

Notochord sheath

Notochord

FIGURE 7-8

Vertebral components in adult lungfish *(Neoceratodus)* and sturgeon. Arrow in lungfish indicates canal occupied by a supportive longitudinal ligament.

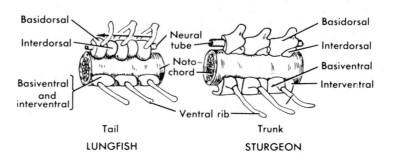

Basidorsal

Interdorsal

Neural tube

Notochord

Basiventral and interventral

Ventral rib

Tail

LUNGFISH

Basidorsal

Interdorsal

Basiventral

Interventral

Trunk

STURGEON

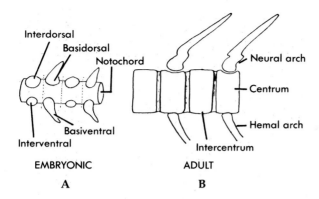

FIGURE 7-9

Tail vertebrae of *Amia*. A centrum and intercentrum develop in each body segment. Embryonic basidorsal and basiventral cartilages contribute to each centrum, and interdorsal and interventral cartilages contribute to each intercentrum.

neural spines are often very tall, and in lower teleosts they are sometimes surmounted by **supraneural bones** (Fig. 7-23, *Perca*). A variety of processes, mostly unlike those of tetrapods, protrude from the arches and centra.

Elasmobranchs and many bony fishes have two centra and two sets of arches in each metamere of the tail, and a few have duplication of centra and arches in the trunk also. Consequently, the number of centra in those regions is double the number of myomeres and spinal nerves. The condition is known as **diplospondyly.** *Amia* has what look like two centra per body segment (Fig. 7-9), but only one bears arches. The other is an intercentrum. The embryonic basidorsal and basiventral cartilages are incorporated in the centra, and the interdorsals and interventrals are incorporated in the intercentra. It may be that diplospondyly in the tail provides this locomotor organ with greater flexibility, but this is pure speculation. Flexibility of the vertebral column is facilitated by the elasticity of the ligaments that interconnect vertebrae, not by intervertebral joints, as in tetrapods.

No movement is possible between the first vertebra and the skull. They are united by cartilage or nonelastic connective tissue. The terminal vertebrae of fishes are part of the caudal fin and are discussed in Chapter 9.

Evolution of Tetrapod Vertebrae

Like the vertebral columns of rhipidistian crossopterygians, those of early tetrapods did not consist of one unit per body segment. A trunk vertebra in rhipidistians and in the earliest known amphibians consisted of a **hypocentrum** (intercentrum), which was a large, anterior, median, wedge-shaped element that was incomplete dorsally, two **pleurocentra,** which were smaller, posterodorsal, intersegmental elements, and a **neural arch** (Fig. 7-10). A vertebra of this kind is **rhachitomous,** and the earliest amphibians are therefore referred to as rhachitomes. All later tetrapod vertebrae appear to be modifications of the rhachitomous variety (Fig. 7-11). Changes leading to modern amniotes were characterized by increased prominence of pleurocentra and concomitant reduction of hypocentra. The evolutionary history of anurans, urodeles, and apodans is not known, hence the history of their vertebrae is unknown.

FIGURE 7-10

Rachitomous vertebrae from the primitive labyrinthodont *Archegosaurus,* left lateral view. **A** and **B,** Trunk. **C-E,** Caudal. Hypocentra are median crescentic bones; pleurocentra are paired. The three encompass the notochord, as shown in Fig. 7-11, *A,* XS. The extra component in **D** is a detached ossification center of the pleurocentrum. (After Goodrich.)

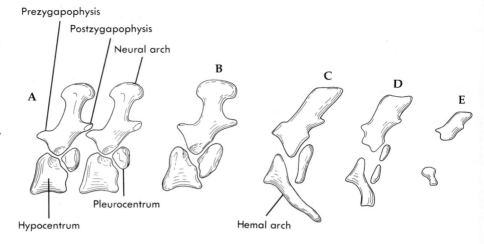

FIGURE 7-11

Postulated modifications leading from labyrinthodont to modern amniote vertebrae. *A, B, B1,* and *B2,* Four varieties of labyrinthodont vertebrae. The rachitomous type occurred in crossopterygians and the earliest tetrapods (*Ichthyostegalia* in Fig. 3-15). *B* is from an amphibian thought to be in the reptile line. *Diagonal lines,* hypocentrum; *stipple,* pleurocentrum.

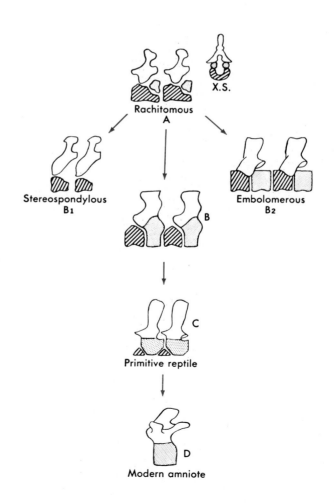

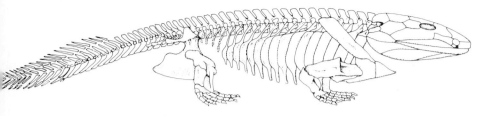

FIGURE 7-12

The three foot long skeleton of *Ichthyostega,* the oldest known tetrapod. Missing ribs indicated by dotted lines. (After Jarvik, E.: Scientific Monthly **80**:152, March 1955.)

Even today, in modern tetrapods each centrum commences ossification at several symmetrical loci in the perichordal blastema and, as the separate ossification centers expand, they coalesce. The centers in the occipital bone, which is thought to be a modified vertebra, are shown in Fig. 8-5. The typical modern tetrapod vertebra is, therefore, a composite structure. The findings of paleontology, comparative anatomy, and embryology all point to the same conclusion: an adult tetrapod vertebra composed of a centrum and an arch in each body segment is not primitive. Primitively, several skeletal components occupied each body segment.

Rhachitomous vertebrae, like the vertebrae of all fishes except gars, were amphicelous; and amphicelous vertebrae are still present in generalized urodeles such as *Necturus,* in apodans, in *Sphenodon,* and in primitive lizards such as geckos, all of which retain vestiges of notochordal tissue in the intervertebral joints. In other tetrapods the centra become flattened at one or both ends—**opisthocelous, procelous,** or **acelous vertebrae**—with attached intercentra or, in mammals, with intervertebral discs (Fig. 7-5). The intervertebral joints of modern amniotes provide more flexibility to the vertebral column than do joints between amphicelous vertebrae.

Regional Specialization in Tetrapod Columns

With the advent of life on land and the accompanying morphological changes in other systems that increased the probability of survival, the vertebral column underwent regional specialization. Tetrapod limbs push against the earth. Initially, they did so purely for locomotion; later they were also supporting the body above the ground. The opposing physical force exerted by the earth in response to the push of a tetrapod appendage is transmitted partly via the ilia of the pelvic girdle to one or more of the hindmost trunk vertebrae (Fig. 7-12). These vertebrae became appropriately modified and are now called **sacral** (Fig. 7-13). The pelvic fins of fishes have no such relationship to the vertebral column.

Survival on land with its irregular contours and scattered obstructions to vision, such as rocks or vegetation, was facilitated in tetrapods by increased mobility of the head with its special sense organs, including eyes that can scan the horizon. This was achieved to a small degree in amphibians by development of a more mobile joint between the first vertebra and the skull. The innovation was improved on in many reptiles by shortening or eliminating ribs on some of the more anterior vertebrae, and by increasing the mobility of the intervertebral joints in that region. Thus **cervical verte-**

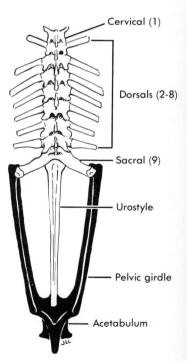

Cervical (1)

Dorsals (2-8)

Sacral (9)

Urostyle

Pelvic girdle

Acetabulum

FIGURE 7-13

Vertebral column and pelvic girdle *(black)* of an anuran. The transverse processes include short ribs. The pelvic girdle is braced against the sacral vertebra.

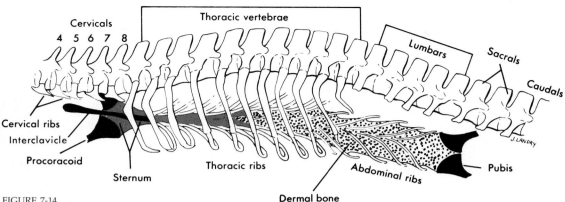

FIGURE 7-14

Vertebrae and ribs of an alligator. Abdominal ribs are also called gastralia. A complete crocodilian column is seen in Fig. 9-33.

brae made an appearance (Figs. 7-14, 7-17, and 9-33, *B*). In crocodilians, lizards, birds, and mammals long ribs have become restricted to the anterior portion of the region between cervical and sacral vertebrae, thereby preserving a thoracic cage that houses the lungs and participates in external respiration. These are **thoracic vertebrae.** The remaining vertebrae anterior to the sacral are **lumbar** (Fig. 7-14). The complete vertebral columns of an alligator, seal, and porpoise are shown in Fig. 9-33.

Among tetrapods, snakes have the longest columns with as many as 400 or more vertebrae. These provide snakes with an elongated, multijointed body that can be coiled into loops or sinuous curves for locomotion without limbs. Caecilians, also limbless, have as many as 250 or more vertebrae, and some urodeles have 100. The awkward leaping anurans have the shortest columns. In turtles and birds only the cervical and caudal segments of the column are flexible. The trunk vertebrae are fused to the synsacrum in birds (Fig. 7-20) and to the carapace in turtles (Fig. 5-37).

THE CRANIOVERTEBRAL JUNCTION AND NECK VERTEBRAE. The first and only cervical vertebra of modern amphibians lacks transverse processes and prezygapophyses, and its cranial end bears two smooth concave facets that articulate with the two knoblike occipital condyles. The condyles of a primitive urodele are illustrated in Fig. 8-12. The absence of processes and the nature of the amphibian craniovertebral joint permits limited dorsoventral rocking of the skull, a function that is not possible in fishes. The two-facet articulation is a change from the earliest amphibians, who had only one occipital condyle.

Amniotes have more cervical vertebrae than amphibians, which results in a long, flexible neck, and the first two vertebrae are modified in such manner as to permit considerably more independent movement of the head. The first vertebra, or **atlas,** named after Atlas of Greek mythology, is ringlike because its centrum is no longer attached (Fig. 7-15, atlas). On its cephalic end are one or two deep concavities for articulation with the single occipital condyle of reptiles and birds or with the two of mammals. These are condyloid joints in which the skull rocks as in nodding ''yes.'' Except in

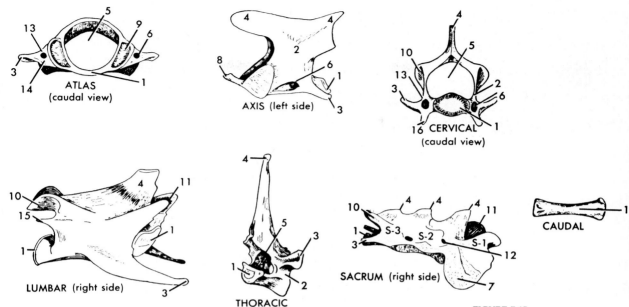

FIGURE 7-15

Cat vertebrae. *1*, Centrum, except in atlas; *2*, pedicle; *3*, transverse process; *4*, neural spine; *5*, vertebral canal; *6*, transverse foramen; *7*, site of articulation with ilium; *8*, odontoid process; *9*, articular facet for axis; *10*, postzygapophysis; *11*, prezygapophysis; *12*, intervertebral foramen transmitting the dorsal ramus of the first sacral spinal nerve; *13*, diapophysis; *14*, parapophysis; *15*, accessory process; *16*, vestige of a cervical rib with two heads, one fused with a diapophysis, the other with a parapophysis to enclose a transverse foramen transmitting the vertebral artery to the brain. *S-1*, *S-2*, *S-3*, sacral vertebrae.

snakes, the centrum of the atlas has become an **odontoid process** of the second vertebra, the **axis.** The process projects forward to rest on the floor of the atlas where it is held in place by a fibrous transverse ligament (Fig. 7-16). (The odontoid process of the relict *Sphenodon* chondrifies as an independent element before uniting with the axis.) The skull and atlas pivot as a unit on the odontoid process, which is the actual axis of rotation. Pivoting is facilitated by the reduction or absence of pre- and postzygapophyses on the first one or two vertebrae. The degree of rotation is maximal in mammals.

In some reptiles and in a few mammals an additional neural arch of uncertain homology, the **proatlas,** is interposed between skull and atlas (Figs. 7-16, *B,* and 7-17). In some lizards, *Lacerta,* for instance, there is a membranous gap instead of a proatlas at that location.

A flexible neck is a necessary accompaniment to appendages that lift the trunk off the ground. It enables these tetrapods to reach food and water on the ground while standing, rather than having to get on their knees or belly to do so. Such a posture would endanger an ungulate in a habitat that includes stealthy carnivores.

Flexibility of the neck is exceptional in birds and turtles because of the way their neck vertebrae articulate. The caudal ends of the centra of birds are saddle-shaped, having a convexity in the right-left axis and a concavity in the dorsoventral axis (Fig. 7-1, *F*). The cephalic end of the next centrum is oppositely shaped to accommodate this configuration. These are **heterocoelous vertebrae.** The joints permit much side-to-side flexion of the neck as well as dorsoventral flexion. Because of the axis-atlas complex, hetero-

FIGURE 7-16

Atlas, axis, and odontoid process of a cat, sagittal section, and of a crocodilian. In **B,** the crocodilian, the vertebrae have been slightly separated for clarity. Note that the second cervical crocodilian rib is bicipital.

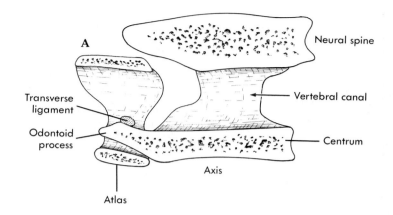

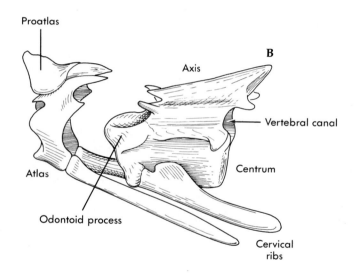

FIGURE 7-17

The eight cervical vertebrae, eight cervical ribs *(red),* and proatlas, *P,* of an alligator, left lateral view. *1,* Atlas and attached first rib. Immediately behind the first rib is the rib of the axis. The ribs are fused to transverse processes.

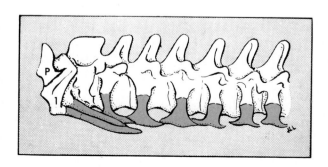

coelous vertebrae, and the largest number of neck vertebrae in any tetrapod—twelve or more in common species, twenty-five in swans—some birds can turn their heads 180 degrees to the rear. Turtles have ball-and-socket joints in the neck, so most turtles can completely retract their head and neck into the protection of the shell by folding the neck dorsoventrally, or, in "side-necked" turtles, sidewise. However, like all other extant reptiles, turtles have only eight cervical vertebrae. In contrast to the high flexibility of the neck, the remainder of the column of birds and reptiles is highly rigid.

Mammals almost always have seven cervical vertebrae. This is as true of the stubby, rigid neck of whales as it is of the neck of the tallest giraffe. The sole exceptions are three edentates (two sloths and the Great Anteater of Central and South America), with six, eight, or nine, and one manatee species with six. In moles several cervical vertebrae ankylose, which no doubt strengthens the neck for burrowing; and all eight cervical vertebrae of cetaceans and armadillos are shortened and more or less fused together.

The cervical vertebrae of birds and mammals have a transverse foramen between the two heads of the vestigial cervical ribs (Fig. 7-15, atlas, axis, and cervical). The successive foramina provide a **vertebroarterial canal** that transmits the vertebral artery and vein to or from the brain. The vessels exist in other amniotes, but they are not enclosed by foramina.

STABILIZING THE HIND LIMBS: SACRUM AND SYNSACRUM. Sacral vertebrae bear short, stout transverse processes that are strong enough to take the thrust of the pelvic girdle as the limbs push against the earth in locomotion (Fig. 7-13). Amphibians have one sacral vertebra, living reptiles and most birds have two. Pelycosaurs, the earliest mammal-like reptiles, had two sacral vertebrae as do today's opossums, but later therapsid reptiles had three. Most modern mammals have three to five. Some perissodactyls have six to eight, and edentates have up to thirteen. When there are more than one, the sacral vertebrae usually ankylose to form a single bone, the **sacrum** (Fig. 7-18). Sacral vertebrae do not differentiate in caecilians, limbless reptiles, or cetaceans, all of which lack hind limbs. In fact, snakes have little differentiation along the entire length of the column, all vertebrae being much alike and all bearing long, slim ribs that are used in locomotion.

In birds the last thoracic vertebra, all lumbars, the two sacrals (three in ostriches), and the first few caudals unite to form one adult complex, the **synsacrum,** and the latter becomes fused to one degree or another with the pelvic girdle (Figs. 7-19, 7-20, and 7-21). The synsacrum provides a rigid framework for the teeter-totter-like, two-legged stance of birds. The thoracic vertebrae anterior to the synsacrum also unite more or less completely, so there is little flexibility in the avian backbone caudal to the neck. This axial rigidity minimizes the number of muscles needed to keep the body streamlined during flight. However, the historical selective forces responsible for a synsacrum may have been quite different from what appear today to be its advantages.

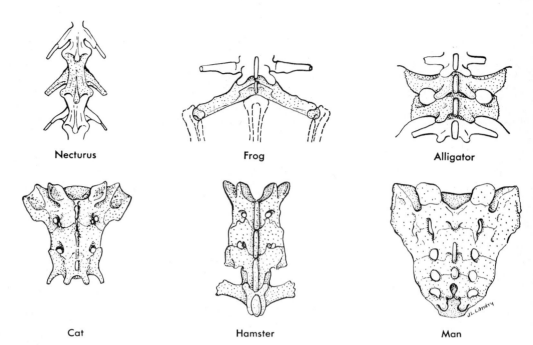

Necturus

Frog

Alligator

Cat

Hamster

Man

FIGURE 7-18

Sacral vertebrae *(stippled)* of selected vertebrates, dorsal views. They have ankylosed to form a sacrum in the amniotes illustrated. The distal end of the transverse processes of frogs incorporate a short rib.

FIGURE 7-19

Diagram of vertebral column of pigeon. *T,* Thoracic; *L,* lumbar; *C,* caudal; *P,* pygostyle, composed of four fused vertebrae.

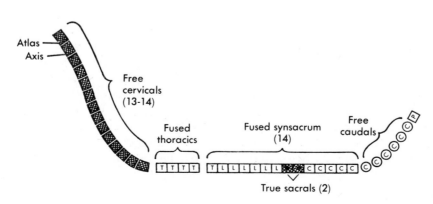

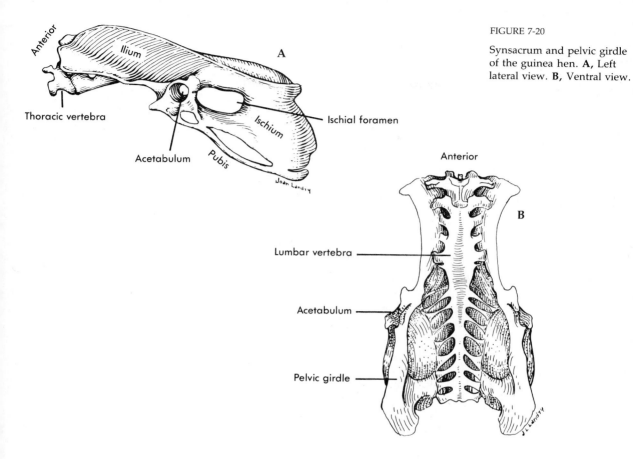

FIGURE 7-20

Synsacrum and pelvic girdle of the guinea hen. **A,** Left lateral view. **B,** Ventral view.

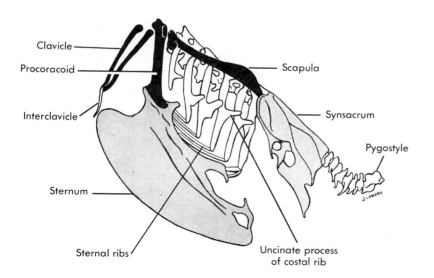

FIGURE 7-21

Skeleton of trunk, tail, and pectoral girdle of a pigeon, head to the left. Bones of the pectoral girdle are black. The ventral keel of the sternum is the carina. The clavicles and interclavicle form the furcula, or wishbone. The first rib illustrated is the second of two cervicals.

Armadillos also have a synsacrum. It consists of up to thirteen fused sacral and anterior caudal vertebrae. In addition, there is extensive fusion of neck vertebrae in these armored mammals.

TAIL VERTEBRAE: UROSTYLE, PYGOSTYLE, AND COCCYX. Caudal vertebrae in early tetrapods probably numbered fifty or more. In modern ones the number is much reduced and highly variable, dictating how effective the tail may be for one use or another. Among mammals, for example, there are as few as three and as many as fifty. The locomotor tail of sperm whales has twenty-four. Toward the end of the tail in all tetrapods the arches and processes become progressively shorter and rudimentary until finally the vertebrae consist solely of small cylindrical centra (Fig. 7-22).

Anurans have a unique **urostyle** at the end of the vertebral column (Fig. 7-13). It arises from a continuous elongated perichordal cartilage at the base of the larval tail, and it grows and ossifies after the tail is lost at metamorphosis. That it is composed of postsacral vertebrae is evident from the one or more coalesced centra, vestigial arches, transverse processes, and nerve foramina that are part of the cranial end of the urostyle in some species.

Among reptiles the caudal vertebrae of lizards deserve mention. Many lizards, when captured by the tail, break off the end distal to the site of capture and scurry away. Thereafter the tail is regenerated. This **autotomy** is implemented by a zone of soft tissue that divides each tail vertebra into cephalic and caudal sections, the location being at the level of a myoseptum. A sudden opposing reflex muscle contraction on each side of the septum causes the break.

Archaeopteryx had a long tail, and modern birds still have a tail, although it is not prominent. Pigeons have fifteen caudal vertebrae. Five are ankylosed with the synsacrum, six are independent, and the last four are fused to form a **pygostyle,** which is the skeleton of the visible part of the tail (Fig. 7-21). The pygostyle develops as four separate cartilaginous centra.

The great apes and human beings have four or five vestigial tail vertebrae comparable to the pygostyle of birds. These vertebrae lack arches, but most of them have rudimentary transverse processes. The last three or four diminish progressively in size and, in humans, they fuse with one another to form a rigid **coccyx** at about twenty-five years of age. The centra of the coccyx are still identifiable, but the last one is a mere nodule of bone. People whose "tailbone" is recurrently sore have fractured the coccyx at some earlier date in a fall that landed them on the end of their spine. In contrast to hominoids, adult rhesus monkeys have a prehensile tail the length of which averages nearly fifty percent of the monkey's sitting height, and prominent chevron bones are present on the more proximal vertebrae.

FIGURE 7-22

Complete set of tail vertebrae from a hamster, left lateral view.

RIBS

Ribs articulate with vertebrae and extend into the body wall. The part proximal to a vertebra is formed intersegmentally by scleroblasts from two successive mesodermal somites. The remainder of the rib arises from lateral plate mesoderm. Most bony ribs are of endochondral origin. However, the **abdominal ribs (gastralia)** in the ventral body wall of crocodilians and some lizards (Fig. 7-14) are membrane bones that might more appropriately be thought of as part of the dermal skeleton.

Fishes

Polypterus and some teleosts have two pairs of ribs, dorsal and ventral, associated with each trunk vertebra (Figs. 7-23, Perca, and 7-24, *A*). **Dorsal ribs** pass laterad into the horizontal skeletogenous septum that separates the epaxial and hypaxial musculature (Figs. 1-2 and 7-6). **Ventral ribs** develop in myosepta and arch ventrad in the lateral body wall just external to the parietal peritoneum. Commencing near the vent, each pair of ventral ribs, or their basal stumps, unite beneath the centrum to form a hemal arch, and this condition continues into the tail.

Most fishes have only ventral ribs. Sharks and a few others have only dorsal ribs. Holocephalans, skates, and some teleosts such as sea horses have no ribs. Neither have cyclostomes, but this is probably correlated with the absence of centra. Fish ribs probably stabilize the intermuscular septa, thereby increasing the effectiveness of the metameric muscles, which are used in locomotion.

Tetrapods

In early tetrapods, ribs were present in most body segments from the atlas to nearly the end of the tail. They became restricted in length and number during the evolution of modern tetrapods, and those in the upper trunk region of amniotes acquired a ventral anchor, the sternum.

Most tetrapod ribs are bicipital, having a dorsal head, or **tuberculum**, and a ventral head, the **capitulum** (Figs. 7-24, *B*, and 7-25). The tuberculum in early tetrapods articulated with the extremity of a diapophysis, and the capitulum articulated with the hypocentrum. With reduction in size of the hypocentrum that took place later in modern amniotes, the articulation of the capitulum became altered. Thereafter, it articulated in one of several fashions: (1) in demifacets on the contiguous lateral ends of two successive centra; (2) with a parapophysis (Fig. 7-24, *B*); or (3) in a facet on a single centrum, in the absence of a parapophysis. Variations along the length of a column are common. For example, capitula of the more anterior ribs of mammals usually articulate in demifacets, and those farther along articulate in a facet on a single centrum. The tuberculum continues to articulate with the diapophysis (transverse process of mammals). It is frequently reduced to a small tubercle on some ribs, especially in mammals, the capitulum becoming the chief site of articulation with the column. In some tet-

FIGURE 7-23

The skeleton in actinopterygian fishes. *Glaucolepis* was a Triassic chondrostean, *Caturus* a Mesozoic holostean, *Clupea* is a generalized teleost, and *Perca* is a specialized teleost. (From Colbert: Evolution of the vertebrates, ed. 3, New York, 1980, John Wiley & Sons. Courtesy of the publishers.)

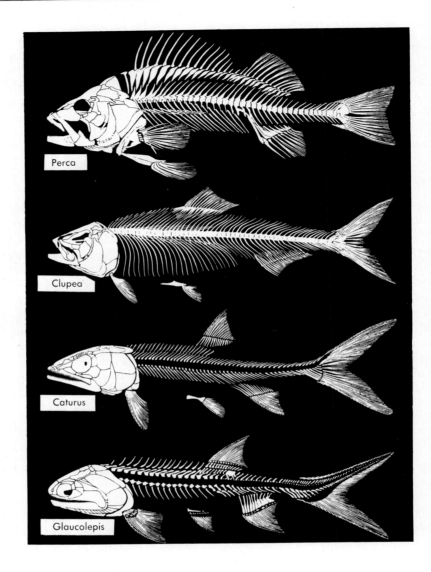

FIGURE 7-24

Dorsal and ventral ribs of some fishes and bicipital rib of a tetrapod. An alternate articulation for the capitulum of tetrapods is seen in Fig. 7-25. The basal stump in **A** and the parapophysis in **B** are not homologues.

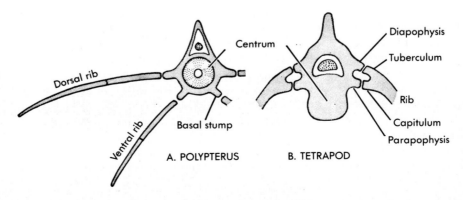

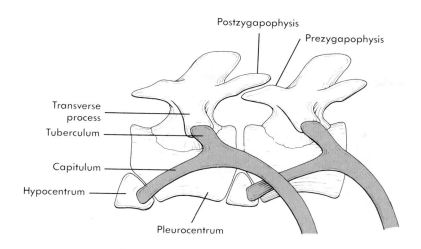

Postzygapophysis

Prezygapophysis

Transverse process

Tuberculum

Capitulum

Hypocentrum

Pleurocentrum

FIGURE 7-25

Two thoracic ribs *(red)* from a primitive tetrapod to show articulation with vertebral column. (After Goodrich.)

rapods, crocodilians, for example, there is a progressive tendency toward fusion of the tuberculum and capitulum until eventually only one head may remain. It articulates with the diapophysis.

Typical thoracic ribs of amniotes consist of two segments, a dorsal or **costal rib,** and a ventral or **sternal rib** (Fig. 7-21). The latter tend to remain unossified, especially in mammals.

AMPHIBIANS. The limbless caecilians have long ribs throughout most of the length of the column, except at its extremities. The ribs of anurans and urodeles have become very short and, in anurans, they are ankylosed to the tips of the transverse processes of the vertebrae, hence appear to be absent (Fig. 7-13). They are seen as separate ossification centers in embryos.

REPTILES. Turtles have no cervical ribs, and the ribs of the trunk vertebrae are fused with the costal plates of the carapace (Fig. 5-37). The two sacral ribs are not fused with the carapace, they are short, and their expanded distal ends are ankylosed to the ilia of the pelvic girdle. They brace the girdle against the vertebral column. Slender, riblike processes may be associated with a few of the caudal vertebrae.

Snakes have long, curved ribs beginning at the axis and continuing far into the tail. There is no sternum for ribs to attach to in snakes, but the ventral ends have ligamentous connections with scutes, and they participate in locomotion. In cobras, the long ribs of the neck can be rotated outward, causing the neck to "spread" (Fig. 7-26).

The ribs of lizards and crocodilians are more "conventional." There are long ribs on many of the trunk vertebrae, short ribs in most of the neck. Crocodilian ribs are illustrated in Figs. 7-14 and 7-17. The primitive geckos have ribs associated with every cervical vertebra; modern lizards usually lack them on the atlas and axis. A half dozen or more of the posterior ribs of the trunk of the gliding lizard *Draco* are greatly elongated and associated

FIGURE 7-26

Cervical ribs of a cobra, a "spreading adder." The first three ribs, associated with the atlas, axis, and third vertebra, are short and hidden by the jaws. *C4,* Neural spine of the fourth cervical vertebra.

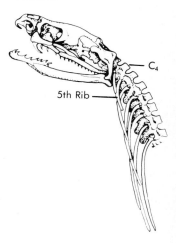

C₄

5th Rib

with a broad fold of skin, the patagium. Like the long ribs of the neck of cobras, these ribs can be rotated outward to elevate the patagium into a winglike membrane on which the lizard soars. When not in use, the patagium folds against the side of the body. In *Sphenodon*, long ribs continue into the tail.

BIRDS. The first two pairs of ribs of many carinates are movable cervical ribs that lack a sternal segment. In some species they are immovable, being fused to cervical vertebrae. The remaining five pairs are thoracic, and the sternal segments are bony. These constitute the major portion of the thoracic basket, ribs that form caudal to them becoming ankylosed into the synsacrum. The functional ribs are thin and flat, and most of them bear **uncinate processes** that are directed caudad and dorsad (Fig. 7-21). Uncinate processes are found also in some lizards, and were present in ichthyostegids (Fig. 7-12). Thin, broad ribs and uncinate processes provide birds with a lightweight but sturdy thoracic body wall skeleton for attachment of powerful flight muscles. Although adult birds have relatively few functional ribs, rudimentary ribs form during embryonic development in all body segments and deep into the tail.

MAMMALS. Mammals typically have only thoracic ribs. The number ranges from nine pairs in some whales to twenty-four in some sloths; twelve pairs are common. Where the number is larger than ten the rest are "floating," that is, the costal cartilages (sternal ribs), if present, fail to reach the sternum. Human beings and apes other than orangutans have twelve pairs of thoracic ribs, and apes have an additional two pairs of lumbars. Humans frequently have an additional rib, cervical or lumbar, as an anomaly, but it usually remains undiscovered unless revealed by an x-ray examination. As described in Chapter 12, thoracic ribs function in respiration in most amniotes.

STERNUM

The sternum is a tetrapod structure, and primarily amniote. It serves in part as a ventral site against which the pectoral girdles of terrestrial vertebrates can be braced. Consequently, its size and morphology are correlated with the extent to which the forelimbs are employed in locomotion. It is absent in limbless amphibians and limbless reptiles. There is little evidence that stem amphibians had one, and among modern amphibians it is well differentiated only in anurans (Fig. 9-7). It is flimsy or absent in urodeles (Fig. 7-27).

Lizards are agile climbers and runners, and the procoracoid bones of the pectoral girdle are firmly braced against a substantial shield-shaped sternum of cartilage or replacement bone (Fig. 7-28). The sternum of crocodilians is continuous with an expanse of a mineralized fibrous membrane (Fig. 7-14). Turtles lack a sternum, their pectoral girdle being united ventrally by fibrous bands or a mere wedge of cartilage.

The sterna of modern birds that are capable of flight have developed an enormous keel or **carina** to which the massive pectoral flight muscles are attached (Fig. 7-21). Pterosaurs also developed a carina—an instance of convergent evolution.

The mammalian sternum is composed of bony segments, or **sternebrae,** except in cetaceans and sirenians (Fig. 7-28, monkey). The last sternebra, **xiphisternum,** bears a cartilaginous or bony **xiphoid process.** In all amniotes a varying number of ribs articulate with the sternum via costal cartilages (sternal ribs), and these, along with the sternum and vertebral column, provide a skeletal enclosure for the thoracic viscera.

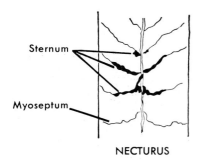

FIGURE 7-27

The "sternum" of a necturus.

NECTURUS

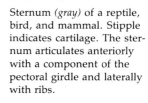

FIGURE 7-28

Sternum *(gray)* of a reptile, bird, and mammal. Stipple indicates cartilage. The sternum articulates anteriorly with a component of the pectoral girdle and laterally with ribs.

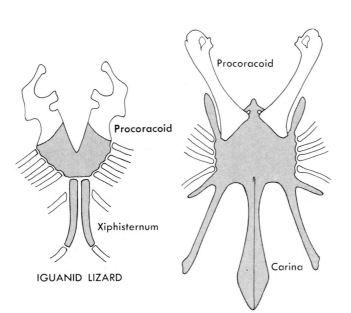

IGUANID LIZARD

CHICKEN

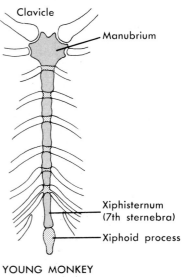

YOUNG MONKEY

The amniote sternum arises as paired mesenchymal bars that later unite and undergo chondrogenesis, and in many mammals **presternal** and **suprasternal** blastemas also develop (Fig. 7-29). The presternal blastema contributes to the manubrium, and the suprasternal centers sometimes do. In a few mammals (insectivores, edentates, rodents, and a few others) the suprasternal centers give rise to independent **suprasternal ossicles** that lie between the clavicle and the manubrium. Some human beings have suprasternal ossicles, but they are not found unless there is an occasion to x-ray the sternoclavicular joint. One might speculate that the presternal and suprasternal centers are vestiges of the median interclavicle and paired coracoids of the pectoral girdle of reptiles.

FIGURE 7-29

Mesenchymal blastemas that contribute to the amniote sternum. Presternal and suprasternal blastemas develop in mammals only. The ventral ends of developing ribs are also shown.

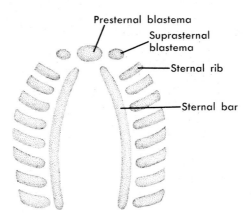

Presternal blastema
Suprasternal blastema
Sternal rib
Sternal bar

Chapter Summary

1. Typical vertebrae consist of centra, neural arches, one or more processes, and, in many tails, hemal arches.

2. Transverse processes (diapophyses) are present in most vertebrates. Tetrapods of one species or another have basapophyses, hypapophyses, parapophyses, and zygapophyses.

3. Fishes have only trunk and tail vertebrae. The trunk vertebrae of tetrapods are subdivided into cervical, dorsal, and sacral vertebrae. Dorsals are further subdivided into thoracic and lumbar when long curved ribs are limited to the anterior trunk.

4. In fishes the notochord persists within the adult vertebral column from skull to tip of tail. It is usually constricted within centra, expanded between centra, producing an hourglass-shaped amphicelous vertebra.

5. The notochord sheath of fishes and primitive urodeles is usually infiltrated with chordal cartilage, and the sheath is surrounded by perichordal cartilage (elasmobranchs), replacement bone (most fishes), or membrane bone (teleosts, apodans, urodeles) to complete the formation of a centrum.

6. The centra of other vertebrates form from multiple ossification centers in perichordal cartilage. All arches are endochondral bone.

7. The axial skeleton of holocephalans consists chiefly of rings of calcified cartilage deposited in the notochord sheath around an unconstricted notochord.

8. Dipnoans and chondrosteans have no centra. Paired basidorsal, interdorsal, basiventral, and interventral cartilages are aligned along the unconstricted notochord.

9. Cyclostomes have no centra. Lateral neural cartilages are perched above the notochord.

10. Diplospondyly is common in the vertebral column of many fishes and early tetrapods.

11. Rhachitomous vertebrae were probably ancestral to all tetrapod vertebrae. They consist of arches, a hypocentrum, and two pleurocentra. The notochord was persistent and the vertebrae were probably amphicelous.

12. Generalized urodeles, caecilians, and some primitive lizards have amphicelous vertebrae and retain much notochordal tissue. In mammals, notochord persists in intercentra as pulpy nuclei.

13. The vertebral column of tetrapods is provided with opisthocelous (specialized urodeles), procelous (anurans and modern reptiles), heterocelous (birds), and acelous (most mammals) vertebrae.

14. Amphibians have one cervical vertebra. The number is higher in reptiles, still higher in birds. Mammals typically have seven. The first two vertebrae in amniotes are the atlas and axis. A proatlas is found in some reptiles and a few mammals.

15. Sacral vertebrae bear stout transverse processes that brace the hind limbs and pelvic girdles against the vertebral column. Amphibians have one, reptiles and birds two, and mammals three to five sacral vertebrae.

16. Sacral vertebrae in amniotes usually unite to form a sacrum. The sacrum of birds unites with adjacent lumbar and caudal vertebrae to form a synsacrum.

17. Caudal vertebrae often bear hemal arches or chevron bones. They are reduced to archless centra near the end of the tail. Caudal vertebrae form a urostyle in anurans, a pygostyle in birds, and a coccyx in humans.

18. Some fishes have dorsal ribs, most have ventral ribs, and a few have both. Agnathans have none. Most ribs are of endochondral bone.

19. Most tetrapod ribs are bicipital with a tuberculum and a capitulum. Most thoracic ribs have costal and sternal segments.

20. Movable ribs are generally confined to the thorax except in limbless species. Short ribs are often fused with transverse processes.

21. Gastralia are the abdominal ribs of crocodilians and some lizards. They arise by intramembranous ossification.

22. Sterna are tetrapod structures. They are absent in limbless amphibians and limbless reptiles, and in turtles, and they are flimsy or absent in urodeles. Flying birds have a carina. Mammals have sternebrae.

23. Amniote sterna arise partly as paired sternal bars that later unite. Embryonic sternal ossicles contribute to the sternum or remain independent.

SELECTED READINGS

Goodrich, E.S.: Studies on the structure and development of vertebrates, London, 1930, The Macmillan Co., Ltd. (Reprinted by The University of Chicago Press, Chicago, 1986.) Although the theoretical discussions are out of date, the work is rich in skeletal morphology and abundantly illustrated.

Hoffstetter, R., and Gasc, J.P.: Vertebrae and ribs of modern reptiles. In Gans, C., Bellairs, A. d'A., and Parsons, T.S., editors: Biology of the reptilia, vol. 1, New York, 1969, Academic Press.

Jarvik, E.: Basic structure and evolution of vertebrates, vol. 1, New York, 1980, Academic Press. An in-depth study of *Amia* and the early fossil vertebrate *Eusthenopteron*.

Laerm, J.: The development, function, and design of amphicelous vertebrae in teleost fishes, Journal of the Linnaean Society of London, Zoology **58**:237, 1976.

Panchen, A.L.: The origin and evolution of early tetrapod vertebrae. In Andrews, S.M., Miles, R.S., and Walker, A.D., editors: Problems in vertebrate evolution, New York, 1977, Academic Press.

Stahl, B.J.: Vertebrate history: problems in evolution, New York, 1974, McGraw-Hill Book Co.

8 SKULL AND VISCERAL SKELETON

In this chapter we will find that vertebrates from fishes to human beings form their head skeleton out of three components: a cartilaginous braincase, ancient dermal armor, and contributions from the branchial skeleton. We will learn how these components are assembled, and note the new functions achieved by the branchial skeleton when vertebrates took up life on land.

The word "skull" is seldom misunderstood by the layman. To him, it is the bony structure that Hamlet held in his hand and gazed at dolefully as he spoke his now famous words, "Alas, poor Yorick." To the morphologist, however, the term poses a problem because of the intimate relationship in fishes between the skeleton that protects the brain and special sense organs of the head and the skeleton of the jaws and branchial arches. The latter has been inherited in modified form by higher vertebrates including, alas, poor Yorick. For this reason, the morphologist may avoid the term "skull" and refer instead to (1) the **neurocranium,** or **primary braincase;** (2) the **dermatocranium;** and (3) the **visceral skeleton,** or **splanchnocranium.** In this chapter, for practical purposes, "skull" will mean what the layman probably thinks it means, but without the lower jaw. We can then classify the parts of the cranial skeleton of vertebrates as follows.

Skull
  Neurocranium
  Dermatocranium
Visceral skeleton
  Embryonic upper jaw cartilage (palatoquadrate) and its replacement
    bones
  Embryonic lower jaw cartilage (Meckel's) and its replacement and
    investing bones
  Skeleton of the branchial arches

The upper jaw is visceral skeleton because it is part of the first visceral arch (Fig. 8-1). However, in bony vertebrates it becomes incorporated into the skull as development progresses.

THE NEUROCRANIUM

The neurocranium (sometimes called endocranium or primary braincase), is the part of the skull that (1) protects the brain and certain special sense organs, (2) arises as cartilage, and (3) is subsequently partly or wholly

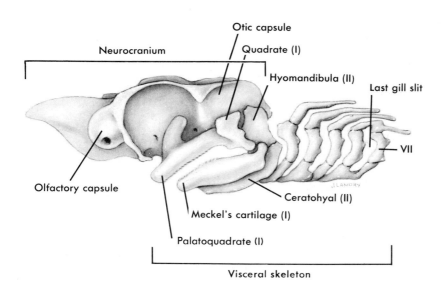

Neurocranium

Otic capsule

Quadrate (I)

Hyomandibula (II)

Last gill slit

Olfactory capsule

VII

Ceratohyal (II)

Meckel's cartilage (I)

Palatoquadrate (I)

Visceral skeleton

FIGURE 8-1

Skull and visceral skeleton of the shark *Squalus acanthias*. *I*, *II*, and *VII*, Skeleton of first, second, and seventh pharyngeal arches. Labial cartilages, gill rakers, and gill rays are omitted.

replaced by bone except in cartilaginous fishes. The neurocrania of all jawed vertebrates develop in a similar manner, as described below. It commences as several independent rudimentary cartilages that later, by expansion, become united to form a definitive cartilaginous braincase.

The Cartilaginous Stage

PARACHORDAL AND PRECHORDAL CARTILAGES AND NOTOCHORD. The neurocranium commences as a pair of parachordal and prechordal cartilages (Fig. 8-2, *A*) underneath the brain. Parachordal cartilages parallel the anterior end of the notochord beneath the midbrain and hindbrain. Prechordal cartilages (also called **trabeculae cranii**) develop anterior to the notochord underneath the forebrain. The parachordal cartilages expand across the midline toward each other and unite. In the process, the notochord and parachordal cartilages are incorporated into a single, broad, cartilaginous **basal plate.** The prechordal cartilages likewise expand and unite across the midline at their anterior ends to form an **ethmoid plate.**

SENSE CAPSULES. While parachordal and prechordal cartilages are forming, cartilage also appears in two other locations: (1) as an **olfactory (nasal) capsule** partially surrounding the olfactory epithelium and (2) as an **otic capsule** completely surrounding the otocyst, which is the developing inner ear (Fig. 8-2, *A* and *B*). The olfactory capsules are incomplete anteriorly, since water (in fishes) or air (in tetrapods) must have access to the olfactory epithelium. The walls of the olfactory and otic capsules are perforated by foramina that transmit nerves and vascular channels.

An **optic capsule** forms around the retina but it is not the orbit, or skeletal socket, in which the eyeball lies. It is the **sclerotic coat** of the eyeball. Although this capsule is fibrous in mammals, cartilaginous or bony plates

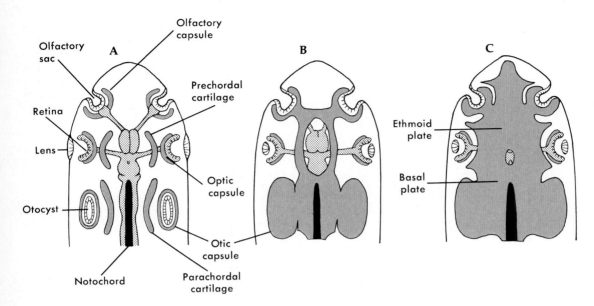

FIGURE 8-2

Early stages in development of a cartilaginous neurocranium, as seen from a ventral view. In **A** the notochord is seen underlying the midbrain and hindbrain. In **B** the notochord has been incorporated into the caudal floor of the neurocranium (basal plate). In **C** a cartilaginous floor has been completed beneath the entire brain, and a hypophyseal fenestra remains between the ethmoid and basal plates. The optic capsule will later become the sclerotic coat of the eyeball.

often form within the sclerotic coat (Fig. 16-17). This is an ancient condition, having been present in crossopterygians and extinct amphibians and reptiles. Because the optic capsule does not fuse with the rest of the neurocranium, the eyeball is free to move independently of the skull. Therefore the sclerotic coat is not conventionally considered part of the neurocranium.

COMPLETION OF FLOOR, WALLS, AND ROOF. The expanding ethmoid plate unites anteriorly with the olfactory capsules, and the expanding basal plate unites with the otic capsules, which are lateral to the hindbrain. The ethmoid and basal plates also expand toward one another until they meet to form a floor upon which the brain thereafter rests (Fig. 8-2, *C*). In the latter process there remains in the midline between the two plates a **hypophyseal fenestra** accommodating the hypophysis and internal carotid arteries, the latter being en route to the brain. Later the fenestra is usually reduced to a pair of foramina transmitting the arteries. Further development of the neurocranium primarily involves construction of cartilaginous walls around the brain and, in lower bony fishes and in amphibians, a partial or complete cartilaginous roof above the brain. The cranial nerves and blood vessels are already present by this time, and cartilage is deposited in such a manner as to leave foramina for these structures. The largest is the foramen magnum in the rear wall of the neurocranium. As the brain, blood vessels, nerves, and sense organs grow, remodeling of the neurocranial cartilage takes place.

The preceding basic pattern of development recurs in all vertebrate classes, producing a cartilaginous neurocranium that houses the brain, olfactory epithelia, and inner ear (Fig. 8-3). The mesenchyme that gives rise to the neurocranium comes from at least two sources. The prechordal car-

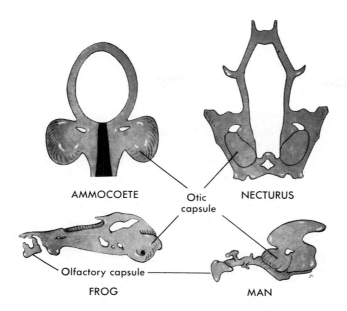

AMMOCOETE   Otic capsule   NECTURUS

Olfactory capsule

FROG   MAN

FIGURE 8-3

Cartilaginous neurocrania from selected embryonic, larval, or immature vertebrates. Dorsal view of ammocoete and *Necturus;* lateral view of frog and embryonic human.

tilages are formed from neural crest ectoderm that streams ventrad in front of the developing eyestalks (optic stalks). The mesenchyme that forms the parachordal cartilages and encapsulates the inner ear comes from mesoderm. The source of the mesenchyme that forms the olfactory and optic capsules has not yet been ascertained.

Cartilaginous Neurocrania of Adult Vertebrates

CYCLOSTOMES. The several cartilaginous components of the embryonic neurocranium remain more or less independent throughout life (Fig. 8-33). An olfactory capsule, otic capsules, a basal plate, and a notochord (not fused with the basal plate) are identifiable, and several other cartilages of unknown homology are also present. The roof above the brain remains unchondrified, hence fibrous.

CARTILAGINOUS FISHES. The neurocranium of *Squalus acanthias* is typical of cartilaginous fishes. It constitutes a high water mark in the development of neurocrania, for the embryonic components unite to form a boxlike adult cartilaginous braincase or **chondrocranium** in which the basic components have almost lost their identity. Walls are fully developed, and the brain is completely roofed by cartilage. The last portion of the roof to chondrify is an area just behind the rostrum, and in young skulls this may still be an unossified fenestra. (Some elasmobranch species have a fenestra at this location throughout life.) The otic capsules are inextricably fused into the posterolateral walls of the braincase, and the olfactory capsules are firmly united to it anteriorly. The notochord is visible on the ventral aspect as a ridge extending cephalad from the base of the foramen magnum. The hypophysis is cradled in a cartilaginous pocket, the sella turcica, beneath the brain. The neurocranium projects forward beyond the olfactory capsules as the rostrum.

On each side of the foramen magnum is an occipital condyle marking the site of the immovable articulation between the occipital region of the neurocranium and the first vertebra. On the posterodorsal aspect of the neurocranium, a depression, the endolymphatic fossa, exhibits two pairs of foramina that transmit endolymphatic and perilymphatic ducts, respectively. The endolymphatic ducts open onto the surface of the head. A well-developed, totally cartilaginous adult skull, devoid of dermal bones, is found only in Chondrichthyes.

LOWER BONY FISHES. The earliest fishes were bony animals. So, too, are almost all modern ones. A cartilaginous neurocranium is primarily an embryonic structure that may persist in restricted areas of an adult skull while becoming replaced by endochondral bone elsewhere. In the chondrosteans *Acipenser* (sturgeon) and *Polyodon* (spoonbill), in the holosteans *Amia* (bowfin) and, to a lesser degree, *Lepisosteus* (garfish), and in dipnoans a largely cartilaginous neurocranium remains throughout life. However, to see it, one must strip away the membrane bones that overlie it (Fig. 8-7). In teleosts and amniotes the embryonic cartilaginous neurocranium is largely replaced by bone during ontogeny. The ossification centers where this takes place will be discussed next.

FIGURE 8-4

Cartilaginous neurocranium of a fetal pig with chief endochondral ossification centers of mammals superimposed *(dots)*. The neurocranium is complete as shown, there being no cartilage above the brain. The ethmoid center becomes the cribriform plate perforated by olfactory foramina. The alisphenoid center is thought to be in the palatoquadrate cartilage in mammals. The otic centers are in the otic capsule.

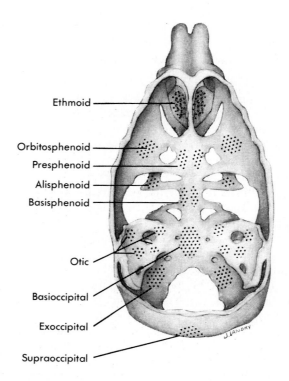

Ethmoid

Orbitosphenoid

Presphenoid

Alisphenoid

Basisphenoid

Otic

Basioccipital

Exoccipital

Supraoccipital

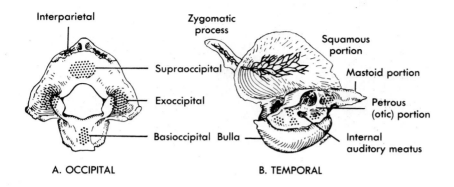

Interparietal

Zygomatic process

Supraoccipital

Exoccipital

Basioccipital Bulla

Squamous portion

Mastoid portion

Petrous (otic) portion

Internal auditory meatus

A. OCCIPITAL   B. TEMPORAL

FIGURE 8-5

Endochondral ossification centers *(dots)* and intramembranous ossification centers *(black networks)* superimposed on the occipital and right temporal bones of an adult cat. **A,** Caudal view. **B,** Medial view. The bulla arises from new cartilage not associated with the earlier neurocranium. The mastoid portion is an outgrowth of the petrous portion.

Neurocranial Ossification Centers

In bony vertebrates the embryonic cartilaginous neurocranium is mostly replaced by replacement bone. The process of endochondral ossification occurs more or less simultaneously at numerous separate ossification centers (Fig. 8-4). Although the specific number of such centers varies in different species, four regional groups are universally involved. These groups—occipital, sphenoid, ethmoid, and otic—will be discussed next.

OCCIPITAL CENTERS. The cartilage surrounding the foramen magnum may be replaced by as many as four bones. One or more endochondral ossification centers ventral to the foramen magnum produce a **basioccipital bone** underlying the hindbrain (Fig. 8-5). Centers in the lateral walls of the foramen magnum produce two **exoccipital bones.** Above the foramen a **supraoccipital bone** may develop. In mammals all four occipital elements usually fuse to form a single **occipital bone.** In modern amphibians, one or more of these may remain cartilaginous, although they were bony in stem amphibians.

The neurocranium of tetrapods articulates with the first vertebra via one or two **occipital condyles.** Stem amphibians had a single condyle borne chiefly on the basioccipital bone. Living reptiles and birds still have a single condyle. Modern amphibians and mammals diverged from the early tetrapod condition, gradually shifting the condyles from the median basioccipital to the two exoccipitals. The selective factors in the environment that caused this shift are unknown, and what the survival value might have been is purely speculative.

SPHENOID CENTERS. The embryonic cartilaginous neurocranium underlying the midbrain and pituitary gland ossifies to form a **basisphenoid bone** (anterior to the basioccipital) and a **presphenoid.** Thus a bony platform consisting of occipital and sphenoid bones underlies the brain. The side walls above the basisphenoid and presphenoid ossification centers form lateral sphenoid elements (**orbitosphenoid, pleurosphenoid,** and others*), and these may remain separate or unite with the basisphenoids and

---

*The alisphenoid of at least some mammals forms in the palatoquadrate cartilage rather than in the neurocranium.

FIGURE 8-6

Bony neurocranium *(red)* of human skull. The calvarium (roof) of the skull has been sawed off, and view is looking down into skull from above. Major endochondral ossification centers are labeled at left. The sphenoid bone exhibits an anterior (lesser) wing enclosing the optic foramina and a posterior (greater) wing between the two label lines for the sphenoid bone. The olfactory foramina are in the cribriform plate of the ethmoid.

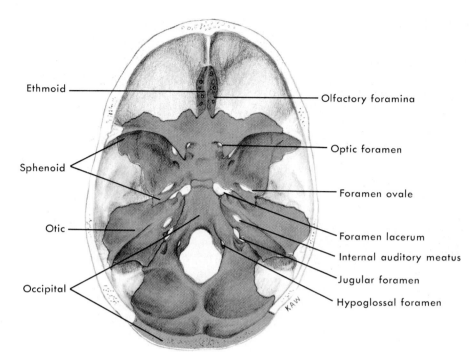

Ethmoid — Olfactory foramina

Optic foramen

Sphenoid — Foramen ovale

Otic — Foramen lacerum

Internal auditory meatus

Jugular foramen

Occipital — Hypoglossal foramen

presphenoids to form a single adult **sphenoid bone** with ''wings'' (Fig. 8-6). The pituitary rests in the sella turcica of the basisphenoid region. No replacement bones develop above the brain.

ETHMOID CENTERS. The ethmoid region lies immediately anterior to the sphenoid and includes the ethmoid plate and olfactory capsules. Of the four major ossification centers in the cartilaginous neurocranium (occipital, sphenoid, ethmoid, otic), the ethmoid more than the others tends to remain cartilaginous. Ossification centers in this region become the **cribriform plate** of the ethmoid, perforated by olfactory foramina (Fig. 8-6), and several of the **conchae,** or turbinal bones **(ethmoturbinals)** in the nasal passageways of crocodilians, birds, and mammals. **Mesethmoid** bones ossify in some mammals and contribute to the otherwise cartilaginous median nasal septum. In anurans the **sphenethmoid** is the sole bone arising in the sphenoid and ethmoid regions.

OTIC CENTERS. The cartilaginous otic capsule is replaced in lower vertebrates by several bones with such names as **prootic, opisthotic,** and **epiotic.** One or more of these may unite with adjacent replacement or membrane bones. For example, in frogs and most reptiles the opisthotics fuse with the exoccipitals and in birds and mammals the prootic, opisthotic, and epiotics all unite to form a single **periotic,** or **petrosal bone.** The petrosal, in turn, may unite with the squamosal, a membrane bone, to form a **temporal bone** (Fig. 8-5, *B*). Six ossification centers have been described in the otic capsule of a human fetus.

THE GENERALIZED DERMATOCRANIUM

The membrane bones of the skull constitute collectively the dermatocranium. After considering how the dermatocranium probably originated, we will examine its basic architecture in generalized vertebrates. The insight thus gained will enable us to relate the dermatocranium to the rest of the skull in modern tetrapods. Because the skulls of modern fishes are highly specialized and vary widely, we will mention them only occasionally.

How It May Have Begun

Much of the body of the earliest vertebrates was encased in bony dermal armor. As far back as ostracoderms this armor varied widely in the extent to which it covered the body—head and anterior trunk only, head and entire trunk, or head, trunk, and tail; in the relative size of the bones or scales making up the armor—large shields, smaller plates, minute scales; and in the relative size of the plates or scales on the head contrasted with those on the rest of the body. There is also evidence that some ostracoderms—cephalaspids, at least—passed through one or more cycles of expansion and reduction of their dermal armor, going from tiny scales to increasingly larger plates, then more or less reversing the trend; and that the reduction came earlier and was more complete on the trunk than on the head. Among jawed fishes, too, there have been those that lost most or all their dermal plates or scales except in the skin overlying the neurocranium and pectoral girdle. Even today, in the relict *Amia* and gars, the membrane bones on the top and sides of the head are still in the skin, where they overlie (invest, ensheath) the neurocranium (Fig. 8-7). The small cheek plates of gars (Fig.

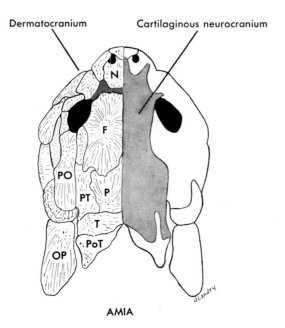

Dermatocranium   Cartilaginous neurocranium

AMIA

FIGURE 8-7

Skull of *Amia calva,* dorsal view. Dermal bones have been removed on the right side to reveal underlying cartilaginous neurocranium. *F,* Frontal bones; *N,* nasal; *OP,* operculum; *P,* parietal; *PO,* postorbital; *PoT,* posttemporal; *PT,* pterotic; *T,* tabular. The bones anterior to the nasals are ethmoids. Premaxillae are not visible in this view.

8-11, *A*) are either skull bones or scales, whichever you prefer to call them. It is as though natural selection has been preserving the armor that protected the brain and the sense organs of the head.

In modern vertebrates, these membrane bones of the head no longer ossify from mesenchyme *within* the dermis but from subdermal mesenchyme that is continuous with that of the developing dermis. It is as though fibroblasts and scleroblasts responsible for these bones differentiated deeper and deeper in the mesenchyme underlying the epidermis until the bones became part of the skull. *It is these bones that constitute the dermatocranium.* The embryonic mesenchyme that gives rise to them migrates into the embryonic head fold from lateral-plate mesoderm of the trunk. Neural crest cells also contribute to one degree or another in at least some vertebrate classes.

Its Basic Structure

For convenience of discussion the generalized dermatocranium can be divided into (1) bones that form above and alongside the brain and neurocranium (roofing bones), (2) dermal bones of the upper jaw, (3) bones of the primary palate, and (4) opercular bones.

ROOFING BONES. An early pattern for roofing bones is seen in crossopterygian fishes and labyrinthodonts, which differ little from ancestral bony fishes (Fig. 8-8). Roofing bones in these vertebrates provided a protective shield over the brain and special sense organs with openings only for the nares and eyes, including the parietal eye. In crossopterygians a series of paired and unpaired scalelike bones extended along the middorsal line from the nares to the occiput, overlying the brain, olfactory capsules, and any other neurocranial cartilage that developed in the area. In labyrinthodonts the unpaired bones were missing and a series of paired **nasals, frontals, parietals,** and **postparietals (dermoccipitals)** took their place. The postparietals disappeared, perhaps by coalescence with the supraoccipital, in phylogenetic lines diverging from the earliest reptiles. A parietal foramen housing the parietal eye is still present in many fishes, amphibians, and lizards.

Forming a ring around the orbit in the generalized skull were a **lacrimal, prefrontal, postfrontal, postorbital,** and **infraorbital (jugal).** The lacrimal bone derives its name from its relationship to the nasolacrimal duct of amniotes, which drains excess fluid, including tears (lachrymae) from the surface of the eyeball to the nasal canal. At the posterior angle of the skull were **intertemporal, supratemporal, tabular,** and, lower down, **squamosal** and **quadratojugal** bones. Labyrinthodonts developed a longer facial area, or snout, than is seen in crossopterygians by elongating the bones between the external nares and orbits. This could have been correlated with altered methods of capturing food or manipulating it in the mouth.

DERMAL BONES OF THE UPPER JAW. The earliest embryonic upper jaw, the palatoquadrate cartilage, is not a dermatocranial element but an endo-

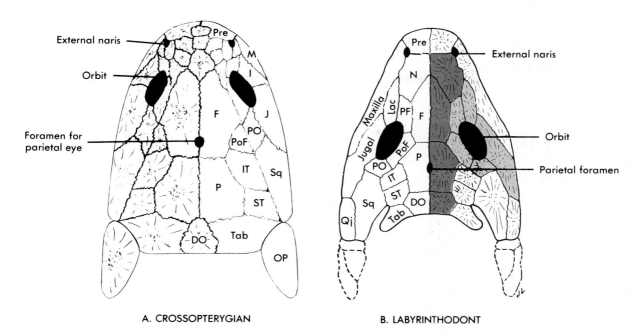

A. CROSSOPTERYGIAN   B. LABYRINTHODONT

FIGURE 8-8

Early dermal bone patterns from which tetrapod dermatocrania probably evolved. **A,** Skull of the crossopterygian fish *Eusthenopteron.* Note midline bones and small, scalelike bones in the rostral region. The location of the parietal foramen is in dispute. **B,** Skull of a Carboniferous labyrinthodont, representing the primitive tetrapod condition. Broken lines indicate deleted opercular bones. *DO,* Dermoccipital (postparietal); *F,* frontal; *I,* infraorbital; *IT,* intertemporal; *J,* jugal; *Lac,* lacrimal; *M,* maxilla; *N,* nasal; *OP,* opercular; *P,* parietal; *PF,* prefrontal; *PO,* postorbital; *PoF,* postfrontal; *Pre,* premaxilla; *Sq,* squamosal; *ST,* supratemporal; *Qj,* quadratojugal; *Tab,* tabular. In B, midline roofing bones are dark red, periorbital bones are light red.

chondral component of the visceral skeleton. It is the only upper jaw that cartilaginous fishes develop, and in many of these it moves independently of the neurocranium (Fig. 8-1). In bony vertebrates this cartilage becomes overlaid, or invested, by tooth-bearing dermal bones, including **premaxillae** and **maxillae,** and these become incorporated into the dermatocranium (Fig. 8-10). Thus the upper jaw, a part of the visceral skeleton, becomes part of the skull. The lower jaw is not incorporated into the skull, and will be discussed with the other components of the visceral skeleton.

PRIMARY PALATAL BONES. The primary palate is the roof of the oropharyngeal cavity of fishes and of the oral cavity of lower tetrapods. In sharks it is cartilaginous, being the floor of the neurocranium on which the brain rests. In bony vertebrates, membrane bones are applied to the underside of the neurocranium and to any part of the palatoquadrate cartilages (upper jaw cartilages) occupying the area. These membrane bones then become the major components of the primary palate.

In crossopterygians and early tetrapods, these membrane bones of the palate were an unpaired **parasphenoid** beneath the sphenoid region of the neurocranium, paired **vomers** beneath the ethmoid region, and paired **palatines, pterygoids,** and **ectopterygoids** laterally (Fig. 8-9). Primitively, teeth formed on all these bones, and several of them still bear teeth in lower vertebrates. Internal nares **(choanae)** pierced the palate anterolaterally. A primary palate is still present with modifications in all tetrapods (Figs. 8-12 and 8-13), but in those that also develop a secondary palate the primary palate remains in the roof of the nasal passageway (Fig. 8-18).

FIGURE 8-9

Primary palate *(red)* of a rhipidistian crossopterygian and a late Paleozoic labyrinthodont.

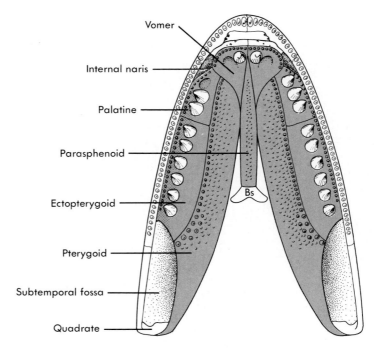

CROSSOPTERYGIAN

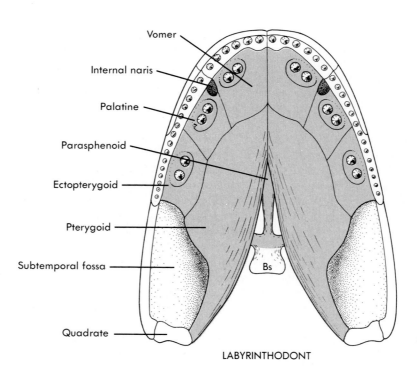

LABYRINTHODONT

OPERCULAR BONES. The operculum is a flap of tissue that arises as an outgrowth of the hyoid arch and extends caudad over the gill slits. It is membranous in holocephalans and absent in elasmobranchs. In bony fishes it is stiffened by squamous plates of dermal bone. Most constant are a large **opercular,** smaller **preoperculars** overlying the site of articulation of the upper and lower jaws, **suboperculars,** and **interoperculars** (Fig. 8-11). In some of the more primitive bony fishes, one or more **gular bones** follow the contour of the gill chambers (Fig. 8-10). They are represented in lungfishes and more specialized ray-fins by **branchiostegal rays,** seven in perches, located in a caudally directed flap, the **branchiostegal membrane** of the operculum. No opercular or gular plates or vestiges of them remain in tetrapods.

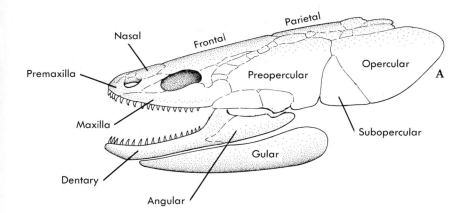

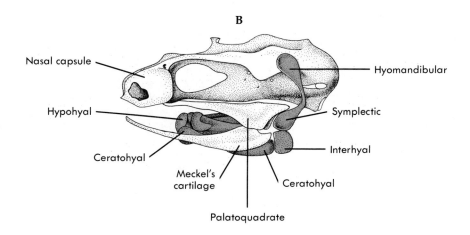

FIGURE 8-10

Skull of *Polypterus.* **A,** Lateral view of intact skull and lower jaw. **B,** Dermal bones stripped away to reveal neurocranium and the endoskeleton of the mandibular and hyoid arches. The premaxilla and maxilla invest the palatoquadrate, incorporating it immovably into the skull. The dentary and angular invest Meckel's cartilage. *Red,* skeleton of second visceral arch. (**A,** modified after Goodrich; **B,** from Kingsley, after Budgett.)

THE NEUROCRANIAL-DERMATOCRANIAL COMPLEX OF BONY FISHES

We will now examine the neurocranium and dermatocranium as they combine architecturally to contribute to the skulls of bony fishes. Later, we will study the skeleton of the pharyngeal arches to complete our picture of the cranial skeleton of fishes. The crossopterygian skull, a prototype for tetrapods, has been discussed earlier.

Chondrosteans

Although remote structurally and in time from early actinopterygians, *Polypterus* retains more features of their skulls than do other chondrosteans (Fig. 8-10). The neurocranium is well ossified, and it is covered along the middorsal line by paired nasals, frontals, and parietals of dermal origin. Lateral to these is a linear series of small, mostly unnamed bony dermal plates. (In instances where names have been assigned to the latter, there is no assurance that they are homologous with others of the same name.) Premaxillae and maxillae develop lateral to the palatoquadrate bones, which are the endoskeletal components of the upper jaw, and a preopercular bone lies lateral to the jaw joint. Other opercular bones complete the dermatocranium laterally, and a primitive gular bone lies in the opercular membrane beneath the gills.

The hinge of the jaw of *Polypterus,* which involves the posterior ends of the palatoquadrate and meckelian cartilages of the first visceral arch, is the basic jaw joint of all vertebrates except mammals (Fig. 8-10, *B*). As in other fishes, components of the hyoid arch skeleton are also involved in this joint.

Quite unlike *Polypterus,* sturgeons and spoonbills retain an almost completely cartilaginous neurocranium throughout life. The only traces of endochondral ossification in the neurocranium of sturgeons are isolated centers in the otic capsule and in the sphenoid bone where it contributes to the wall of the orbit. The bill of the spoonbill is an extension of the cartilaginous rostrum.

Holosteans

The skulls of *Amia* and garfishes have many of the characteristics of chondrostean skulls and bear little resemblance to the skulls of modern bony fishes (Figs. 8-7 and 8-11, *A*). The dermal bones are grooved and pitted, exhibiting the sculpturing effects of the underside of the dermis, rather than being smooth as in modern fish skulls; they can be characterized equally well as large and small scales and plates. The cheek plates of gars particularly are mute but compelling evidence that dermatocrania are derived from bony dermal scales that sank beneath the skin to invest the underlying endoskeleton. In *Amia,* the neurocranium remains highly cartilaginous, with only isolated centers of endochondral ossification. In gars it is ossified to a greater degree, although not matching the fully ossified neurocrania of teleosts. A detailed account of the anatomy of the skull of *Amia* has been provided by E. Jarvik.

Teleosts

The skulls of modern ray-finned fishes are highly specialized and architecturally diverse. This is correlated with the diverse feeding habits of the group, which includes species that harvest plant tissues or capture animal life in every conceivable niche of the streams, lakes, and oceans of the world. A combination of highly maneuverable jaws and palates, coupled with the largest number of bones in the skull of any vertebrate, accounts for the anatomical diversity.

The skull of a carp, a minnow-like cyprinid in the order Cypriniformes, is illustrated in Fig. 8-11. Like many teleost skulls, it is compressed laterally and vaulted dorsally. Most of the bones will be recognized from earlier discussions. The maxillae, premaxillae, dentary, articular, quadrate, and symplectic are associated with the jaws and hyoid arch and will be discussed with the visceral skeleton. The posttemporal is a common roofing bone of the dermatocranium of fishes. The mesethmoid, epiotic, and pterotic in teleosts in general usually incorporate replacement bone from the neurocranium and bone of intramembranous origin.

The neurocranium in most teleosts is fully ossified except for the olfactory capsules, which remain partly cartilaginous in all vertebrates. In cyprinids, however, neurocranial ossification is less than complete, and the islands of cartilage seen on the surface of the carp skull are unossified regions of the neurocranium or, at some sites, new cartilage.

Not seen in the illustration is the branchiostegal membrane beneath the gill chambers. In teleosts it contains branchiostegal rays instead of the more primitive gular bones of early ray-finned fishes.

Many of the names given to bones of teleost skulls have been assigned arbitrarily on the basis of location (epiotic, palatine), shape (pterygoid), or some other characteristic (squamosal). This presents a problem when a bone with the same name turns out to be an ossification center within cartilage—the palatoquadrate, for instance—in one vertebrate and a membrane bone in another. In such instances, the bones bearing the same name are not homologues. On the other hand, homologues have been given different names in different species, which poses problems of synonymy. Harrington (1955) has provided an annotated synonymy of teleost skull bones. However, it is in need of updating. Bones with the term pterygoid incorporated in their name are usually palatal bones. The bones so labelled on the illustration of the carp skull are of questionable homology.

A parasphenoid and vomer bone of intramembranous origin and a palatine that ossifies in the palatoquadrate cartilage are among common contributions to teleost palates.

Dipnoans

Following their divergence from the evolutionary line leading to tetrapods, dipnoans did not experience the burst of speciation undergone by teleosts, and changes in the architecture of their skulls have been more conservative. With respect to the dermatocranium, time seemingly has taken lungfishes in a direction opposite that of teleosts, whose dermal bones

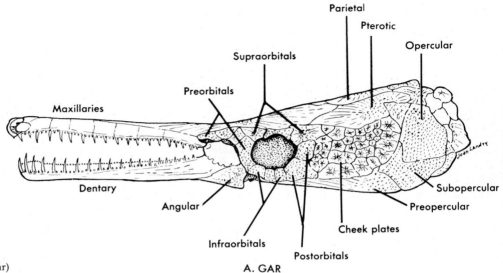

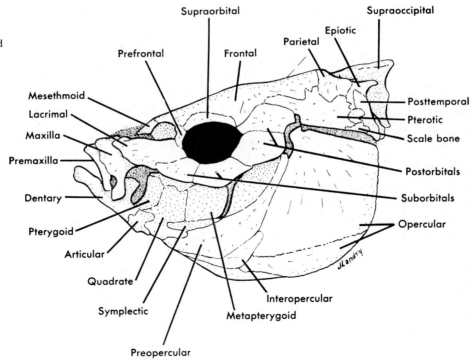

FIGURE 8-11

Skull of a holostean (gar) and of a modern fish (carp). Note the scalelike nature of the dermal bones in the gar and the series of maxillary bones. In the carp the dark stipple represents unossified cartilage.

A. GAR

B. CARP

increased in number and decreased in size. The dermatocranium of lung-fishes has evolved from a large number of scalelike dermal bones in fossils to a relatively few broad bony plates in today's species. The neurocranium is largely cartilaginous like that of some chondrosteans and holosteans, rather than bony, as in teleosts, and the palate has been reconstructed to accommodate an opening of the nasal canal into the oropharynx.

THE NEUROCRANIAL-DERMATOCRANIAL COMPLEX IN MODERN TETRAPODS

We can now look at the neurocranial-dermatocranial complex as it has evolved in modern tetrapods. A disproportionate amount of time will be spent on reptiles because they have diversified more than the other tetrapods and because during the Mesozoic the skulls of some of them underwent architectural changes that were incorporated into the skulls of birds and mammals.

Amphibians

The skulls of modern amphibians are considerably modified from those of labyrinthodonts (Fig. 8-8, *B*), although they are still **platybasic** (flattened) compared with the vaulted skulls of higher amniotes. Much of the neurocranium remains cartilaginous, except in apodans, in which rigidity may be correlated with burrowing. The only replacement bones in anurans and urodeles are a sphenethmoid, two prootics, and two exoccipitals, the latter bearing condyles. In perennibranchiate urodeles even the sphenethmoid fails to ossify.

The dermatocranium consists of fewer membrane bones than that of labyrinthodonts. The bones that once surrounded the orbit have been lost except for the lacrimals and prefrontals that are still present in primitive urodeles (Fig. 8-12, *A*). Also missing are the primitive bones of the temporal region from the orbit caudad (intertemporals, supratemporals, tabulars, postparietal). This leaves the otic capsule (prootic) exposed dorsally and laterally. Only the squamosal and, sometimes, quadratojugal remain in this region. Premaxillae and maxillae are usually represented in the upper jaw (the margin of the dermatocranium), but in perennibranchiates even maxillae fail to develop.

The primary palate has been altered. Ectopterygoids have been lost, and the pterygoids have been reduced to a pair of bipartite bones that brace the upper jaw against the braincase posteriorly (Fig. 8-12, *B*). In anurans, the palatines have been reduced to transverse splinters that brace the upper jaw against the palate anteriorly, and enormous palatal vacuities have developed (Fig. 8-13, *Rana*). As a result, the large eyeballs, which monitor the environment just above the water line when the frog is nearly submerged, can be retracted into the roof of the oral cavity. In urodeles, on the contrary, the parasphenoid has become exceptionally broad (Figs. 8-12, *B*, and 8-13, *Necturus*). Changes associated with loss of gills are discussed under "The visceral skeleton."

FIGURE 8-12

Skull of *Ranodon*, a primitive urodele in the family Hynobiidae. **A,** Dorsal view; **B,** palatal view. *Light red*, dermal bones; *dark red*, neurocranial bones; *stipple*, cartilage. *col*, Columella; *con*, occipital condyle; *eth*, sphenethmoid; *ex*, exoccipital; *fr*, frontal; *hy*, a dorsal segment of the hyoid skeleton; *lac*, lacrimal; *mx*, maxilla; *na*, nasal; *pal*, palatine; *par*, parietal; *pf*, prefrontal; *pmx*, premaxilla; *pq*, palatoquadrate cartilage; *pro*, prootic; *ps*, parasphenoid; *pt*, pterygoid; *qd*, quadrate; *sq*, squamosal; *vo*, vomer. *I* and *II* indicate origin from the first or second visceral arch. (After Schmalhausen, courtesy Academic Press. From Lebedkina.)

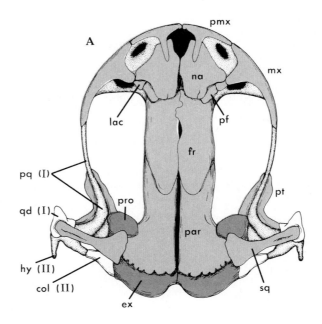

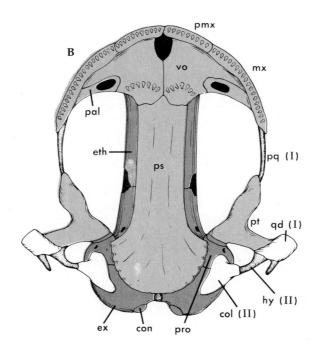

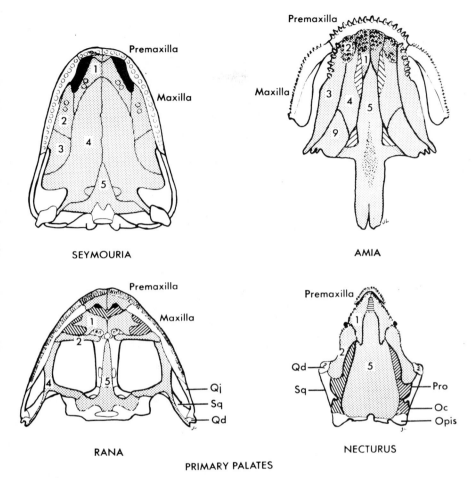

SEYMOURIA

AMIA

RANA

NECTURUS

PRIMARY PALATES

FIGURE 8-13

Primary palates of a reptile-like amphibian *(Seymouria)*, a lower bony fish *(Amia)*, and two amphibians *(Rana and Necturus)*. Cartilage is indicated by diagonal lines; internal nares are black, and palatal bones are stippled. *1*, Vomer; *2*, palatine (in *Necturus*, palatopterygoid); *3*, ectopterygoid; *4*, endopterygoid; *5*, parasphenoid; *9*, epipterygoid (of endochondral origin). *Oc*, Cartilaginous portion of otic capsule; *Opis*, opisthotic; *Pro*, prootic; *Qd*, quadrate; *Qj*, quadratojugal; *Sq*, squamosal.

Reptiles

The skulls of cotylosaurs were little changed from those of labyrinthodonts, and some of the primitive features are still present in modern reptiles. Among these are a full complement of neurocranial bones, a single occipital condyle, a larger complement of membrane bones than remains in other modern tetrapods (Fig. 8-14, alligator), and, in *Sphenodon* and many lizards, a parietal foramen housing a third eye. Among major modifications of early tetrapod skulls that are seen in reptiles are temporal fossae, a partial or complete secondary palate, and, in therapsid reptiles, two occipital condyles and increased prominence of the dentary bone. Fossae have been transmitted to birds through archosaurs, and all four modifications have been transmitted to mammals through therapsid reptiles.

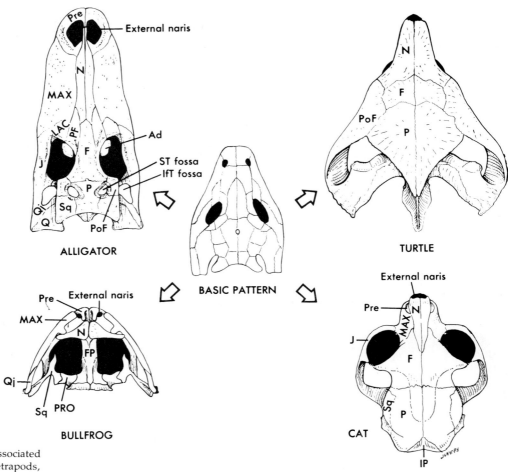

FIGURE 8-14

Roofing area and associated bones in selected tetrapods, dorsal views. The basic pattern represents a labyrinthodont. The turtle is an alligator snapping turtle, *Macroclemys temminckii*. The alligator skull has two temporal fossae, hence is diapsid. *F,* Frontal; *LAC,* lacrimal; *N,* nasal; *P,* parietal; *PF,* prefrontal; *PoF,* postfrontal; *Pre,* premaxilla; *Sq,* squamosal; *ST,* supratemporal fossa; *Qj,* quadratojugal; *Ad,* adlacrimal; *FP,* frontoparietal; *IfT,* infratemporal fossa; *IP,* interparietal; *J,* jugal; *MAX,* maxilla; *PRO,* prootic; *Q,* quadrate. As a study aid you may wish to color homologous bones on the different skulls with the same colors.

ROOFING BONES

TEMPORAL FOSSAE. A temporal fossa (temporal fenestra) is a cavity in the temporal region of the amniote skull bounded by one or more bony arches (Fig. 8-15). Stem reptiles had none, so their skulls are **anapsid,** that is, lacking a temporal arch. Today, only turtles among living amniotes lack temporal fossae (Fig. 8-16). Synapsida, the extinct mammal-like reptiles, developed a single **lateral temporal fossa** surrounded by the postorbital, squamosal, and jugal bones, the last two forming an underlying **zygomatic arch.** This **synapsid skull** was transmitted to mammals.

Archosaurs and the ancestors of *Sphenodon,* lizards, and snakes had **superior and inferior temporal fossae.** When there are two arches, the skull is **diapsid.** The lower arch corresponds to the zygomatic arch of synapsids. The upper arch, beneath the superior temporal fossa, is formed by the postorbital and squamosal bones. Crocodilians and *Sphenodon* still have diapsid skulls (Figs. 8-14, alligator, and 8-17, *A*), but modern lizards have

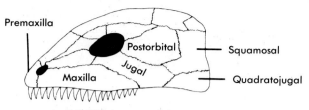

A. ANAPSID (stem reptile)

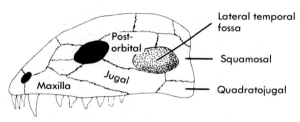

B. SYNAPSID (mammal stock)

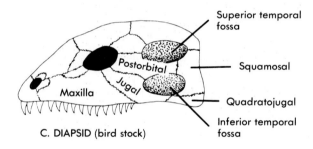

C. DIAPSID (bird stock)

FIGURE 8-15

Temporal fossae in reptiles leading to birds and mammals. The squamosal and postorbital bone in the diapsid skull form the superior temporal arch. The squamosal and jugals form the zygomatic arch in the synapsid skull.

FIGURE 8-16

Skull of small sea turtle, lateral view, lower jaw removed. *1*, Premaxilla; *2*, maxilla; *3*, jugal; *4*, quadratojugal; *5*, quadrate; *6*, prefrontal; *7*, frontal; *8*, postfrontal; *9*, parietal; *10*, squamosal; *11*, supraoccipital; *12*, middle ear cavity.

FIGURE 8-17
Diapsid **(A)** and modified diapsid **(B,C)** skulls of three reptiles.

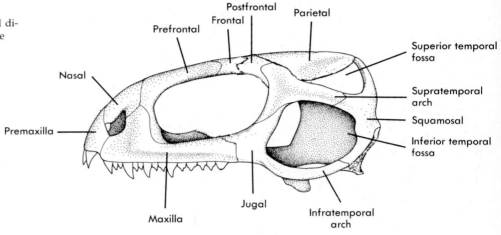

SPHENODON

A

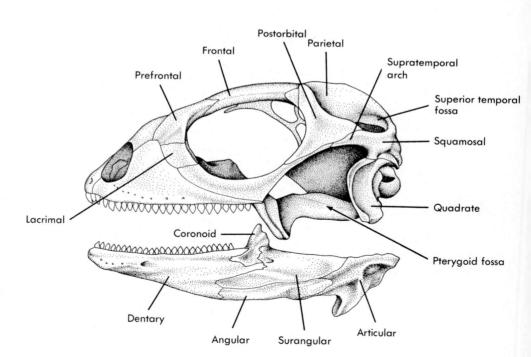

IGUANA

B

lost part or all of the lower arch, and snakes lost both arches, leaving a void in the lateral wall of the dermatocranium (Fig. 8-17, *B* and *C*). These are **modified diapsid skulls.** Correlated with erosion of the dermatocranium was an increase in cranial kinesis.

IIchthyosaurs and plesiosaurs had one dorsal fossa that may have been equivalent to the superior fossa of diapsids, since the postorbital bone met the squamosal below the fossa. This **euryapsid** condition no longer exists.

The temporal region of a turtle skull is an enigma. It has no fossa, which suggests a primitive reptilian condition. Nevertheless, there appears to have been considerable excavation at the rear. Supratemporal, tabular, and postparietal bones are missing, the postorbital bone has united with the postfrontal, and the parietal and squamosal bones have receded from the rear, leaving a wide vacuity that is seen best from a caudal or ventral view (Figs. 8-14, turtle, and 8-16). This raises a question as to whether turtle skulls are truly anapsid, or secondarily so. The question is unanswered.

Temporal fossae provide space and surfaces in functionally advantageous positions for accommodating the powerful adductor muscles needed to operate the lower jaws of amniotes. In labyrinthodonts and cotylosaurs the adductor mandibulae (chief levator muscle of the lower jaw) was confined to cramped quarters internal to the temporal region of the dermatocranium. (In cartilaginous fishes, as seen in Fig. 10-22, *A*, it lies just under the skin of the head because there is no dermatocranium.) At best, this muscle enabled bony fishes and primitive tetrapods to seize food, bite off pieces, and close the mouth to prevent return of the food to the environment. The food was swallowed whole. Temporal fossae provided space for

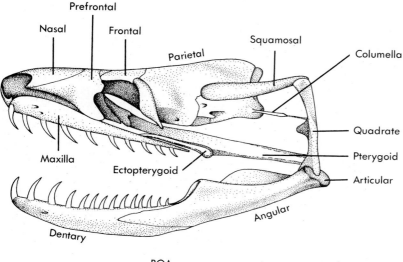

BOA

C

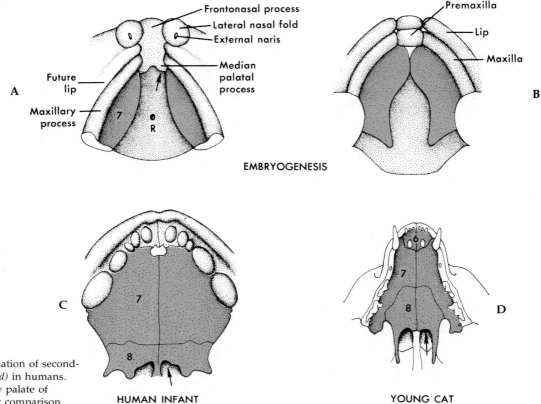

EMBRYOGENESIS

SECONDARY PALATES

HUMAN INFANT

YOUNG CAT

FIGURE 8-18

A to C, Formation of secondary palate *(red)* in humans. D, Secondary palate of young cat for comparison. Arrows indicate nasal passageways. *6,* Palatal process of premaxilla; *7,* palatal process of maxilla; *8,* palatal process of palatine bone. In A (fetus approximately 18 weeks old) the palatal processes of the maxillae are growing toward the midline, forming a secondary roof in the oral cavity. In B the processes have met anteriorly. In C the palate is complete. *R,* Rathke's pouch in primary palate.

an expanding adductor muscle to shorten and thicken during contraction; they provided exits to the surface of the dermatocranium so that one portion of the adductor, the temporalis muscle, could spread upward onto the temporal region of the skull; and they provided a bony infratemporal (zygomatic) arch to which a second portion of the adductor, the masseter, could acquire an anchorage. Fig. 10-22, C, shows the temporalis and masseter in relation to the temporal fossa of a primate. These powerful new adductors, assisted by other muscles of the mandibular and hyoid arches, make possible the complex side-to-side, forward-backward, and rotary chewing movements seen in herbivorous mammals that grind grasses or chew their cud and in carnivores that macerate flesh and crush bones.

SECONDARY PALATES. A secondary palate appears first in reptiles. It is a horizontal partition that divides the primitive oral cavity into separate oral and nasal passageways, thereby displacing the internal nares (posterior choanae) caudad (Fig. 11-3, cat). Embryonic development of a secondary palate in a mammal is illustrated in Fig. 8-18. In vertebrates with a secondary palate the primary palate remains in the roof of the nasal passageway,

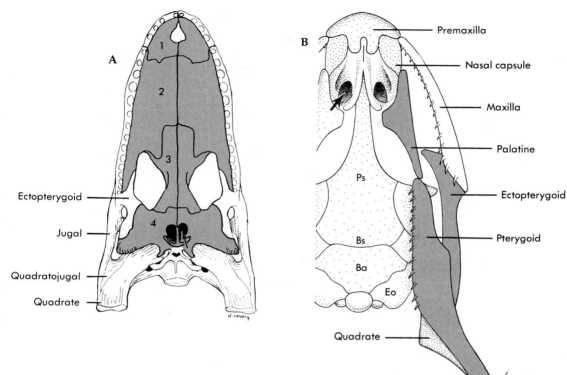

FIGURE 8-19

A, The long rigid secondary palate of an alligator *(red).* **B,** Kinetic palatal complex of the snake *Tropidonotus (red).* Arrows indicate location of internal nares. In **A:** *1, 2,* and *3,* Palatal processes of premaxilla, maxilla, and palatine bones, respectively; *4,* pterygoid. In **B:** *Ba,* basioccipital; *Bs* and *Ps,* ankylosed basisphenoid and parasphenoid contributions to primary palate; *Eo,* exoccipital. In **B,** the left and right palatine and pterygoid bones bound a median palatal fissure. A vomer bone, not visible from a ventral view, lies anterior to the parasphenoid. (**B,** from Kingsley, after W.K. Parker.)

usually with a reduced complement of membrane bones. For example, no parasphenoid bone is present in crocodilians or mammals.

In **crocodilians,** medially directed, shelflike, **palatal processes** of the premaxillae, maxillae, palatine, and pterygoid bones meet in the midline to form a completely bony secondary palate with internal nares far to the rear (Fig. 8-19, *A*). Crocodilians can complete the separation by operating curtainlike vela that hang down from the palate just anterior to the internal nares. **Mammals** are the only vertebrates other than crocodilians having a secondary palate that extends all the way to the pharynx, but in mammals the last part of it—the soft palate—lacks bone. The advantage of separating the air passageway from the food and water channel in crocodilians, whales, and suckling mammalian babies is discussed on p. 408. The advantage to any mammal that masticates its food is readily demonstrable. Try eating a hamburger while holding your nose closed!

In reptiles other than crocodilians, not all the palatal processes reach the midline, so the secondary palate is incomplete. Among **turtles,** varying degrees of completion can be observed by examining different species (Fig. 8-20). When the secondary palate is incomplete, the respiratory airstream is channelized in a fairly deep longitudinal groove, the **palatal fissure,** situated between the two palatine bones in the roof of the oral cavity (Fig. 8-20, *C*). The edges of the bones are covered by fleshy **palatal folds** seen also in birds (Fig. 11-4).

In modern **lizards** the quadrate bones, upper jaw, palate, and frontal and parietal bones have become movable as a unit independently of the rest

FIGURE 8-20

Species differences in the secondary palates (*gray*) of turtles. **A,** *Chelydra serpentina.* **B,** *Macroclemys temminckii.* **C,** *Lepidochelys olivacea*, temporal and supraoccipital regions omitted. *1,* Vomer; *2,* palatine of primary palate; *4,* pterygoid; *6, 7,* and *8,* palatal process of premaxilla, maxilla, and palatine bones, respectively. Note that, in **A,** only the maxilla has palatal processes and these are rudimentary. The palatal fissure is best illustrated in **C,** where it is bounded laterally by the palatine bones and roofed over by the vomer. Arrows point to internal nares.

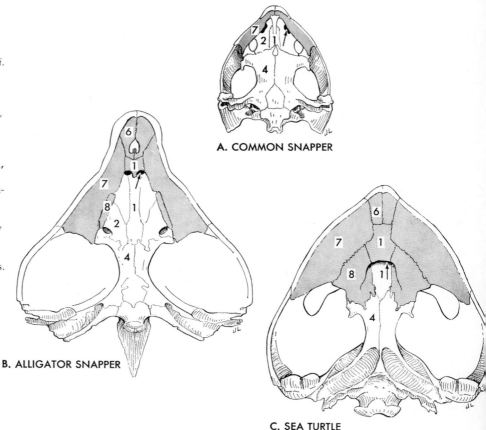

A. COMMON SNAPPER

B. ALLIGATOR SNAPPER

C. SEA TURTLE

FIGURE 8-21

A and **B,** One kind of kinetism in the palate of a snake; **C,** kinetism in a bird's skull. Note the hinge in front of the orbit between nasal bone and braincase. Note the relationship of the quadrate, pterygoid, and palatine in Fig. 8-22. Arrows indicate direction of movement as the mouth opens.

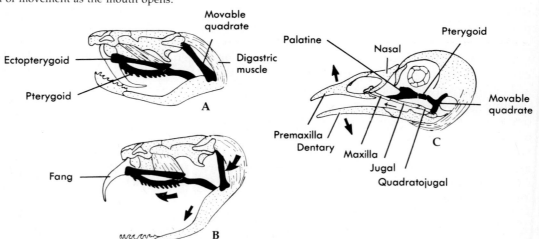

of the skull, a condition referred to as **cranial kinesis.** Joints within this complex and between the complex and the rest of the skull enable lizards to raise their upper jaw when they open their mouth, a movement that cannot be accomplished by an animal in which the upper jaw is fused with the braincase. Structural details vary among the species. There is a common belief that crocodiles can raise their upper jaws—it appears that way when a crocodile is in the water—but that is not the case. The crocodilian skull is not kinetic. Nevertheless, cranial kinesis of one variety or another is a common phenomenon in fishes, and it is seen also in caecilians, snakes, and birds. It was present in crossopterygians, and in some labyrinthodonts and extinct reptiles. Its advantage is not always clear.

Snakes have even more highly specialized palates than lizards. The palatine, pterygoid, and ectopterygoid of the primary palate and the quadrate of the upper jaw have become movable as a unit, with species variations, while the vomer and parasphenoid continue to undergird the neurocranium (Figs. 8-19, *B,* and 8-21). Cranial kinesis enables many snakes—especially vipers—to open their mouths wide enough to swallow animals larger than their own head. Whether snakes may be said to have a secondary palate is a matter of definition. As has already been mentioned, secondary palates arise as medially directed palatal processes of one or more of the primary palatal bones. What may be the beginning of, or a vestige of, a medially directed palatal process is seen on the palatine bone of the snake illustrated in Fig. 8-19, *B.* The fact is, the jaws, palate, and pharyngeal musculature of many snakes have become so highly specialized for admitting large prey into the oral cavity, for moving it caudad in the cavity from mouth to esophagus, and for swallowing it, that the generalized relationships of the dermatocranial-splanchnocranial complex have become greatly disrupted. It should also be added, parenthetically, that secondary palates probably evolved several times during the course of reptilian speciation.

TABLE 8-1   Skulls of Early Tetrapods Contrasted with Those of Modern Amphibians and Reptiles with Reference to a Few Selected Characteristics

	Early Tetrapods	Modern Reptiles	Modern Amphibians
Neurocranium	Well ossified	Well ossified	Mostly cartilage
	One condyle	One condyle	Two condyles
	Platybasic	Tropibasic	Platybasic
Primary palate	Complete complement of dermal bones	Relatively complete	Fewer
	Parasphenoid small	Small	Large in urodeles
	Vacuity small	Small	Large in anurans
	Internal nares lateral	Medial	Lateral
Secondary palate	None	Partial or complete	None
Dermal roofing bones	Complete complement	Some reduction	Extensive reduction
Parietal foramen	Present	Present in some	Confined to larvae
Marginal bones	Complete complement	Usually complete	Fewer
Bones ensheathing Meckel's cartilage	Numerous	Numerous	Fewer

Table 8-1 contrasts the skulls of early tetrapods and modern amphibians and reptiles with respect to the palates and a few other traits.

Birds

Differences between birds' skulls and those of their archosaur ancestors are correlated with altered feeding habits, a larger brain, reduction in thickness of all bones, loss of sutures, and massive orbits (Fig. 8-21, C). The neurocranium is well ossified, incomplete dorsally, and bears a single occipital condyle. There is a reptilian complement of dermal bones, but these have become thin, and most of the sutures between them have been entirely eliminated in adult skulls. Despite this, the architecture of the skull is such as to make it sturdy.

One might think of the avian skull as comprising two functional regions: at the rear, a solid bony box (neurocranium and dermatocranium) that houses the brain, olfactory organ, eyeball, and ear complex, and which therefore protects all structures needed for input and processing of information required during navigation; and at the front a visceral area, the elongated palate and beak, the primary role of which is procurement of food. The nonvisceral part of the skull, unlike that of reptiles, is vaulted **(tropibasic),** bulging outward and arching upward alongside the greatly enlarged brain. The orbits are at the anterolateral edges of the braincase. They reflect the size of the eyeball, which is uniquely shaped, massive, and has exceptional visual acuity. The orbit is open posteriorly to a cavernous fenestra resulting from loss of the temporal arch that separated the superior and inferior temporal fossae of archosaurs. The zygomatic arch is still intact, although very slim. Thus birds have a modified diapsid skull. The interorbital septum consists of a perpendicular, partly cartilaginous, mesethmoid bone—the only cartilage in the skull other than that of the olfactory capsule.

The visceral portion of the skull consists chiefly of a beak and palate. Premaxillae and dentary bones usually form the major part of the beak. The premaxillae, maxillae, and nasals are frequently fused to form a unit that articulates with the frontal bone in a hinge joint. Consequently this part of the beak is able to be raised independently of the rest of the skull (Fig. 8-21, C). The palate, too, is kinetic to one degree or another, resembling that of snakes except that the ectopterygoids have been lost (Fig. 8-22). When a bird with a kinetic palate opens its mouth by lowering the lower jaw, the quadrate bone is pushed forward and the motion is transmitted to the upper beak via either a movable palate, a movable zygomatic arch, or a combination of these. The parasphenoid is immobile, being fused to the basisphenoid. There are a multitude of variations in the structure of beaks and palates, and in their anatomical relationships to one another and to the braincase. For this reason, palatal structure has been used as one basis for establishing major avian taxa.

Not all bird skulls are equally kinetic. Birds such as woodpeckers that subject the beak to rough usage usually have minimal kinetism. The ethmoid region of the neurocranium and a narrow arch of roofing bones

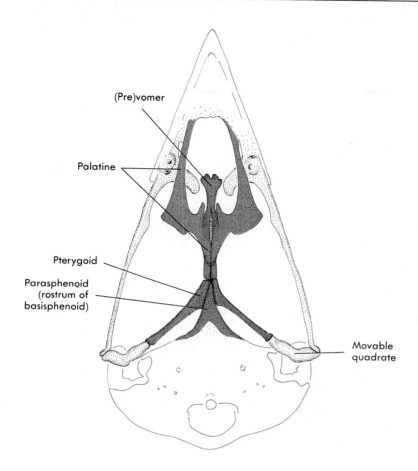

(Pre)vomer

Palatine

Pterygoid

Parasphenoid
(rostrum of
basisphenoid)

Movable
quadrate

FIGURE 8-22

Primary palate *(red)* of a bird, the cotinga, ventral view. The parasphenoid is immovable. The reptilian homologue of the prevomer (vomer?) is conjectural.

between the orbits bear the shocks generated when woodpeckers drill in solid wood. Some adaptations of the beak for procuring food are illustrated in Fig. 1-14.

Mammals

Major features that differentiate mammalian skulls from those of reptiles other than synapsids are emergence of the dentary as the sole bone of the lower jaw, modifications in the temporal region of the dermatocranium, an altered site of articulation of the lower jaw with the braincase, alterations in the secondary palate, and the presence of three bones (ear ossicles) in the middle ear cavity. The skull has become increasingly domed as the cerebral hemispheres ballooned dorsally, laterally, and caudad. The volume of the cranial cavity of modern man is larger than that of as recent a hominid as Neanderthal man.

The neurocranium is incomplete dorsally with the result that membranous soft spots, or **fontanels,** can be felt in the heads of newborn babies until ossification of the roofing bones is complete (Fig. 8-23). Fontanels

FIGURE 8-23

Two stages in the development of the human skull. **A,** Intramembranous ossification is under way. The neurocranium *(black)* is incomplete lateral to and above the brain. **B,** Intramembranous ossification has progressed, but "soft spots" (fontanels) remain where there is no cartilage or bone. The exoccipital and supraoccipital are neurocranial bones.

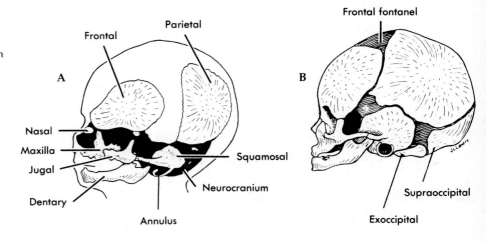

FIGURE 8-24

Chief endochondral *(red)* and dermatocranial components in a typical mammalian skull. The vomer, palatine, and pterygoid are parts of the primary palate. The premaxilla and maxilla contribute horizontal processes to the secondary palate. The dentary is a membrane bone of the visceral skeleton. The alisphenoid is said to be derived from the palatoquadrate cartilage or, occasionally, to be intramembranous in origin.

As, Pleurosphenoid (alisphenoid)
Ba, Basioccipital
Bs, Basisphenoid
C, Cribriform plate of ethmoid
E, Ethmoid, perpendicular plate
Ex, Exoccipital
F, Frontal
I, Interparietal
N, Nasal
Os, Orbitosphenoid
Ot, Otic (petrous)
P, Parietal
Pa, Palatine
PM, Premaxilla
Ps, Presphenoid
Pt, Pterygoid
So, Supraoccipital
Sq, Squamosal
Vo, Vomer

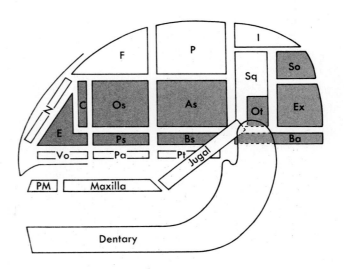

enable the fetal skull to be molded appropriately during delivery through the narrow birth canal. A small **bregmatic bone** ossifies in the frontal fontanel in some species and occurs as an anomaly in human beings. Paracelsus called it the "antiepileptic bone" because he believed it served as a pop-up valve for relieving pressure in the head. Occipital bones bear two occipital condyles inherited from therapsid reptiles. The basioccipital and sphenoid bones form a floor on which the brain stem rests, and ethmoid cartilages or bones underlie the olfactory bulbs, house the olfactory epithelium, and transmit olfactory nerve bundles. Ossification centers in the otic capsules form a pair of **periotic (petrosal) bones** that lie buried beneath the overgrown temporal lobes of the cerebral hemispheres. The endoskeletal components of a typical mammalian skull are diagrammed in Fig. 8-24.

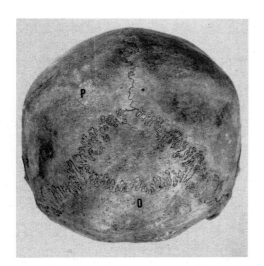

FIGURE 8-25

Inca bone in a human skull from the Aleutian Islands. *P*, Parietal; *O*, occipital. (Courtesy William S. Laughlin.)

The dermatocranium is usually represented in mammals by paired premaxillae, maxillae, jugals (malars), nasals, lacrimals, and squamosals; by paired or unpaired frontals and parietals; and by an unpaired interparietal, all of which are paired in embryos and, in some species, in neonates. A postparietal was found in *Homo erectus* and is still present in some human populations, chiefly Mongolians. It is sometimes called an **Inca bone** because it was common among Inca Indians (Fig. 8-25). Premaxillae are not identifiable in human skulls because they unite with the maxillae early in embryonic life, a discovery made by the poet-biologist Goethe, much to his "unspeakable joy." The zygomatic arch varies from massive to slender, depending on the power exerted by the masseter muscle. It has become extremely delicate, or even incomplete, in insectivores that do little chewing of their food.

The temporal complex of mammals consists of numerous components of intramembranous and endochondral origin (Figs. 8-26 and 8-27). The **squamous portion** is the squamosal of lower tetrapods. The **petrous portion** is the ossified otic capsule. The **tympanic portion,** new in mammals, surrounds the middle ear cavity and sometimes expands to form a **tympanic bulla** (Fig. 8-28). Associated with the tympanic portion is a bony ring, the **annulus tympanicus** (Fig. 8-37) derived, according to evidence from embryonic opossums, from the angular bone of reptiles.* The tympanic membrane, or eardrum, is attached to it. A mastoid portion of endochondral origin is new in mammals. Although the tympanic and petrous portions are separate bones in some mammals, they frequently unite to form a **petrotympanic bone,** as in rabbits, and the petrotympanic may unite with the squamosal to form a **temporal bone,** as in cats and man. The dorsal tip of

---

*The annulus of anurans is not a homologue. It is thought to be derived from the palatoquadrate cartilage.

FIGURE 8-26

Skull of modern man. The temporal bone is red. The parietal bone is separated from the frontal by the coronal suture, from the occipital by the lambdoidal suture, and is bounded ventrally by the squamosal suture.

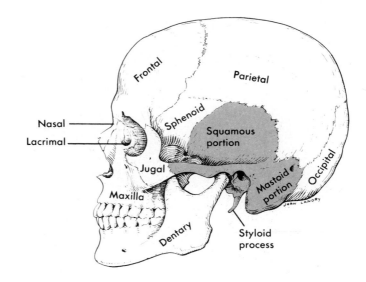

FIGURE 8-27

Schematic representation of the multiple nature of the temporal bone of mammals. Note reduction in number of separate elements from the condition in reptiles (outer circle) to mammals (other circles). The two dermal elements have asterisks. The mastoid portion and tympanic bulla are mammalian innovations. The ossicles are within the temporal bone but not part of it.

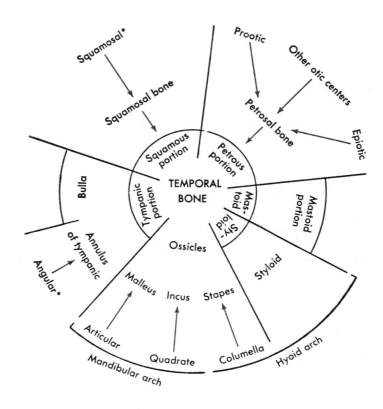

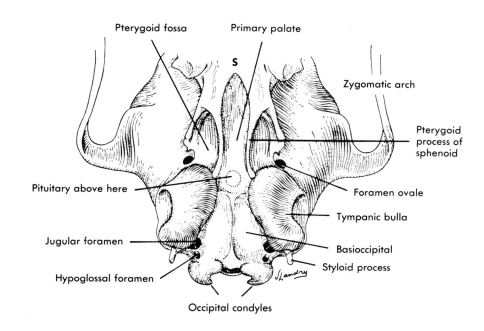

FIGURE 8-28

Hamster skull, caudal part, ventral view. *S*, Secondary palate. The primary palate is the roof of the nasopharynx.

Pterygoid fossa

Primary palate

S

Zygomatic arch

Pterygoid process of sphenoid

Pituitary above here

Foramen ovale

Tympanic bulla

Jugular foramen

Basioccipital

Hypoglossal foramen

Styloid process

Occipital condyles

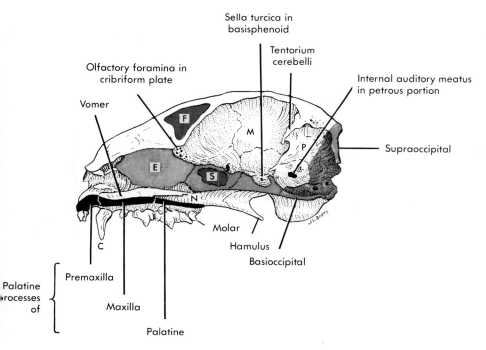

FIGURE 8-29

Sagittal section, cat skull, bony part of a secondary palate in black. *C*, Canine tooth; *E*, mesethmoid (perpendicular plate of ethmoid) in nasal septum; *F*, frontal sinus in frontal bone; *M*, middle cranial fossa housing cerebral hemispheres; *N*, nasal passageway; *P*, posterior cranial fossa housing cerebellum; *S*, sphenoidal sinus in presphenoid bone. Light gray designates ethmoid, sphenoid, and occipital components of the neurocranium.

Sella turcica in basisphenoid

Tentorium cerebelli

Olfactory foramina in cribriform plate

Internal auditory meatus in petrous portion

Vomer

F

M

P

Supraoccipital

E

S

N

Molar

C

Hamulus

Basioccipital

Palatine processes of

Premaxilla

Maxilla

Palatine

the skeleton of the hyoid arch sometimes coalesces with the temporal bone to become a **styloid process** (Fig. 8-26).

The squamosal bone has become a new site of articulation of the lower jaw with the skull in mammals. The shift from the quadrate-articular joint of other vertebrates is correlated with changes in the structure of the mammalian lower jaw (pp. 255 and 256).

Air-filled cranial sinuses are often found within the maxilla, frontal, sphenoid, and ethmoid bones (Fig. 8-29, *F* and *S*). The frontal sinuses of sheep and goats extend into the horns. When male goats butt heads at speeds of up to 35 miles (60 km) per hour as part of the mating ritual, the walls of the sinus act as a bony brace that shunts shock waves away from the brain to the vertebral column via the bones of the skull. Inflammation of the sinuses (sinusitis) is a common ailment in human beings.

Of the primary palate inherited from reptiles, an unpaired vomer lies at the base of the nasal septum, which consists of a **mesethmoid** bone and more or less cartilage (Fig. 8-29). A **nasal process of the palatine** bone now lies in the lateral wall of the nasopharynx where it contributes to the wall of the orbit, and a **palatal process** contributes to the secondary palate. The pterygoids are reduced to small, winglike, **pterygoid processes** of the sphe-

TABLE 8-2   Chief Foramina and Bony Canals in the Mammalian Skull

Foramen or Canal	Transmits
Incisive (anterior palatal)	Duct from oral cavity to vomeronasal organ
Infraorbital	Infraorbital branch of maxillary nerve and of internal maxillary artery
Jugular	Nerves IX, X, and XI; jugular vein
Lacerum	Internal carotid artery
Magnum	Spinal cord, vertebral arteries, and spinal roots of XI and XII
Mandibular	Inferior alveolar artery and nerve
Mental	Mental artery and nerve
Olfactory	Filia olfactoria
Optic	Optic nerve and ophthalmic artery from carotid plexus
Orbital fissure	Nerves III, IV, and VI and ophthalmic division of nerve V; in rabbits also transmits maxillary division of nerve V
Ovale	Mandibular division of nerve V
Rotundum	Maxillary division of nerve V; foramen is missing in rabbits
Stylomastoid	Exit of facial canal; transmits nerve VII
Facial canal	Nerve VII
Canal for auditory tube	Auditory tube enters tympanic bulla just lateral to foramen lacerum
Hypoglossal canal	Opens in caudal wall of jugular foramen; transmits nerve XII
Internal auditory meatus	Nerve VII to facial canal and nerve VIII to inner ear
Nasolacrimal canal	Nasolacrimal duct from orbit to nasal cavity

noid complex that serve as the anatomical origin of the pterygoid muscle, a derivative of the adductor mandibulae of lower vertebrates (Fig. 8-28). The parasphenoid and ectopterygoids have been lost. The mammalian secondary palate is illustrated in Figs. 8-18, 8-29, 11-3, and 12-12, *A*. The bony part (hard palate) consists of palatal processes of the premaxilla, maxilla, and palatine bones. Caudal to the bony part is a membranous soft palate.

Mammals have three pairs of scroll-like **turbinal bones,** or **nasal conchae,** extending into the nasal cavity from the maxillary, nasal, and ethmoid bones (Fig. 12-12, *A*). The maxillary and nasal conchae are covered with *nasal epithelium,* the venous plexuses of which warm inspired air en route to the lungs. The third pair of conchae, ethmoturbinals, are high in the nasal cavity and are covered with *olfactory epithelium.* Their turbinal structure traps some of the inspired air so that weak or highly diluted odorants are more likely to be detected. Most reptiles other than turtles have one pair of turbinals. Birds have two.

Unique in mammals is the presence of the posterior tips of the palatoquadrate and meckelian cartilages in the middle ear cavity, where they now serve as ear ossicles along with the dorsal tip of the hyoid arch skeleton. How they got there will be discussed shortly.

The chief foramina and bony canals in the mammalian skull are listed in Table 8-2.

Reduction in Number of Bones During Phylogeny

The number of individual bones, especially membrane bones, has tended to be reduced during the evolution of tetrapods. With reference to Fig. 3-3, any tetrapod group at the end of an arrow has fewer bones in the skull than the group preceding it in the phylogenetic line. Labyrinthodonts had fewer than crossopterygians, cotylosaurs had fewer than labyrinthodonts, modern reptiles have fewer than cotylosaurs, and mammals have fewer than mammal-like reptiles. Modern amphibians have fewer than their ancestors, the labyrinthodonts. This generalization does not mean that *modern* reptiles have fewer bones than *modern* amphibians. In fact, reptiles have more. But then, modern reptiles were not derived from modern amphibians.

The reduction is a result of fusion of adjacent embryonic ossification centers, phylogenetic loss of ossification centers, and obliteration of sutures in young animals. Reduction in the number of membrane bones in the mandible during phylogeny illustrates this trend (Fig. 8-36 and Table 8-3).

Membrane bones often unite with adjacent replacement bones, giving rise to a single bone with a dual history. Postfrontals and supratemporals sometimes unite with replacement bones of the otic capsule to form sphenotic and pterotic bones; the squamosal unites with otic and other elements to contribute to a temporal bone. The mammalian interparietal, a membrane bone, may unite with the supraoccipital. Unions such as these have reduced the number of bones in the skulls of recent tetrapods.

TABLE 8-3   Reduction in Number of Dermal Bones Investing Meckel's Cartilage When Early Vertebrates Are Contrasted with Later Ones

	Fishes			Tetrapods			
				Primitive	Modern		
Primitive	Crossopterygians	Modern	Labyrinthodonts	Reptiles and Birds	Amphibians	Mammals
Dentary	Dentary	Dentary*	Dentary	Dentary	Dentary	Dentary
Angular	Angular	Angular†	Angular	Angular	Angular‡	
Surangular	Surangular		Surangular	Surangular		
Infradentary§	Splenial		Splenial	Splenial	Splenial‡	
Infradentary	Coronoid	Derm-articular‖	Coronoid	Coronoid		
Infradentary	Prearticular		Prearticular			
Infradentary			Intercoronoid			
Infradentary			Precoronoid			
Infradentary			Postsplenial			

Primitive forms had a larger number of bones than modern ones. Reptiles have retained more of the primitive elements than other modern tetrapods.

*Dentary incorporates mentomeckelian of endochondral origin in some teleosts.
†May be absent. Sometimes named surangular.
‡Sometimes incorporated in an angulosplenial.
§Variable number.
‖May include articular of cartilage origin.

THE VISCERAL SKELETON

The visceral skeleton, or **splanchnocranium,** is the skeleton of the pharyngeal arches. In fishes, therefore, it is the skeleton of the jaws and gill arches. In tetrapods, this skeleton has become modified to perform new functions on land.

The blastemas that give rise to the visceral skeleton come from neural crests, and they first secrete cartilage. Later, the cartilage may be partly or wholly *replaced* by bone. Only in the first arch is it *ensheathed* by dermal bone. We will look first at a shark, in which no bone forms and in which the visceral skeleton is seen in its primitive role, that of supporting the jaws and gills.

Sharks

Squalus acanthias is a generalized vertebrate except that it lacks bones. The visceral skeleton consists of cartilages in each pharyngeal arch (Fig. 8-1) and median **basihyal** and **basibranchial cartilages** in the pharyngeal floor (Fig. 8-30). The skeleton in each arch conforms fairly closely to a basic pattern (Fig. 8-31, *A*), and all but the first and last in *Squalus* support gills. The first arch and, to a degree, the second, are modified for procuring food.

The skeleton of the mandibular arch consists of two cartilages on each side, a **palatoquadrate cartilage** dorsally and **Meckel's cartilage** ventrally

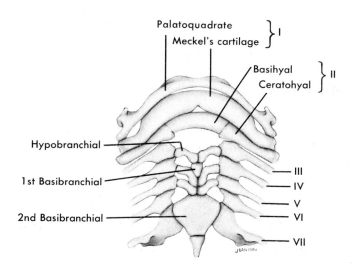

FIGURE 8-30

Visceral skeleton of *Squalus acanthias*, ventral view. *III* to *VII*, Ceratobranchial cartilages of the third to seventh pharyngeal arches.

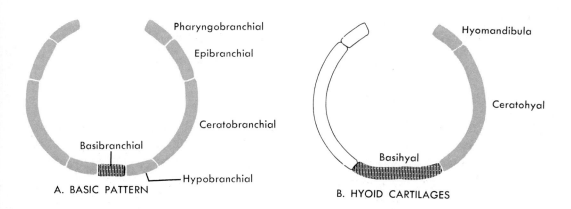

A. BASIC PATTERN

B. HYOID CARTILAGES

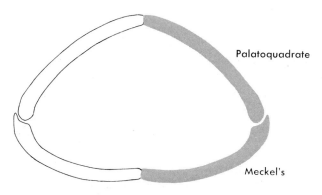

C. MANDIBULAR CARTILAGES

FIGURE 8-31

Skeletal components of a typical branchial arch, **A**, and modifications in the hyoid and mandibular arches of *Squalus acanthias*, **B** and **C**. Midventral elements in the pharyngeal floor are shown in white on black. The basihyal is paired in embryos.

(Fig. 8-31, C). The left and right cartilages meet in the midline to form the upper and lower jaws. Slender **labial cartilages** of unknown significance (not illustrated) extend from the angles of the mouth into a position where lips would be. The skeleton of the hyoid arch consists of paired **hyomandibular cartilages** dorsally and gill-bearing **ceratohyals** laterally (Figs. 8-1 and 8-31, B).

At the angle of the mouth, Meckel's cartilage and the palatoquadrate cartilage articulate with one another and with the hyomandibula in a movable joint united by ligaments (Fig. 8-1). The dorsal end of the hyomandibula is bound by ligaments to the otic capsule and suspends the jaws and the entire branchial skeleton from the neurocranium. This is **hyostylic jaw suspension.**

Bony Fishes

The visceral skeleton of bony fishes resembles that of sharks in general morphology (Fig. 8-32). Chief differences are that the embryonic palatoquadrate and Meckel's cartilages in bony fishes become invested by membrane bones, the hyoid skeleton may consist of more segments, the embryonic cartilages of the gill arches are replaced by bone, and the upper jaw becomes incorporated into the dermatocranial complex.

To gain a general impression of the palatoquadrate cartilage and its fate in bony fishes, the reader is urged to look once again at Fig. 8-10. In most bony fishes (1) this cartilage becomes *invested* laterally by two membrane bones, a premaxilla and maxilla (but see the relict gar, Fig. 8-11); (2) the portion in the roof of the oropharyngeal cavity, the palatal portion, develops two or three sites of endochondral ossification including **palatine** and **ectopterygoid** bones (Fig. 8-9, A); (3) the posterior tip ossifies to become a **quadrate** bone; and (4) the entire complex of ensheathing and replacement bones fuses with the rest of the dermatocranium to one degree or another, depending on the species.

Meckel's cartilages ossify at their caudal ends to become **articular bones.** The remainder of the cartilage becomes invested by several membrane bones including, always, a **dentary** and **angular,** and sometimes a **surangular.** The surangular was present in fossil teleosts but has since been lost. The mandible of the relict *Amia* consists of paired **dentaryinfradentaries** (a single composite bone), angulars, surangulars, **prearticulars,** and four pairs of **coronoids.** All but the angular and surangular bear teeth. In some fishes a short segment of Meckel's cartilage on either side of the mandibular symphysis ossifies to become a **mentomeckelian bone.**

The replacement bones of the hyoid arch of a relatively primitive teleost are shown in Fig. 8-32. The **symplectic** and **interhyal** are ossification centers in the embryonic hyomandibular cartilage, and the **epihyal** is an ossification center in the ceratohyal. The symplectic usually articulates with the quadrate bone, or with the quadrate and lower jaw. Its role in a kinetic fish skull will be discussed shortly.

The skeleton of a typical bony gill arch, like the gill arches of a shark, consists of four segments, **hypobranchial, ceratobranchial, epibranchial,**

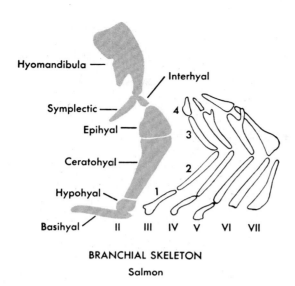

BRANCHIAL SKELETON
Salmon

FIGURE 8-32

Visceral skeleton of a salmon, upper and lower jaws removed. Hyoid cartilages are in red. The basihyal is unpaired. *1* to *4*, Hypobranchial, ceratobranchial, epibranchial, and pharyngobranchial elements of the third arch.

and **pharyngobranchial,** the latter in the pharyngeal roof (Fig. 8-31, *A*). Only the middle two segments bear gills. The pharyngeal borders of the branchial arches are covered with bony dermal plates with flattened or pointed denticles. The size and arrangement of the plates and denticles depend on the nature of the diet and the method of food procurement.

Feeding Mechanisms in Bony Fishes

Early voracious jawed fishes had wide mouths, the hinge of the jaws was far back under the skull, and the upper jaw was fused with the braincase and therefore incapable of independent movement. As a result, in feeding, the mandible was simply lowered and then snapped shut on prey in the manner of sharks. Whatever portion of the prey was inside the orobranchial chamber was swallowed.

With the appearance of the type of kinetism seen in the upper jaw, hyoid arch, and palate of teleosts, more complex mechanical means of procuring food became possible. As an accompaniment of these changes, the large adductor mandibulae inserted farther and farther forward on the mandible, an additional array of jaw muscles developed, and the mouth became narrower and more oval. The effect of these changes was that the upper and lower jaws of the most highly specialized teleosts can be thrust forward independently of the skull and employed in feeding by inertial suction. The mechanism works equally well in herbivorous and carnivorous fishes, although not in precisely the same manner.

Feeding by inertial suction in the herbivorous freshwater teleost *Petrotilapia* has been described by Liem. In its natural habitat this fish feeds on algae attached to submerged objects; in an aquarium it feeds equally well on commercial fish food. Abduction of the hyomandibula expands the orobranchial chamber, the lower jaw falls open, and the opercular cavity is then expanded, in that sequence, creating forward suction. As a result, zooplankton in the water immediately ahead are slowly drawn into the mouth. The lower jaw is then raised, the upper jaw remains briefly protruded, and the hyomandibula is adducted, which compresses the chamber. The two phases of the pharyngeal pump, expansion and compression, occupy together about 600 milliseconds. Activity of the muscles was recorded by electromyography. By manipulating the hyoid skeleton and jaws appropriately, the protruded jaws can be directed upward for collecting food on the surface, or downward for food on the substrate.

A similar mechanism operates in predatory fishes. When the hyomandibula is drawn forward, the symplectic forces the kinetic elements of the upper jaw and palate to slide forward, and the premaxillae and dentary bones close on the prey. A process of the maxilla usually becomes attached by a ligament to the dentary bone, so that movement of one participates in displacing the other. The details of the anatomical relationships and the mechanics of their operation are as numerous as the taxa that exhibit them. It should be pointed out, however, that snapping and biting has not gone out of style. One half or more of the teleosts have nonprotrusible jaws.

Many years ago it was proposed that the jaws are branchial arches that became modified for predatorial feeding as an alternative to filter feeding. There is some evidence for this in that the innervation, blood supply, and muscles of the mandibular arch are the most anterior of a series of structures of similar ontogeny and relationships that are repeated in each gill arch, as will be seen in later chapters. The concept, although generally accepted, is based for the most part on conjecture.

Hyostyly, Amphistyly, and Autostyly

The jaw-hyoid complex, whether of a shark, teleost, or tetrapod, must be braced against some support, and the nearest in fishes is the braincase. In elasmobranchs and most actinopterygians the hyomandibular cartilage is braced against the otic capsule and the jaws are braced against the hyomandibula (Fig. 8-1). As mentioned earlier, the condition is known as **hyostyly.** A more primitive condition is seen in some older sharks in which jaws and hyoid are both braced directly against the braincase, a condition known as **amphistyly.** A third variant is seen in lungfishes and chimaeras, in which the upper jaw is incorporated immovably into the braincase and the hyomandibula plays no role in jaw suspension. This condition has been called **autostyly.** All tetrapods are autostylic. More specialized terminologies are employed by specialists in fish morphology.

Cyclostomes

The visceral skeleton of cyclostomes is totally unlike that of jawed fishes (Fig. 8-33). There are no identifiable palatoquadrate or Meckel's cartilages and, of course, cyclostomes lack bone. A V-shaped **dental plate (lingual cartilage)** bearing horny teeth and located in the floor of the buccal cavity

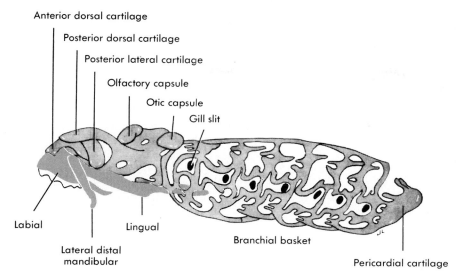

Anterior dorsal cartilage

Posterior dorsal cartilage

Posterior lateral cartilage

Olfactory capsule

Otic capsule

Gill slit

Labial

Lingual

Lateral distal mandibular

Branchial basket

Pericardial cartilage

NEUROCRANIUM and VISCERAL SKELETON
Lamprey

FIGURE 8-33

Neurocranium and visceral skeleton of a lamprey. Red elements are the "jaws." The olfactory capsule is a midline structure, otic capsules are paired. The lingual cartilage is sometimes referred to as a tongue.

moves in and out of the buccal funnel, serving as a rasping tonguelike organ. It is operated by protractor and retractor muscles that are attached to a narrow, segmented, immovable **basal plate** cartilage lying immediately beneath it (not illustrated). There is insufficient evidence that any of these cartilages are derived from the first visceral arch. As for an upper jaw, an early study of the visceral skeleton of hagfishes led to the conclusion that a rudimentary upper jaw may be fused with the neurocranium, but this remains hypothetical. The dental plate, or "tongue," and associated cartilages are adaptations for a unique method of feeding. The rest of the visceral skeleton consists of a branchial basket, less well developed in hagfishes, which is simply a fenestrated framework of continuous cartilage immediately under the skin surrounding the pharynx. The caudal end of it in lampreys is located at the level of the heart.

Tetrapods

With life on land, the visceral skeleton, and particularly the branchial skeleton, underwent profound changes from that of jawed fishes. Some previously functional parts were deleted, and those that persisted perform new and sometimes unexpected functions. It is these modifications, which took place over the course of millions of years, that we will examine in the pages that follow. But first, we will look at some of these changes in a frog tadpole, where they occur during the course of two or three days!

Larval frogs have six pairs of visceral cartilages, the last four of which bear gills. The cartilages of the gill arches meet ventrally in a **hypobranchial plate** (Fig. 8-34, *A*). During metamorphosis (Fig. 8-34, *B* and *C*) the third,

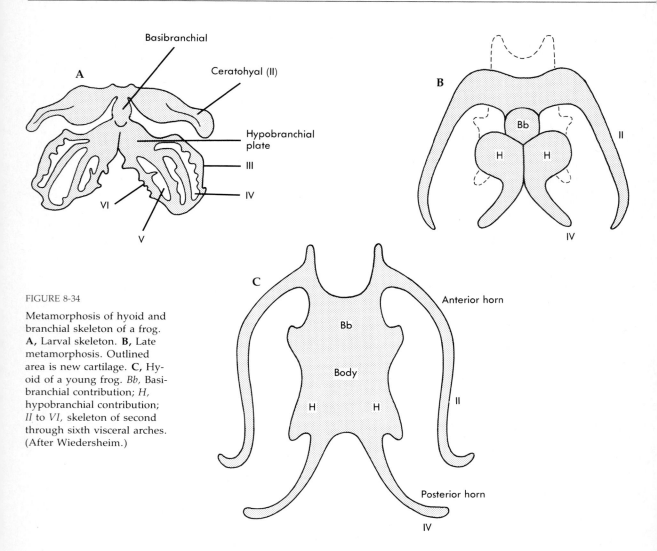

FIGURE 8-34

Metamorphosis of hyoid and branchial skeleton of a frog. **A,** Larval skeleton. **B,** Late metamorphosis. Outlined area is new cartilage. **C,** Hyoid of a young frog. *Bb,* Basibranchial contribution; *H,* hypobranchial contribution; *II* to *VI,* skeleton of second through sixth visceral arches. (After Wiedersheim.)

fifth, and sixth visceral cartilages regress. The hypobranchial plate enlarges, and, along with the first basibranchial, becomes incorporated into the body of the hyoid. The latter thereafter is a broad cartilaginous and bony plate in the buccopharyngeal floor. The ceratohyal cartilage of the second arch is reduced to a slender anterior horn, or cornu, of the hyoid apparatus, and the cartilage of the fourth arch becomes a posterior horn. As a result of these and other changes, a visceral skeleton initially adapted for branchial respiration becomes converted, in the span of a few days, to one adapted for life on land. Among its new roles is serving as an anchorage for the muscular tetrapod tongue. Perennibranchiate amphibians, of which *Necturus* is an example, retain a reduced gill-bearing branchial skeleton throughout life.

FATE OF THE PALATOQUADRATE AND MECKEL'S CARTILAGES. We have already seen in bony fishes that the embryonic palatoquadrate cartilage becomes ensheathed by dermal bones (premaxillae and maxillae), that the palatine portion contributes in one degree or another to the primary palate, and that the posterior end of this cartilage ossifies to become a quadrate bone at the hinge of the jaws. The story is the same for amphibians, reptiles, and birds (Figs. 8-12, *A,* 8-16, 8-19, *B,* and 8-21, *C*). The probable fate of the quadrate region in mammals—that it becomes the incus, an ear ossicle—will be discussed shortly.

The embryonic Meckel's cartilage of tetrapods may continue to grow and become a prominent core of cartilage within the mandible, as in turtles and crocodilians (Fig. 8-35), but more often little remains in adults. However, it becomes invested by dermal bones which included, primitively, a **dentary, angular, surangular, splenial,** one or more **coronoids,** and a **prearticular.** Modern tetrapods have a reduced number, and mammals have only the dentary (Fig. 8-36).* As in bony fishes, the posterior end of Meckel's cartilage ossifies to become an **articular bone** at the hinge of the jaws, except in mammals. In mammals it becomes the malleus, a middle ear ossicle (p. 257).

EXPANSION OF THE DENTARY AND A NEW JAW JOINT IN MAMMALS. As explained earlier, temporal fossae enabled the adductor mandibulae muscles of synapsid reptiles to enlarge, subdivide, and acquire anatomical origins on the temporal region of the skull and on the zygomatic arch. The increased mass of these new muscles resulted in expansion of the dentary bone where the muscles inserted. The changes in the dentary can be followed in successive generations of therapsid reptiles (Fig. 8-36, *B* and *C*). A ramus developed on the dentary and then expanded increasingly upward in the temporal fossa toward the origin of the temporalis muscle (Figs. 8-36, *D,* and 10-22). Eventually the other dermal bones of the mandible disappeared, the articular bone became an ear ossicle, and mammals were left with a mandible consisting solely of two dentary bones.

---

*The mentomeckelian bone at the mandibular symphysis of anurans is intramembranous in some species, endochondral in others.

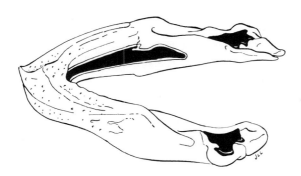

FIGURE 8-35

Mandible of an adult sea turtle, from the left and above, showing core of Meckel's cartilage *(black)* ensheathed by membrane bone.

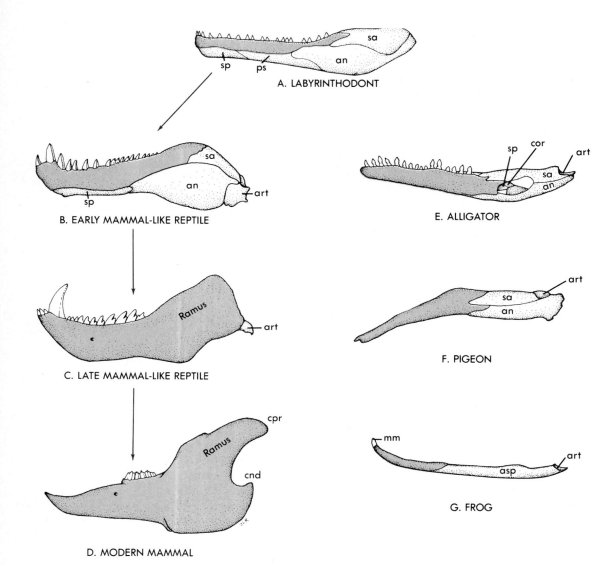

FIGURE 8-36

The mandibles of tetrapods, left lateral views. **A** to **D,** Probable evolutionary stages leading to modern mammals. The dentary *(red)* became increasingly larger, whereas other bones were reduced and finally lost. Arrows indicate phylogenetic pathways. **E** to **G,** Lower jaws of three modern tetrapods for comparison with primitive pattern. All are dermal bones except the mentomeckelian and articular. *an,* Angular; *art,* articular (cartilage in frog); *asp,* angulosplenial; *cnd,* condyle for articulation with squamosal; *cor,* coronoid; *cpr,* coronoid process; *mm,* mentomeckelian; *ps,* postsplenial; *sa,* surangular; *sp,* splenial. **B,** Pelycosaur; **C,** late therapsid; **D,** rabbit.

Expansion of the dentary brought it close to the squamosal, and there it eventually established a new site for articulating with the skull (Fig. 8-26). For a while the lower jaw of evolving mammals articulated at two sites, the reptilian one (articular against the quadrate) and the new one (condyloid process of the dentary against the squamosal). The two existed side by side in *Eozostrodon,* one of the oldest known mammals of the late Triassic. Eventually the articular and quadrate were "captured" by the middle ear cavity and only the new articulation remained. The shape, slant, and relationships of the condyloid process of the mandible differ with the demands made on it by the feeding habits of the various mammalian orders.

EAR OSSICLES FROM THE HYOMANDIBULA AND JAWS. It will be recalled that the hyomandibular cartilage of sharks is interposed between the quadrate cartilage and the otic capsule, and that the latter houses the inner ear (Fig. 8-1). In an autostylic jaw the hyomandibular bone can be dispensed with, and that is what happened to it in extant lungfishes. It persisted in tetrapods, but lost its articulation with the quadrate except in urodeles, becoming attached instead to the newly formed tympanic membrane (eardrum). The other end continued to abut against the otic capsule, and the entire hyomandibular became surrounded during embryogenesis by an evagination of the first pharyngeal pouch, the middle ear cavity (Fig. 17-19). The hyomandibula thereby became the first ear ossicle, or **columella,** conducting sound waves from the eardrum to the inner ear in tetrapods below mammals. Fig. 8-12 shows the columella still interposed between the quadrate and the otic capsule (prootic bone) in a primitive urodele. These amphibians lack tympanic membranes, and their middle ear cavities are vestigial.

When the dentary bone of therapsid reptiles acquired an articulation with the squamosal the articular and quadrate bones were free to perform a new function or to disappear. There is no doubt that the caudal end of Meckel's cartilage becomes the **malleus** in mammals. The embryonic Meckel's cartilage in one mammalian species after another can be seen projecting into the area where middle ear cavitation is proceeding, and its cartilaginous tip can be observed to separate, ossify, and become the malleus (Figs. 8-37 and 16-13).

The evidence that the quadrate bone of synapsid reptiles became the incus of the middle ear of mammals is less direct, but the facts are: (1) the articular and quadrate bones have been articulating in a diarthrosis since the time of the earliest jawed fishes, and they were doing so in therapsid reptiles; (2) the articular has separated from the lower jaw and is now the malleus; (3) the quadrate has disappeared from the posterior tip of the upper jaw; and (4) the articular still articulates in a diarthrosis with a bone, the **incus,** an ear ossicle which, if it is not the quadrate, is of unknown origin and homology (Fig. 16-13). In 1837, C. Reichert proposed the theory that mammalian ear ossicles are derived from the jaws, and in 1913 the theory was modified by E. Gaupp to include the origin of the stapes from the hyomandibula. Subsequent knowledge from paleontology and embry-

FIGURE 8-37

Posterior tip of Meckel's cartilage and associated ear ossicles surrounded by the developing middle ear cavity *(gray)* in a mammalian embryo.

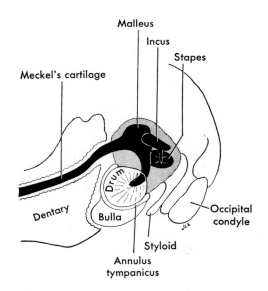

ology has strengthened the Reichert-Gaupp theory, and it is now generally accepted.*

THE TETRAPOD HYOID. The hyoid consists of a median plate, or **body,** at the base of the tongue in the pharyngeal floor just anterior to the larynx, and two or three paired **horns** in the pharyngeal walls (Figs. 8-34, 8-38, *B* to *H,* and 8-39). The body is derived from basihyal and basibranchial cartilages; the anterior horns develop from the ceratohyal of the second pharyngeal arch; and the more caudal horns arise from the third and, frequently, the fourth arches.

In lizards and birds the body of the hyoid is narrow, and an elongated process extends forward into the long darting tongue as an **entoglossal bone** (Fig. 8-38, *C* and *E*). In some male lizards (anoles and related genera) a similar process extends caudad into the dewlap (gular pouch). In snakes the entire branchial skeleton is vestigial.

The hyoid of mammals has two horns, an anterior pair from the second arch and a posterior pair from the third. In cats the anterior horns are the longer ones (greater horns) and are composed of four segments (Fig. 8-38, *G*). The dorsalmost, or **tympanohyal,** ends in a notch in the tympanic bulla. (The stapes, which became detached from it during phylogeny, is inside the bulla.) In humans the anterior horns are the shorter ones (lesser horns), an unossified **stylohyoid ligament** represents the middle segment, and the tympanohyal is attached to the temporal bone as a **styloid process** (Figs. 8-26 and 8-40). In rabbits, too, the anterior horn is shorter (Fig. 8-39), and the tympanohyal is represented by a slender **stylohyal bone** embedded in the tendon of insertion of the stylohyoid minor muscle.

*For a critique of the theory and an alternate one see Jarvik, vol. 2, p. 161.

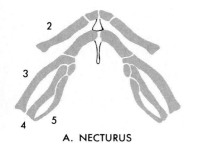

A. NECTURUS

2
3
4   5

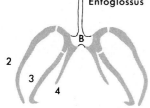

2
3
4

C. LIZARD

Entoglossus

B

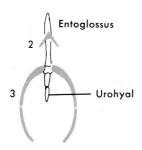

Entoglossus

2

3   —— Urohyal

E. CHICKEN

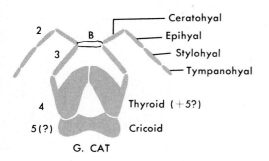

2
3
4
5(?)

B

Ceratohyal

Epihyal

Stylohyal

Tympanohyal

Thyroid (+5?)

Cricoid

G. CAT

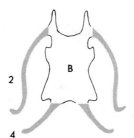

2
4

B. FROG

B

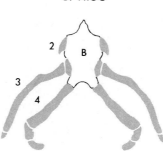

2
3
4

B

D. TURTLE

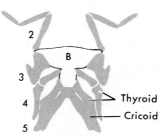

2
3
4
5

B

Thyroid

Cricoid

F. MONOTREME

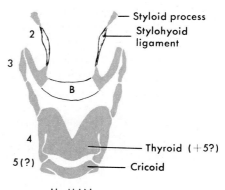

2
3
4
5(?)

Styloid process
Stylohyoid
ligament

B

Thyroid (+5?)

Cricoid

H. MAN

FIGURE 8-38

Skeletal derivatives of the second through fifth pharyngeal arches in selected tetrapods. **B,** Body of hyoid. *2* to *5,* Derivatives of arches 2 through 5. The projections from the body in **B** to **H** are the horns of the hyoid. In **E,** the body of the hyoid extends forward into the tongue as an entoglossus to which are attached two paraglossals *(2).*

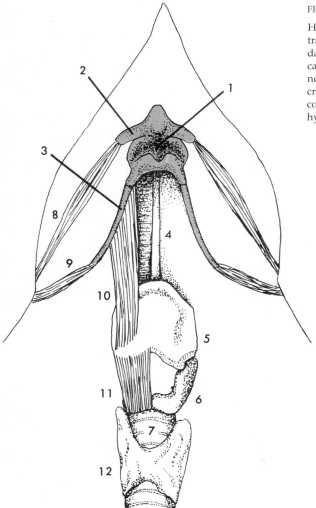

FIGURE 8-39

Hyoid *(red)*, larynx, and associated structures of a rabbit, ventral view. *1*, Body of hyoid; *2*, anterior horn of hyoid; *3*, caudal horn of hyoid; *4*, median thyrohyoid ligament; *5*, thyroid cartilage; *6*, cricoid cartilage; *7*, trachea; *8*, stylohyoideus minor muscle; *9*, stylohyoideus major; *10*, thyrohyoideus; *11*, cricothyroideus; *12*, thyroid gland. The stylohyoidei and cricothyroideus are branchiomeric muscles; the thyrohyoideus is hypobranchial.

FIGURE 8-40

Derivatives of the visceral skeleton *(red)* in a human being. *Ia*, Broken line connects derivatives of palatoquadrate cartilage; *Ib*, broken line connects vestiges and derivatives of Meckel's cartilage; *II*, broken line from lesser horn of hyoid to styloid process to stapes connects derivatives of hyoid arch. The section between hyoid and styloid process is the stylohyoid ligament; *III-V*, derivatives of the third, fourth, and fifth visceral arches. *III* is at the tip of the posterior (greater) horn of the hyoid, illustrated in Fig. 12-12, *B.*

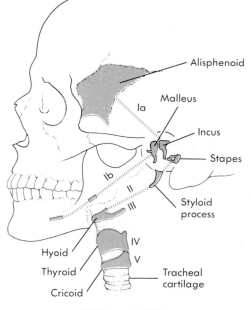

VISCERAL SKELETON

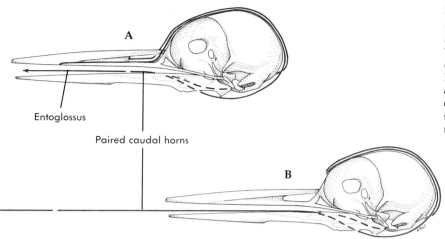

A

Entoglossus

Paired caudal horns

B

FIGURE 8-41

Caudal horns of the hyoid of a hairy woodpecker. **A**, Tongue retracted. **B**, Tongue extended. The paired caudal horns lie close together and appear as one when extended. Broken lines indicate position of horns medial to the mandible.

The hyoid anchors the highly mobile tongue of tetrapods, provides attachment for some of the extrinsic muscles of the larynx, is the skeleton for the buccopharyngeal pressure pump used in respiration in anurans, has subtle effects on lower jaw movements, and is the site of attachment of muscles that participate in swallowing. The associated hypobranchial and branchiomeric muscles approach the hyoid from many directions—lower jaw, larynx, sternum, clavicle, the temporal region of the skull, and elsewhere. These muscles stabilize the hyoid in a single position, or move it forward, backward, up, or down.

The hyoid of woodpeckers, in association with the tongue, is a remarkable tool for impaling grubs (Figs. 8-41 and 11-6). From the base of the entoglossal bone, two long, flexible, posterior horns loop caudad, then dorsad around the occipital region of the skull and forward under the scalp all the way to the lore, where both horns dip into the right nasal passageway, continue forward a short distance farther, and terminate. The termination of the two horns in one nasal passageway is totally unexpected and lacks a functional explanation. When impaling grubs, the horns are straightened by an accelerator muscle that shoots the tongue into the prey. Elastic recoil of the horns immediately returns the tongue to the mouth with impaled food.

THE LARYNGEAL SKELETON. Nearly all tetrapods have **cricoid** and **arytenoid cartilages** or replacing bones, and mammals have **thyroid cartilages** or **bones** in addition (Figs. 12-11, 12-12, and 12-14). Thyroid cartilages arise from mesenchyme of the fourth pharyngeal arch and, perhaps, the fifth. Cricoid and arytenoid cartilages appear to be products of the fifth arch. Since the caudal end of the pharyngeal arch series has been subject to reduction during evolution, it is not surprising that problems are encountered in relating the caudalmost laryngeal cartilages to specific arches.

Perspective

It is evident that the visceral skeleton is a complex that was associated primitively with feeding and branchial respiration. In tetrapods it has been modified in part for transmission of airborne sounds, for attachment of tongue muscles, and for support of vocal cords. These adaptations illustrate how mutations alter ancient structures for new functions. Some of the skeletal derivatives of the visceral arches in representative living vertebrates are listed in Table 8-4.

TABLE 8-4   Skeletal Derivatives of Pharyngeal Arches in Sharks and *Approximate* Homologues in Bony Vertebrates

Arch	Shark	Teleost	Necturus	Frog	Reptile and Bird	Mammal
I	Meckel's cartilage Pterygoquadrate	Articular* Quadrate Epipterygoid Metapterygoid	Articular Quadrate Palatal cartilage	Articular Mentomeckelian† Quadrate Annulus tympanicus (?)	Articular Quadrate Epipterygoid	Malleus Incus Alisphenoid
II	Hyomandibula Ceratohyal Basihyal	Hyomandibula Symplectic Interhyal Epihyal Ceratohyal Hypohyal Basihyal	Rudimentary Ceratohyal Hypohyals	Columella (stapes) Styloid process in mammals Anterior horn of hyoid Body of hyoid Entoglossus in reptiles and birds	2nd horn of hyoid Body of hyoid	
III	Pharyngobranchial Epibranchial Ceratobranchial Hypobranchial	Pharyngobranchial Epibranchial Ceratobranchial Hypobranchial	Epibranchial Ceratobranchial	Body of hyoid	Body of hyoid	
IV	Branchial skeleton	Branchial skeleton		Last horn and body of hyoid Laryngeal cartilages (?)	Last horn and body of hyoid	Thyroid cartilages
V	Branchial skeleton	Branchial skeleton		Laryngeal cartilages (?) (precise homologies unknown)		Thyroid cartilages
VI	Branchial skeleton	Branchial skeleton		Not present		
VII	Branchial skeleton	Branchial skeleton		Not present		

*Sometimes part of derm-articular.

†Of intramembranous origin in some species.

Chapter Summary

1. During ontogeny the neurocranium is constructed of prechordal and parachordal cartilages, notochord, and cartilaginous olfactory and otic capsules knit together and completed by cartilaginous walls and, in lower vertebrates, a roof above the brain. An optic capsule remains independent of the rest of the neurocranium and sometimes forms scleral rings.

2. In agnathans and cartilaginous jawed fishes the neurocranium remains cartilaginous throughout life. In agnathans the components are loosely articulated.

3. The chief centers of ossification in the neurocranium of bony vertebrates are occipital, sphenoid, ethmoid, and otic. Basic replacement bones are exoccipital, basioccipital, supraoccipital; basisphenoid, presphenoid, orbitosphenoid, pleurosphenoid; mesethmoid, cribriform plate, ethmoturbinals; prootic, opisthotic, epiotic, or petrosal. These may be reduced in number by fusion.

4. The neurocrania of teleosts are well ossified. Those of dipnoans, and of other ray-fins except *Polypterus,* retain considerable cartilage.

5. A single occipital condyle is found in ancient tetrapods and in modern reptiles and birds. Modern amphibians, synapsid reptiles, and mammals have two.

6. The membrane bones of the skull constitute the dermatocranium. They are vestiges of ancient dermal armor.

7. The chief dermal bones of primitive tetrapod skulls were (a) **roofing bones**—nasal, frontal, parietal, postparietal; intertemporal, supratemporal, tabular, squamosal, quadratojugal; lacrimal, prefrontal, postfrontal, postorbital, infraorbital (jugal); (b) **upper jaw bones**—premaxilla and maxilla; (c) **primary palatal bones**—parasphenoid, vomer, palatine, endopterygoid, ectopterygoid; and (d) **opercular bones.** A parietal foramen was present.

8. Teleosts have the largest number of named dermatocranial bones; dipnoans have a smaller number of large dermal bony plates. The dermatocrania of the remaining bony fishes resemble those of their Devonian ancestors.

9. Modern amphibian skulls have lost many membrane bones, and much of the neurocranium remains unossified except in apodans. Anurans have large palatal vacuities. Apodan skulls have been modified the least.

10. Modern reptiles retain well-ossified neurocrania, a single occipital condyle, many membrane bones, and, in lizards, a parietal foramen.

Among specializations are temporal fossae, and a partial or complete secondary palate (turtles, some lizards, crocodilians).

11. Temporal fossae result in diapsid (crocodilians and *Sphenodon*), synapsid (therapsids), or euryapsid skulls (ichthyosaurs and plesiosaurs). Turtle skulls are anapsid; those of lizards, snakes, and birds are modified diapsid. Mammals have synapsid skulls.

12. Secondary palates arise as horizontal processes of premaxillae, maxillae, and palatine bones. They are complete only in crocodilians and mammals.

13. Cranial kinesis is found in many fishes including the crossopterygian precursors of tetrapods, and in caecilians, snakes, some lizards, and birds. In amniotes at least it is correlated with modes of procuring food.

14. Bird skulls are modified diapsid and reptilian. Dermal bones are numerous, but sutures have been obliterated except in ratites. The skulls are thin, domed, and have large orbits. Jaws are elongated to form a beak.

15. Mammalian skulls have a single temporal fossa. The dentary is the sole bone of the lower jaw, and it articulates with the squamosal. The quadrate and articular have become middle ear ossicles. The braincase is expanded, and the dermatocranial bones are reduced in number. There is a full complement of neurocranial bones with some fusions. A temporal complex combines numerous separate components.

16. The cyclostome visceral skeleton is unlike that of any other vertebrate. There are no palatoquadrate or Meckel's cartilages, and no hyoid or branchial arches. A cartilaginous branchial basket is located under the skin of the pharyngeal region.

17. The visceral skeleton of elasmobranchs consists of palatoquadrate and Meckel's cartilages, hyoid cartilages, branchial cartilages, and median ventral basihyal and basibranchial cartilages.

18. Jaw suspension in fishes is mostly hyostylic. Some older sharks were amphistylic. Chimaeras, lungfishes, and tetrapods have autostylic jaw suspension.

19. The visceral skeleton of bony fishes resembles in general that of sharks except that it ossifies. The hyoid arch skeleton consists of a larger number of components.

20. The palatoquadrate cartilages in all bony vertebrates are invested by membrane bones. Their caudal ends become quadrate bones, except in mammals. Meckel's cartilages are invested, and their caudal ends become articular bones, except in mammals.

21. In mammals the articular and quadrate become middle ear ossicles (malleus, incus).

22. Commencing with anurans, the hyomandibula becomes a columella (stapes). The remainder of the second arch skeleton, that of the third, and sometimes part of the fourth give rise to horns of the hyoid. The remainder of the fourth arch and perhaps the fifth give rise to the skeleton of the larynx.

23. The neurocranium and dermatocranium arise from neurectoderm and from head. The splanchnocranium arises from neurectoderm.

SELECTED READINGS

Readings on feeding mechanisms will be found also in Selected Readings, Chapter 11.

Allin, E.F.: Evolution of the mammalian ear, Journal of Morphology **147**:403, 1975.

Brodal, A., and Fänge, R., editors: The biology of Myxine, Oslo, 1963, Norway Universitetsforlaget.

Carroll, R.L.: The hyomandibular as a supporting element in the skull of primitive tetrapods. In Panchen, A.L., editor: The terrestrial environment and the origin of land vertebrates, New York, 1980, Academic Press.

Crompton, A.W., and Parker, P.: Evolution of the mammalian masticatory apparatus, American Scientist **66**:192, 1978.

de Beer, G.R.: The development of the vertebrate skull, Oxford, 1937, The Clarendon Press. Reprinted with a Foreword by Brian K. Hall and James Hanken by The University of Chicago Press, Chicago, 1985.

Gans, C., and Parsons, T.S., editors: Biology of the reptilia, vol. 4, New York, 1973, Academic Press.

Goodrich, E.S.: Studies on the structure and development of vertebrates, London, 1930, The Macmillan Co., Ltd. Reprinted by The University of Chicago Press, Chicago, 1986.

Greaves, W.S.: The mammalian jaw mechanism—the high glenoid cavity, American Naturalist **116**:432, 1980.

Harrington, R.W., Jr.: The osteocranium of the American cyprinid fish, Notropis bifrenatus, with an annotated synonymy of teleost skull bones, Copeia, no. 4, p. 267, 1955.

Jarvik, E.: Basic structure and evolution of vertebrates, 2 vols., New York, 1980, Academic Press.

Liem, K.F.: Adaptive significance of intra- and interspecific differences in the feeding repertoires of cichlid fishes, American Zoologist **20**:295, 1980.

Lombard, R.E., and Bolt, J.R.: Evolution of the tetrapod ear: an analysis and reinterpretation, Biological Journal of the Linnaean Society **11**:19, 1979.

Moore, W.J.: The mammalian skull, Cambridge, 1981, Cambridge University Press.

Ørvig, T.: The dermal skeleton: General considerations. In Ørvig, T., editor: Current problems in vertebrate phylogeny, New York, 1968, Interscience-Wiley.

Radinsky, L.B.: Evolution of skull shape in carnivores. 1. Representative modern carnivores, Biological Journal of the Linnaean Society **15**:369, 1981.

Schmalhausen, I.I.: The origin of terrestrial vertebrates (Translated from the Russian by Leon Kelso), New York, 1968, Academic Press.

Stahl, B.J.: Vertebrate history: problems in evolution, New York, 1974, McGraw-Hill Book Co.

Symposia in American Zoologist

Mammalian mastication: an overview, **25**:289, 1985.

Evolutionary morphology of the actinopterygian fishes, **22**:237, 1982.

GIRDLES, FINS, LIMBS, AND LOCOMOTION

In this chapter we will focus on skeletal structures that, along with the axial skeleton, participate in locomotion. We will find that the endoskeleton of fins varies widely but that all tetrapod limbs reflect a basic pattern. We'll see how the tetrapod pelvic girdle has been altered to facilitate delivering large cleidoic eggs or mammalian young and how limbs have been modified for life in the water, on land, or in the air. We'll also learn how tetrapods propel themselves in the absence of limbs. Along the way we'll examine some hypotheses concerning the origin of fins and a plausible theory of the origin of limbs.

The girdles and the skeleton of fins and limbs constitute the **appendicular skeleton.** Girdles brace fins and limbs against the force that these appendages transmit from the substrate. The forces are strongest in amniotes, because their limbs hold the body above the ground. The pectoral girdle, in turn, is braced against the skull in many bony fishes and against the sternum in tetrapods, and the pelvic girdle is braced against the vertebral column in tetrapods.

Most vertebrates that lack one or both pairs of appendages are either aquatic or live in burrows. They also usually have an elongated trunk. Agnathans, caecilians, snakes, amphisbaenians, and some lizards have no paired appendages. Lacking posterior paired appendages are a number of teleosts, including eels, urodeles in the family Sirenidae, the lizard *Bipes*, cetaceans, and sirenians. There is also a lizard with hind limbs only. Locomotion without fins or limbs in water is accomplished by exaggerated fishlike swimming movements—lateral undulation of the trunk and tail—but limbless locomotion on land necessitates additional adaptations for making forward progress. It is not unusual for an embryonic limb bud to appear transitorily in limbless species.

PECTORAL GIRDLES

In its basic form a pectoral girdle is a U-shaped skeletal complex in the body wall that articulates with the anterior fins or limbs. It is stabilized by the multidirectional forces exerted on it by muscles that originate on the axial skeleton and insert on the girdle and by other muscles that arise on the girdle and insert on the fin or limb.

Pectoral girdles in all vertebrates are modifications of a basic pattern seen in **early jawed fishes.** In these vertebrates the girdle consisted of three pairs of replacement bones that constituted an endoskeleton and at least four pairs of investing bones derived from dermal armor (Fig. 9-1, basic pattern).

FIGURE 9-1

Pectoral girdle in selected phylogenetic lines. Dermal bones are red, and cartilage and replacement bones are black. Triangles represent interclavicle. *P*, Postcleithrum. Only one half of each girdle is illustrated, and relationships have been distorted when necessary to emphasize homologies. The basic pattern is based on *Eusthenopteron*. In alligator and bird only, *1* is the procoracoid.

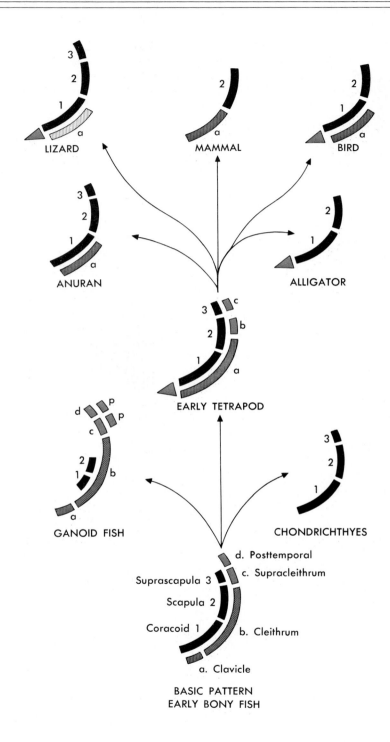

LIZARD

MAMMAL

BIRD

ANURAN

ALLIGATOR

EARLY TETRAPOD

GANOID FISH

CHONDRICHTHYES

d. Posttemporal
c. Supracleithrum
Suprascapula 3
Scapula 2
Coracoid 1
b. Cleithrum
a. Clavicle

BASIC PATTERN
EARLY BONY FISH

As in today's fishes, it was located immediately behind and directly in line with the branchial skeleton, resembling in this respect the skeleton of a gill. (This inspired the hypothesis that the pectoral girdle is a modified gill arch.) The *replacement bones* were, on each side, a ventral **coracoid,** a **scapula** that received the force transmitted to the body by the fin or limb, and a **suprascapula.** The *dermal bones* were a small ventral **clavicle** that met the opposite clavicle in a midventral symphysis, a large **cleithrum** overlying the scapula, a smaller **supracleithrum,** and a **posttemporal** bone that anchored the girdle to the tabular region of the skull. **Postcleithral** bones sometimes accompanied the supracleithrum (Fig. 9-1, ganoid fish).

In **later bony fishes** the coracoid and scapula were reduced in size and the cleithrum became the major bone of the girdle (Fig. 9-2). In teleosts the embryonic coracoid and scapula unite to form an adult coracoscapula (Fig. 9-3). A full complement of dermal bones has remained except in holosteans

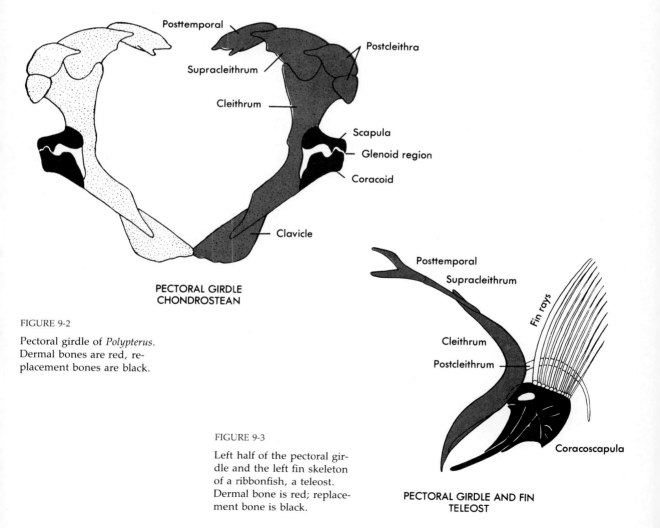

PECTORAL GIRDLE
CHONDROSTEAN

FIGURE 9-2

Pectoral girdle of *Polypterus.* Dermal bones are red, replacement bones are black.

FIGURE 9-3

Left half of the pectoral girdle and the left fin skeleton of a ribbonfish, a teleost. Dermal bone is red; replacement bone is black.

PECTORAL GIRDLE AND FIN
TELEOST

FIGURE 9-4

Cartilaginous pectoral girdle of *Squalus*, anterior view.

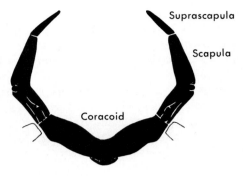

PECTORAL GIRDLE

FIGURE 9-5

Left half of the pectoral girdle of selected tetrapods, lateral views. Dermal bones are red, replacement bones are stippled. *1*, Coracoid or procoracoid; *2*, scapula. *a*, Clavicle; *b*, cleithrum; *in*, interclavicle. In turtles the clavicles and interclavicle are fused with the shell.

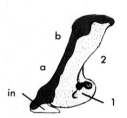

STEM AMPHIBIAN

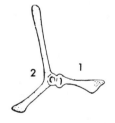

TURTLE

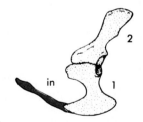

ALLIGATOR

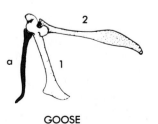

GOOSE

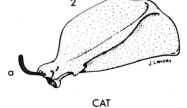

CAT

PECTORAL GIRDLES

and teleosts, who lost the clavicle. **Cartilaginous fishes** have no bones dermal or otherwise, in their girdle, and the coracoid, scapula, and suprascapula remain cartilaginous throughout life (Fig. 9-4).

The pectoral girdle of **early tetrapods** was hardly different from that of early jawed fishes. If you compare the early bony fish and the early tetrapod diagrams in Fig. 9-1 you will note that tetrapods have acquired an additional membrane bone, the **interclavicle,** and that they have lost the posttemporal bone, which braced the girdle against the skull in fishes. The suprascapula and supracleithrum are not significant in phylogeny, because they either drop out or, more often, simply become part of the scapula and cleithrum, respectively. Fig. 9-5, stem amphibian, illustrates the left half of an early tetrapod girdle from a lateral view. Present were an interclavicle, clavicle, cleithrum, all of which were dermal bones, and a coracoid and scapula, which were replacement bones. These are the legacy of all later tetrapods. We will follow their fate during subsequent tetrapod evolution.

The interclavicle, once it appeared, was persistent. Although not found in modern amphibians, it is present in most amniotes other than placental mammals (Figs. 9-1, alligator, lizard, bird, and 9-6).

The coracoids in tetrapods arise from an embryonic coracoid plate having an anterior and posterior pair of ossification centers. Bones that arise from the anterior center have been called **procoracoids, precoracoids,** or **anterior coracoids.** Those of the posterior centers are simply coracoids. With the development of a midventral sternum in tetrapods, the coracoid or procoracoid began to brace the glenoid region of the scapula against this new structure (Figs. 7-28, iguanid lizard, chicken; 9-6; and 9-7, C). This stabilized the scapula, which until then received unassisted the entire force exerted through the forelimbs as they pushed against the earth. The bracing may partly account for the fact that, as time went by, tetrapod forelimbs became more supportive of the trunk than they had been in labyrinthodonts.

The fate of the coracoids and clavicle seems interrelated. In some tetrapods the clavicle assisted the coracoid or even replaced it as the ventral brace. In others, such as crocodilians, some lizards, and some mammals, the clavicle was lost, although it appears transitorily in crocodilian embryos. The mammalian clavicle is mentioned in more detail later.

The cleithra and supracleithra have been lost in all living tetrapods. A scapula, bearing part or all of a glenoid fossa, is always present.

The mammalian pectoral girdle is the outcome of evolutionary changes that gave rise to therapsids and their mammalian descendants. A therapsid girdle is illustrated in Fig. 9-6, A. The bones illustrated there are the legacy of the mammals. Monotremes have the same bones today (Fig. 9-6, B and C). In other mammals the only bones that remain are the scapula, a clavicle in some species (Fig. 9-5, cat), and the coracoid process of the scapula, a vestige of the old coracoid plate that overhangs the glenoid fossa (Fig. 10-20). This process, like several other portions of the scapula, develops as a separate ossification center. In humans it remains independent of the scapula until about age 15 years.

The mammalian scapula is divided by a scapular spine into supraspinous and infraspinous fossae. The fossae are the anatomic origins of strong muscles that insert on the humerus (Fig. 10-20), and the spine is an insertion site for some of the appendicular muscles that arise on the vertebral column.

The clavicle is large in insectivores and primates, which are generalized mammals, and in mammals with strong forelimbs that are used for digging, climbing, or flying. In the latter the strong clavicle braces the scapula against the sternum. At the opposite extreme are mammals with no clavicle: cetaceans, ungulates, subungulates, and some carnivores. In other carnivores, including the cat family, the clavicle has been reduced to a slender splinter that reaches neither the sternum ventrally nor the scapula dorsally.

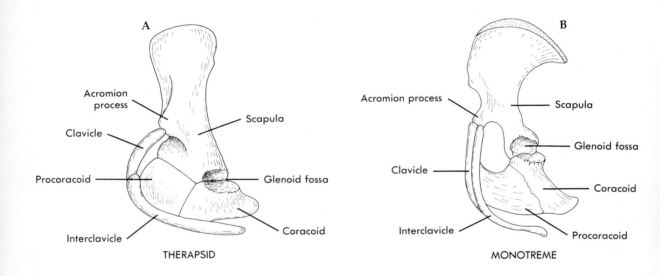

FIGURE 9-6

A and **B**, Components of therapsid and monotreme pectoral girdles oriented to show similarities. **C**, Pelvic girdle and associated bones of *Ornithorhynchus*, ventral view. Note the large coracoid bracing the scapula against the sternum.

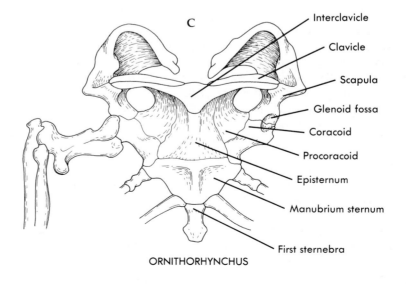

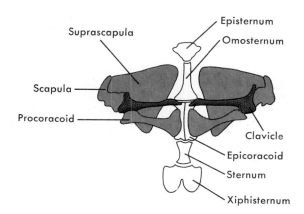

Suprascapula
Episternum
Omosternum
Scapula
Procoracoid
Clavicle
Epicoracoid
Sternum
Xiphisternum

FIGURE 9-7

Sternum and pectoral girdle of frog, ventral view. Membrane bone is red, and replacement bones except epicoracoid are gray.

The smaller the clavicle the more freedom of movement is afforded the mammalian scapula and the better the shoulder is able to absorb shock transmitted from the foot. The vestigial clavicle is partly responsible for the agility of cats, and its total absence facilitates grazing in ungulates.

The phylogenetic history of the components of the pectoral girdle of vertebrates inspires this observation: *In general, in bony fishes there was a reduction in the number of replacement bones in the pectoral girdle, whereas in tetrapods it is the dermal bones that have been reduced or lost.* This is illustrated graphically in Fig. 9-1.

Two loose ends are worth calling attention to: Urodeles fail to develop any dermal bones in their girdles. And have you noticed that tetrapods almost never brace their *pectoral* girdles against the axial skeleton? It is said to have happened in a few pterosaurs. The story is quite different with respect to the pelvic girdle.

PELVIC GIRDLES

Pelvic girdles in most fishes consist of a pair of simple cartilaginous or bony **pelvic** or **ischiopubic, plates** that meet in a midventral **pelvic symphysis** and provide a brace for the pelvic fins (Fig. 9-8, herring). In cartilaginous fishes and lungfishes, two embryonic cartilages unite to form one adult plate (Figs. 9-8), shark; and 9-9). In teleosts that have a short trunk the pelvic plate lies immediately behind, or sometimes below, the pectoral girdle and is often attached to it. For this reason, pelvic fins may project below or even anterior to pectorals. There are no dermal bone components in the pelvic girdles of either fishes or tetrapods.

Tetrapod embryos also develop cartilaginous pelvic plates. Each plate ossifies at two centers to form a **pubic bone, or pubis,** and a more posterior **ischial bone, or ischium** (Fig. 9-10). (In *Necturus* the pelvic plate remains cartilaginous except for a small ossification center.) Dorsal to the pelvic plate an additional blastema gives rise to an **ilium.** At the junction of the pubis, ischium, and ilium a socket, the **acetabulum,** accommodates the head of the femur.

FIGURE 9-8

Pelvic plates *(black)* of a bony and cartilaginous fish.

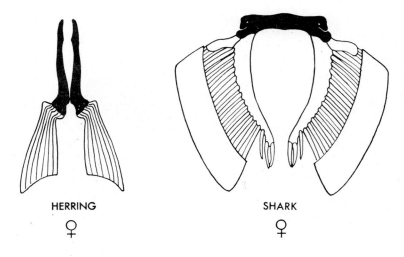

HERRING ♀

SHARK ♀

FIGURE 9-9

Pelvic plate *(black)* and fin of a male shark, showing basal fin cartilages modified as claspers *(gray)*. Compare female shark, Fig. 9-8.

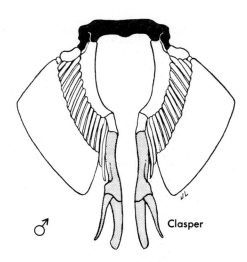

♂

Clasper

Dorsally, the ilium is braced against the stout transverse processes of sacral vertebrae, with short sacral ribs intervening. The ribs are usually ankylosed to the transverse processes and are not evident except in embryos or larvae (Fig. 9-10, *Necturus*). Ventrally, except in birds, there is a symphysis between either the two pubic bones **(pubic symphysis),** the two ischia **(ischial symphysis),** or both **(ischiopubic symphysis)** (Fig. 9-11). The symphysis is in the midventral coelomic wall immediately anterior to the cloaca. The architecture of the region is such that the force transmitted to the two acetabula as a result of gravity (weight bearing) or locomotion is distributed in two directions: to the sacrum dorsally via the ilia and to the symphysis ventrally via the ischia and pubic bones. The proportion distributed in each direction depends on the posture of the animal at rest and in motion. The joint between the head of the femur and the girdle is stabilized by muscles that approach the femur from opposing directions. The sacrum and the girdle, usually united in rigid sacroiliac joints, form a bony enclosure, the **pelvis,** that encircles the caudal end of the coelom. The resulting **pelvic cavity** contains the urogenital organs and the terminal portion of the large intestine.

Posture and mode of locomotion are correlated with the shape of the ilium, ischium, and pubis, the anatomical relationships of these bones, and their proportional size. A squatting, jumping frog has a different set of vectors affecting the pelvic girdle than does a bird, deer, kangaroo, or

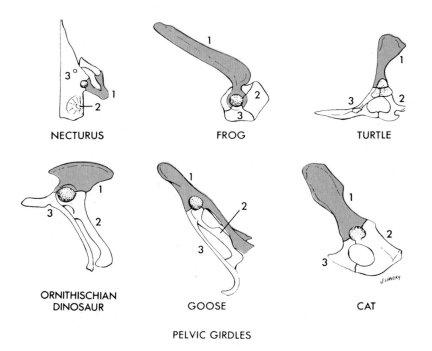

NECTURUS   FROG   TURTLE

ORNITHISCHIAN DINOSAUR   GOOSE   CAT

PELVIC GIRDLES

FIGURE 9-10

Left halves of pelvic girdles of selected tetrapods, lateral views, except *Necturus,* which is a ventral view. *1,* Ilium *(red); 2,* ischium; *3,* pubis (unossified pelvic plate in *Necturus*). In *Necturus,* a sacral rib is seen attached to the dorsal end of the bony ilium. There are no dermal bones in pelvic girdles.

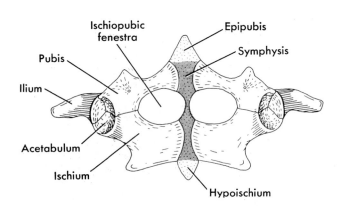

Ischiopubic fenestra   Epipubis
Pubis   Symphysis
Ilium
Acetabulum
Ischium
Hypoischium

FIGURE 9-11

Pelvic girdle of *Sphenodon,* ventral view, showing ischiopubic symphysis.

FIGURE 9-12

Half of pelvic girdle of a primitive tetrapod, selected reptiles, and a modern mammal, lateral views, head to the left. The pubis and ischia are differentially colored. **A,** *Seymouria,* a borderline reptilelike amphibian. **B,** *Alligator.* **C,** *Allosaurus.* **D,** *Varanus.* **E,** *Ophiacodon,* a primitive reptile on the mammalian line. **F,** Coxal (innominate) bone of cat. In **B,** the anterior acetabular wall is partly cartilaginous *(gray)*, and in **B** and **C** the medial wall of the acetabulum (not to be confused with the mammalian obturator foramen) is incomplete. Note the craniad reorientation of the ilium in mammals. (**A** after White; **C** after Gilmore; **D** after Butschli; **E,** redrawn from Osteology of the reptiles by A.S. Romer by permission of The University of Chicago Press [copyright 1956 by the International Copyright Union].)

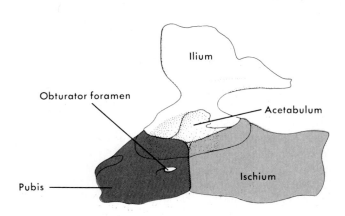

Ilium

Obturator foramen

Acetabulum

Pubis

Ischium

A. PRIMITIVE TETRAPOD

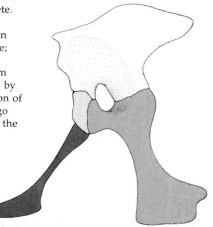

B. ALLIGATOR

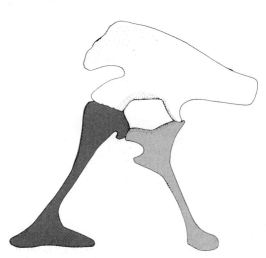

C. SAURISCHIAN DINOSAUR

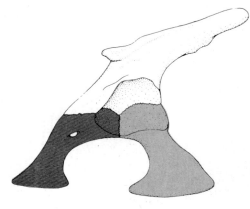

D. MONITOR LIZARD

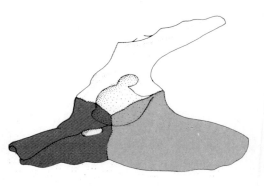

E. EARLY SYNAPSID

marine turtle. In frogs the ilia are slender and greatly elongated, and they extend from the sacral vertebra to the end of the urostyle, where they meet the ischia and pubic bones and where the acetabulum is located (Fig. 7-13). The joint between ilium and sacral vertebra is free to move when a frog pushes off at the start of a leap. As the frog lands, the joint, along with others in the leg, helps dissipate the force of the impact, acting as a shock absorber. In many tetrapods this **sacroiliac joint** is not freely movable.

Urodeles have limbs that can scarcely lift the sagging belly off the substrate without the buoying effect of water. The limbs, therefore, bear only part of the weight, whether the animal is resting motionless on the bottom of a pond or is on land. Most of the force exerted against the girdle of a urodele is a result of pushing against the substrate when moving about, not when resting. The pelvic girdle of urodeles differs little from that of fishes, except that small ilia are braced against the sacral vertebra. A median cartilage, the **prepubic,** or **ypsiloid,** extends from the girdle forward in the ventral body wall.

A variety of pelvic girdle architecture is found among reptiles, which is correlated with their divergent body structure, stances, and modes of locomotion. An expanded ilium accommodates the additional hind limb muscles necessary for more efficient locomotion on land than is seen in amphibians. In expanding, the ilium has become braced against an additional vertebra. In most reptiles the pubis is directed cephalad and the ischium caudad, and because the ilium is directed dorsad the girdle is triradiate (Fig. 9-12, *B, C,* and *D*). In ornithischian dinosaurs, however, the pubis was directed caudad, paralleling the ischium, as in birds (Fig. 9-10, ornithischian dinosaur and goose). Turtles and the generalized *Sphenodon* have an ischiopubic symphysis (Fig. 9-11), but some of the more specialized reptiles have only an ischial symphysis. A wide **ischiopubic fenestra** develops on each side between the ischium and pubis in *Sphenodon* (Fig. 9-11), in turtles (Fig. 9-10), and in lizards. It had not yet evolved in early reptiles. It is called the **obturator fenestra** (or **foramen**) in mammals because in these it transmits the obturator nerve that supplies some of the hind limb muscles.

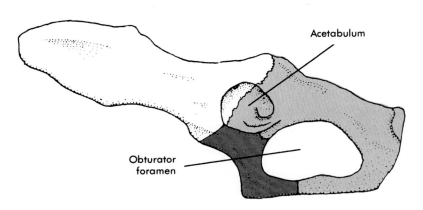

Acetabulum

Obturator
foramen

F. CAT

FIGURE 9-12, cont'd

For legend see opposite page.

(In other reptiles the nerve passes through a small foramen near the ischiopubic fenestra.) An **epipubic bone** and a **hypoischial bone** usually develop in association with the pelvic girdle of reptiles (Fig. 9-11), and one or both are present in monotremes and marsupials. In the latter it is also called **marsupial bone** because it supports the marsupial pouch.

The ilia and ischia of birds are enormously expanded, providing a broad site for attachment of the muscles used in bipedal locomotion, and the girdle is braced against lumbar and sacral vertebrae (Fig. 7-20). The pubic bones are reduced to long splinters that are directed caudad (Fig. 9-10, goose). The absence of a pelvic symphysis provides a large pelvic outlet for laying massive eggs.

In mammals the ilium, ischium, and pubis ankylose early in postnatal life to form a left and right innominate (coxal) bone (Fig. 9-12, *F*). Dorsally, each innominate joins the sacrum in an immobile sacroiliac joint. Ventrally, the two innominates meet in a symphysis to complete the bony pelvis. The ischium does not always contribute to the symphysis. In some species a small acetabular (cotyloid) bone ossifies in the acetabular wall.

Mammalian young are delivered through a pelvic outlet bounded ventrally by the pubic symphysis and dorsally by the first few caudal vertebrae (coccyx in human beings). In late pregnancy the fibrocartilage separating the bones at the symphysis is softened by hormones, which permits expansion of the pelvic outlet for delivery. In mice six days pregnant the gap between the two bones at the symphysis was shown on x-ray films to be only 0.25 mm. Thirteen days later, on the day of birth, the gap had widened to 5.6 mm.

FINS

Most fins are simple steering devices that control the direction of movement, or stabilizers that prevent rolling (as a result of torque), side-to-side wobbling (yaw), and the inclination, or pitch, of the body. Paired fins also serve as brakes to slow or halt forward motion, but they play little or no role in propulsion. Locomotion in fishes is chiefly by lateral undulation of the trunk and tail with its vertical caudal fin.

Fins, whether paired, median, or caudal, consist mostly of two surfaces of skin, the dermis of each surface being stiffened by flexible **fin rays (lepidotrichia or ceratotrichia)** supported at the base by **pterygiophores,** which are small cartilages or bones (Fig. 9-13). A lepidotrichium is a ray consisting of jointed *bony* dermal scales aligned end to end in *bony* fishes. Ceratotrichia are the long horny rays of *cartilaginous* fishes (Fig. 9-17, *A*). Short delicate **actinotrichia,** similar to ceratotrichia, develop distally in the fins of both groups. In elasmobranchs, dipnoans, and some lower ray-finned fishes, body scales continue onto the fin, growing smaller distally. These provide additional support to the fin membrane (Fig. 9-13).

Although most fins are supported by rows of **basal and radial pterygiophores** (Fig. 9-17), there is no morphologic difference between proximal and distal rows in some generalized actinopterygians (Fig. 9-17, *C*). It is likely that basals resulted from coalescence of proximal radials. An instance

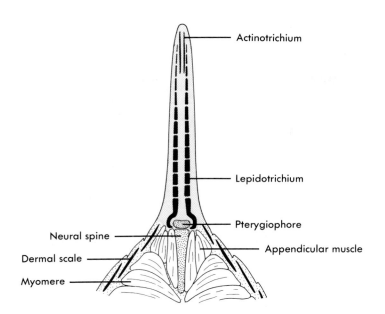

FIGURE 9-13

Supporting elements of the dorsal fin of a teleost in diagrammatic cross section. (After Goodrich.)

of what might then be extreme coalescence is seen in the single large basal pterygiophore of *Heptanchus* (Fig. 9-17, *A*). The pectoral fins of most sharks have three basals, to which the terms **pro-, meso-,** and **metapterygia** have been applied (Fig. 9-15, *D*). In the most specialized teleosts, on the other hand, basals have been lost and mere vestiges of radials remain to provide a base for a pair of fin rays (Fig. 9-16, *B*). Caudal fins have the added support of the vertebral column. Striated muscle masses from several successive myomeres extend into the base of a fin and insert on the pterygiophores.

Paired Fins

Paired fins of living and fossil fishes may be grouped in three general categories, **lobed fins, fin fold fins,** and **ray fins** (Fig. 9-14). Not unexpectedly, these correspond to the three groups of jawed fishes that emerged from the Devonian and have persisted to the present day: sarcopterygians, chondrichthyeans, and actinopterygians. The fin skeletons of representative members of these groups show distinctive morphologic patterns. Other skeletons in these same groups, particularly fin fold and ray fin types, have undergone so much modification that it is difficult to relate them to a primitive condition. Considering the enormous variety of fossil and living species, this is not surprising, but as mentioned in Chapter 3, the result has been a lack of agreement among competent paleontologists as to the kinships of early jawed fishes, hence of fish fins.

The skeleton of paired fins is braced against the corresponding girdle, the site being in the glenoid region or glenoid fossa of the scapula (or scapula and coracoid) of the pectoral girdle (Figs. 9-2, 9-16, and 9-43, *A*), and on

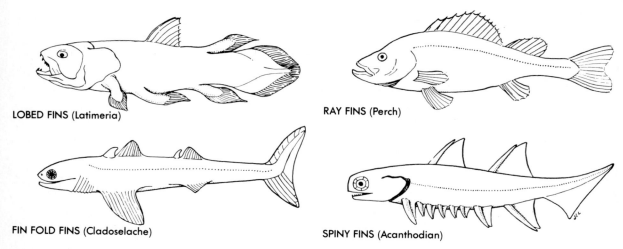

LOBED FINS (Latimeria)

RAY FINS (Perch)

FIN FOLD FINS (Cladoselache)

SPINY FINS (Acanthodian)

FIGURE 9-14

External appearances of some fins.

a prominence on the lateral or posterior aspect of the pubic plate of the pelvic girdle (Fig. 9-8).

Lobed fins consist of a fleshy, muscular proximal portion containing the endoskeleton and a membranous portion that is stiffened by fin rays (Fig. 9-14, *Latimeria*). The fin arises from a narrow base and is shaped like a paddle or an oar. These fins are characteristic of dipnoans and crossopterygians.

The fin skeleton of the dipnoan *Neoceratodus* consists of a bony jointed central axis, the bones of which function like basals, and a series of preaxial and postaxial radials (Fig. 9-15, *A*). The fin is said to be **biserial** because of the two series of radials. Crossopterygians have a variant in which postaxial radials are less numerous than in dipnoans (Fig. 9-43, *A*). The biserial fin was given the name **archipterygium** to indicate that it may have been ancestral to all other fin types. This is no longer considered likely.

Fin fold fins, characteristic of cartilaginous fishes, have a broad base. It is broad in modern sharks, and it was even broader in the Paleozoic (Fig. 9-14, *Cladoselache*). The *pelvic fins* of Chondrichthyes, Paleozoic and modern, have a basal or axial element, but radials are confined to the preaxial border (Fig. 9-15, *B* and *C*). In males the pelvic fin skeleton has been modified to become an intromittent organ, or clasper (Fig. 9-9). The *pectoral fins* of these same fishes exhibit a wide diversity of skeletal morphology. The pectoral fin of a modern shark, *Heterodontus*, is illustrated in Fig. 9-15, *D*. To what extent these variants are strictly adaptive would be a subject for heated argument among students of evolutionary biology, but all will agree that pectoral fins bear most of the burden of regulating pitch and acting as brakes to forward motion. The evolutionary origins of fin fold fins will not be settled until the evolutionary history of Chondrichthyes is established beyond doubt.

Ray fins are characteristic of actinopterygians (Fig. 9-14, perch). The most specialized teleosts have no basals, and only vestigial radials, which commence at or close to the endoskeletal component of the girdle (Fig. 9-16, *B*). The result is a fin that is highly flexible all the way to the body wall. The fin skeletons of chondrosteans, however, are more like those of Paleozoic actinopterygians, having prominent complements of basals and radials (Fig. 9-16, *A*). As pointed out earlier, the pelvic girdle of teleosts with short trunks lies immediately behind, or sometimes below, the pectoral girdle. In these species the pelvic fins lie below, or even below and anterior to, the pectorals.

It should be mentioned that a few fishes fly, although not as far as the Wright brothers. Characins, which are primitive voracious teleosts inhabiting fresh tropical waters, get an initial thrust out of the water with the

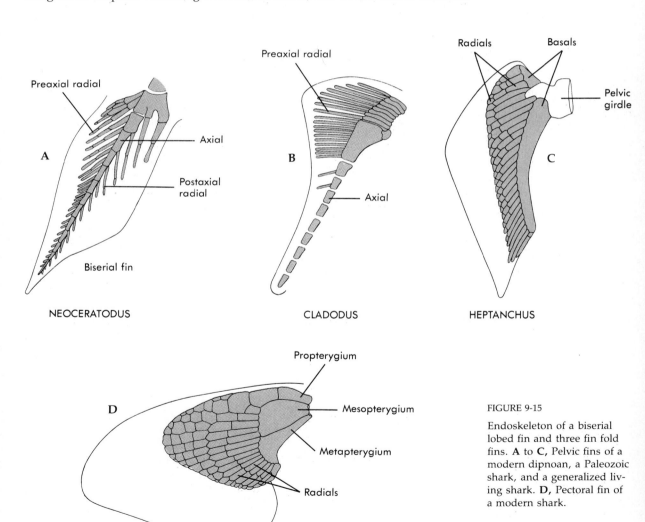

FIGURE 9-15

Endoskeleton of a biserial lobed fin and three fin fold fins. **A** to **C**, Pelvic fins of a modern dipnoan, a Paleozoic shark, and a generalized living shark. **D**, Pectoral fin of a modern shark.

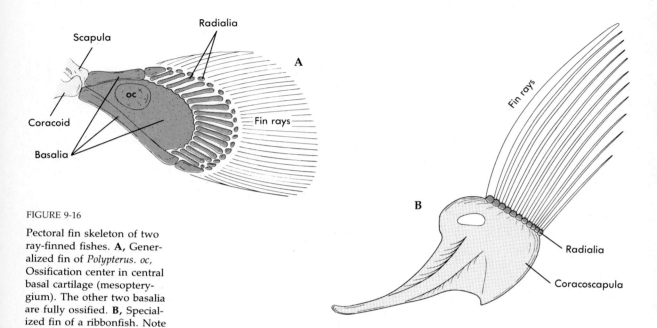

FIGURE 9-16

Pectoral fin skeleton of two ray-finned fishes. **A,** Generalized fin of *Polypterus. oc,* Ossification center in central basal cartilage (mesopterygium). The other two basalia are fully ossified. **B,** Specialized fin of a ribbonfish. Note loss of basalia and proximal radialia.

caudal fin and, beating winglike pectorals, fly several yards using appendicular muscles that, for a fish, are exceptionally large. This behavior is said to occur only when the fish is alarmed. If so, it's a fishy way of performing a disappearing act.

Origin of Paired Fins

What was the nature of structures that might have given rise to paired fins? This is one of the more puzzling questions in the study of phylogeny, because to answer it requires more knowledge than we have of early vertebrates. This being so, the best we can do to satisfy our curiosity, which is the motivation for basic science, is to look for whatever clues we might find among ostracoderms and early jawed fishes, and then speculate.

The **fin fold hypothesis** states that paired fins are derived from a pair of continuous fleshy folds of lateral body wall analogous to the metapleural folds of an amphioxus. Some anapsid ostracoderms had such folds, although they were higher on the body wall. If such a structure were to become interrupted in the middle of the trunk and the remaining sections were to be invaded by muscle buds at the base and endowed with an endoskeleton, the result would be paired fins like those of *Cladoselache* (Fig. 9-14). However, there is no evidence that this happened, and the hypothesis is of historical interest only.

According to the **gill arch hypothesis** of Gegenbaur, pectoral and pelvic girdles are modified gill arches and the skeleton within the fin is an expansion of gill rays. It is true that many students of comparative anatomy, seeing a shark skeleton suspended in a museum jar for the first time, think that the pectoral girdle is part of the pharyngeal skeleton because of its location immediately behind the last pharyngeal arch, and its U shape. It looks as though it might have been a gill arch. Whether it was is speculative.

The most promising hypothesis is the **fin spine hypothesis** proposed near the middle of the twentieth century by Gregory and Raven. In early acanthodians (Fig. 9-14, spiny fins) pectoral and pelvic appendages were the largest of a series of lateral hollow spiny appendages that extended the length of the trunk and supported a fleshy membrane. Evidently they were practically immobile, and in no way resembled the pectoral and pelvic fins of Chondrichthyes or Osteichthyes. In later acanthodians, weak fin rays were present in the membranes of the two pairs at the pectoral and pelvic levels only, and small radial elements supported the membranes at their base. In time, acanthodians tended to lose all except the pairs containing fin rays. This may be a clue to the origin of paired fins, but it is not known that acanthodians were ancestral to later fishes. Among placoderms, arthrodires had a pair of fixed spines projecting behind the head, antiarchs had jointed, spiny, armored pectoral appendages (Fig. 3-7), and some very early placoderms, probably bottom-dwellers, had pectoral fins resembling those of modern rays, an instance of evolutionary convergence. Although it is uncertain that placoderms are a natural group, they seem to have been "experimenting" with designs for later paired appendages.

No known protochordate has structures that conceivably could have given rise to paired vertebrate fins, even allowing the imagination free rein, and the earliest known vertebrates had none. Reliable clues to the origin of paired fins may be hidden forever in the obscurity of elapsed time.

Median Fins

In addition to paired fins, most fishes have one or two median **dorsal fins,** an **anal fin** just behind the vent or anus and a **caudal fin** at the end of the tail (Fig. 9-14, perch). Median fins act like keels, keeping fishes from rolling to the left and right. During forward movement they minimize yaw that results from the side-to-side thrusts of the locomotor tail. Lack of a stabilizing structure of one kind or another would not only increase the cost of locomotion in terms of energy expended, but would probably place a species in jeopardy. Bottom dwellers might be less disadvantaged. Because of the body shape of Rajiformes we might expect that they would not be subject to appreciable yaw; nevertheless, rays other than sting rays have two dorsal fins far back on the tail, but no anal fin.

Dorsal fins are short in some species, but in lampreys and eels they extend the entire length of the trunk, providing mechanical stability to the elongated body and compensating, no doubt, for the lack of a lobed caudal fin and a full complement of paired fins. It may also augment the lateral thrust of the elongated body against the water. The anal fin of males of some viviparous teleost species has been modified to become an intromittent organ, or **gonopodium,** fertilization being internal. The endoskeleton of dorsal fins rests on the axial skeleton (Fig. 9-17, B and C).

FIGURE 9-17

Median dorsal fin skeletons of a generalized shark, a ray, and a sturgeon. The basal pterygiophores of median dorsal fins rest on the vertebral column. (**A** after Daniels; **B** and **C** after Goodrich.)

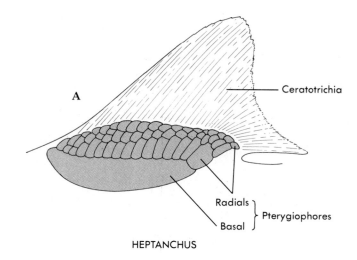

HEPTANCHUS

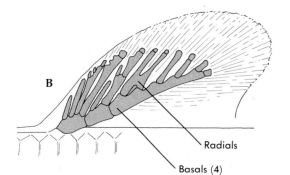

RAJA

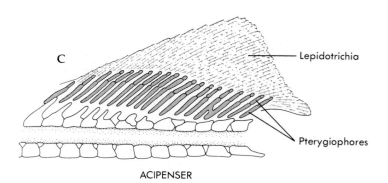

ACIPENSER

Caudal Fins

Caudal fins increase the effectiveness of the tail as a locomotor organ. Those of fishes are classified on the basis of the direction taken by the terminal portion of the notochord and vertebral column, and their shape (Fig. 9-18, *A* to *C*). A tail that contains dorsal and ventral lobes and in which the notochord turns upward into a larger dorsal lobe is said to be **heterocercal.** This was the predominant type in the Paleozoic, having been present in placoderms, Paleozoic sharks, and at least some acanthodians, and it is still seen in modern sharks and two relict chondrosteans, namely, sturgeons and spoonbills (Fig. 3-9). There is also a rare condition, **hypocercal,** in which the vertebral column turns downward.

FIGURE 9-18

Caudal fins of fishes. **A** to **C,** Major morphologic varieties, notochord in black. The heterocercal fin is probably the most primitive. **D,** Successive developmental stages in the flounder *Pleuronectes,* notochord in red, showing transition from heterocercal to homocercal.

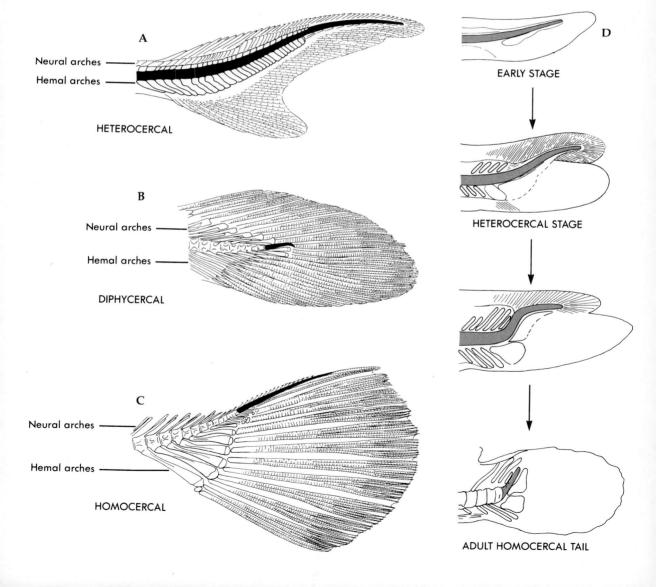

FIGURE 9-19

Skeleton of the homocercal caudal fin of the primitive teleost *Clupea.* The bony urostyle encloses all but the terminal segment of the upturned notochord. Dorsal radials tend to disappear in more specialized teleosts. Fin rays extend to the margin of the fin from the hypurals and dorsal radials. *oc,* Opisthoural cartilage.

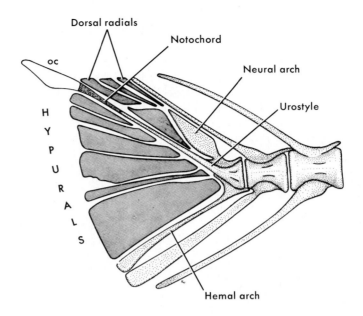

Derived from the heterocercal condition were innumerable variants that make classification of caudal tails less than definitive. Two terms that have been used conventionally are diphycercal and homocercal. Both types are symmetrical externally, but the notochord and vertebral column terminate differently in the two. In **diphycercal** tails the vertebral column ends with very little upbending, as in dipnoans and *Latimeria* (Figs. 3-12 and 9-18, *B*). In **homocercal tails** the notochord, encased within a bony sheath, or urostyle (Fig. 9-19), turns far dorsad, and the fin membrane is supported by hemal arches or hypural bones, the two probably being homologous (Figs. 9-18, *C*, and 9-19). Homocercal tails are common in teleosts but not restricted to them. Several hypural bones are braced against the underside of the teleost urostyle, and fin rays extend from these, there being no ventral radials. Specialized teleosts have only two hypurals and no dorsal radials.

During embryonic development a homocercal tail is initially heterocercal (Fig. 9-18, *D*). Applying von Baer's theorem, we could conclude that modern caudal fins are modifications of a heterocercal condition. That conclusion is reinforced by the fact that heterocercal tails predominated during the Paleozoic.

That so many permutations have taken place in the architecture of fish tails since the Paleozoic, resulting in innumerable gradations between heterocercal and other types, has given rise to disagreement among morphologists as to which term should be applied in a given species. In the ultimate analysis it is a matter of lack of specificity of the definitions, understandable inasmuch as a truly definitive list would be interminable. There have been attempts to use the morphologic features of caudal fins in unraveling the phylogenetic relationships among early and subsequent fishes. The approach has been thwarted by enormous diversity, an incomplete fossil record, and the realization that similarities may be the result of convergent evolution.

Caudal fins have developed as adaptations to aquatic life in a few amniotes. The fishlike ichthyosaurs developed them, and cetaceans and sirenians acquired horizontal rather than vertical fins, the former being more often referred to as flukes. The vertebral column provided internal support for the caudal fins of ichthyosaurs, turning ventrad into the ventral lobe of their bilobed tails, but there were no fin rays. The flukes of marine mammals have neither fin rays nor endoskeletal support. The same is true of the caudal fins of larval amphibians.

TETRAPOD LIMBS

Although tetrapods typically have four limbs, some have lost one or both pairs and in others the forelimbs have been modified as wings or paddles. By employing limbs with appropriate modifications, tetrapods swim, crawl, walk, run, hop, jump, dig, climb, glide, or fly to avoid enemies, seek food and shelter, and find a mate. Each life style necessitates some modification of the appendicular skeleton.

Tetrapod limb skeletons consist of five segments: **propodium, epipodium, mesopodium, metapodium,** and **phalanges.** In the forelimb these correspond to the bones of the upper arm, forearm, wrist, palm, and digits, the last three constituting the **manus,** or hand (Fig. 9-20). Table 9-1 lists these segments and their corresponding parts in the hind limb by common names. The skeleton within homologous segments in the various tetrapods is remarkably similar despite outward appearances; it is the orientation of the bones, the relative mobility of the joints, and the complexity of the appendicular muscles as much as the skeleton per se that makes possible the variety of locomotor activities of tetrapods. The most striking differences in skeletons are at the distal ends of the appendages.

The limbs of early tetrapods were short, the first segment extended nearly horizontally from the trunk, and the second segment was perpendicular to the first, directed downward (Fig. 9-21, *D*). This posture persists to a considerable degree in urodeles, tortoises, and primitive lizards. In other reptiles and in birds and mammals there has been a rotation of the entire appendage toward the body, so that the long axes of the humerus and femur more nearly parallel the vertebral column, the elbow being directed caudad and the knee cephalad (Fig. 9-33, *A*). Limbs so oriented are excellent shock absorbers. They also develop greater leverage for pulling (forelimbs) and pushing (hind limbs) against the ground.

There is no structure known that could have given rise to tetrapod limbs except the paired fins of fishes. A hypothetical stage in the transition is illustrated in Fig. 9-21, *B*. To produce only the externally observable differences between a fin and a limb necessitated the following modifications, not necessarily in the sequence listed, not simultaneously, and certainly not exclusively, inasmuch as concomitant adaptive modifications had to be made in the entire skeletomuscular system subserving each appendage: (1)

TABLE 9-1   Homologous Segments in Anterior and Posterior Limbs of Tetrapods

	Name of Segment		Skeleton
Anterior Limb	1 Upper arm (brachium)		Humerus
	2 Forearm (antebrachium)		Radius and ulna
	3 Wrist (carpus)	}	Carpals
	4 Palm (metacarpus)	} Manus	Metacarpals
	5 Digits	}	Phalanges
Posterior Limb	1 Thigh (femur)		Femur
	2 Shank (crus)		Tibia and fibula
	3 Ankle (tarsus)	}	Tarsals
	4 Instep (metatarsus)	} Pes	Metatarsal
	5 Digits	}	Phalanges

FIGURE 9-20

Generalized pattern of a right anterior limb, viewed from above, palm down. *1* to *5*, First to fifth digits.

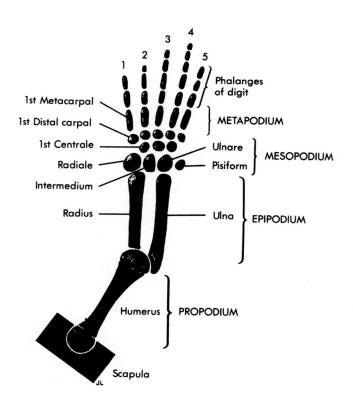

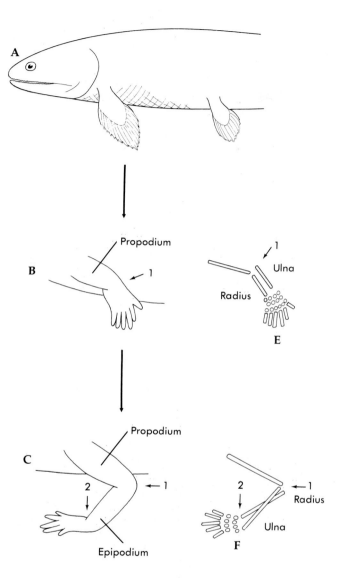

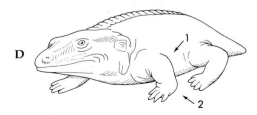

FIGURE 9-21

From fin to limb in the Devonian. **A,** Rachitomous crossopterygian. **B,** Hypothetical transitional stage. Note location of future elbow joint *(arrow 1)*. **C** and **D,** Temnospondylous labyrinthodonts. In **C,** note elongation of forearm and formation of a wrist joint *(arrow 2)*. **E** and **F,** Orientation of skeletal elements of **B** and **C.** In **F** note that some reorientation of the radius and ulna may have been necessary for the manus to lie flat on the ground.

elongation of the two bones of the epipodium; (2) formation of hinge joints at the elbow and knee and of joints between epipodium and wrist or ankle, and sometimes within the ankle, as in birds (p. 305); (3) rotation of the long axes of the humerus and femur to parallel the vertebral column (Fig. 9-21, *F*); and (4) emergence of a definitive manus and pes. These modifications are illustrated in Fig. 9-21.

Generally, hind limbs have stouter muscles, are used to a greater degree in powering locomotion, and are statistically longer than forelimbs. The hind limbs of fleet amniotes, such as cursorial lizards and deer, are considerably longer than the forelimbs, and in leaping species such as frogs and kangaroos hind limbs may be twice as long. An optional bipedal posture is usually associated with short arms and a stout muscular tail that is used for a prop (Fig. 3-22, *Tyrannosaurus*). Brachiating primates—primates that swing from branch to branch—have exceptionally long arms. The ability of any of these species to survive depends partly on these adaptations. Accounts of how various appendages are used in locomotor activities will be found in some of the Selected Readings at the end of this chapter and of Chapter 10.

Propodium and Epipodium

The humerus is the bone of the upper arm. The similarity of the humeri of all tetrapods is more striking than any differences (Fig. 9-22). Variations in length, diameter, and shape are adaptive modifications. The odd humer-

FIGURE 9-22

Humerus, radius, and ulna of the left forelimb, lateral views. *H*, Humerus; *R*, radius; *U*, ulna. In the frog the radius and ulna have united to form a radioulna, *RU*. In the bat the ulna is vestigial.

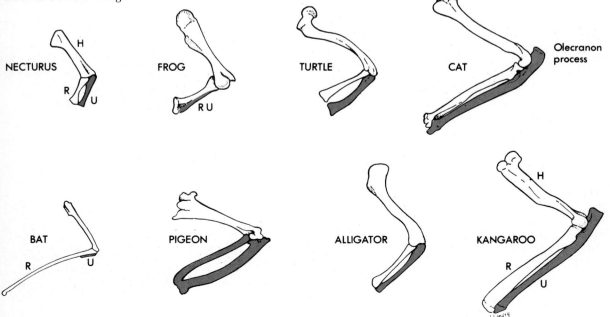

us of the mole (Fig. 9-23), for example, has expansions for insertion of massive shoulder muscles for digging. The humeri of carinate birds have a slender central cavity containing diverticula from the lungs.

The radius and ulna are bones of the forearm. The radius is a preaxial bone articulating proximally with the humerus and distally with wrist bones on the thumb side of the hand. It bears most of the force being transmitted from wrist to humerus. The ulna is a longer, postaxial bone articulating proximally with the humerus and radius and distally with wrist bones on the side opposite the thumb. The ulna sometimes fuses with the radius, or it may be vestigial, as in frogs and bats (Fig. 9-22).

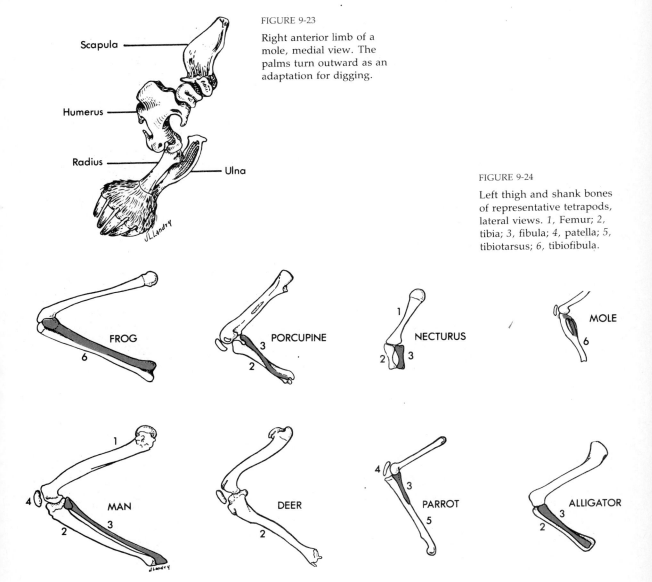

FIGURE 9-23

Right anterior limb of a mole, medial view. The palms turn outward as an adaptation for digging.

Scapula

Humerus

Radius

Ulna

FIGURE 9-24

Left thigh and shank bones of representative tetrapods, lateral views. *1*, Femur; *2*, tibia; *3*, fibula; *4*, patella; *5*, tibiotarsus; *6*, tibiofibula.

FROG

PORCUPINE

NECTURUS

MOLE

MAN

DEER

PARROT

ALLIGATOR

The femur is the propodial bone of the thigh, and the tibia and fibula are bones of the lower leg (shank). The three bones differ relatively little from one tetrapod to another. A sesamoid bone, the **patella,** or kneecap, develops in birds and mammals. It ossifies in the tendon of insertion of the powerful extensor muscle of the thigh where the tendon passes over the complicated knee joint to insert on the tibia. The patella protects the joint from the abrasive action of the tendon. The fibula may unite partially or completely with the tibia, it may be reduced to a splinter, as in birds, or it may be lost, as in ungulates (Fig. 9-24, mole, frog, bird, deer). In birds the tibia fuses with the proximal row of tarsals to form a **tibiotarsus.**

Manus: The Hand

Wrist, palm, and digits constitute a functional unit: the hand, or manus. Considering the wide variety of modifications that have appeared since the labyrinthodonts, the skeletons of the hands of all tetrapods are remarkably similar.

In generalized hands the wrist consists of three more or less regular rows of carpal bones (Fig. 9-20, mesopodium). The proximal row has a radial carpal **(radiale)** at the end of the radius, an ulnar carpal **(ulnare)** at the end of the ulna, and an **intermedium** between the two. At the ulnar end of the proximal row in most reptiles and mammals is a sesamoid bone, the **pisiform.** The middle row of carpals in a generalized hand consists of **centralia**—three or four in early tetrapods, two in early reptiles—one of which is sometimes displaced to the proximal or distal row of carpals. The distal row is composed of five **distal carpals** numbered 1 to 5 commencing on the thumb side. Table 9-2 lists the names of the carpal bones.

The metacarpals are the skeleton of the palm. Primitively, there were probably as many distal carpals and metacarpals as there were digits.

Each digit consists of phalanges. The generalized phalangeal formula commencing with the thumb is thought to have been 2-3-4-5-3, the formula

TABLE 9-2   Synonymy of Carpal Bones

Terms Preferred by Comparative Anatomists	Nomina Anatomica*	Anglicized Names and Synonyms
Radiale	Os scaphoideum	Scaphoid, navicular
Intermedium	Os lunatum	Lunate, lunar, semilunar
Ulnare	Os triquetrum	Triquetral, cuneiform
Pisiform	Os pisiforme	Pisiform, ulnar sesamoid
Centralia (0 to 4)	Os centrale	Central carpal(s)
Distal carpal 1	Os trapezium	Trapezium, greater multangular
Distal carpal 2	Os trapezoideum	Trapezoid, lesser multangular
Distal carpal 3	Os capitatum	Capitate, magnum
Distal carpal 4 ⎫ Distal carpal 5 ⎭	Os hamatum	Hamate, unciform, uncinate

*Terms approved by the Eighth International Congress of Anatomists at Wiesbaden in 1965.

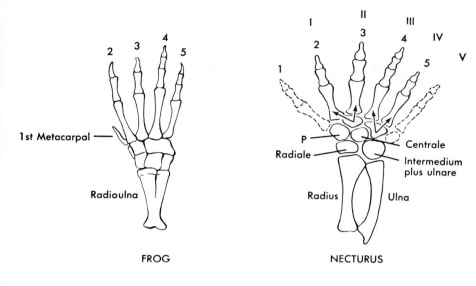

FROG

NECTURUS

FIGURE 9-25

Hands of *Rana catesbeiana* and *Necturus*, dorsal views. Which finger is missing in *Necturus?* Arabic numerals suggest that in a necturus the thumb, *1*, is missing and the little finger, *5*, is present. Roman numerals suggest that the thumb, *I*, is present and the little finger, *V*, and last distal carpal and meta-carpal are missing. Arrows indicate existing muscle attachments. Broken lines represent nonexistent elements, one of which has been lost. *P*, Distal carpal number two or prepollex, depending on interpretation. In the frog shown, the bone at the base of the first metacarpal is probably a displaced centrale.

in generalized reptiles. In late therapsids it had become 2-3-3-3-3, an almost universal formula for modern mammals with five fingers (Figs. 9-26, man, and 9-28, *B*).

Modifications of the manus with few exceptions involve *reduction in the number of bones by evolutionary loss or fusion.* A less common modification is the *disproportionate lengthening or shortening of some of the bones.* Least common is an *increase in the number of phalanges.* Centralia frequently unite with one of the proximal carpals or disappear. As a result, most reptiles and numerous mammals have a single centrale, and it is sometimes found among the proximal row of carpals. Fusion of distal carpals 4 and 5 is common and results in a **hamate bone.** Phalanges or entire digits may be lost. In the latter event the corresponding metacarpal becomes vestigial or lost.

Most modern amphibians have lost one *digit* in the hand, but retain five in the foot, and the corresponding *metacarpal* has been reduced or lost (Fig. 9-25). Several *wrist bones* in amphibians have been lost as independent bones because the embryonic intermedium and ulnare often unite, a proximal carpal often unites with an adjacent carpal, and fusion between centralia and proximal or distal carpals is common, as observed in embryos. Members of the urodele genus *Amphiuma* have one to three fingers. The three distal carpals of *Necturus* have been considered to be, commencing on the radial side (Fig. 9-25), carpal 2, carpal 3, and hamate (fused carpals 4 and 5). However, it could be that the fifth finger rather than the thumb is missing. The three distal carpals would then represent a **prepollex** (an extra bone that sometimes occurs near the thumb, or **pollex**), carpals 1 + 2, and carpals 3 + 4. Carpal 5 would be missing. This interpretation takes into account that the bone labeled *P* in Fig. 9-25 has no muscle connecting it with a finger and that the muscle from the first and second digit attaches to the second of the three distal carpals. This carpal also frequently has double ossification centers. This is one approach used in attempting to determine

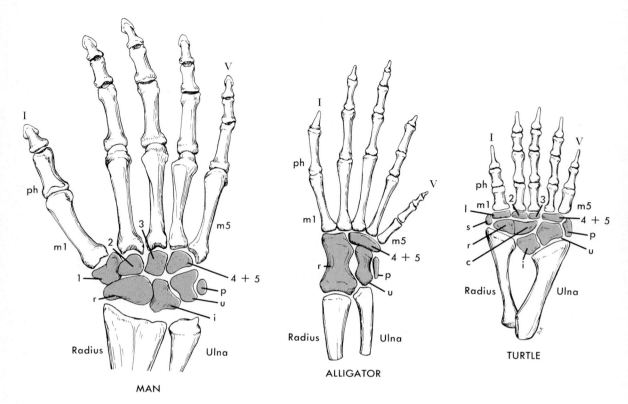

FIGURE 9-26

Right manus of man, alligator, and turtle, dorsal views, carpal bones in red. *c,* Centrale; *I,* intermedium; *m1* and *m5,* first and fifth metacarpals; *p,* pisiform; *ph,* proximal phalanx; *r,* radiale; *s,* radial sesamoid; *u,* ulnare; *1* to *5,* distal carpals; *I* and *V,* first and fifth digits. The alligator has an additional carpal that cannot be seen from this view. Wrist bones are red.

homologies of the hand or foot. The technique used by Alberch and Gale (1985) is another.

The muscles that insert on the hands and feet of urodeles are neither strong nor well differentiated, and the joints between the epipodia and wrists or ankles and between the wrists or ankles and the metapodia are capable of little mobility. Thus neither the hands nor the feet of urodeles generate locomotor thrust. They are chiefly platforms, or podia, which, pressing on the substrate, provide friction while muscles higher on the limbs extend the legs. The same is true of the hands, but not the feet, of anurans.

The hands of living reptiles and primitive mammals such as insectivores and primates tend to remain pentadactyl and to have five metacarpals and a nearly full complement of carpals except central carpals, or centralia (Fig. 9-26, turtle and man). In crocodilians, however, the wrist has been reduced to five adult bones (Fig. 9-26, alligator), and in birds the entire manus has been reduced (Fig. 9-27). When present in mammals, the centrale may lie in the distal row of carpals, as in rabbits, or it may unite with the radiale and intermedium to form a scapholunar bone of triple origin, as in cats (Fig. 9-28). Human fetuses have a centrale that remains as an independent center of ossification until the third fetal month, when it fuses with the radiale. Among major modifications of the hand are those for flight, life in the ocean, swift-footedness, and grasping.

FIGURE 9-27

Left manus of a bird. *I* to *III*, Digits; *M₁* to *M₃*, metacarpals fused with three distal carpals to form a carpometacarpus.

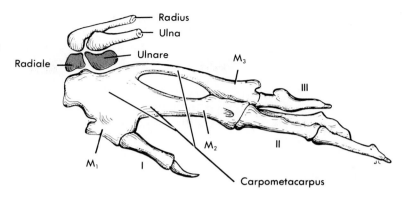

FIGURE 9-28

Left manus of cat and rabbit, anterior view, carpal bones in red. The pisiform bone is on the palmar side and cannot be seen. The scapholunar of cats combines radiale, intermedium, and central carpal, which are separate bones in kittens. The phalangeal formula of both animals is 2-3-3-3-3. *1* to *3*, First through third distal carpals. *I* and *V*, First and fifth metacarpals. Hamate incorporates fourth and fifth distal carpals. (From Kent, G.C.: Anatomy of the vertebrates: a laboratory guide, ed. 3, St. Louis, 1978, The C.V. Mosby Co.).

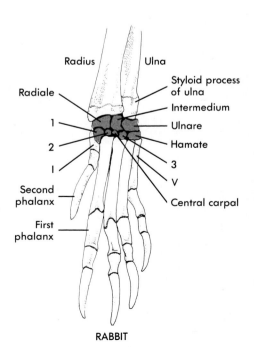

CAT

RABBIT

ADAPTATIONS FOR FLIGHT. In many birds the hand has little independent role in propulsion in air, but being at the end of an airfoil, it has an aerodynamic effect. Loss and fusion of bones have reduced the hand to a rigid, tapering structure (Fig. 9-27). Despite this, most of the basic components of tetrapod hands are identifiable in avian embryos. Two carpals (radiale and ulnare) form in the proximal row, and three in the distal row. As development progresses the three distal carpals unite with the three metacarpals to form a rigid **carpometacarpus.** Three fingers are usually present, and the number of phalanges has been reduced. (Terns often develop four embryonic digits, but only three persist.) The fingers often bear claws that are used in nonaerial locomotion, and like the rest of the hand, they are covered by feathers.

The first finger of birds that maneuver, alight, and take off in limited spaces is elongated, prominent, and independently movable and is called an **alula.** Songbirds have short, broad wings, and the feathered alula serves as an accessory airfoil at the leading edge of the wing as the bird flits among tree branches. Carnivorous birds have moderately long, broad wings adapted for slow speed flight and quick landings in limited spaces. When these birds are braking, the alula is moved away from the rest of the wing, which creates a slot through which air rushes, making it possible to maintain stable low-speed headway until additional braking is applied with the main wings and tail feathers. There is a joint with considerable flexibility at the elbow, and another between the forearm and wrist, except in birds that hover. When flexed at the wrist, the hands exert a strong braking effect for landing, especially in birds with a large wingspread. Birds that hover have high-speed wings (up to 50 strokes per second in hummingbirds), the hand is as long as the arm or longer, and the entire wing is quite rigid.

Contrary to the condition in birds, the hand in pterosaurs and bats was, or is, the main part of the wing. Pterosaurs (Fig. 9-29, *B*) had four fingers, three of which were normal and bore claws. The fourth was embedded in the wing membrane **(patagium)** and consisted of four enormously elongated phalanges that made this finger as long as the entire body. The associated metacarpal was not elongated, but it was much enlarged. Bats have five fingers (Fig. 9-29, *A*). The thumb is normal and bears a claw. The other four fingers are elongated, and are associated with four greatly elongated metacarpals. These and the phalanges constitute the skeleton of the patagium. The three proximal carpals are united in a single bone. Movement of the hand is responsible for takeoff and true flight in bats. No one has ever seen a pterosaur take off!

Flying lemurs have a patagium, but it is less well developed than in bats and pterosaurs, and the fingers, although embedded in it, are not elongated. Flying lemurs soar but are not capable of true flight. Patagiums in such unrelated animals as pterosaurs, bats, lemurs, and even flying lizards are instances of convergent evolution.

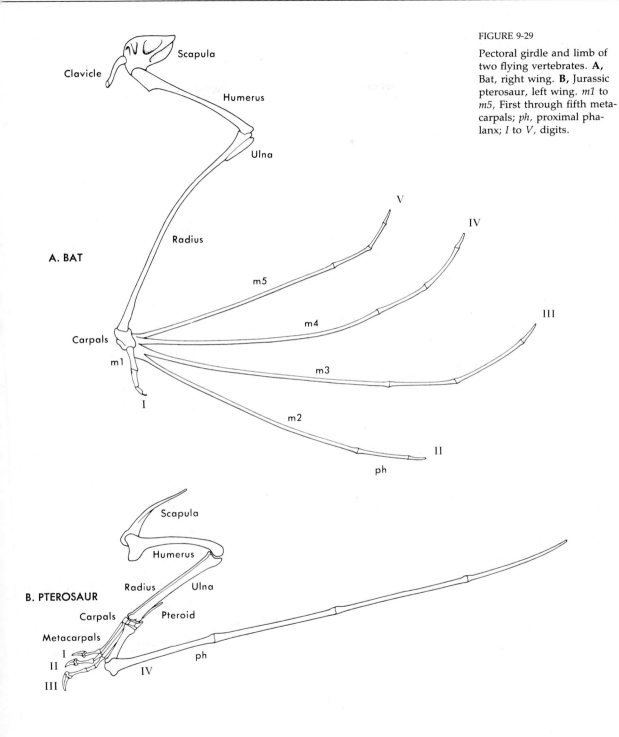

FIGURE 9-29

Pectoral girdle and limb of two flying vertebrates. **A,** Bat, right wing. **B,** Jurassic pterosaur, left wing. *m1* to *m5,* First through fifth metacarpals; *ph,* proximal phalanx; *I* to *V,* digits.

ADAPTATIONS FOR LIFE IN THE OCEAN. The hands of ichthyosaurs (Figs. 9-30 and 9-31), plesiosaurs, some sea turtles, penguins, cetaceans, sirenians, seals, and seal lions have become paddlelike flippers. They are flattened, short, and stout, and in several groups the number of phalanges has greatly increased. In some ichthyosaurs there were as many as 26 phalanges per digit, or more than 100 in a single hand. Dolphins show the same modification (Fig. 9-31). Within the flippers of most of the other swimmers, however, the bones conform closely to the generalized tetrapod pattern (Fig. 9-32). Some aquatic mammals have lost all traces of hind limbs (Fig. 9-33, C).

FIGURE 9-30

Jurassic and Cretaceous ocean-dwelling reptile, ranging up to 3 m in length. (From Colbert, E.H.: Evolution of the vertebrates, ed. 2, New York, 1969, John Wiley & Sons, Inc.)

Ichthyosaurus

FIGURE 9-31

Convergent evolution in anterior limbs. **A,** Extinct, water-dwelling reptile. **B,** Water-dwelling mammal. h, Humerus; r, radius; u, ulna.

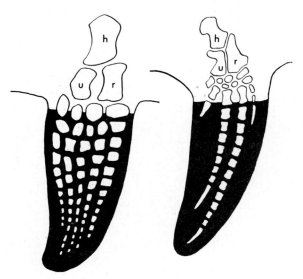

A. ICHTHYOSAUR   B. DOLPHIN

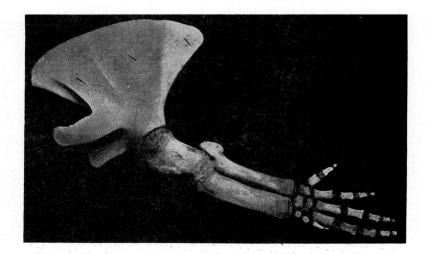

FIGURE 9-32

Right forelimb and pectoral girdle of a beaked whale. A remarkable resemblance to basic pattern remains, despite the fact that the limb has become paddlelike. (Courtesy American Museum of Natural History, New York.)

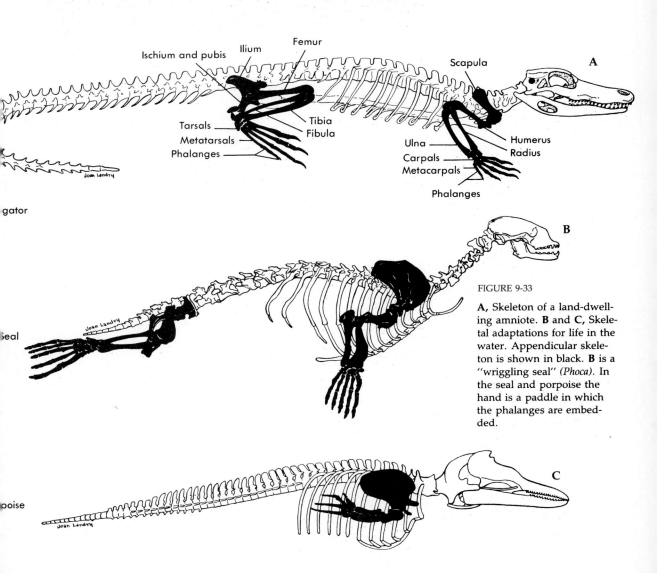

Ischium and pubis

Ilium

Femur

Scapula

A

Tarsals

Metatarsals

Phalanges

Tibia
Fibula

Ulna

Carpals

Metacarpals

Humerus
Radius

Phalanges

Joan Landry

gator

B

FIGURE 9-33

A, Skeleton of a land-dwelling amniote. B and C, Skeletal adaptations for life in the water. Appendicular skeleton is shown in black. B is a "wriggling seal" *(Phoca)*. In the seal and porpoise the hand is a paddle in which the phalanges are embedded.

Seal

Joan Landry

poise

C

Joan Landry

ADAPTATIONS FOR SWIFT-FOOTEDNESS. Mammals with pentadactyl hands and feet are usually **plantigrade,** which means that the wrists, ankles, and digits all rest on the ground. This is a primitive tetrapod stance not associated with fleetness. Generalized mammals—monotremes, marsupials, insectivores, primates—are plantigrade, as are some specialized mammals such as bears and the arboreal raccoons (Fig. 9-34, monkey). Mammals in which only the first digit has been reduced or lost tend to be **digitigrade,** which means that they bear their weight on digital arches with wrist and ankle elevated (Fig. 9-34, dog). Among digitigrade mammals are rabbits, rodents, and most carnivores. These mammals run faster, walk more silently, and are more agile than plantigrade species, and some, such as cats of various kinds, are the fastest runners of all (Fig. 10-10). Cheetahs, for example, can sprint at an estimated 70 miles (112 km) per hour (Hildebrand, 1980). To a carnivore, speed makes eating more likely; to a herbivore such as a hare, it may mean reaching a burrow instead of being eaten.

The extreme modification of reducing the number of digits and walking on the tips of the remaining digits is seen in ungulates. Unguligrade mammals walk on four, three, two, or even one digit (Fig. 9-34, deer), with wrists and ankles elevated well above the ground. The claws have become thick hoofs that bear the body weight and protect the living tissues of the toes from the abrasive action of the substrate (Fig. 5-20, hoof). The metacarpals and metatarsals that correspond to the missing digits are vestigial or lost, and those that remain are much elongated and frequently united (Fig. 9-35, horse, deer, camel). This has the effect of providing an additional functional segment to the limb. Although excelled in sprinting speed by

FIGURE 9-34

Plantigrade, digitigrade, and unguligrade feet, from left to right. Ankle bones are black; metatarsals are gray.

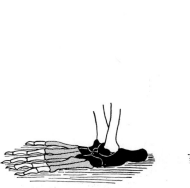
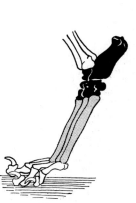

MONKEY   DOG   DEER

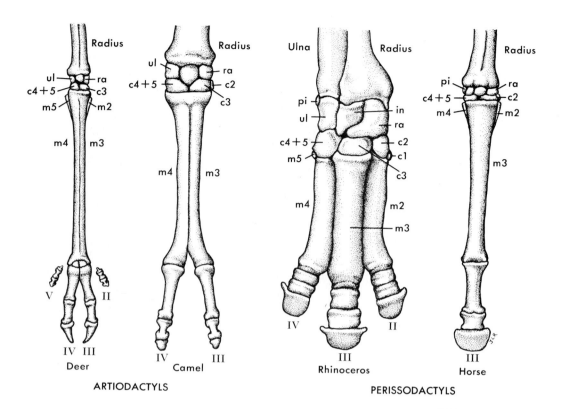

ARTIODACTYLS   PERISSODACTYLS

FIGURE 9-35

Right manus of several ungulates as seen from the front. *c2* to *c5*, Distal carpals 2 to 5; *in*, intermedium; *m2* to *m5*, metacarpals 2 to 5; *pi*, pisiform; *ra*, radiale; *ul*, ulnare; *II* to *V*, digits. In horses, *m3* is the "cannon bone."

some carnivorous digitigrade mammals, ungulates can sustain their speed for much longer. Not only are the feet of ungulates well suited for running, but they also function extremely well in rugged mountainous terrains. However, specialization has made their fingers and toes useless for anything else. This is the price of specialization.

Sequential evolutionary steps leading to the most specialized unguligrade stance may be illustrated by placing the fingers and palm flat on a tabletop, with the forearm perpendicular to the surface. This represents roughly the plantigrade position. Raising the palm off the table while keeping the fingers flat on the table illustrates roughly the digitigrade position. Unguligrade conditions may be illustrated by placing only the fingertips on the table and then raising the thumb, the little finger, the second finger, and finally the fourth finger, leaving only the third finger to bear the body weight, as in modern horses. Those fingers that fail to reach the table represent digits that have been successively reduced or lost in ungulates.

The horse underwent these successive changes commencing with the early *Eohippus*, which had four digits on the manus, and culminating in the modern *Equus*, which has a single digit. Despite extreme specialization of the manus of the modern horse, the proximal row of carpals (Fig. 9-36) is intact, and the distal row lacks only the first carpal. With the loss of digits I, II, IV, and V, metacarpals 1 and 5 have been lost and 2 and 4 have been

FIGURE 9-36

Mesaxonic and paraxonic manus of representative ungulates, showing bones lost (*white*) and retained (*black*) and indicating distribution of body weight through wrist and digits. The horse and camel are used as specific examples, and the number of bony elements is correct for these animals. *I,* Intermedium; m_2 to m_4, second, third, and fourth metacarpals; *p,* pisiform; *ph,* first phalanx; *r,* radiale; *u,* ulnare; *1* to *5,* distal carpals *I* to *V,* digits.

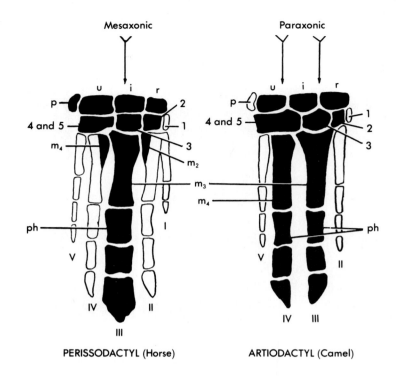

PERISSODACTYL (Horse)   ARTIODACTYL (Camel)

reduced to splinters. Metacarpal 3, associated with digit III, has elongated (Fig. 9-35).

Evolution among the ungulates seems to have progressed along two independent lines. In the line leading to **artiodactyls,** the weight of the body tended to be distributed equally between digits III and IV (Fig. 9-36, camel). Thus arose the "cloven" hoof (Fig. 9-37). Such a foot is said to be **paraxonic** because the body weight is borne on two parallel axes. Artiodactyls of today have an even number of digits. In the evolutionary line leading to **perissodactyls,** the body weight increasingly tended to be borne on digit III, the middle digit. This is a **mesaxonic** foot (Figs. 9-35 and 9-36, horse). Most perissodactyls have an odd number of digits; tapirs, however, have a small fourth digit on the front feet, three on the rear. It is the mesaxonic foot and not the number of digits that defines the perissodactyls.

ADAPTATIONS FOR GRASPING. Many mammals are able to flex their hand at the joint between palm and fingers. Rodents, for example, sit on their haunches and nibble on food held between *two hands,* which are flexed in this manner and which *face one another.* A further specialization is the ability to wrap the fingers *around* an object so that it is held securely in *one hand.* This is accomplished by flexing the fingers at each interphalangeal joint. Only primates can do that.

Another step in the evolution of the mammalian hand was development of an opposable thumb—one that can be made to touch the tips of each of the other digits. This was accomplished by formation of a saddle joint at the

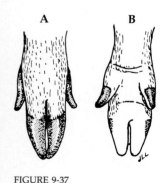

FIGURE 9-37

Foot of fetal pig. There is a cleft between digits. Thus the hoof is "cloven." **A,** As seen from in front; **B,** as seen from behind.

base of the thumb where it meets the palm, by setting the thumb at increasingly wider and wider angles to the index finger, and by the evolution of strong adductor pollicis (thumb) muscles. True opposability is found in Old World monkeys, but even there the hand does not have the full range of functional capability that has evolved in human beings. Neither New World monkeys nor anthropoid apes have a perfectly opposable thumb. With such a hand humans were able to fashion increasingly sophisticated instruments, commencing with rocks chipped by design and continuing to the electronic computer. As John Napier has said, "The implements of early man were as good (or as bad) as the hands that made them." Of course, evolution of the brain was an essential concomitant. Still, it seems impossible that any species on the planet earth lacking a prehensile hand, even having the brain, could evolve so sophisticated an existence.

Pes: The Hind Foot

The pes is comparable bone for bone with the manus except that there is no consistent equivalent of the pisiform bone (Table 9-3). Tetrapods primitively had four centralia in the pes. The number was reduced to one or two in primitive reptiles and to one in primitive mammals.

The pes of three amphibians is illustrated in Fig. 9-38. In anurans two elongated stout proximal tarsals—tibiale and fibulare—are firmly united at each end to provide a strong mechanical brace terminating at an intratarsal joint. At the joint are two or three small distal tarsals, one of which may be a displaced centrale. This architecture provides a broad palm with webbed fingers for pushing off when jumping, the sole means of anuran locomotion on land.

Living reptiles display considerable phylogenetic loss and embryonic fusion of ankle bones, turtles to a lesser degree than others (Fig. 9-39). The

---

TABLE 9-3   Comparison of Skeletal Elements of Manus and Pes

Manus*	Pes, with Synonyms	
Radiale	Tibiale	Talus or astragalus†
Intermedium	Intermedium	
Ulnare	Fibulare	Calcaneus
Pisiform		
Centralia (0 to 4)	Centralia (0 to 4)	Navicular
Distal carpal 1	Distal tarsal 1	Entocuneiform
Distal carpal 2	Distal tarsal 2	Mesocuneiform
Distal carpal 3	Distal tarsal 3	Ectocuneiform
Distal carpal 4 } Hamate Distal carpal 5 }	Distal tarsal 4 } Distal tarsal 5 }	Cuboid
Metacarpals (1 to 5)	Metatarsals (1 to 5)	
Digits (I to V)	Digits (I to V)	

*For synonyms, see Table 9-2.
†Often incorporates the intermedium and a centrale.

proximal row of tarsals is reduced to a single bone in *Sphenodon* and many lizards and has been given the name **astragalocalcaneus**. It incorporates all proximal tarsal bones and a centrale. A highly flexible intratarsal joint between proximal and distal tarsals helps bipedal lizards to run rapidly in a digitigrade manner, the long tail maintaining balance. The primitive joint between leg and ankle in these lizards is immobilized by ligaments. Most reptiles have five toes, although alligators and some lizards have four and some freshwater turtles have three. The phalangeal formula for *Sphenodon*, 2-3-4-5-4, is the generalized formula for reptiles. Alligators have reduced it to 2-3-4-4-0, and turtles to 2-3-3-3-2.

FIGURE 9-38

Left pes of amphibians. **A,** Rachitomous labyrinthodont *Trematops*. **B,** Generalized plethodontid salamander *Hydromantes*. **C,** *Rana catesbeiana*. Tarsal bones are red, and centralia are darker. *I* and *V,* First and fifth digits; *1* to *5,* distal tarsals; *m,* metatarsal.

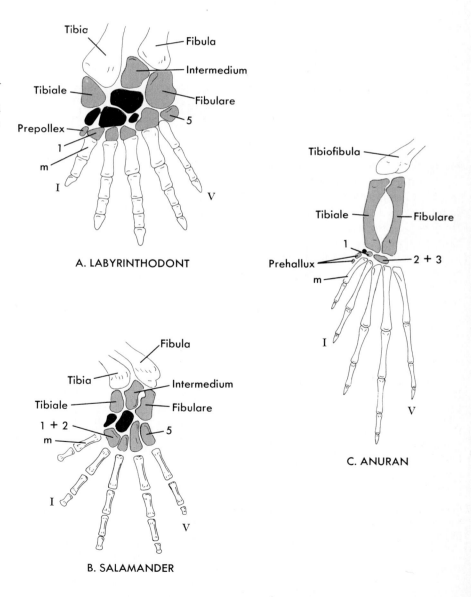

A. LABYRINTHODONT

B. SALAMANDER

C. ANURAN

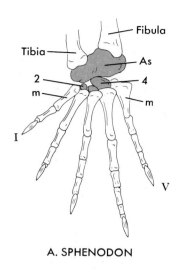

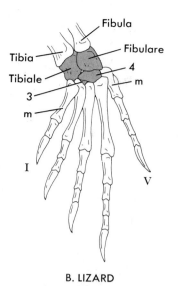

A. SPHENODON

B. LIZARD

FIGURE 9-39

Left pes of *Sphenodon* and the lizard *Uromastix*. *I* and *V*, First and fifth digits; *2* to *4*, distal tarsals; *As*, astragalo-calcaneus; *m*, metatarsal.

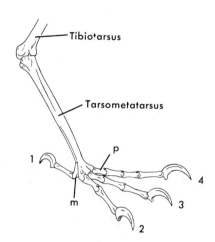

FIGURE 9-40

Left ankle and digits of a passerine bird, medial view, head to the right. *1* to *4*, Digits; *m*, first metatarsal; *p*, phalanx. The claws obscure the terminal phalanges.

A bird's foot is highly modified (Fig. 9-40). The proximal tarsals are united with the lower end of the tibia to form a **tibiotarsus,** there are no centralia, and the distal tarsals are united with the second, third, and fourth metatarsals to form a long, rigid **tarsometatarsus.** A first metatarsal remains in some species. *There is an intratarsal joint between the tibiotarsus and tarsometatarsus and a joint between the tarsometatarsus and the toes.* The latter permits a digitigrade stance (Fig. 3-26, *A*). This stance keeps the bird "ready" for takeoff, because extending the leg at the knee and intratarsal joint gives the initial thrust for becoming airborne. The thigh of a standing bird is not visible because it is quite short and, along with the knee, tucked away under the feathers or incorporated in the body wall.

FIGURE 9-41

A, Left pes of cat, anterior view. The phalangeal formula is 0-3-3-3-3. **B,** Left ankle and associated bones of rhesus monkey, lateral view. *1* to *5,* Distal tarsals; *I, II,* and *V,* metatarsals; *s,* sesamoid in the peroneus longus muscle. For synonyms for mammalian ankle bones (red) see Table 9-3.

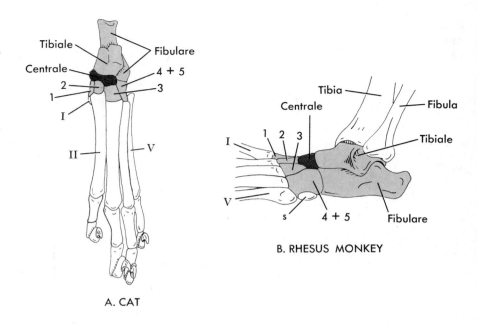

A. CAT

B. RHESUS MONKEY

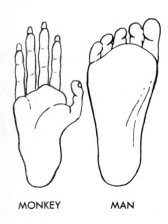

MONKEY   MAN

FIGURE 9-42

The partially opposable big toe of an Old World monkey and the nonopposable toe of man.

Most birds have four toes, a few have three; ostriches alone have two. The phalangeal formula is the same as that for the first four toes of *Sphenodon.* Usually three toes are directed radially forward and one comes off the back of the foot; but a few birds, including woodpeckers and parrots, have two toes at the back, the four forming an X **(zygodactyly).** The arrangement enables woodpeckers to obtain a firm grip on rough bark on a vertical tree trunk while drilling, braced against the trunk by stiff tail feathers. Birds can sleep while on a perch because the long tendons of the flexor muscles of the lower leg pass along the posterior aspect of the ankle to insert on the claw-bearing digits. The weight of the body pulls on the tendons, keeping the claws flexed on the perch.

Mammals, like their therapsid ancestors, lack an intratarsal joint but have a large hinge joint between the shank and the ankle. The tibiale is the principal weight-bearing bone of the ankle (Fig. 9-41, *B*). The other proximal tarsal, the fibulare, is elongated backward in plantigrade mammals, upward in digitigrades and unguligrades (Fig. 9-34). It is the heelbone of plantigrades. Except for a reduction in the number of centralia, the number of mammalian ankle bones has not changed appreciably since the first therapsids. In hominoids a **metatarsal arch,** or instep, distributes the body weight over four solid bases: the two heels and the ball of each foot. This has been compared with the four-pedestal architecture of the Eifel Tower. The arch also absorbs some of the shock generated by bipedal locomotion and provides "spring" for walking and running.

The phalangeal formula for the pes of early therapsids was 2-3-3-3-3, and this is still the typical formula for modern pentadactyl mammals. The great toe, or **hallux,** is opposable in many primates but not in human beings (Fig. 9-42). The condition is correlated with brachiation.

Origin of Limbs

Although the problem of the origin of paired fins may never be satisfactorily resolved, this is not true of the origin of tetrapod limbs. As mentioned, the paired fins of some ancient fish must have been the precursors of tetrapod limbs. The question then arises, "Did the skeleton of the fin of any known Devonian fish evince the potential for becoming a limb?" For an answer we turn to the lobe-finned crossopterygians, who resembled the first tetrapods in many respects.

The skeleton in the basal lobe of rhipidistian crossopterygian fins bears a striking resemblance to that of a tetrapod limb (Fig. 9-43). In the pectoral fins of crossopterygians a single pterygiophore, which we will call a humerus, articulates proximally with the scapula and distally with a pair of pterygiophores we will call radius and ulna. Loss of fin rays and modifications of the pterygiophores distal to the radius and ulna could have produced the skeleton of the first tetrapod limb. Westoll proposed that the first two rows of labyrinthodont carpals may be modified radials and that the more distal

FIGURE 9-43

Left pectoral fin of a Devonian crossopterygian, and left forelimb of an early tetrapod oriented to show similarities of endoskeleton. *Dark gray,* dermal bone of pectoral girdle. The fin is viewed from above; the limb is viewed from in front.

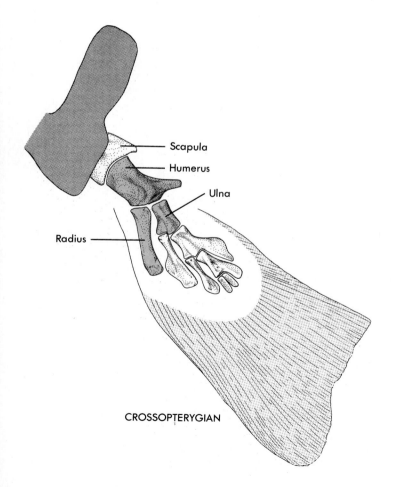

CROSSOPTERYGIAN

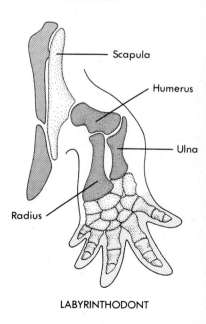

LABYRINTHODONT

bones of the manus may have evolved from new ossification centers. Meanwhile, the girdles of labyrinthodonts remained fishlike.

It is probable that crossopterygian fins were used at times for resting on the water bottom. They would not have borne appreciable weight because of the buoyancy of water. Minor modifications would have permitted "walking" on the muddy bottoms close to shore. Several hundred living species of fishes do this today, including the Australian lungfish. Using pectoral fins, some living fishes move several feet inland nightly. Some climb inclined planes with little fringed fins that have a remarkable resemblance to hands. Fig. 9-21 illustrates the changes in external appearance that must have taken place in the conversion of fins to limbs.

The pressures that drove vertebrates onto the land must, of necessity, be conjectural. It may have been the absence of predators on land, or that there was less competition for food, or simply that food was abundant on land. Or perhaps it was simply a manifestation of the tendency of organisms to invade a contiguous environment whenever nothing prevents the invasion. Whatever the explanation, it was inevitable that a limb more suitable for life on land would evolve from the fin of a fish.

LOCOMOTION WITHOUT LIMBS

Limbs provide traction and leverage for locomotion on land. Snakes and limbless lizards cannot gain traction in this manner; yet they are highly successful on land because modifications of the vertebral column, ribs, body wall muscles, and skin have provided them with alternative methods of moving about.

The most common method of locomotion in snakes and in limbless lizards is by forming irregular loops that become propped against, and push against, any available stationary object on their path, such as a clump of vegetation, a rock, a root, or simply an irregularity of the surface (Fig. 9-44). While the animal is moving, the eyes, which are close to the ground, constantly scan the substrate for the next contact point. When one looms, the head is thrust in that direction. On contact a wave of muscular contraction commences at the head and travels to the last segment of the tail, which is pulled forward in its turn. A minimum of three contact sites is necessary for this mode of locomotion, and no appreciable force is exerted downward against the ground. This mode of locomotion, referred to as **serpentine,** or **lateral undulation,** is a modification of the swimming movements of fishes. A basic necessity is a body that is metameric with respect to the skeletomuscular system.

Some snakes glide—seemingly flow—forward while keeping the entire body in a straight line, a spectacle that some viewers find unnerving! The ventral skin acts like a conveyor belt, sections of which stop and go. This has been termed **rectilinear locomotion,** and it depends on generating friction between sections of the ventral skin and the substrate. Unlike snakes that press laterally against contact points for leverage, these snakes press against the ground, using bunched groups of belly scutes as intermittent holdfasts. After a brief interval of holding, during which time the next more

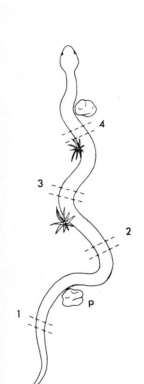

FIGURE 9-44

Serpentine locomotion. Forward progression is achieved by exerting pressure against contact points, or props, *p,* on the path. The head moves forward to a new contact point, and the rest of the body follows the established looping path, the segment at position *1* moving successively to positions *2, 3,* and *4* while the loops remain stationary with reference to the contact points. Eventually, each loop will be at the end of the tail and new loops will have formed anteriorly. Any track consists of sinuous waves. (Based on studies by Gans.)

caudal group of scutes is catching up, the downward pressure exerted by the first group is released, and the group flows forward, stretched out and elevated from the substrate, until it catches up with the group ahead. Meanwhile, the group immediately behind is holding fast. Thus rhythmic waves of forward flowing and resting scutes sweep along the body from head to tail. The skin, like that of many other snakes, is only loosely attached to the underlying tissues; there is much elastic tissue in the dermis, and the scutes, approximately one body segment long, overlap and are connected to the next by a pleat of cutaneous membrane that unfolds during movement. This combination of factors enables each scute to be released from the ground and commence its forward movement an instant before the next. Two sets of striated muscles are responsible for rectilinear locomotion. A pair of slender costocutaneus muscles extends downward and backward from high on each rib to the dermis at the edges of each scute. When these contract they lift the scute off the substrate so that it no longer serves as a holdfast, and they draw it forward while stretching the interscutal membrane. A second, more powerful pair extends less obliquely from a scute to the lower end of a more caudal rib. Contraction of the latter maintains the forward movement of the body mass within the skin envelope. Amphisbaenians use rectilinear progression, but the entire skin moves, not just the ventral skin.

Sidewinding (Fig. 9-45) enables rattlesnakes and other snakes to occupy sandy deserts, where the ground offers too few stationary contact points for lateral undulation and is too unstable to provide sustained friction for rectilinear locomotion. Sidewinding is also used by many snakes in other environments when they are temporarily in a situation where, because of the nature of the surface, other methods of locomotion would be clumsy or

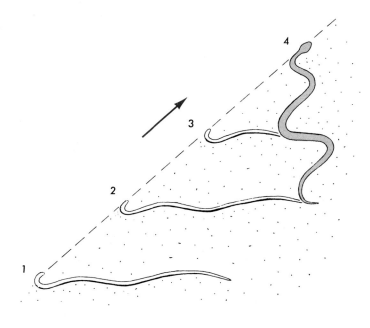

FIGURE 9-45

Locomotion in sidewinder snakes. *1* to *4*, Successive tracks in sand left when the head and neck are lifted from the ground, thrust forward, and set down firmly at the next position. The body does not touch the ground between tracks. The rest of the body is "flowing" above ground to the new position. Dotted lines and arrow indicate direction of movement.

ineffectual. In sand the body usually occupies two or three tracks at a time, and more or less of the anterior quarter of the snake is thrust forward to start a new one.

Burrowing snakes propel themselves in the burrow by modified lateral undulation, bracing S-shaped loops against the burrow wall and exerting horizontal force against the wall while thrusting the head and forebody forward in the burrow where new loops and braces are established. The more caudal sections then advance and establish new loops. The process, reminiscent of the operation of the bellows of a concertina (accordion), has been called **concertina** movement. The method in combination with constrictor movements and others is used in climbing trees, especially when following channels in the bark. According to Carl Gans, a long snake often combines lateral undulation with concertina movement in appropriate habitats as a common method of going from place to place.

All the foregoing methods of locomotion are made possible by vertebral columns consisting of up to 400 or more highly flexible intervertebral joints; by exceptional ribs that extend from the atlas to the tip of the tail and reach almost to the midventral line; by the unusually large number of muscle bundles that interconnect vertebrae and connect vertebrae with ribs; by wide, smooth, overlapping horny ventral scutes interconnected by pleated membranes; by the exceptional elasticity of the dermis; and by loose skin that allows independent movement of skin and enclosed body mass. These are the more obvious features that facilitate locomotion without limbs on land. Many systems had to undergo modification for the foregoing adaptations to be effective.

Chapter Summary

1. The appendicular skeleton consists of pectoral and pelvic girdles and the skeleton of the fins or limbs.

2. All pectoral girdles are modifications of a basic morphologic pattern seen in early bony fishes, consisting of three cartilages or endochondral bones and of four or five membrane bones derived from dermal armor.

3. Coracoids, scapulae, and suprascapulae are endoskeletal. Coracoids are absent above monotremes. Scapulae are universal except in some limbless species.

4. Clavicles, cleithra, supracleithra, posttemporals, and postcleithra are dermal bones. Cleithra and supracleithra are confined to fishes. Clavicles occur in all classes of bony vertebrates but are best developed in tetrapods. A midventral interclavicle of intramembranous origin is found in reptiles, birds, and monotremes.

5. Recent bony fishes have tended to lose the replacement bones of the pectoral girdle. Tetrapods have tended to lose dermal bones.

6. There are no dermal bones in the pelvic girdle of any vertebrate.

7. The pelvic girdle of fishes consists of two pelvic (ischiopubic) plates. These articulate laterally with a pelvic fin, and usually meet ventrally in a pubic symphysis. The plates unite to form a median bar in sharks and lungfishes.

8. In tetrapods two ossification centers in each pelvic plate become ischial and pubic bones that may form midline symphyses. An ilium braces the girdle against the vertebral column.

9. Additional small bones commonly associated with the amniote pelvic girdle include epipubics and hypoischials.

10. The ilium, ischium, and pubis unite to form an innominate bone (coxa) in mammals. It encloses an obturator foramen. The sacrum and pelvic girdle constitute a pelvis that encloses a pelvic cavity.

11. Excepting sesamoids, the bones of fins and limbs are endochondral.

12. Paired fins are of three types: lobed fins (Sarcopterygii) with a central jointed axis and pre- and postaxial radials; fin fold fins (Chondrichthyes) with a proximal row of basals and distal rows of radials; and ray fins (Actinopterygii). Ray fins are reduced to a few short radials, or none.

13. Fin rays are ceratotrichia, lepidotrichia, and actinotrichia.

14. Median fins are dorsal and anal. They have a skeleton of pterygiophores and rays. The caudal fins of teleosts also contain a urostyle and hypural bones. Caudal fins include diphycercal, heterocercal, homocercal, and hypocercal varieties.

15. The origins of fins is unknown. Fin fold, gill arch, and fin spine hypotheses have been proposed.

16. Tetrapod limbs are modifications of crossopterygian fins.

17. Tetrapod limb skeletons consist of a propodium, epipodium, mesopodium, metapodium, and phalanges. The last three constitute the manus or pes.

18. The orientation of the bones, relative mobility of the joints, and complexity of the muscles determine the nature of the locomotor activities of tetrapods.

19. Structural modifications of the manus and pes involve reduction in the number of bones, disproportionate lengthening or shortening of segments, and in some aquatic tetrapods increase in number of phalanges.

20. Digits have been reduced to four or fewer in modern amphibians, three in birds, and as few as one in some ungulates. Loss of digits is accompanied by loss or reduction of associated carpals and metacarpals or tarsals and metatarsals.

21. The most striking modifications of the manus are in flying tetrapods (pterosaurs, birds, bats), water-adapted reptiles and mammals (ichthyosaurs, plesiosaurs, cetaceans, sirenians), and ungulates. A few species lack one or both pairs of limbs.

22. Mammalian stances are plantigrade, digitigrade, and unguligrade. Ungulate feet are hoofed and either mesaxonic (perissodactyls) or paraxonic (artiodactyls).

23. Birds have no free ankle bones, one free metatarsal, and two to four toes, and are digitigrade. Some swift lizards are digitigrade while running.

24. Snakes move by lateral undulation, sidewinding, or rectilinear or concertina movements.

SELECTED READINGS

Additional references to locomotion will be found in Selected Readings at the end of Chapter 10.

Alberch, P., and Gale, E.A.: A developmental analysis of an evolutionary trend: digital reduction in amphibians, Evolution 39(1):8, 1985.

Daniel, J.F.: The elasmobranch fishes, Berkeley, 1934, University of California Press.

Daniel, T.L.: Unsteady aspects of aquatic locomotion, American Zoologist 24:121, 1984.

Elder, H.Y., and Trueman, E.R., editors: Aspects of animal movement, Cambridge, England, 1980, Cambridge University Press.

Gans, C.: How snakes move. In Vertebrate structure and functions: readings from *Scientific American* with introductions by Normal K. Wessells, San Francisco, 1955-1974, W.H. Freeman and Co., Publishers.

Goodrich, E.S.: Studies on the structure and development of vertebrates, London, 1930, The Macmillan Co., Ltd. (Reprinted by The University of Chicago Press, Chicago, 1986.)

Hildebrand, M.: The adaptive significance of tetrapod gait selection, American Zoologist 20:255, 1980.

Jarvik, E.: Basic structure and evolution of vertebrates, vol. 1, New York, 1980, Academic Press.

Lande, R.: Evolutionary mechanisms of limb loss in tetrapods, Evolution 32:73, 1978.

Lindsey, C.S.: Form, function, and locomotory habits in fish. In Hoar, W.S., and Randall, D.J., editors: Fish physiology, vol. 7, Locomotion, New York, 1978, Academic Press.

Napier, J.: The evolution of the hand. In Vertebrate structure and functions: readings from *Scientific American* with introductions by Norman K. Wessells, San Francisco, 1955-1974, W.H. Freeman and Co., Publishers.

Ostrom, J.H.: Bird flight: How did it begin? American Scientist 67:46, 1979.

Rackoff, J.S.: The origin of the tetrapod limb and the ancestry of tetrapods. In Panchen, A.L., editor: The terrestrial environment and the origin of land vertebrates, New York, 1980, Academic Press.

Romer, A.S.: Osteology of the reptiles, Chicago, 1956, University of Chicago Press.

Walker, A.D.: Evolution of the pelvis in birds and dinosaurs. In Andrews, S.M., Miles, R.A., and Walker, A.D., editors: Problems in vertebrate evolution, New York, 1977, Academic Press.

Webb, P.W., and Blake, R.W.: Swimming. In Hildebrand, M., editor: Vertebrate functional morphology, Cambridge, Massachusetts, 1983, Harvard University Press.

Webb, P.W., and Smith, G.R.: Function of the caudal fin in early fishes, Copeia, 1980, p. 559.

Westoll, T.S.: The origin of the primitive tetrapod limb. Proceedings of the Royal Society of London, Series B, Biological Sciences, 131:373, 1943.

10

MUSCLES

In this chapter we will classify muscles from several viewpoints and then examine skeletal muscles of fishes and tetrapods in functional groups, noting the primitive pattern and seeing how it has been subsequently modified.

Muscle is a tissue; muscles are organs. The tissue, hence the organ, is specialized to perform a single function: to shorten when stimulated, and thereafter to recover. Shortening is a result of chemical changes in two muscle proteins, actin and myosin. Shortening of a sufficient number of contractile units causes a corresponding shortening and fattening of the entire muscle mass. If the muscle surrounds a lumen, the lumen is compressed. If it extends between two structures, one of these is drawn toward the other. The usual stimulus for muscle contraction is a nerve impulse, although this is not true of cardiac muscle, as will be explained shortly.

This chapter is concerned primarily with the muscles that stabilize or operate the skeleton. These muscles are the final determinants of posture, locomotion, and orientation of the body in the environment.

CLASSIFICATION OF MUSCLE TISSUES AND MUSCLES

Muscle tissues and muscles may be classified according to any criterion that serves a useful purpose. When classified histologically, they are either **striated, cardiac** (striated with unique characteristics), or **smooth.** This and other classifications useful in functional morphology are discussed below.

Striated Muscle Tissue. Striated muscle is a tissue composed of long, cylindrical, multinucleate **muscle fibers** with transverse bands and longitudinal striae (Figs. 10-1, *A*, and 10-2). Parallel **myofibrils** within the fiber are responsible for the longitudinal striations. Each myofibril is composed of repeated **sarcomeres,** and each sarcomere is composed of two species of proteinaceous filaments, **myosin** and **actin** (Fig. 10-1, *A*). The striated appearance of a muscle fiber is a result of the perfect alignment in lateral register of all the sarcomeres in the muscle fiber. Stimuli are provided in the form of neurotransmitters, which are short-lived amines secreted at nerve endings in motor end plates (Fig. 10-2).

The striated muscle fiber is not a cell but a syncytium formed during histogenesis when myoblasts become aligned end to end and unite, contributing their nuclei and protoplasm to the cylindrical syncytium contained within a thin membrane, the **sarcolemma.** Muscle fibers are assembled to form skeletal muscles consisting also of the connective, vascular, and neural tissues required to provide mechanical support, tensile strength, metabolic needs, and stimuli for contraction.

The fibers of striated muscles are of several functional varieties that can be identified histochemically as fast-twitch glycolytic fibers (highly fatigable), fast-twitch oxidative-glycolytic fibers (relatively fatigue resistant), slow-twitch oxidative fibers (nonfatigable), and slow tonic fibers. Their ratios vary in different regions of a single muscle and in one muscle compared to another. They are related to the role the muscle plays in locomotion, as opposed, for instance, to maintenance of posture. Muscles that are called on for bursts of activity, such as short periods of rapid swimming in sharks, swift but not sustainable running in cats, or flapping flight in birds have a mix of fiber types different from the mix in muscles used for posturing. For example, up to 99% of the fibers in the pectoral muscles of pigeons are fast-twitch fibers, whereas in ratites, most of which are flightless, up to 40% are of the slow tonic variety. The slow fibers in the pectoral muscle of carinate birds are often located in the deep backbelly of the muscle. The relationship between histochemical fiber types and motor function is under current investigation in several vertebrate classes. It is too early to generalize concerning the specific role of each type.

Cardiac Muscle. The musculature (myocardium) of the heart is a special category of striated muscle tissue (Fig. 10-1, *B*). It contains myofibrils and filaments of actin and myosin arrayed similarly to those in skeletal muscle fibers, and the mechanisms of contraction are essentially the same. It differs in that it is a branching system of cells, a functional but not a cytoplasmic **syncytium,** in which successive cells are separated by specialized dual junctions, the **intercalated disks.** Cardiac muscle differs dramatically from skeletal muscle in its mode of excitation. It contracts automatically, without rhythmicity, when the interstitial fluids include chlorides of sodium, potassium, and calcium in precise proportions. The autonomic nervous system imposes rhthymicity.

Smooth Muscle Tissue. Unlike striated muscle, smooth muscle occurs most often in sheets as part of an organ. The cells are spindle shaped, uninucleate, have myofibrils, but lack striations (Fig. 10-1, *C*). Smooth muscle, like cardiac muscle, is innervated by the autonomic nervous system.

Voluntary and Involuntary Muscles. Muscles may be either voluntary or involuntary according to whether they can be operated at will. Voluntary muscles contract at will unless fatigued. This does not mean that they may not be operated unintentionally (reflexly) if the body is endangered, as when the skin comes in contact unexpectedly with a pin, or when holding one's breath becomes detrimental to physical welfare. Involuntary muscles, on the contrary, are not ordinarily subject to volitional control. The terms have less significance when applied to lower vertebrates.

Skeletal and Nonskeletal Muscles. Muscles attached to the skeleton are **skeletal muscles** and are striated and voluntary. They are disposed in accordance with a generalized pattern that is obvious in fishes and discernible in higher vertebrates, as will be seen shortly. Evolutionary modifications of the basic pattern are correlated chiefly with assumption of life on land. One variety of skeletal muscles, **branchiomeric,** differs from others in that they arise from neural crests rather than from myotomes of mesodermal somites, as described in Chapter 4.

FIGURE 10-1

Muscle tissues. **A,** Section of a striated muscle fiber, as seen with light microscopy, and its contractile units. **B,** Cardiac muscle tissue. Dark bars are intercalated disks at cell boundaries. **C,** Isolated smooth muscle cells from intestinal wall. **A** depicts, respectively, a section of one muscle fiber, a section of one myofibril, and muscle contraction as a function of sliding protein filaments. The left and right boundaries of the sarcomere are too fine to be seen with light microscopy.

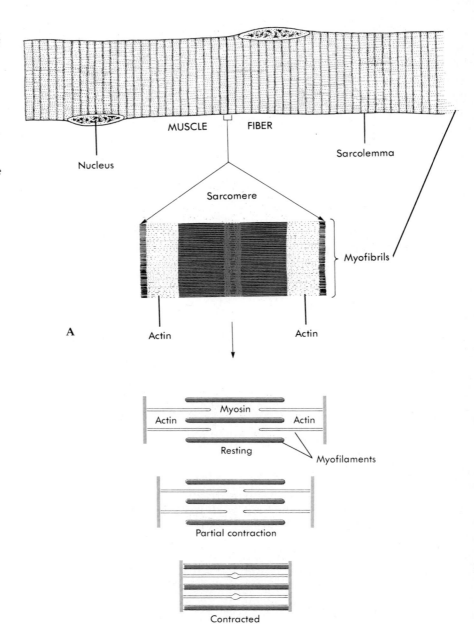

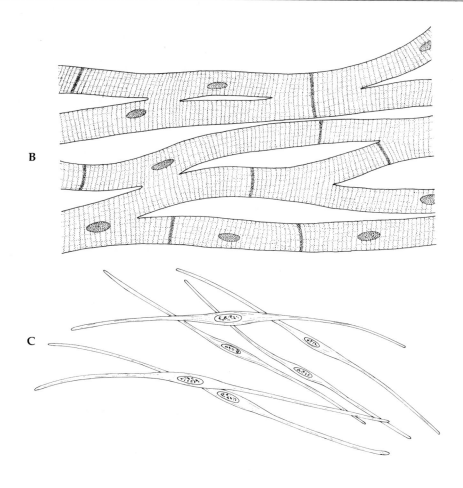

B

C

FIGURE 10-1, cont'd

For legend see opposite page.

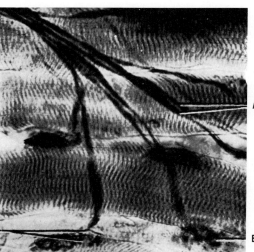

Myelinated fibers

Naked fibers

End plate

FIGURE 10-2

Innervation of skeletal muscle fibers. (From Bevelander, G., and Ramalay, J.A.: Essentials of histology, ed. 8, St. Louis, 1979, The C.V. Mosby Co.)

Muscles that do not attach to the skeleton must be referred to as **non-skeletal,** because no other designation (for example, smooth, involuntary, visceral) is an antonym for skeletal. Having made this clear, it can be added that most nonskeletal muscles are smooth and involuntary.

If we use the foregoing criteria most of the muscles of the vertebrate body can be classified according to the following scheme.

Skeletal, striated, voluntary muscles
    Axial
        Body wall and tail
        Hypobranchial and tongue
        Extrinsic eyeball
    Appendicular
    Branchiomeric
    Integumentary
Nonskeletal, smooth, involuntary muscles
        Muscles of tubes, vessels, and hollow organs
        Intrinsic eyeball muscles
        Erectors of feathers and hairs
    Cardiac muscle
    Electric organs

The foregoing categories are not mutually exclusive, and no such scheme can be devised. Some branchiomeric muscles have become secondarily associated with paired appendages, the ciliary muscles of the eyeball of reptiles and birds are often striated, and erectors of feathers and hairs are smooth integumentary muscles.

Somatic and Visceral Muscles. Somatic muscles orient the animal—its body, or soma—with respect to the external environment. They enable an

TABLE 10-1   Some Contrasts Between Somatic and Visceral Muscles

Somatic Muscles	Visceral Muscles Except Branchiomeric
Striated, skeletal, voluntary	Smooth, nonskeletal, involuntary*
Primitively segmented	Unsegmented
Myotomal	Arise mostly from lateral mesoderm†
Mostly in body wall and appendages	Mostly in splanchnopleure‡
Primarily for orientation in external environment	Regulate internal environment
Innervated directly by spinal nerves and cranial nerves III, IV, VI, and XII	Innervated by postganglionic fibers of autonomic nervous system§

*Cardiac muscle is striated.
†Branchiomeric muscles arise from neural crests.
‡Those in the body wall erect hairs or feathers or constrict blood vessels.
§Branchiomeric muscles are innervated directly by cranial nerves V, VII, IX, and X.

animal to go deeper into an environment to pursue food or a mate and to withdraw from an environment that is hazardous. Skeletal muscles other than those of the visceral arches are somatic. Because somatic muscles are either ontogenetic or phylogenetic derivatives of myotomes, they are also referred to as **myotomal muscles.**

Visceral muscles are those that generally maintain an appropriate internal environment. They are the muscles of hollow organs, vessels, tubes, and ducts; the intrinsic musculature of the eyeball, the erector muscles of hairs and feathers, and the striated muscles of the jaws and the remaining visceral arches. Visceral muscles compress lumens, cause peristalsis, and serve as sphincter and dilator muscles in such diverse locations as gill pouches, the iris diaphragm, and the pyloric sphincter. Branchiomeric muscles are a special variety of visceral muscles. They are attached to the skeleton of the visceral arches, and function as muscles of the jaws and gills of fishes. They have acquired some new functions in tetrapods. Reasons for classifying them as visceral are given on p. 349.

If we decide to classify musculature on the basis of whether it is somatic or visceral, a scheme somewhat like the following would be useful.

Somatic muscles
    Muscles of the body wall and tail
    Hypobranchial and tongue muscles
    Extrinsic eyeball muscles
    Appendicular muscles
Visceral muscles
    Branchiomeric muscles
    Muscles of tubes, vessels, hollow organs
    Intrinsic eyeball muscles
    Erectors of feathers and hairs
    Cardiac muscle

As in the first classification scheme, the above categories are not mutually exclusive. A few appendicular muscles are branchiomeric, erectors of feathers and hairs are in the body wall, and electric organs have been omitted because some are somatic and some are visceral.

Except for branchiomeric muscles, visceral muscles exhibit few evolutionary changes throughout the vertebrate series and will receive little attention in this chapter. Some contrasts between somatic and visceral muscles are presented in Table 10-1.

INTRODUCTION TO SKELETAL MUSCLES
Skeletal Muscles as Organs

Skeletal muscles consist of muscular and tendinous portions. The **muscular portion** consists of parallel fascicles of relatively few striated muscle fibers, each fascicle enveloped in and pervaded by fine connective tissue, the **endomysium.** Fascicles are combined into larger parallel bundles, which are encompassed collectively by a continuous reticulum of dense

FIGURE 10-3

Origin, insertion, and direction of tensile force exerted by two muscles of the right upper arm of a rabbit, lateral view (i.e., head is to the right). Arrows within muscles indicate direction of pull on the forearm; arrows at right indicate direction of displacement of forearm when muscles contract. *Bi,* biceps brachii; *h,* humerus at elbow; *r,* radius; *s,* scapula; *Tr,* long head of triceps brachii; *u,* ulna. *Tr* provides the motive power for a first-class lever, and *Bi* for a third-class lever. The two muscles must operate synergistically, monitored by proprioceptive feedback to the central nervous system to effect smooth operation of the forearm against a load.

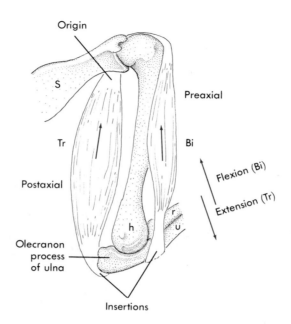

connective tissue, the **perimysium.** Surrounding the entire muscle is a tough glistening fibrous sheath, the muscle fascia, or **epimysium.** The perimysium and epimysium form a single continuum of high tensile strength encompassing the contractile components of the organ.

Tendons are continuations of the organ beyond the site where muscle fascicles end, the collagenous bundles of the perimysium and epimysium passing over directly into those of the tendon. In the same manner, at the site of attachment of the tendon to the skeleton the connective tissue fibers of the perichondrium or periosteum pass directly into the tendon. Consequently, tension produced by muscle contraction is spread throughout the entire organ, internally and externally, and from one skeletal attachment to its opposite.

The **anatomical origin** of a muscle, as opposed to its phylogenetic or ontogenetic origin, is the site of attachment that remains fixed under most functional conditions; that is, the structure on which it originates is not displaced when the muscle contracts. For example, when the biceps muscle of the upper arm contracts, the forearm is drawn toward the upper arm. The origin is, therefore, somewhere above the elbow (Fig. 10-3). The **insertion** of a muscle is the site of attachment that is normally displaced by contraction of the muscle. The biceps inserts on a bone of the forearm. A muscle may cause displacement of the origin instead of the insertion if the insertion becomes immobilized by other muscles. For example, the geniohyoid muscle, which is attached to the hyoid bone and to the lower jaw at the chin, either lowers the jaw or draws the hyoid forward, depending on which one is immobilized at the time.

Muscles with a pronounced bulge in the middle are said to have a **belly.** Some muscles are straplike, or broad and flat, and have no belly. Broad, flat muscles usually have broad tendons. Such a tendon is an **aponeurosis.** Broad muscles that insert on connective tissue **raphes** (seams) such as the linea alba (Fig. 10-8) or the raphe in the floor of the buccal cavity may compress a cavity and the organs within it.

During dissection of adult mammals, students frequently notice that skeletal muscles vary in size in the two sexes. Androgens, the predominant gonadal hormones of males, cause amino acids to be linked together into polypeptides and proteins. Inasmuch as muscle is 80% protein, androgen results in statistically demonstrable larger muscles in males. This confers on the male an added advantage during mating, which increases the likelihood of impregnation of all ovulated eggs in a population and therefore the probability of survival of the species.

Actions of Skeletal Muscles

Skeletal muscles may be categorized according to their function. **Extensors** tend to straighten two segments of a limb or vertebral column at a joint. **Flexors** tend to draw one segment toward another. **Adductors** draw a part toward the midline. **Abductors** cause displacement away from the midline. **Protractors** cause a part, such as the tongue or hyoid, to be thrust forward or outward; **retractors** pull it back. **Levators** raise a part; **depressors** lower it. **Rotators** cause rotation of a part on its axis. **Supinators** are rotators that turn the palm upward, **pronators** make it prone (turn it downward). **Tensors** make a part such as the eardrum more taut. **Constrictors** compress internal parts. **Sphincters** are constrictors that make an opening smaller; **dilators** have the opposite effect. Most sphincter and dilator muscles are nonskeletal.

It is exceptional for a single skeletal muscle to act independently; muscles nearly always act synergistically in functional groups. While one group is contracting, a group of so-called antagonistic muscles must relax simultaneously, otherwise a stalemate results. For all muscle groups to function smoothly they must be under reflex control of the cerebellum, which dispatches motor impulses to appropriate muscles on receiving sensory feedback from proprioceptive receptors located in the muscles, in their tendons, and in the bursae and capsules of affected joints (Chapter 16).

Names and Homologies of Skeletal Muscles

Skeletal muscles have been named for the direction of their fibers (oblique, rectus), location or position (thoracis, supraspinatus, superficial), number of subdivisions (triceps), shape (deltoid, teres, serratus), origin or insertion (xiphihumeralis, stapedius), action (levator scapulae, risorius), size (major, longissimus), and for still other features including a combination of these. Insight into the significance of a muscle's name should aid in recalling other information about the muscle.

Names were given originally to muscles in accordance with one or the other of the preceding criteria without regard to homology. Therefore, the fact that two muscles have the same name is no assurance that they are homologous, and the more distantly related the two animals are, the more likely it is that the muscles are not homologous. Similarity of location, origin, and insertion are a good starting point, but not a reliable basis for establishing homologies. One reason is, muscles sometimes alter their sites of attachment during evolution as they spread and produce new slips. A possible example is the genioglossus muscle that inserts on the mammalian tongue. It was inherited, of course, from reptiles. The probable homologue in some birds inserts on the sublingual seed pouch. The only way to be sure these are the same muscle is through a combination of methods involving, although not exclusively, their embryogenesis and nerve supply. One method of seeking homologues, therefore, is to compare the early developmental stages of two muscles suspected of being homologous. Do the anlagen appear to be identical? If so, the two should be homologues, even though in one species the anlage expands in one direction or another to assume new relationships, whereas the anlage of the other does not do so. If the anlagen are similar, the muscles may be homologous. On this premise the coracoscapular muscle of reptiles is homologous with the supraspinatus and infraspinatus muscles of mammals. Neurological studies often provide substantiating evidence. If the motor cell bodies that innervate two muscles suspected of being homologues are located in homologous nuclei within the cord or brain, this is substantiating evidence. Of course, there must also be some confidence that the two motor nuclei are homologous. The latter method is able to identify muscles of the mammalian head and neck that operated the gills of fishes, because most of the motor cell bodies of branchiomeric muscles are in a different motor column than the motor cell bodies supplying any other group of muscles. Embryological and neurological evidence is the most reliable at present for establishing muscle homologies. However, such data are available for relatively few individual muscles. On the other hand, homologies between functional groups of muscles may be deduced with a much greater degree of reliability.

AXIAL MUSCLES

Axial muscles are the skeletal muscles of the trunk and tail. They extend forward beneath the pharynx as hypobranchial and tongue muscles, and they are present in the orbits as extrinsic muscles of the eyeballs. They do not include branchiomeric or appendicular muscles.

The immediately evident feature of the axial muscles of generalized vertebrates is their metamerism (Fig. 10-4). This primitive condition along with a flexible vertebral column enables fishes and many aquatic tetrapods to propel themselves in water by rhythmical lateral undulatory movements that sweep from head to tail (Fig. 10-5). These muscles perform the same function for limbless tetrapods living on land. In higher tetrapods the metamerism became partly obscured when paired appendages requiring an extensive extrinsic appendicular musculature assumed responsibility for locomotion on land. Nevertheless, even in mammals much of the axial musculature remains metameric.

The axial muscles are segmental because of their embryonic origin: they arise from segmental mesodermal somites (Figs. 10-6 and 15-6). Mesenchyme cells from the myotome of each somite stream into the embryonic lateral body wall and migrate ventrad while undergoing repeated cell division (Figs. 4-9 and 10-7). They cease migrating when they reach the midventral line, where a longitudinal connective tissue raphe, the **linea alba,** forms. These myotomal cells give rise to blastemas for body wall muscles; and because the somites are metameric, the blastemas are initially metameric. Blastemal cells, having become myoblasts, unite to form striated muscle fibers, and the body wall muscles commence to take shape. The metamerism of the embryonic somites is expressed in adults as **myomeres**

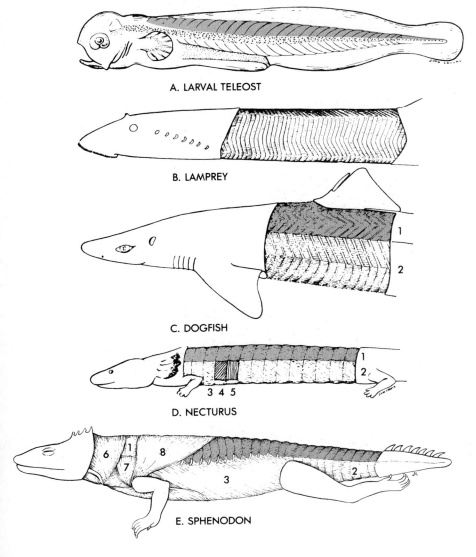

A. LARVAL TELEOST

B. LAMPREY

C. DOGFISH

D. NECTURUS

E. SPHENODON

FIGURE 10-4

Trunk muscles of selected vertebrates. *1,* Expaxials (dorsalis trunci, *red*); *2,* hypaxials; *3,* external oblique; *4,* internal oblique; *5,* transverse muscle of abdomen. The notochord in the larval teleost is located just behind the eye. In *Sphenodon* the appendicular muscles are *6,* trapezius; *7,* dorsalis scapulae; *8,* latissimus dorsi. The horizontal skeletogenous septum separates *1* and *2* in C and **D.**

FIGURE 10-5

Locomotion by lateral undulation in a fish.

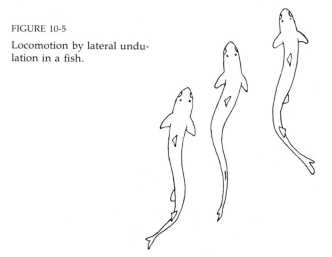

FIGURE 10-6

Axial muscle origins and innervation in a generalized vertebrate embryo (diagrammatic). The eyeball muscles organize near three preotic somitomeres at the level of the oculomotor, trochlear, and abducens nuclei, respectively. Segmental muscles of the trunk arise from trunk somites and are supplied by corresponding segmental nerves. Hypobranchial musculature migrates forward into the floor of the pharynx accompanied by a nerve supply. Motor fibers innervating myotomal muscle have their cell bodies in the somatic motor column of the cord and brain. The postotic somitomeres (*dotted outlines*), which vary in number in different species, make no contribution to the musculature. The central nervous system has been projected above the embryo for clarity. Amniotes have three somitomeres anterior to the three preotic somitomeres seen here.

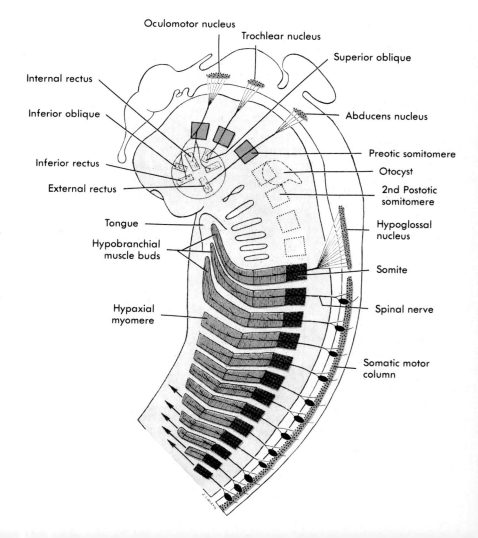

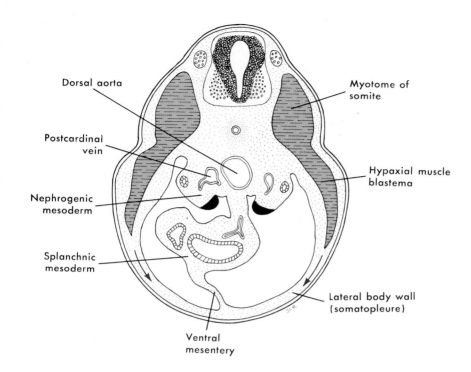

FIGURE 10-7

Cross section of mammalian embryo showing invasion of myotomal cells *(red)* into lateral body wall to form hypaxial muscles.

Dorsal aorta

Postcardinal vein

Nephrogenic mesoderm

Splanchnic mesoderm

Myotome of somite

Hypaxial muscle blastema

Lateral body wall (somatopleure)

Ventral mesentery

wherever myosepta (myocommata) separate the muscle of one body segment from the next.

Myosepta do not form in the abdominal region of higher tetrapods (Fig. 10-4, *E*). As a result their abdominal musculature consists of broad sheets rather than serially repeated myomeres. Nevertheless, these sheets are innervated by as many spinal nerves as there were somites that contributed to them.

Trunk and Tail Muscles of Fishes

The musculature of the body wall and tail of fishes consists of metameric **myomeres** separated by zigzag **myosepta** to which the muscle fibers attach (Fig. 10-4, *A* to *C*). Typically, there is a myomere for each vertebra, and a spinal nerve for each myomere (Fig. 10-6). The myosepta are anchored to transverse processes; and because the vertebrae are intersegmental, each myomere pulls on two successive vertebrae and, when present, on two ribs, causing lateral flexion between two centra (Fig. 7-4). Brook lampreys *(Lampetra)* lack this segmentation.

Except in agnathans, the myomeres are divided into a dorsal and a ventral mass by a horizontal sheet of connective tissue, the **lateral (horizontal) skeletogenous septum,** that extends between the transverse processes and the skin throughout the length of the trunk and tail (Figs. 7-6 and 10-4, *C* and *D*). The absence of this septum in agnathans may be correlated with the absence of vertebrae. Dorsal to the septum (lateral to the neural arches) are

FIGURE 10-8

Hypobranchial and ventral branchiomeric muscles of a shark. At left, anterior to the first hypaxial myomere, are three hypobranchial muscles. At right are branchiomerics. The intermandibular is superficial to the coracohyoid and coracomandibular muscles and has been partly removed on the left to reveal the coracomandibular muscles.

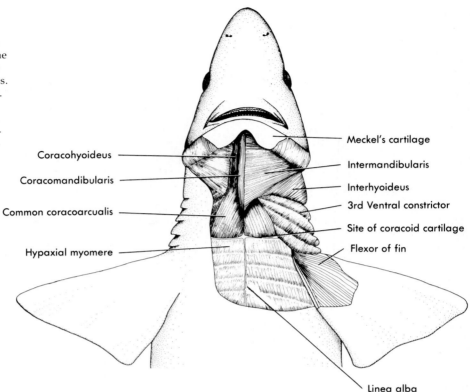

Coracohyoideus

Coracomandibularis

Common coracoarcualis

Hypaxial myomere

Meckel's cartilage

Intermandibularis

Interhyoideus

3rd Ventral constrictor

Site of coracoid cartilage

Flexor of fin

Linea alba

epaxial muscles. Beneath the septum in the lateral body wall are **hypaxial muscles.** The linea alba and a middorsal raphe separate the myomeres of the two sides of the body. The myosepta of fishes are seen to zigzag from dorsal to ventral when the skin is removed, but deeper within the body wall and tail they are elaborately folded. In response to unilateral waves of motor impulses that sweep caudad over successive spinal nerves, the wave on one side being in advance of the other, successive myomeres exert a pull on the vertebral column, first on one side, then the other, evoking rhythmical lateral undulations of the trunk and tail. These propel the fish during locomotion.

A thin sheet of **oblique fibers** lies superficial to the main hypaxial mass ventrolaterally in many fishes, and a thin ventralmost ribbon of still more superficial fibers parallels the linea alba on each side. The latter resembles the rectus abdominis muscle of tetrapods (Fig. 10-13).

The metamerism of the hypaxial muscles of fishes is interrupted where the pectoral and pelvic girdles are built into the body wall, and by the gills. Dorsal to the gills the epaxial muscles continue to the skull as **epibranchial muscles** (Fig. 10-22, *A*). Beneath the gills the hypaxials extend to the lower jaw as specialized **hypobranchial muscles** that have lost their metamerism (Fig. 10-8).

Trunk and Tail Muscles of Tetrapods

Urodeles have metameric epaxial and hypaxial muscles that produce swimming movements in water by lateral undulation of the trunk and tail, as in fishes (Fig. 10-4, *D*). Because urodele limbs are not strong weight bearers, these same muscles produce squirming, swimminglike locomotion on land. Amniotes, on the other hand, except limbless ones, have evolved strong supportive limbs operated by complex appendicular muscles, and this was accompanied by loss of much of the primitive metamerism of the axial musculature. Except for the deepest slips, the epaxial myomeres of the more specialized amniotes have coalesced to become long unsegmented epaxial bundles lying on the transverse processes of the vertebrae from the occiput to the tip of the tail (Fig. 10-9). These long bundles extend (straighten) the vertebral column, keeping it supportive of the weight of the viscera. A similar modification occurred in the dorsalmost hypaxial region, where long unsegmented **subvertebral** muscles formed beneath the transverse processes. Subvertebral muscles flex (arch) the column. In no tetrapod is the effectiveness of this combination of extensors and flexors of the vertebral column better demonstrated than in fleet digitigrade carnivorous mam-

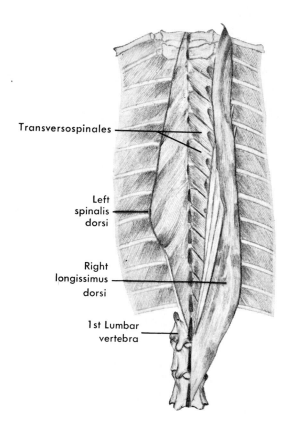

Transversospinales

Left spinalis dorsi

Right longissimus dorsi

1st Lumbar vertebra

FIGURE 10-9

Three epaxial muscles of the thorax of a hamster, dorsal view. The right spinalis dorsi has been removed to reveal the deeper transversospinales that connect transverse processes with the neural spine on the second vertebra forward. The longissimus dorsi, which lies lateral to the spinalis dorsi, has been removed on the left.

mals, which usually must "hump" to capture food, or in jackrabbits, who must scamper to avoid being eaten (Fig. 10-10).

The remaining hypaxial musculature, that in the lateral body wall, also changed, losing the myosepta and becoming split into several superficial and deep layers, the fibers of each layer being disposed obliquely to those of the others. This modification is seen in its simplest manifestation in urodeles (Fig. 10-4, *D*). As an accompaniment to these various changes, the horizontal septum ultimately became insignificant and disappeared. Figure

FIGURE 10-10

Extension and flexion of the vertebral column of a fleet digitigrade mammal.

FIGURE 10-11

Scheme of axial muscle disposition in the abdominal region of a mammal, based on cross section of a rabbit, cephalic view. The cutaneous maximus muscle has been removed.

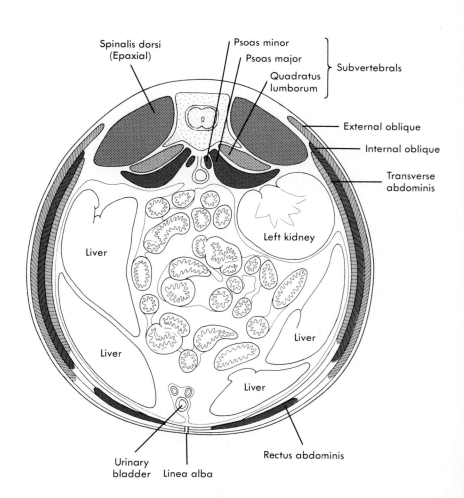

10-11 illustrates the components of the axial musculature as they are disposed in the abdominal region of a mammal.

EPAXIAL MUSCLES OF THE TRUNK. The epaxial muscles of tetrapods extend from the base of the skull to the tip of the tail (Fig. 10-12). In lower tetrapods these retain their primitive metamerism, extending between successive epaxial myosepta and transverse processes as the dorsalis trunci (Fig. 10-4, *D* and *E*, in color). The most anterior epaxial muscles attach to the occipital region of the skull. In amniotes the epaxial muscles of the trunk have become increasingly specialized into nonsegmental functional bundles that extend over several or many body segments, and are divisible arbitrarily into four groups: intervertebrals, longissimus, spinales, and iliocostales.

Intervertebrals are the deepest epaxial muscles and the only ones to retain their primitive metamerism in mammals. They extend between two successive transverse processes (intertransversarii), two neural spines (interspinales), two neural arches (interarcuales), or two successive zyg-

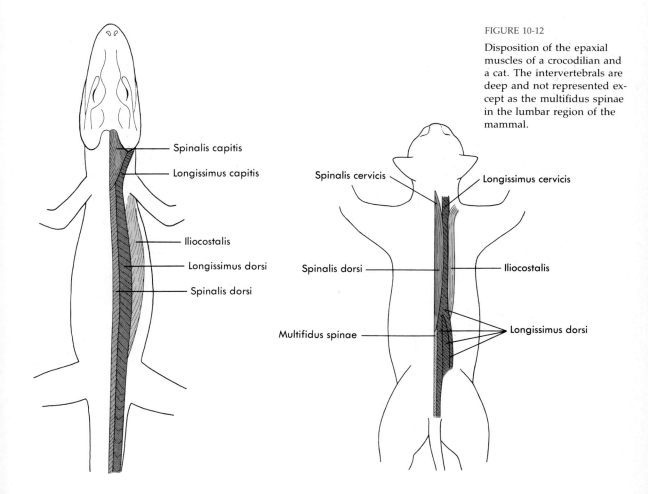

FIGURE 10-12

Disposition of the epaxial muscles of a crocodilian and a cat. The intervertebrals are deep and not represented except as the multifidus spinae in the lumbar region of the mammal.

Spinalis capitis
Longissimus capitis
Iliocostalis
Longissimus dorsi
Spinalis dorsi

Spinalis cervicis
Longissimus cervicis
Spinalis dorsi
Iliocostalis
Multifidus spinae
Longissimus dorsi

apophyses (interarticulares), and participate with longer epaxials in exten-
tion and lateral flexion of the vertebral column. In mammals they are sometimes grouped with short spinales bundles and designated collectively, the
multifidus spinae.

The **longissimus** and **spinales** muscles occupy, respectively, lateral and
medial positions above the vertebral column and are named according to

TABLE 10-2   Representative Somatic (Myotomal) Muscles of Head, Trunk, and Forelimbs
of Mammals

	Eyeball	Tongue	Hypobranchial
Head and neck	Superior oblique Inferior oblique Medial rectus Lateral rectus Superior rectus Inferior rectus	Genioglossus Hyoglossus Styloglossus Lingualis	Geniohyoideus Sternohyoideus Sternothyroideus Thyrohyoideus Omohyoideus

	Epaxial Muscles	Hypaxial Muscles

Trunk	Intervertebrals
	  Intertransversarii
	  Interspinales
	  Interarcuales
	  Interarticulares
	Longissimus
	  L. dorsi
	  L. cervicis
	  L. capitis
	  Extensor caudae
	    lateralis
	Spinales
	  S. dorsi
	  S. cervicis
	  S. capitis
	  Transversospinales
	Iliocostales
		Subvertebrals
		  Longus colli
		  Quadratus lumborum†
		  Psoas minor
		Oblique group (parietals)
		  Internal and external in-
		    tercostals
		  Internal and external
		    obliques of abdomen
		  Cremaster
		  Supracostals
		    Scalenus*
		    Serratus dorsalis*
		    Levatores costarum*
		    Transversus costarum
		  Diaphragm
		Transverse group (parietals)
		  Transversus thoracis (subcos-
		    tal)
		  Transversus abdominis
		Rectus muscles
		  Rectus abdominis
		  Pyramidalis

	Extrinsic	Intrinsic

Forelimb	Secondary appendicular
	  Levator scapulae
	  Rhomboideus
	  Serratus ventralis
	Primary appendicular
	  Latissimus dorsi
	  Pectorales
		See Table 10-4

*May be epaxial by derivation.
†Secondary appendicular muscle.

their location (Table 10-2). They continue into the tail as lateral and medial extensors of the tail. The most anterior divisions (capitis) insert on the skull and assist in movements of the head. The longissimus is so named because in all tetrapods it is the longest epaxial bundle. In the lumbar region of mammals the longissimus dorsi consists of three distinct bundles (Fig. 10-12, *B*). Spinales include long, medium, and short bundles that connect neural spines or transverse processes with neural spines several or many segments cephalad. Transversospinales are short spinales bundles, two of which are labeled in Fig. 10-9.

The **iliocostales** lie lateral to the longissimus, arise on the ilium, and pass forward to insert on the upper ends of ribs and on uncinate processes. Extensions continue forward into the neck, but not caudad into the tail.

The epaxial muscles become increasingly hidden by the expanding appendicular muscles and lumbodorsal fascia of higher amniotes, which must be removed to reveal the underlying epaxials (compare Fig. 10-18, *A*, *B*, and *C*).

In turtles and birds the vertebral column in the trunk is rigid, and in that region epaxials are poorly represented. The longissimus, for example, is mostly ligamentous, having few fleshy components. In the neck it is a different story. The intervertebrals, longissimus, and spinales are well represented, providing the cervical region of these tetrapods with exceptional flexibility. In birds one component, the complexus muscle, which arises on several of the anteriormost transverse processes and inserts on the interparietal bone, provides power for cracking the eggshell with the beak during hatching. In contrast to turtles and birds, the epaxials of the trunk of limbless tetrapods provide a considerable share of the power for locomotion.

HYPAXIAL MUSCLES OF THE TRUNK. The hypaxial muscles of the trunk of tetrapods can be divided arbitrarily into **subvertebrals** (longitudinal bundles close to the vertebral column), **oblique** and **transverse sheets** in the lateral body wall (parietals), and **rectus** muscles (longitudinal straplike muscles close to the linea alba). (Representatives of these groups are listed in Table 10-2 and represented schematically in Fig. 10-11.)

Subvertebral Muscles. Bundles of subvertebral muscles lie immediately under the transverse processes of the vertebrae, where their primary function is to flex the vertebral column. They are present in the neck of amniotes as the **longus colli.** These are the most prominent subvertebrals in turtles and birds because the remainder of the vertebral column is rigid. Subvertebrals have become meager in the thoracic region of most amniotes, but they are prominent in the lumbar region, where they are represented chiefly by the **quadratus lumborum,** and in mammals by a **psoas minor** in the coelomic roof. (The psoas major of primates and some other mammals is an extrinsic appendicular muscle). Portions of the lumbar subvertebral musculature of cattle are sold as tenderloin in markets and restaurants in the United States. The subvertebral musculature does not extend into the tail.

Oblique and Transverse Muscles. The parietal muscles have become stratified into superficial and deep layers, typically **external oblique, internal oblique,** and **transverse muscles of the abdomen,** and in amniotes **external intercostal, internal intercostal,** and **transverse muscles of the thorax.** The fibers of all these pass more or less obliquely from origin to insertion. One or another of these sheets may either be split in two or lost. In aquatic urodeles, for example, in which the parietal muscles participate in locomotion, the external oblique muscle is split into superficial and deep parts. In crocodilians and some lizards all three layers consist of two sheets each. In anurans the internal oblique is sometimes missing, and in birds all sheets are thin. In turtles they are less than thin—they are vestigial! But because of the rigid shell a turtle couldn't use them anyway. (If you would like some turtle soup, it will have to come from the neck, tail, or appendicular muscles!)

Muscle slips from the inferior border of the internal oblique, and sometimes the transverse abdominal muscle, of male mammals form a **cremaster muscle** that loops around the spermatic cord commencing at the inguinal ring and inserts on a fibrous sheath in the wall of the scrotum below the testes (Fig. 14-21). It is best developed in mammals with permanently open inguinal canals, such as rabbits, because it retracts the testes into the canals in those species.

The oblique and transverse muscles, particularly the intercostals, play a major role in external respiration in most amniotes other than turtles, and an accessory role in mammals. The intercostals are assisted by specialized subdivisions of **supracostal muscles** that differentiate on the surface of the rib cage, chiefly **scalenus, serratus dorsalis, levatores costarum,** and **transversus costarum.** Along with the rectus, parietal muscles support the abdominal viscera in a muscular sling, and they compress the viscera for such functions as egg laying, delivery of mammalian young, and emptying of the digestive tract. Although they lack segmental myosepta in amniotes, their metameric history is evidenced by their embryonic origin from successive somites and by their innervation via ventral rami of successive spinal nerves.

Rectus Muscles. The rectus muscles of most tetrapods, from amphibians to mammals, extend between the pubic symphysis caudally and the pectoral girdle, first rib, or manubrium sterni, then continue, usually without interruption, to the lower jaw as hypobranchial muscles (Figs. 10-11, and 10-13). In birds and in some mammals, including higher primates, the rectus muscles are interrupted in the thorax. A **pyramidal muscle** in the ventral wall of the marsupial pouch of metatherians is a slip of the rectus abdominis. Some higher mammals have vestiges of the pyramidal muscle, either normally or as an anomaly. The rectus muscles assist in flexing the trunk if the vertebral column is supple, and they aid in supporting the abdominal viscera.

In urodeles the rectus muscles are strictly segmental, but in anurans and amniotes they exhibit at varying intervals irregular transverse tendinous lines, or **inscriptions,** reminiscent of metameric myosepta (see rectus abdominis, Fig. 10-19, *B*). It is not certain that these inscriptions are phylogenetic remnants of myosepta, but they perform the same mechanical function, enabling what would otherwise be excessively long muscles to remain equally taut over their entire length by providing intermediate sites for attachment. Similar inscriptions are present in the oblique and transverse muscles of the abdomen of amniotes.

Mammalian Diaphragm. In mammals, mesenchyme from several embryonic somites at the level of the developing fourth, fifth, and sixth cervical spinal nerves, depending on the species, migrates caudad in the somatopleure to invade the embryonic septum transversum, which is converted into a dome-shaped muscular diaphragm separating the thoracic and abdominal cavities. When completed, the diaphragm consists of a central **tendinous portion** with a pair of semilunar extensions, and a **muscular** portion. The latter converges on the tendinous portion from all directions, arising on the xiphoid process ventrally and on the caudalmost ribs or their costal cartilages laterally **(sternocostal portion),** and from several lumbar vertebrae dorsally **(vertebral portion).** The vertebral portion consists of a pair of muscular triangular masses, the **crura,** that are firmly anchored to the lumbar vertebrae by short, tough cylindrical tendons. The mammalian diaphragm has become more important than the parietal musculature in creating a suction pump for mammalian respiration. Innervation of the diaphragm by *ventral* rami of several (cervical) spinal nerves indicates its hypaxial status. The diaphragms of reptiles and birds (Chapter 12) are only slightly muscular, and have a different embryonic origin from that of mammals.

MUSCLES OF THE TAIL. The musculature of the tail of tetrapods, like that of fishes, is a direct continuation of the axial musculature of the trunk, having arisen from postcloacal mesodermal somites. Consequently it is divided into epaxial and hypaxial masses. This is most evident in urodeles and generalized reptiles such as *Sphenodon* (Fig. 10-4, *E*). That at the junction between trunk and tail anchors the tail to the more posterior trunk vertebrae and to components of the pelvic girdle, especially the ilium and ischium. In general the epaxials are extensors, straightening the tail, and the hypaxials are flexors or abductors, bending the tail downward or laterad.

Needless to say, the volume of the caudal musculature varies among tetrapods. In snakes, for example, which employ the tail in locomotion, epaxial and hypaxial muscle masses are well developed almost to the tip of the tail. In some others—modern lizards and turtles, for instance—the musculature is meager distally. As mentioned, caudal centra, arches, and processes usually become increasingly vestigial along the length of the tail. As a correlate, the tail musculature in that species does likewise.

In mammals, the spinales and longissimus dorsi of the trunk, which are epaxials, continue into the tail as long medial and lateral extensors lying on the dorsal aspect of the vertebral column. These are supplemented by additional intrinsic epaxial muscle slips arising from caudal vertebrae, particularly in the proximal portion of the tail. The muscles insert by many long slender tendons on the dorsal aspect of more distal vertebrae. As the vertebrae become increasingly vestigial distally, the muscles dwindle until only their long tendons of insertion remain.

The hypaxial musculature of mammals continues into the tail as flexor and abductor muscles. They arise from the medial surface and caudal border of the wing of the ilium, from beneath the transverse processes of the last lumbar vertebrae, or from the dorsal aspect of the sacrum, and pass into the tail as long bundles paralleling the vertebral column laterally and ventrally. Most of the lateral muscles are abductors that insert laterally on more distal centra or on transverse processes. The ventral muscles are flexors that insert ventrally on centra. These extrinsic bundles in mammals extend only part way into the tail—in cats, for example, to about the tenth caudal vertebra. They are supplemented in the tail by long and short intrinsic flexors that arise on proximal centra and insert ventrally and more distally by long tough tendons. The intrinsic bundles, like epaxials, dwindle distally. The major fleshy mass of the mammalian tail is therefore in the proximal portion of the tail near its base; distally, the tail is mostly skin, stringlike tendons, and cylindrical bones (Fig. 7-22) united by interosseus ligaments. The meatiest oxtail soup comes from the proximal portion of the tail!

The **dilator and sphincter muscles of the cloaca** of lower vertebrates and the **sphincter ani of mammals** arise from myoblasts that contribute to the hypaxial musculature anchoring the tail to the sacral vertebrae and pelvic girdle. In turtles, for instance, these muscles arise on sacral and proximal caudal vertebrae and insert on the circumferential skin of the vent. In mammals they have abandoned their skeletal attachments.

In urodeles, reptiles, and birds, a **caudofemoralis muscle** contributes to the fleshy part of the tail, arising on the base of the tail and inserting by tendons on the femur. In crocodilians this muscle is highly specialized, arising from transverse processes and neural arches halfway to the end of the tail and inserting by two strong tendons, one on the femur and the other extending beyond the knee into the shank. Whether this is a primary or secondary appendicular muscle has not been established.

Hypobranchial and Tongue Muscles

The hypobranchial muscles develop from cephalic extensions of hypaxial blastemas beneath the pharynx (Fig. 10-6). In fishes and urodeles these extend forward from the pectoral girdle to the branchial skeleton, hyoid, and mandible (Figs. 10-8 and 10-13). In one species or another they strengthen the floor of the pharynx and pericardial cavity, aid the branchiomeric muscles in changing the shape of the pharyngeal floor, and open the mouth. With development of a long neck in amniotes they have become longer and straplike and now stabilize the hyoid and larynx or draw these forward or backward, depending on their attachments and on what other muscles are doing at the time. They bear such names as **sternohyoid, sternothyroid, thyrohyoid, omohyoid,** and **geniohyoid.**

The tongue of tetrapods has become essentially a protrusible mucosal sac anchored to the hyoid apparatus and stuffed with hypobranchial muscle. This explains why, in bats for instance, muscles extend into the tongue from as far back as the sternum. Among the most common extrinsic muscles of the protrusible tongue are the paired **hyoglossus, styloglossus,** and

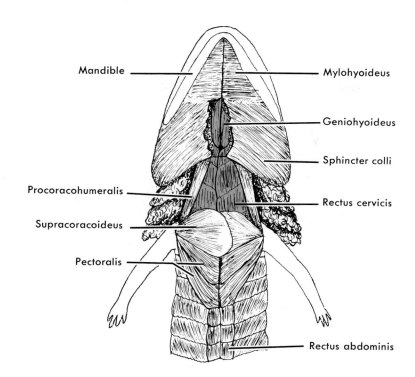

Mandible

Mylohyoideus

Geniohyoideus

Sphincter colli

Procoracohumeralis

Rectus cervicis

Supracoracoideus

Pectoralis

Rectus abdominis

FIGURE 10-13

Anteroventral muscles of *Necturus*. The sphincter colli is a very thin muscular sheet on the surface of the powerful branchiohyoideus. The hypobranchials are shown in red.

genioglossus. Mammals and some reptiles also develop an intrinsic tongue muscle, the **lingualis,** that has no skeletal attachments.

Because of the derivation of hypobranchial and tongue muscles from the anteriormost trunk somites they are supplied by the most anterior spinal nerves or, with respect to the tongue muscles of amniotes by the last cranial nerve. The motor fibers that innervate them have their cell bodies in the **somatic motor column (somatic efferent column)** of the central nervous system (Figs. 10-6 and 15-34, *SE*). This column supplies myotomal muscles everywhere in the body.

Extrinsic Eyeball Muscles

The extrinsic eyeball muscles are striated, skeletal, voluntary muscles that have their anatomic origins on the axial skeleton and are innervated by somatic motor nerve fibers. In all these respects they are identical with the myomeres of the trunk and tail of fishes, which, it should be recalled, arise from the myotomes of somites. In elasmobranch embryos the extrinsic eyeball muscles arise from the vicinity of three pairs of **preotic somitomeres,** which are part of a series of about a dozen similar segmental mesenchymal aggregations that are reminiscent of somites and are located in the head, anterior and posterior to the embryonic otocysts (Fig. 10-6). Whether these preotic and postotic somitomeres are vestigial somites, rudimentary somites, or neither, cannot be known for certainty; they do not differentiate

FIGURE 10-14

Extrinsic eyeball muscles of a generalized vertebrate, left eyeball, viewed from back of the eye. The rectus muscles of all vertebrates arise from a single location on the orbital wall near the optic foramen. The origin of the obliques varies slightly.

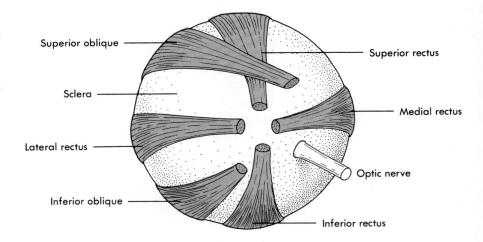

to the extent of exhibiting recognizable myotomes. They are found in fishes, amphibians, and amniotes, and demonstrate unequivocally the marked tendency toward segmentation of the vertebrate head.

The most anterior somitomere in the vicinity of the developing eyeball is situated close to where the third cranial nerve emerges from the midbrain. This nerve innervates four eyeball muscles, the **superior, medial,** and **inferior rectus,** and the **inferior oblique** (Fig. 10-14 and Table 10-3). The next more posterior somitomere is located at the level where the fourth cranial nerve emerges from the midbrain. This nerve innervates the **superior oblique** eyeball muscle. The last preotic somitomere is located close to where the sixth cranial nerve emerges. This nerve innervates the **lateral rectus** eyeball muscle. (The fifth cranial nerve, which intervenes in the eyeball muscle series, is a branchiomeric nerve that grows downward into the mandibular arch.)

Although the myoblasts that give rise to eyeball muscles cannot be traced to specific somitomeres, hence are not myotomal by embryonic origin, they are preceded in development by somitomeres; the nerves that supply them are, like spinal nerves, serially disposed; and the cell bodies of the motor fibers that innervate them are in the motor column of the central nervous system, as are the cell bodies of motor fibers that innervate myotomal muscle elsewhere in the body. In view of this, and in consideration of the early segmentation of head mesenchyme, it appears that the rectus and oblique eyeball muscles, like the myomeres of fishes, are derivatives of an ancient segmental axial musculature that, in the head, has been "conscripted" to operate lateral visual organs. As might be expected, vertebrates with vestigial eyeballs have vestigial eyeball muscles.

In many amniotes, muscles insert on the upper lids and nictitating membrane (**pyramidalis** of reptiles and birds, **quadratus** of birds, **levator palpebrae superioris** of reptiles and mammals). The specific source of their myoblasts has not been ascertained, but their innervation by somatic motor fibers with cell bodies in the somatic efferent column indicates that they are

TABLE 10-3   Chief Extrinsic Eyeball and Eyelid Muscles and Their Innervation

Cranial Nerve Supply	Extrinsic Eyeball Muscles	Eyelid Muscles
III (oculomotor)	Superior rectus Inferior rectus* Medial (internal) rectus Inferior oblique	Levator palpebrae superioris
IV (trochlear)	Superior oblique	
VI (abducens)	External (lateral) rectus Retractor bulbi	Pyramidalis of eye Quadratus of eye

*In lampreys the inferior rectus is innervated by cranial nerve VI.

somatic muscles (Table 10-3). **Protractors and retractors of the eyeballs** of reptiles and birds, and **depressors of the lower lids,** when present, are evidently not of axial origin, because they are innervated by the fifth cranial nerve.

APPENDICULAR MUSCLES

Appendicular muscles are those that insert on the girdles, fins, or limbs, moving the appendage. As mentioned, most fishes use axial, not appendicular muscles, for locomotion. As a correlate, the appendicular muscles of fishes are uncomplicated, exhibit little variety, have little mass, and perform a restricted function. Those of tetrapods have become increasingly numerous and complex as a consequence of adaptations for life on land.

Fishes

Paired fins arise in embryos as fin folds that protrude from the lateral body wall (Fig. 10-15). Thereafter, hollow **muscle buds** sprout from the lower edges of a series of embryonic myomeres near the base of each fin fold (Fig. 10-16). The buds split into dorsal and ventral moities, invade the developing fin, and establish blastemas from which dorsal and ventral muscle masses are formed. Dorsal blastemas form extensors, or elevators, of the fin; ventral blastemas form the flexors, or depressors (Fig. 10-17). The resulting musculature establishes attachments to the girdles, basalia, radialia (if any), and to the fascia overlying the base of the fin rays. Any muscle tissue that may develop distal to these masses is insignificant except in the claspers of male Chondrichthyes and in the undulating winglike fins of rays and skates. As a result, most paired fins are operated as unjointed unitary appendages.

The muscles of dorsal fins organize from myotomal mesenchyme that is giving rise to epaxial myomeres. Those of ventral fins arise from hypaxial myotomal mesenchyme. The musculature of median fins is meager, with the exception of those anal fins that have become gonopodia.

FIGURE 10-15

Top, Origin of shark appendicular muscles from somites, and their innervation by corresponding spinal nerves. *Bottom,* Probable phylogenetic derivation of appendicular muscles of a mammal from six somites based on their innervation. *C4* and *T1,* Dorsal root ganglia of the fourth cervical and first thoracic spinal nerves of the brachial plexus.

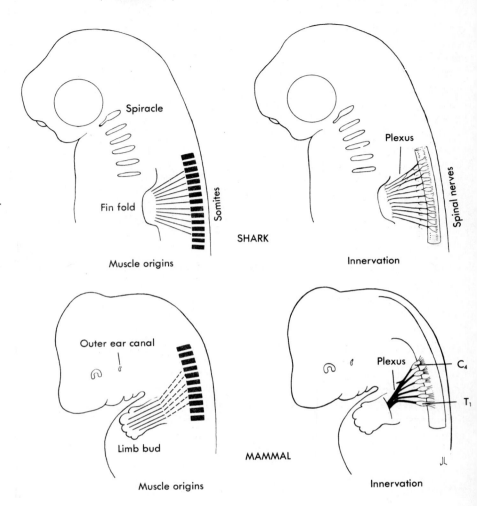

FIGURE 10-16

Budding of fin musculature from embryonic body wall myomeres in the shark *Pristiurus.* In the electric ray 26 myotomes participate. *b,* Muscle buds; *ff,* fin fold; *m,* myomeres.

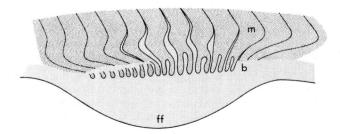

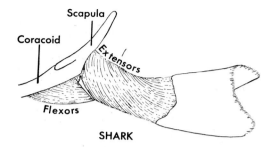

FIGURE 10-17

Appendicular muscles of pectoral fin of shark. Extensors, which are dorsal, or postaxial, are also called levators. Flexors, which are ventral, or preaxial, are adductors.

Tetrapods

The appendicular muscles of tetrapods in general constitute a larger proportion of the total body weight than do those of fishes, and they are far more complex because of the many leveraged joints within tetrapod limbs. As in the fins of fishes, the appendicular muscles of tetrapods are disposed in opposing functional groups, including extensors and flexors. Primitively, the postaxial (posterior, dorsal) muscles extended the appendage, and preaxials (anterior, ventral) flexed it (Fig. 10-17). But that was before the days of the tetrapod limb. With the advent of the five-segment limb and its rotation toward the long axis of the body, where elbows point backward and knees point forward (Fig. 9-21), these terms become anachronisms except to research scholars. Not only have the bones of the pro-, epi-, meso-, and metapodia become reoriented, but so have the muscles, often encircling a joint to insert on the opposite surface of the next segment. For this reason we will discuss tetrapod muscles with reference primarily to their anatomical relationship—dorsal or ventral—when this is feasible, rather than by reference to their phylogenetic history (pre- and postaxial). In attempting to visualize the location of dorsal and ventral muscles one should not attempt to use the human arm as a reference. It has been greatly reoriented, and in addition the independent mobility of the radius enables the palm to be turned up or down while the elbow is immobilized. The biceps and triceps brachii of mammals (Fig. 10-3) is a classic example of the relationship between preaxial, anterior, and flexion on the one hand, and postaxial, posterior, and extension on the other. These muscles have retained a primitive relationship with respect to the long axis of the humerus. The reorientation of the tetrapod appendage has played havoc with students of vertebrate evolution, but it has made possible the flight of an eagle, the agility of a mountain goat, and the marvel of the human arm and hand.

Anatomists find it useful to divide appendicular muscles into two groups according to their anatomical origins. **Extrinsic appendicular muscles** have their origins on the axial skeleton or fascia of the trunk and insert on the girdles or on the skeleton of the limbs, thereby moving the girdle, particularly the pectoral girdle, and the limb. **Intrinsic appendicular muscles** have their origins on the girdle or proximal skeletal elements of the limb and insert on more distal elements, thereby moving, more or less independently, the segments of the tetrapod limb, pro-, epi-, meso-, metapodia, and phalanges. All extrinsic muscles must be completely severed for removal of the pectoral girdle and limb from the body as a unit. In the process intrinsic muscles remain intact. The pelvic girdle and limb cannot be removed as a unit because the girdle is ankylosed to the sacrum.

Morphogenesis. Most extrinsic muscles of tetrapod limbs commence development from blastemas *within the embryonic body wall;* and just as in fishes, buds from these blastemas grow toward and establish an insertion on a girdle or proximal bone of the limb. Muscles that develop in this manner are called **secondary appendicular muscles** by embryologists because of their embryonic origin from the *axial* musculature. The levator scapulae, serratus ventralis, and rhomboideus, all of myotomal derivation, and the trapezius, sternomastoid, and cleidomastoid, of branchiomeric origin, are secondary appendicular muscles of the forelimbs. They will be discussed shortly. In species in which a muscle bearing the name quadratus lumborum attaches to the femur, the muscle, if named with regard to its homologies, is a secondary appendicular muscle of the hind limb. Secondary appendicular muscles are the most primitive, being extensions of the locomotor hypaxial

FIGURE 10-18

Superficial shoulder muscles of an amphibian, reptile, and mammal, illustrating increased extent of shoulder muscles in successively higher tetrapods. The anterior and posterior slips of the trapezius in *Necturus* are also named supracoracoideus and procoracohumeralis, respectively.

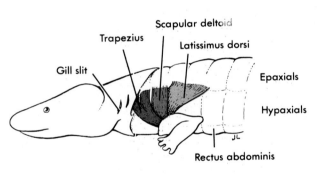

A. NECTURUS

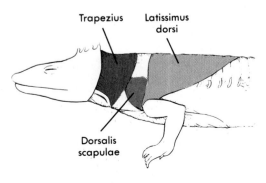

B. SPHENODON

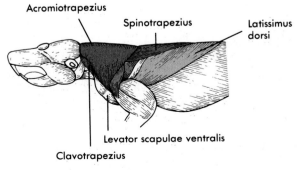

C. HAMSTER

muscles of ancestral fishes. And in fishes, these are the only kind of appendicular muscles. They deserve better than "secondary" billing!

Intrinsic appendicular muscles of tetrapods, unlike those of fishes, organize from blastemas *within the developing limb,* rather than as buds from hypaxial myomeres. Muscles that form in this manner have been called **primary appendicular muscles.** They include a few extrinsic muscles that form when these blastemas spread trunkward to establish an attachment on the axial skeleton. The latissimus dorsi of the forelimb (Fig. 10-18) and the iliopsoas of the hind limb of generalized mammals are in the latter category.

The source of the myoblasts in these intrinsic blastemas is difficult to trace. It has not been shown that they originate in somites; therefore it cannot be said with certainty that they are myotomal by embryonic origin. However, their innervation is such as to leave little doubt that they are myotomal by phylogenetic derivation (Fig. 10-15). Formation of intrinsic appendicular muscles in this manner, that is, in situ, skipping an ancestral ontogenetic stage of development, may be a time-saving modification of selective value. It is one of many instances of developmental precocity in vertebrates. No biological scientist today expects ontogeny to recapitulate phylogeny.

EXTRINSIC MUSCLES OF THE PECTORAL GIRDLES AND FORELIMBS

Dorsal Group. The most constant dorsal extrinsic appendicular muscle of tetrapods is the **latissimus dorsi,** a primary appendicular muscle that inserts on the humerus. In urodeles (Fig. 10-18, *A*) it is a delicate triangular muscle arising from the superficial fascia that overlies the epaxial myomeres of the shoulder region. In reptiles it became stronger, it spread dorsad to acquire a firm attachment to the tough fascia that is anchored to neural spines, and it broadened its axial origins by spreading still farther caudad (Fig. 10-18, *B*). In mammals the trend toward a broader dorsal anchorage continued. It now arises from the neural spines of most of the thoracic vertebrae caudal to the first few, and from the tough fibrous lumbodorsal fascia that overlies the lumbar vertebrae and extends all the way to the base of the tail (Fig. 10-18, *C*). The result of these dorsocaudad expansions was a continuing increment in the force the muscle was able to exert on the forelimb.

Deep to the latissimus dorsi (and to the trapezius, to be discussed next) in most amniotes are three additional extrinsic muscles or muscle groups that insert on the scapula: the levatores scapulae, the rhomboideus group (found only in crocodilians among living reptiles), and the serratus ventralis (serratus anterior). The **levators of the scapulae** have their origins on the transverse processes of the atlas or on the basioccipital bone (levator scapulae ventralis) and on the transverse processes of a number of posterior cervical vertebrae (levator scapulae dorsalis). The **rhomboideus group** arises collectively from the occiput and from the neural spines of a series of cervical and anterior thoracic vertebrae. The **serratus ventralis,** as the name implies, is serrate, because it arises by many separate prominent slips from a series of ribs, and in some mammals from a series of cervical vertebrae. All the foregoing insert on the dorsal border of the scapula except the mammalian levator scapulae ventralis, which inserts on a process of the scapular spine near the glenoid fossa.*

The pharyngeal arches make a contribution to the dorsal extrinsic musculature of the forelimb. The **trapezius,** a superficial muscle of the shoulder region that inserts primitively on the scapula, is a survivor of the **cucullaris** muscle of fishes (Figs. 10-18 and 10-22, *A*). Like branchiomeric muscles, the trapezius forms from mesenchyme that streams from neural crests, and it receives its

---

*The levator scapulae dorsalis and the serratus ventralis of mammals are sometimes considered a single muscle, the serratus ventralis, which is only one of the frustrations encountered by students attempting to ascertain muscle homologues from a chart. Attention is directed to the earlier discussion of muscle homologies.

FIGURE 10-19

Pectoral musculature of selected tetrapods *(red).* **A,** *Necturus.* **B,** Frog. **C,** Pigeon. **D,** Crocodilian. **E,** Rabbit. In **B,** the cutaneous pectoris is an integumentary slip of the pectoral musculature that has acquired an insertion on the skin between the forelimbs. It is superficial to the other pectoral muscles. In **C** the pectoral muscle has been removed on the right to reveal the supracoracoid. In **E,** the pectoralis primus and pectoralis major have been removed on the right to expose deeper muscles. A pectoscapularis lies deep to the cephalic border of the pectoralis minor.

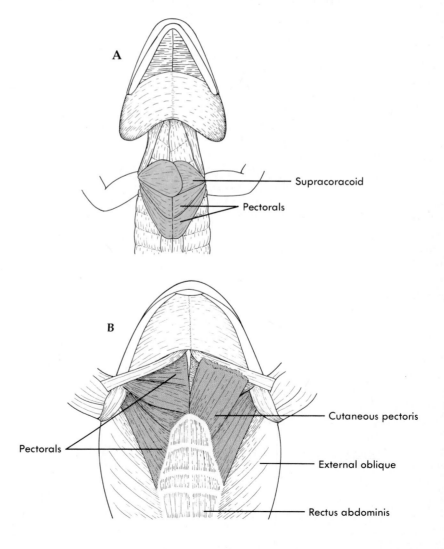

motor innervation via branchiomeric rather than spinal nerves. The trapezius has undergone the same dorsal and caudal extension as the latissimus dorsi, and in mammals it has become subdivided into several components, including **clavotrapezius, acromiotrapezius,** and **spinotrapezius.** Two amniote neck muscles that arise on the temporal region of the skull, the **cleidomastoid,** which inserts on the clavicle, and the **sternomastoid,** which inserts on the sternum, are also derivatives of the trapezius (Figs. 10-19, *D* and *E*, and 10-22, *C*).

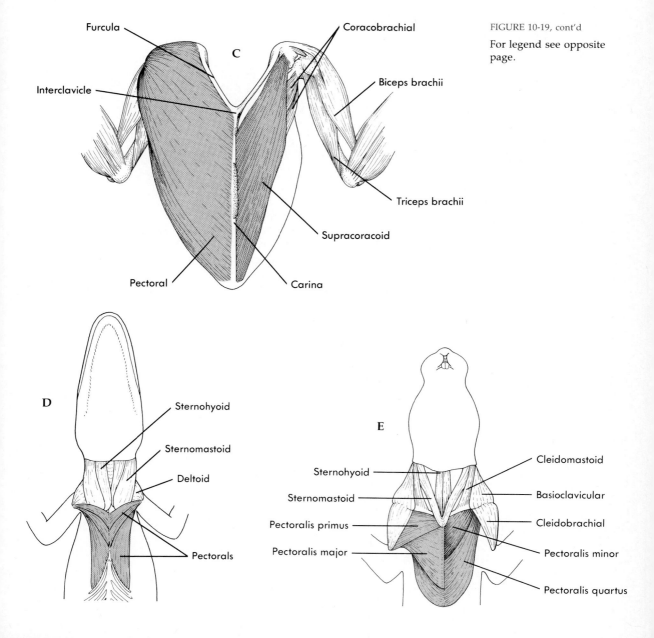

FIGURE 10-19, cont'd

For legend see opposite page.

Ventral Group. The ventral extrinsic musculature of the forelimb, subsumed under the general term **pectoral muscles,** has undergone the same expansive changes as the dorsal group (Fig. 10-19, *A* to *E*). Originating primitively from the coracoid cartilages or bones and the adjacent midventral raphe as a fan-shaped muscle with two or three subdivisions, they have extended their origins to include, in one species or another, the epicoracoid, entire sternum, some of the costal cartilages, and part of the midventral raphe of the abdomen and neck; and in mammals they have become divided into a varying number of individual muscles with superficial and deep locations. They converge to insert on the humerus near the glenoid fossa. In amniotes the insertion site is marked by prominent tuberosities or ridges below the head of the humerus. The pectoral muscles are the weak adductors of the forelimbs of a necturus that help, but not greatly, in keeping the arms from spreading; and they are the powerful ventral flight muscles of birds that elevate and depress the wings in flight. The **panniculus carnosus** of mammals (Fig. 10-23) is a "displaced" superficial sheet of pectoral musculature that became separated from the remainder of the pectoral musculature to insert on the skin. A similar muscle in anurans, the **cutaneous pectoris,** maintains a more primitive orientation (Fig. 10-19, *B*).

INTRINSIC MUSCLES OF THE PECTORAL GIRDLES AND FORELIMBS

Dorsal Group. Arising from the scapula of modern mammals are five postaxial muscles inherited with modifications from reptiles. They are the **deltoids, teres major, teres minor** (deeply located), **subscapular** (on the medial surface), and the **long head of the triceps brachii** (Fig. 10-20). With the exception of the triceps, these muscles insert on the humerus a short distance below its head and align it with the scapula, rotate it around its long axis, or adduct it. The triceps, which has two additional heads that arise on the humerus, inserts on the olecranon process of the ulna (Fig. 10-3), where the three heads apply a powerful pull that extends the forearm. To test your own triceps, try a few push-ups! (This exercise extends the arm against resistance.) Distal to the triceps two **supinators** of the manus (one sometimes named **brachioradialis**) connect the distal end of the humerus with the radius, and an assortment of **extensors of the hand** with long distal tendons (Fig. 10-20) insert on the skeleton of the wrist and digits, some as far distad as the terminal phalanx. The shortest ones are intrinsic to the manus.

The triceps is a classic example of an extensor muscle, paralleling the postaxial surface of the humerus in every vertebrate class. (Some students who have dissected a frog may wonder why they saw no triceps brachii. It was there. It is more often called **anconeus** in laboratory manuals.) Muscles of lower tetrapods that originate on the suprascapula, insert on the humerus, and bear the name **dorsalis scapulae (scapular deltoid)** are likely homologues of the deltoid muscles of higher amniotes. The mammalian teres major appears to be a dissociated portion of the reptilian latissimus dorsi that has acquired an attachment to the scapula. The mammalian teres minor, a short deep muscle along the postaxial border of the scapula proximal to the glenoid fossa, may have been derived from the reptilian **scap-**

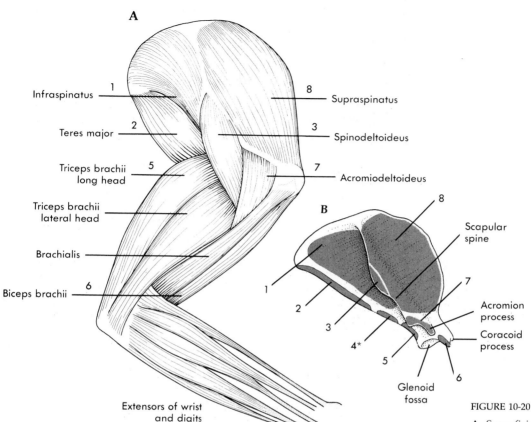

A, Superficial muscles of the right scapula and proximal part of the forelimb of a cat, lateral view.

Infraspinatus — 1
Teres major — 2
Triceps brachii long head — 5
Triceps brachii lateral head
Brachialis
Biceps brachii — 6

8 — Supraspinatus
3 — Spinodeltoideus
7 — Acromiodeltoideus

B

8 — Scapular spine
7 — Acromion process
— Coracoid process
Glenoid fossa

1
2
3
4*
5
6

Extensors of wrist and digits

FIGURE 10-20

A, Superficial muscles of the right scapula and proximal part of the forelimb of a cat, lateral view. **B,** Sites of attachment of correspondingly numbered muscles originating on the scapula. *4,* Origin of teres minor, a deep muscle not visible in this view. Also not seen is the subscapularis muscle (origin, medial surface of scapula) and the tiny coracobrachialis (origin, coracoid process of scapula).

ulohumeralis anterior, which occupies approximately the same position in iguanids. The reptilian subscapular muscle, readily identifiable on the basis of its location in iguanids, has expanded to occupy a larger area of the subscapular fossa in mammals. It is sometimes called **subcoracoscapularis** in reptiles.

Ventral Group. The preaxial musculature of the forelimb of modern mammals is represented proximally by the **supraspinatus** and **infraspinatus** (Fig. 10-20) and by a small deep **coracobrachialis,** all three of which extend from the mammalian scapula to the humerus. The first two are derivatives of the **supracoracoid** muscle of reptiles, an expansive fleshy muscle that in iguanids arises over a broad area of the procoracoid bone near the glenoid fossa at the base of the scapula and inserts on the proximal end of the humerus. Embryological studies have shown that this reptilian muscle in embryonic mammals spreads upward onto the lateral aspect of the developing mammalian scapula during ontogeny, pushes its way beneath the deltoid, and establishes itself on both sides of the scapular spine to become the supraspinatus and infraspinatus muscles. The coracobrachialis, also present in reptiles, is a short muscle in mammals that originates on the coracoid process, a vestige of the reptilian coracoid plate. A **biceps brachii** (scapula to radius) and a **brachialis** (humerus to ulna, Fig. 10-20) are the major flexors of the forearm of reptiles and of mammals. A small deep mammalian **anconeus** (not homologous with the anconeus of frogs) extends between the distal extremity of the humerus and the proximal end of the ulna, and a small transverse **epitrochleoanconeus** partially encircles the elbow joint medially. Distal to these, **supinators of the mammalian hand** insert on the radius, and **flexors of the hand,** with long tendons and with origins chiefly on the humerus, insert on the carpals, metacarpals, and phalanges. Intrinsic to the manus are very short flexors.

Not mentioned thus far is a superficial **cleidobrachialis (clavobrachialis, clavohumeralis)** that extends from the mammalian clavicle, when present, to the ulna. It has the superficial appearance of being a continuation of the basioclavicular muscle of rabbits (Fig. 10-19, *E*) and of the clavotrapezius of cats, because many fibers of these two muscles continue uninterruptedly into the cleidobrachial. However, the latter is innervated by spinal nerves, so it is unlikely that it is an extension of the branchiomeric musculature.

The musculature of birds is essentially reptilian. A large proportion of the weight of a bird is represented by the extrinsic musculature of the fore-limb and the intrinsic musculature of the hind limb. The intrinsic muscles of the wing have been reduced, the epaxial musculature of the trunk is vestigial, and the parietal muscles of birds are thin.

Table 10-4 lists the chief intrinsic muscles of the pectoral girdle and fore-limbs of mammals and their probable homologues in reptiles.

MUSCLES OF THE PELVIC GIRDLES AND HIND LIMBS

Pelvic girdles, unlike pectorals, are incapable of independent mobility, being firmly ankylosed to the axial skeleton dorsally and united in a symphysis ventrally. Consequently, the extrinsic locomotor muscles of the posterior appendages insert on the femur. In mammals these include the iliopsoas, the gluteus, and the pyriform, which are discussed below.

The psoas portion of the mammalian **iliopsoas** arises from an extensive area of the fascia covering the psoas minor muscle and from a series of lumbar vertebrae; the iliac portion arises from the border of the ilium. The two portions ultimately unite and insert by a powerful tendon on a small rough process, the lesser trochanter, located on the femur near its head. The muscle rotates the femur, and in doing so turns the foot outward. In primates and some other mammals the iliopsoas is represented by two separate muscles, the **psoas major** and the **iliacus.** The internal **puboischiofemoral** muscle of reptiles may be a homologue.

A functional group of three mammalian hip muscles, **gluteals, pyriform,** and **gemelli,** has, collectively, a wide origin on sacral and caudal vertebrae and on the ilium and ischium, and inserts on a large prominence, the greater trochanter, near the head of the femur. The group, of which the gluteus is the most powerful, abducts the thigh and, like the iliopsoas, rotates the femur to carry the foot outward. The gluteal and the pyriform (**iliofemoralis** and **caudifemoralis brevis** of iguanids) are primary appendicular muscles that have increased their leverage by spreading onto the axial skeleton.

TABLE 10-4 Chief Intrinsic Muscles of the Pectoral Girdle and Forelimbs of Mammals and Their Probable Homologues in Reptiles

	Mammals	Reptiles
Muscles of girdle		
Girdle to humerus, proximally	Deltoideus	{ Deltoideus clavicularis { Dorsalis scapulae
	Subscapularis Teres minor	Subcoracoscapularis Scapulohumeralis anterior
	Supraspinatus ⎫ Infraspinatus ⎭	Supracoracoideus
	Coracobrachialis Teres major	Coracobrachialis Slip of latissimus dorsi
Muscles of upper arm		
Girdle or humerus to proximal end of radius or ulna	Triceps brachii Biceps brachii Brachialis Epitrochleoanconeus Anconeus	Triceps brachii Biceps brachii Brachialis Epitrochleoanconeus Anconeus
Muscles of forearm*		
Humerus and proximal end of radius and ulna to hand	Extensors and flexors of carpus and digits (Brachioradialis may be a slip of the extensor carpi radialis) Supinators and pronators of hand	
Muscles of hand†	Extensors, flexors, abductors, adductors of digits	

*Frequently have long tendons of insertion.
†Reduced in species in which digits are reduced.

A **quadratus femoris** complex, consisting of four muscles (three **vasti** and a **rectus femoris**), arises collectively on the ilium and on the trochanters and upper shaft of the femur and inserts on the ligament in which the patella is embedded. The vasti extend the leg, and the rectus femoris adducts the thigh, rotating it to carry the foot inward. Other extensors or adductors of the thigh include the **semimembranosus, adductor femoris, adductor longus, pectineus, sartorius,** and **gracilis.** They arise on the ilium, ischium, or pubis, and insert on the shaft of the femur or on the patellar ligament and tibia. The sartorius is a large muscular strap that arises on the ilium and passes diagonally down the medial aspect of the thigh. It is the muscle that may become irritated when an adult human who is not used to doing so sits too long on the floor with the legs crossed in the manner of oldtime tailors who, in creating sartorial splendor, sewed by hand.

Two **obturator** muscles flex, rotate, and abduct the thigh. They arise from the lip of the obturator foramen (Fig. 9-12, *F*) and from the ischium and pubis, and insert on the proximal end of the femur.

A **biceps femoris** and **semitendinosus** are primarily flexors of the leg, although the former also abducts the thigh and the latter extends it. They arise on the ischium and insert on the patella or tibia.

The foregoing muscles either insert on the femur or parallel it to insert just beyond it. Distal to these are **extensors and flexors of the foot,** similar to those of the hand. Among these is the **gastrocnemius,** an extensor of the foot made famous by Achilles. Its long postaxial tendon inserts on the calcaneous bone. Notable also are the very long postaxial tendons that pass behind a bird's heel and insert on the phalanges. The weight of the perching bird stretches the tendons and retracts the claw-bearing toes so that a sleeping bird, muscularly relaxed and with minimal expenditure of energy, tightens its grip on the perch.

A few comments about urodele muscles is appropriate, inasmuch as necturus is frequently dissected in courses in vertebrate morphology. The only extrinsic pelvic muscle of urodeles is a deep **caudifemoralis** that passes from some of the caudal vertebrae to the femur and draws the leg backward during locomotion on land, and a weak slip or two from the tail that are probably subvertebrals. The caudifemoralis is identifiable in reptiles and birds, being especially powerful in crocodilians, who lash out with their tail. Other hind limb muscles of urodeles are intrinsic. For instance, the superficial fan-shaped muscles seen ventrally in necturus between the posterior appendages, namely, the **puboischiofemoralis externus** and **puboischiotibialis,** are primitive intrinsic muscles arising on the pelvic plate.

As mentioned, the pectoral girdles and forelimbs are joined to the trunk by extrinsic appendicular muscles, vessels, nerves, and skin alone, whereas the pelvic girdles are firmly ankylosed to the axial skeleton. This accounts in part for the fact that cats jump by pushing off with their hind limbs but land on their forelimbs, where much of the shock is absorbed by the extrinsic and intrinsic musculature. Among the reasons frogs can't do this is lack of a cerebellum competent to direct precision landings. The result is that frogs land awkwardly—almost abashedly— on whatever part of the body touches down first. The hind limbs of ungulates are good shock absorbers because elongated metatarsals provide what amounts to an extra shock-absorbing joint at the end of the shank.

Innervation of Appendicular Muscles

Branches of the ventral rami of several successive spinal nerves grow into the fin folds and limb buds to innervate the muscles that are developing within the fin or limb. In fishes these nerves come from the same body segments that contribute muscle buds to the appendage (Fig. 10-15). Although in tetrapods the muscles within the appendage cannot be traced directly to myotomes, it seems probable that the number of spinal nerves entering a tetrapod limb is indicative of the number of somites that contributed mesenchyme to the limb phylogenetically.

BRANCHIOMERIC MUSCLES

Associated with the pharyngeal arches of vertebrates (Fig. 10-21) is a series of striated, skeletal, voluntary, visceral, branchiomeric muscles. The basic architectural pattern of these muscles is illustrated in fishes, in which they operate the jaws and successive gill arches (Fig. 10-22, *A*). In tetrapods they still operate the jaws. However, with loss of gills, the more posterior ones have acquired new functions.

That branchiomeric muscles must be classified as visceral is evident from the following facts:

1. Their position in the wall of the alimentary canal identifies them as visceral.
2. The motor nerve fibers innervating them are in a visceral motor column of the brain (Fig. 15-34, SVE) close to the column that innervates smooth muscles and glands (Fig. 15-34, GVE). They are not in the column that innervates other skeletal muscles (SE).
3. Functionally, they are associated with two visceral processes, food handling and respiration.
4. Their embryonic origin is not from myotomes.

The foregoing criteria set branchiomeric muscles apart from other striated skeletal muscles.

Muscles of the Mandibular Arch

In all vertebrates the muscles of the first arch are primarily jaw muscles. In *Squalus*, whose branchial muscles are relatively generalized (Figs. 10-8 and 10-22, *A*), a **levator palatoquadrati** arises on the otic capsule and inserts on the quadrate end of the upper jaw cartilage; a powerful **adductor mandibulae** arises on the quadrate process and inserts on Meckel's cartilage; and an **intermandibularis** extends between Meckel's cartilage and a strong midventral raphe in the pharyngeal floor. A slender **craniomaxillaris** inserting on the upper jaw completes the first arch muscles of *Squalus*. The levator palatoquadrati raises the upper jaw, which is possible because *Squalus* has hyostylic jaw suspension. The craniomaxillaris assists in this. The

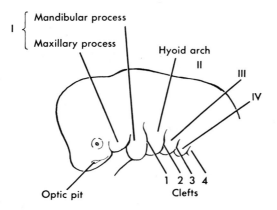

FIGURE 10-21

Pharyngeal arches of a vertebrate.

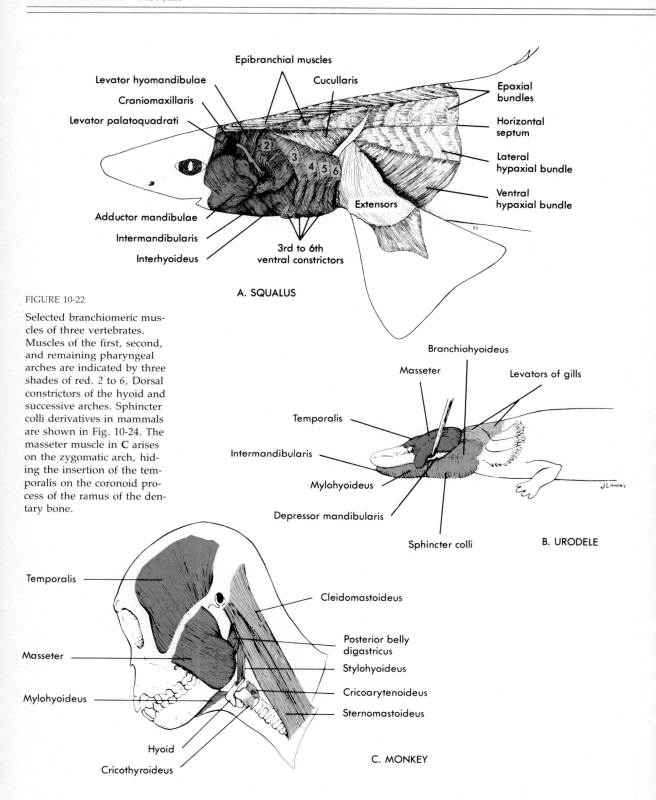

Epibranchial muscles

Levator hyomandibulae

Craniomaxillaris

Levator palatoquadrati

Cucullaris

Epaxial bundles

Horizontal septum

Lateral hypaxial bundle

Ventral hypaxial bundle

Adductor mandibulae

Intermandibularis

Interhyoideus

Extensors

3rd to 6th ventral constrictors

A. SQUALUS

FIGURE 10-22

Selected branchiomeric muscles of three vertebrates. Muscles of the first, second, and remaining pharyngeal arches are indicated by three shades of red. 2 to 6, Dorsal constrictors of the hyoid and successive arches. Sphincter colli derivatives in mammals are shown in Fig. 10-24. The masseter muscle in **C** arises on the zygomatic arch, hiding the insertion of the temporalis on the coronoid process of the ramus of the dentary bone.

Branchiohyoideus

Masseter

Levators of gills

Temporalis

Intermandibularis

Mylohyoideus

Depressor mandibularis

Sphincter colli

B. URODELE

Temporalis

Cleidomastoideus

Posterior belly digastricus

Stylohyoideus

Cricoarytenoideus

Sternomastoideus

Masseter

Mylohyoideus

Hyoid

Cricothyroideus

C. MONKEY

adductor mandibulae raises the lower jaw, thereby closing the mouth during the phase of the respiratory cycle when the spiracle is also closed and water is being forced over the gills by constriction of the walls of the orobranchial chamber. The adductor also enables the shark to hold in a viselike grip any prey unlucky enough to be caught. The intermandibularis is a ventral constrictor that elevates the anterior pharyngeal floor during respiration. The lower jaw is lowered by the coracomandibularis, a hypobranchial muscle.

The adductor mandibulae is the most powerful muscle of the first arch from fish to man. In **tetrapods** it has become split into several muscles—many in snakes, commonly three in amniotes. These are the **masseter, temporalis,** and **pterygoideus.** They have spread in three directions on the skull, and approach the mandible from these three directions, supporting the lower jaw in a muscular sling and providing most of the multidirectional tensile forces that produce the sidewise, up-and-down, forward-and-back, and rotatory chewing movements of such different mammals as herbivores, carnivores, and rodents. The masseter and temporalis muscles are illustrated in Fig. 10-22, C. The pterygoids are medial to the ramus of the mandible, arising in and near the pterygoid fossa of the skull (Fig. 8-28), and cannot be seen until the mandibular symphysis is separated and a dentary bone is retracted laterad. The role of temporal fossae in the evolution of chewing muscles is discussed in Chapter 8.

The intermandibular muscle is called **mylohyoid** in tetrapods (Fig. 10-13). One slip of it has probably given rise to the **digastric muscle (anterior belly** of the digastric when there are two bellies). One slip of the first arch muscle that was attached to the articular bone of therapsid reptiles remained attached there when the bone became the malleus in the middle ear, and that muscle, now the **tensor tympani,** puts tension on the eardrum. All mandibular arch muscles are innervated by cranial nerve V.

Muscles of the Hyoid Arch

The principal muscles of the hyoid, arch of *Squalus* elevate the arch or constrict the pharyngeal cavity, partly as a function of respiration and feeding. A **levator hyomandibulae** and a **dorsal constrictor** arise on the neurocranium and insert on the hyomandibula and ceratohyal cartilage. An **interhyoideus** is a ventral constrictor (Figs. 10-8 and 10-22, A). In bony fishes the dorsal constrictor is subdivided, and part of it operates the operculum. Because of the joint between the upper jaw, lower jaw, and hyomandibula, contraction of hyoid arch muscles results in movements of the jaws.

In **tetrapods** hyoid arch muscles continue to serve some of the functions seen in fishes, but they have also acquired some quite different roles. In *Necturus* a **branchiohyoid muscle** of the hyoid arch arises on the ceratohyal cartilage and inserts on the epibranchial cartilage of the first gill. Along with levators of the gills it waves the gills back and forth in the water for respiration. A **depressor mandibulae** opens the mouth of urodeles and many reptiles (Fig. 10-22, B); and in mammals a **digastric muscle (posterior belly** when there are two) participates in chewing movements (Fig. 10-22, C). A slender **stylohyoid muscle** connects the styloid or jugular process of the skull with the anterior horn or body of the hyoid in mammals (Fig. 8-39). The muscle is so slim in cats that it is almost vestigial. (A posterior portion of the stylohyoid, when present, is derived from the third arch.)

A thin **sphincter colli** lies like a collar (for which it was named) under and closely adherent to the skin of the neck of lower tetrapods. It is superficial to the branchiohyoid muscle of *Necturus* (Fig. 10-22, *B*). In reptiles and birds it spreads upward around the rear of the skull to insert on the skin of the head, and is called **platysma.** In mammals the platysma spreads forward onto the face to become muscles of facial expression, or **mimetic muscles.** The sphincter colli is thought to be derived from a portion of the interhyoideus of fishes.

The **stapedial muscle** of the middle ear of mammals attaches to the stapes, which is a derivative of the hyomandibula. It responds reflexly to extra loud sounds (p. 580).

All these hyoid arch muscles are innervated by the facial nerve (cranial nerve VII), so named because of its wide distribution to the mimetic muscles under the skin of the face of humans. The innervation of the two muscles of the middle ear by two different branchiomeric nerves, V and VII, is not surprising to those who know the phylogenetic history of the ear ossicles to which the muscles attach.

Muscles of the Third and Successive Pharyngeal Arches

The muscles of arches III through VI are **constrictors, levators, adductors,** and **interarcuals** that compress or expand the pharyngeal cavity and gill pouches during respiration. Constrictors in sharks lie just under the skin, covered by tough subcutaneous fascia that is not readily removed, and they attach to strong fascia above and below the gill pouches (Fig. 10-22, *A*). They compress the gill pouches, expelling respiratory water. The levators of these arches make up a strong muscle sheet, the **cucullaris,** which raises the pharyngeal wall assisted by the levator hyomandibulae of the second arch. The cucullaris is thought to be the precursor of the **trapezius, cleidomastoid,** and **sternomastoid** muscles of tetrapods. Adductors deep in the gill arches connect epibranchial and ceratobranchial cartilages and cause the lateral pharyngeal walls to bow outward when the muscles contract, thereby expanding the pharyngeal chamber. Two sets of interarcual muscles in the roof of the pharynx just above the mucosa connect successive pharyngobranchial cartilages with epibranchials and draw these together. Their action assists in further expanding the pharynx. The interarcuals in the floor of the pharynx are not branchiomeric; they are hypobranchial muscles, the most superficial of which is the common coracoarcualis (Fig. 10-8). They assist in expanding the gill pouches during respiration. In bony fishes the branchiomeric muscles caudal to the hyoid arch are much reduced as a consequence of the role of the operculum in moving respiratory water across the gills.

In tetrapods, branchiomeric muscle has pretty much disappeared from what had been the gill-bearing arches. Remaining from arch III are a **sty-**

lopharyngeus that is used in swallowing, and a **posterior belly of the sty-lohyoideus** in some mammals (Fig. 8-39, stylohyoideus major). Remaining from arch IV are the intrinsic muscles of the larynx: **cricothyroid, cricoary-tenoid,** and **thyroarytenoid.** The chief branchiomeric muscles and their innervations are listed in Table 10-5.

.BLE 10-5   Chief Branchiomeric Muscles and Their Innervation in *Squalus* and in Tetrapods

aryngeal Arch	Pharyngeal Skeleton in *Squalus*	Chief Branchiomeric Muscles		Cranial Nerve Innervation
		Squalus	Tetrapods*	

Mandibular arch	Meckel's cartilage	Intermandibularis	Intermandibularis
			  Mylohyoideus (anterior part)
			  Digastricus (anterior part)
				V

		Adductor mandibulae	Adductor mandibulae
			  Masseter
			  Temporalis
			  Pterygoidei
			  Tensor tympani	

	Pterygoquadrate cartilage	Levator palatoquadrati Craniomaxillaris		
Hyoid arch	Hyomandibula	Levator hyomandibulae	Stapedius	VII
	Ceratohyal	Dorsal constrictor	Stylohyoideus (anterior part)	

	Basihyal	Interhyoideus	Depressor mandibulae
			Digastricus (posterior belly)
			Sphincter colli
			  Platysma
			  Mimetics	

[	Gill cartilages	Constrictors Levators Adductors Interarcuals	Stylopharyngeus Stylohyoideus (posterior part)	IX
' to VI	Gill cartilages	Constrictors Levators Adductors Interarcuals	Thyroarytenoideus Cricoarytenoideus Cricothyroideus	X
			Striated pharyngeal muscles	X and XI
		Cucullaris (derived also from dorsal constrictor 3)	Trapezius Sternomastoideus Cleidomastoideus	Occipitospinal nerves in shark; spinal roots of XI in amniotes

ndented muscles in this column may be derivatives of the preceding muscle.

INTEGUMENTARY MUSCLES

In fishes and amphibians, slips of branchiomeric or myotomal muscles insert one place or another on the dermis, attaching the skin firmly to the underlying muscle at those locations but causing little movement of the skin. **Costocutaneous muscles** of snakes are hypaxial muscles used in locomotion. But it is only in mammals that integumentary muscles are well differentiated. A **panniculus carnosus (cutaneous maximus)** wraps around the entire trunk of some mammals, enabling armadillos to roll into a ball when endangered, forms a sphincter around the entrance to marsupial pouches, and vigorously shakes flies off horses (Fig. 10-23). It is poorly developed in monkeys, and absent in humans. In bats slips of the pectoral muscles insert on the skin of the wing membrane as **patagial muscles.**

The most notable integumentary muscles are **mimetics**—those that express emotion (Fig. 10-24). More than 30 different muscles in humans depress the corners of the mouth in grief, raise them for smiling, wrinkle the forehead, raise the eyebrows quizzically, tightly close the lids, pucker the lips, draw milk when suckling, dilate the nostrils, and direct the pinnas of the ears toward faint sounds. One muscle, the **caninus,** elevates the part of the upper lip that hides the spearlike canine tooth used for ripping into flesh. When a human uses it, he or she is said to be sneering.

The preceding are extrinsic integumentary muscles that arise elsewhere and insert on the skin. Intrinsic integumentary muscles are mostly smooth muscles **(arrectores plumarum** or **pilorum)** that attach to feather or hair follicles and ruffle the feathers or raise the fur for insulation or in response to danger. Unlike extrinsic integumentary muscles, they are innervated by visceral motor fibers.

FIGURE 10-23

Panniculus carnosus (cutaneous maximus) of a cat and primate. Note difference in the extent of the muscle in these two animals.

CAT   MONKEY

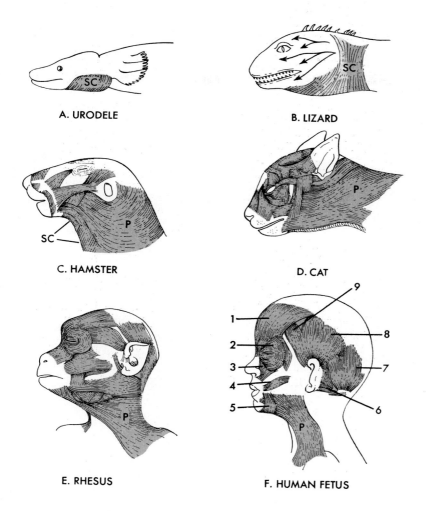

A. URODELE

B. LIZARD

C. HAMSTER

D. CAT

E. RHESUS

F. HUMAN FETUS

FIGURE 10-24

Evolution of mammalian mimetic muscles from hyoid arch muscles of lower tetrapods. The sphincter colli, *SC,* spreads onto the neck in reptiles and becomes the platysma, *P,* in mammals. It then spreads forward onto the head and face, as indicated by *arrows* in **B,** to become muscles of facial expression. Note increasing differentiation of mimetics in the mammals shown. *1,* Frontalis; *2,* orbicularis oculi; *3,* quadratus labii superioris; *4,* risorius; *5,* triangularis; *6,* posterior auricular; *7,* occipital; *8,* superior auricular; *9,* anterior auricular.

ELECTRIC ORGANS

In many fishes certain muscle masses are modified to produce, store, and discharge electricity. In *Torpedo,* the electric ray, an electric organ lies in each pectoral fin near the gills. It is of branchiomeric origin, being supplied by motor fibers of cranial nerves VII and IX. In *Raia,* a skate, and *Electrophorus,* a South American electric eel, electric organs lie in the tail and are modified hypaxial muscles (Fig. 10-25). The potential produced by these organs in eels amounts to 600 V and is used to paralyze or kill prey or to repel other organisms. Some other fishes have electric organs with a lower electric potential that serve as radarlike mechanisms or for communication. Certain neuromast organs are receptors for these signals.

Electric organs consist of a large number of electric disks, up to 200,000 in the tail of one ray, piled in either vertical or horizontal columns. Each disk, or **electroplax,** is a modified multinucleate muscle fiber embedded in a vascular jellylike extracellular material surrounded by connective tissue. Nerve endings terminating on each disk induce the discharge.

Electric organs seem to have no systematic distribution among fishes, and various types probably resulted from convergent evolution. The electric organ of *Malapterurus,* a freshwater catfish from the Nile, envelops nearly the entire body—head, trunk, and tail—in a jellylike mass located subcutaneously rather than among muscles. It appears to be composed of modified skin glands rather than muscle tissue.

FIGURE 10-25

Electric organs in tail of electric eel. Each nucleated horizontal disk (electroplax) is a modified hypaxial muscle cell. *C,* Centrum; *M,* epaxial myomere.

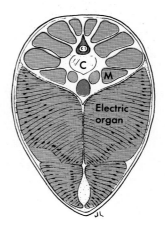

Chapter Summary

1. Muscle may be classified as striated, smooth, or cardiac; voluntary or involuntary; skeletal or nonskeletal; somatic (myotomal) or visceral (nonmyotomal).

2. Skeletal muscles are striated and voluntary. Excepting branchiomeric muscles, they are somatic, arising from mesodermal somites directly or indirectly.

3. Smooth muscle is found in the walls of hollow viscera, tubes, and vessels, and in miscellaneous sites including the eyeball and skin.

4. Cardiac muscle is striated visceral muscle constituting the myocardium of the heart.

5. Classified according to their actions, muscles may be flexors, extensors, abductors, adductors, protractors, retractors, levators, depressors, rotators (including pronators and supinators), constrictors, dilators, or tensors.

6. Skeletal muscles may be named on the basis of the direction of their fibers, location, number of parts, shape, attachments, action, size, or miscellaneous considerations.

7. Homologies between muscle groups are more readily ascertained than homologies between individual muscles. Embryonic origin and nerve supply are the most reliable criteria.

8. The trunk and tail muscles of fishes and tailed amphibians are locomotor, arranged as myomeres, and divided into epaxial and hypaxial masses by a horizontal skeletogenous septum. Loss of myosepta tends to obscure the primitive metamerism.

9. The metamerism of axial muscles is an expression of the metamerism of embryonic somites.

10. Epaxial and hypaxial masses are innervated by dorsal and ventral rami of spinal nerves, respectively.

11. In higher tetrapods epaxial muscles form long bundles, and hypaxial muscles of the trunk have lost their metamerism except where there are long ribs. The latter have also become stratified into oblique, rectus, and transverse sheets.

12. Hypaxial muscles immediately ventral to transverse processes in amniotes are disposed in long bundles (subvertebral muscles).

13. Hypaxial muscles in all vertebrates extend forward under the pharynx as hypobranchial and tongue muscles.

14. The mammalian diaphragm and cremaster muscles are hypaxial muscles that have migrated accompanied by their spinal nerves.

15. Eyeball muscles arise from three preotic somitomeres in elasmobranchs, develop in situ in higher vertebrates. They are innervated by cranial nerves III, IV, and VI and are myotomal by phylogenetic derivation.

16. In fishes and lower tetrapods appendicular muscles are disposed as ventral (preaxial) flexors and dorsal (postaxial) extensors. Reorientation in higher tetrapods has to one degree or another obscured the primitive condition.

17. Primary appendicular muscles organize from blastemas within the limb bud. They sometimes spread into the trunk to acquire origins on the axial skeleton.

18. Secondary appendicular muscles organize from blastemas within the body wall and achieve an attachment to the girdle or limb. They constitute most of the extrinsic limb musculature.

19. Extrinsic appendicular muscles have anatomical origins on the axial skeleton and insertions on the girdles and limbs. Intrinsic appendicular muscles have no axial attachment.

20. Branchiomeric muscles are striated, skeletal, voluntary, visceral muscles that arise from neurectoderm.

21. The muscles of the first visceral arch operate the jaws. In mammals a derivative operates the malleus. They are innervated by cranial nerve V.

22. The muscles of the second visceral arch attach to the hyoid skeleton, lower jaw, and in fishes, operculum. A sphincter colli muscle spreads onto the head of amniotes to become platysma and mimetic muscles. Second arch muscle also attaches to the stapes. The muscles are innervated by cranial nerve VII.

23. Muscles of the third and remaining arches operate gills in fishes. In higher vertebrates they have new functions, including swallowing and vocalization. The muscles of the third arch are innervated by cranial nerve IX, those of successive arches by X.

24. The cucullaris has become subdivided into trapezius, sternomastoid, and cleidomastoid muscles in amniotes.

25. Extrinsic integumentary muscles reach peak development as the panniculus carnosus (cutaneous maximus) of mammals and mimetics of primates. Intrinsic integumentary muscles are chiefly smooth muscles inserting on feathers and hairs.

26. Electric organs are columns of modified axial, appendicular, or branchiomeric muscle fibers (electroplaxes) capable of producing, storing, and discharging electric potential. A few appear to be modified skin glands.

SELECTED READINGS

Alexander, R.M.: Functional design in fishes, ed. 2, London, 1970, Hutchinson University Library.

Alexander, R.M.: Elastic energy stores in running vertebrates, American Zoologist **24:**85, 1984.

Allen, E.R.: Development of vertebrate skeletal muscle, American Zoologist **18:**101, 1978.

Bennett, M.V.L.: Electric organs. In Hoar, W.S., and Randall, D.J., editors: Fish physiology, vol. 5, New York, 1971, Academic Press.

Bone, Q.: Locomotor muscle. In Hoar, W.S., and Randall, D.J., editors: Fish physiology, vol. 7, New York, 1978, Academic Press.

Cundall, D.: Activity of head muscles during feeding by snakes: a comparative study, American Zoologist **23:**383, 1983.

Ellsworth, A.H.F.: Reassessment of muscle homologies and nomenclature in conservative amniotes: the echidna, Tachyglossus; the opossum, Didelphis; and the tuatara, Sphenodon, Huntington, N.Y., 1975, R.E. Krieger Publishing Co.

George, J.C., and Berger, A.J.: Avian myology, New York, 1966, Academic Press.

Hildebrand, M.: How animals run. In Vertebrate structure and functions: readings from Scientific American, with introduction by Norman K. Wessells, San Francisco, 1955-1974, W.H. Freeman and Co., Publishers.

Lauder, G.V.: On the relationship of the myotome to the axial skeleton in vertebrate evolution, Paleobiology **6:**51, 1980.

Raikow, R.J., Boreckt, S.R., and Berman, S.L.: The evolutionary reestablishment of a lost ancestral muscle in the bowbird assemblage, Condor **81:**203, 1979.

Sacks, R.D., and Roy, R.R.: Architecture of the hind limb muscles of cats: functional significance, Journal of Morphology **173:**185, 1982.

Wardle, C.S., and Videler, J.J.: Fish swimming. In Elder, H.Y., and Truman, E.R., editors: Aspects of animal movement, Cambridge, England, 1980, Cambridge University Press.

Symposia in American Zoologist
Skeletal muscle tissue, **18:**95, 1978.

Mammalian mastication: an overview, **25:**289, 1985.

11

DIGESTIVE SYSTEM

In this chapter we will examine the alimentary canal, the glands and other evaginations that arise from it, and some spectacular modifications associated with varying food habits, such as tongues that are stored under the scalp, tooth plates that crush molluscs, gizzards that macerate, stomachs that return cuds to the oral cavity, and ceca that are the home of cellulose-digesting bacteria. We will start by recalling some of the means employed by vertebrates for obtaining energy in the form of food from their environment.

PROCURING FOOD

Any observant person with an interest in animal life can cite numerous means by which animals obtain energy in the form of food from their environment. Of the techniques employed, some are more obvious than others. A few will be mentioned in anticipation of better appreciating the structural adaptations associated with feeding that are described in this and other chapters.

Ancestral chordates and ostracoderms were filter feeders. This process, which can be employed only by aquatic organisms, consists of passively filtering organic matter out of the incoming respiratory stream and propelling the food particles to the rear of the pharynx for swallowing. Filter feeding in sea squirts, amphioxus, and larval cyclostomes has been described in Chapter 2. A more active version of filter feeding is employed by some fishes, such as spoonbills, tiny clupeids, and huge basking sharks. Plankton and small fishes are strained out of the respiratory water stream by long filamentous gill rakers that hang into the pharyngeal chamber from the gill arches. The largest mammals, baleen whales, strain tons of small fish, jellyfish, and other invertebrates from the sea each day through sieves of whalebone, or baleen, that hang into their oral cavity. However, whales take water into the oral cavity solely for the food it contains, since they do not breathe with gills.

With the advent in arthrodires and placoderms of jaws and muscular body walls that can be used for locomotion and pursuit, more aggressive methods of obtaining food became possible. Jaws, at first invested with bony dermal armor, ultimately were furnished with small, often sharp, single denticles that we recognize as teeth. Many organisms thereupon became predators, and their method of feeding, as in modern sharks, was a bite-tear-swallow technique that required no tongue, oral glands, or other specializations of the oral cavity.

A less energy-consuming procedure evolved as a result of further adaptations of the skull and hyoid arches. The modifications enabled some teleosts to approach close to organisms that are small enough to be swallowed, extend protrusible jaws, create suction, close the mouth, retract the jaws, and swallow. Anyone who has watched a goldfish feeding on flakes of fish food has observed this technique. Lampreys, being parasitic, have a different feeding technique. They rasp the tissues of the host with their spiny "tongue" and suck the debris into their pharynx.

On land, long sticky tongues are found among amphibians, squamate reptiles, and many birds and mammals. Some snakes impale prey on their upper jaw teeth. Winged tetrapods pick up grubs, seeds, and grains and perform other food-getting acts with appropriately shaped beaks, and shore birds pierce fish with them, or scoop fish from the sea, pelican fashion. Sanguinivorous bats suck whole blood, meanwhile preventing clotting by coagulants in their saliva. And baby mammals suck milk, using muscular cheeks and lips.

Herbivorous ungulates crop grasses, and carnivorous mammals use a snap-bite-tear technique that often includes the piercing effect of a saberlike tooth. Because of appropriate joints in the wrist or digits, primates can grasp food and convey it to the mouth, and rodents can hold food between their hands and nibble. Other methods of food getting will come to mind.

Food taking depends on food finding. This is accomplished by sense organs that monitor the external environment. Chemical receptors such as olfactory organs, mechanical receptors such as inner ears and lateral line organs of fishes and amphibians, thermal receptors such as the loreal pits of some snakes, capsulated touch receptors such as those on the sensitive snout of pigs, visual receptors, and electroreceptors—one or more of these alert one vertebrate or another to the presence and location of food. Once the energy-containing food is within the body the digestive tract can process it, extract needed molecules, and return unassimilated matter to the environment.

THE DIGESTIVE TRACT: AN OVERVIEW

The digestive tract is a tube, seldom straight and often tortuously coiled, commencing at the mouth and emptying either into a cloaca or to the exterior via the anus (Fig. 11-1). It functions in the ingestion, digestion, and absorption of foodstuffs and in elimination of undigested wastes. In most fishes respiratory water also enters the mouth, but this water is quickly shunted through gill slits and out of the digestive tract.

Major subdivisions of the tract are oral cavity, pharynx (oropharyngeal cavity in fishes), esophagus, stomach, and small and large intestines. Associated with the tract are accessory organs such as tongue, teeth, oral glands, pancreas, liver, and gallbladder. Blind evaginations, or ceca, are common. The digestive tract and accessory organs constitute the digestive system. The digestive organs beyond the esophagus are in the coelom and, for the most part, attached to the dorsal body wall by a median dorsal mesentery that conducts blood vessels and nerves to the organs (Fig. 1-2). The embryonic origin of the coelom as a cavity within the lateral mesoderm is discussed in Chapter 4.

Differences in the anatomy of digestive tracts caudal to the pharynx are correlated not so much with whether an animal lives on land or in the water as with the nature and abundance of food. Is it readily absorbable when ingested, as in vampire bats, or does it require extensive enzymatic activity or mechanical maceration, as in carnivores? Is the food supply constant, so that whenever an animal is hungry the food is likely to be there, or does it have to be stalked? If the latter, the meal is probably bulky and there has to be room for it in an appropriately expandable stomach until it can be digest-

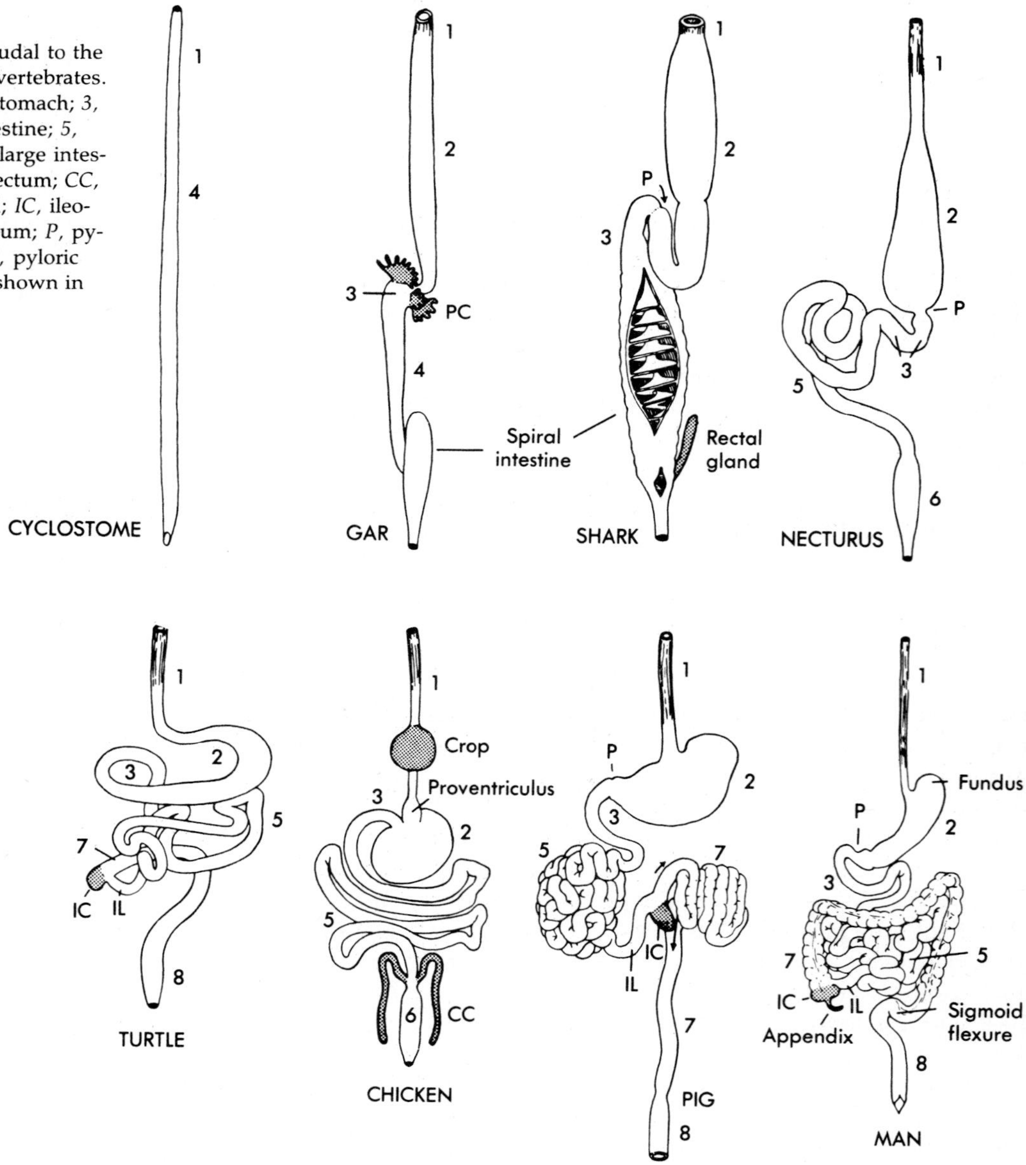

FIGURE 11-1

Digestive tracts caudal to the pharynx of a few vertebrates. *1*, Esophagus; *2*, stomach; *3*, duodenum; *4*, intestine; *5*, small intestine; *6*, large intestine; *7*, colon; *8*, rectum; *CC*, paired ceca of bird; *IC*, ileocolic cecum; *IL*, ileum; *P*, pyloric sphincter; *PC*, pyloric ceca. All ceca are shown in gray.

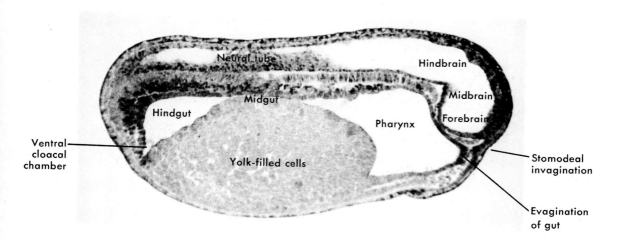

Ventral cloacal chamber

FIGURE 11-2

Sagittal section of 3.5 mm frog tadpole. (From Phillips, J.B.: Developments of vertebrate anatomy, St. Louis, 1975, The C.V. Mosby Co.)

ed. And what is the shape of the animal's body? If it is long, like that of a cyclostome or snake, the tract will probably be straight. If the trunk is short, as in turtles and anurans, the intestine must be coiled in order to provide a sufficient absorptive area. Some fishes have a spiral valve, or typhlosole, that increases the absorptive area.

The entire digestive tract is ciliated in many larval vertebrates, and there are cilia in the stomach of many teleosts, in the oral cavity, pharynx, esophagus, and stomach of some adult amphibians, in the ceca of some birds, and in other locations in various species. Cilia are present for a time in the stomach of a human fetus. However, peristalsis is chiefly responsible for moving foodstuffs along the alimentary canal. The tract is lined by endoderm, as are all evaginations from it. Surrounding the endodermal lining caudal to the pharynx is splanchnic mesoderm (Fig. 4-9), which provides the muscular and connective tissue layers of the tract and the blood vessels and lymphatics, and supports the base of multicellular glands and the plexi of visceral nerves.

The embryonic digestive tract consists of three regions. The part containing the yolk, or to which the yolk sac is attached, is the **midgut.** Anterior to the midgut is **foregut** and caudal to it is the **hindgut** (Fig. 11-2). The foregut elongates to become the posterior portion of the oral cavity, the pharynx, esophagus, stomach, and most of the small intestine. The hindgut becomes the large intestine and cloaca. Little of the midgut remains in adults. From stomach to cloaca the embryonic gut is attached to the dorsal body wall of the trunk by a continuous dorsal mesentery and to the ventral body wall by a ventral mesentery (Fig. 11-24, *A*). Much of the dorsal mesentery remains throughout life, but the ventral mesentery disappears except at two sites. It remains in all vertebrates at the liver where it is called **falciform ligament,** and it remains in tetrapods as the **ventral mesentery of the urinary bladder.**

The anterior part of the oral cavity arises from the **stomodeum,** a midventral invagination of the ectoderm of the head (Figs. 1-1 and 11-2). A thin membrane, or **oral plate,** temporarily separates the two. It ultimately rup-

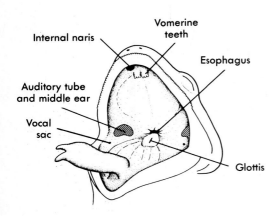

Vomerine teeth

Internal naris

Esophagus

Auditory tube and middle ear

Vocal sac

Glottis

A. MALE FROG

FIGURE 11-3

Oral cavity of an amphibian and a mammal. The roof of the oral cavity in a frog is a primary palate. In the cat it is a secondary palate *(red).* The latter consists of a hard (bony) palate and a soft palate *(light red).* Crossed arrows indicate pharyngeal chiasma where food and water streams cross.

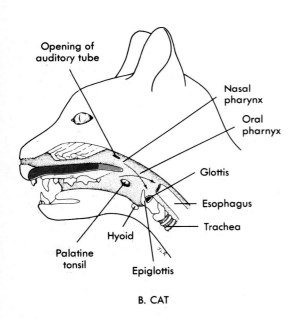

Opening of auditory tube

Nasal pharynx

Oral pharnyx

Glottis

Esophagus

Trachea

Hyoid

Palatine tonsil

Epiglottis

B. CAT

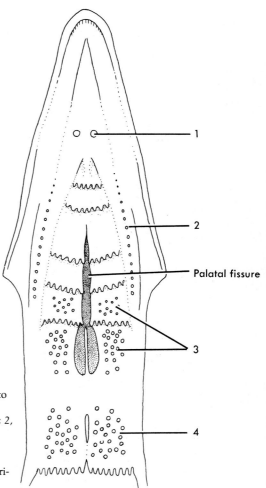

1

2

Palatal fissure

3

4

FIGURE 11-4

Palate of domestic hen. *1* to *4,* Gland openings as follows; *1,* paired maxillaries; *2,* lateral palatines; *3,* medial palatines; *4,* sphenopterygoids. Internal nares are above palatal folds at anterior end of palatal fissure.

tures to provide an entrance to the foregut. A similar invagination, the **proctodeum,** provides an exit from the hindgut when the **cloacal plate** ruptures (Fig. 1-1).

MOUTH AND ORAL CAVITY

The mouth is the entrance to the digestive tract. In gnathostome fishes it opens into the **oropharyngeal cavity,** which ends at the esophagus. In tetrapods it opens into the **oral cavity,** which terminates at the pharynx.

As we have seen earlier, the roof of these cavities is a primary palate in fishes and amphibians, a partial or complete secondary palate in amniotes. The roof is pierced far anteriorly in lungfishes and amphibians by internal nares (Fig. 11-3, A). In snakes, lizards, and birds a deep groove, the palatal fissure, channelizes respiratory air (Fig. 11-4). In crocodilians and mammals the entire roof of the oral cavity is a secondary palate that provides nasal passageways all the way to the pharynx (Fig. 11-3, B).

In mammals a trench, the **oral vestibule,** separates the gums, or **alveolar ridges,** from the cheeks and lips. From the vestibule **cheek pouches** extend backward under the skin and integumentary muscles on each side of the head in some rodents. When these pouches are filled with grain that is being transported from fields to a burrow for storage, a hamster looks as though it had the mumps! Birds have a median **sublingual seed pouch** that is employed in the same manner. When full, the pouch hangs beneath the rear of the oral cavity suspended in a sling formed by the mylohyoid muscle. It is emptied by shaking the head vigorously. Salivary and other glands also empty into the oral cavity in one species or another.

Tongue

The tongue of jawed fishes and perennibranchiate amphibians is a mere crescentic or angular elevation in the floor of the pharynx caused by the underlying hyoid skeleton. This lean hyoid elevation is a **primary tongue** (Fig. 11-5). The tongue of most amphibians consists of the primary tongue and an additional contribution, the **glandular field,** from the pharyngeal

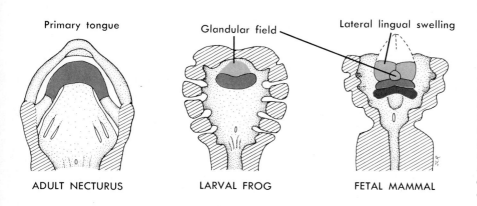

Primary tongue Glandular field Lateral lingual swelling

ADULT NECTURUS   LARVAL FROG   FETAL MAMMAL

FIGURE 11-5

Floor of oral cavity depicting stages in evolution of the mammalian tongue. First arch derivatives, *light red;* second arch, *medium red;* third arch, *dark red.* Dotted outline in mammal shows final extent of lateral lingual swellings. In mammals the glandular field is also known as the tuberculum impar, and arch II endoderm is later covered by arch III endoderm.

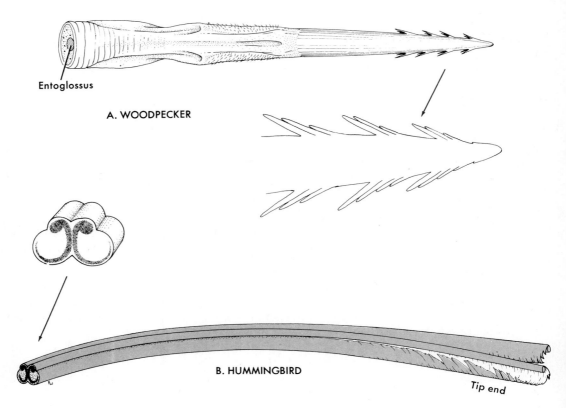

Entoglossus

A. WOODPECKER

B. HUMMINGBIRD

Tip end

FIGURE 11-6

Tongues of two birds with different feeding habits. **A,** Ventral surface. **B,** Lateral view.

floor anterior to the hyoid arch (Fig. 11-5, larval frog). The tongue of reptiles and mammals has a primary tongue contribution from arch II and a pair of **lateral lingual swellings** from arch I that become stuffed with myotomal muscle. In addition, arch III contributes by growing forward over the second arch mucosa and excluding it from the surface. In birds the lateral lingual swellings are suppressed. The "tongue" of agnathans is not really a tongue but a lingual cartilage of unknown homology that is operated by protractor and retractor muscles (Fig. 12-2).

The tongue is widely used for capturing or gathering food. Long, sticky tongues of insectivorous or nectar-feeding tetrapods from salamanders to bats and anteaters dart in and out at lightning speed. The tongue of the four-foot-long Great Anteater is two feet in length! When the mouth of a toad is closed, the tongue lies folded back over itself so that the tip is directed toward the esophagus. Flipping the tongue out of the mouth occurs when the long fibers of the genioglossus medialis muscle, which extend to the tip of the tongue, stiffen to form a complex of intrinsic rods, and the genioglossus basalis muscle, which forms a wedge beneath the anchored anterior end of the tongue, suddenly swells. The combination flips the rigid tongue over the mandibular symphysis. As it returns to the mouth by contraction of the hypoglossal muscles, the captured insect or other food is propelled caudad toward the esophagus (Gans and Gorniak).

Woodpeckers have a tongue that shoots like an arrow into dark crevices in tree trunks, impaling grubs and carrying them to the mouth (Figs. 8-41 and 11-6, *A*). The tiny tongue of hummingbirds darts rapidly back and forth between flower and mouth, collecting droplets of nectar at the hollow frayed tip (Fig. 11-6, *B*). A parrot's tongue is armed with two flexible horny shields in the walls of a seedcup that is used for feeding on seeds, grain, and fruits (Fig. 11-7). Like fingernails, the shields are composed of keratinized epithelial cells that grow forward from a nail-like bed halfway back on the tongue. The anterior edge is constantly being worn away and replaced. Embedded within the tongue of birds and lizards is an entoglossal bone, an anteriorly directed process of the hyoid (Fig. 8-38, *E*). In many birds a long pair of paraglossal bones are attached to the entoglossus near the tip of the tongue and extend caudad embedded in its edges.

The huge immobilized muscular tongue of some baleen whales directs tons of incoming sea water into huge reservoirs under the throat and chest.

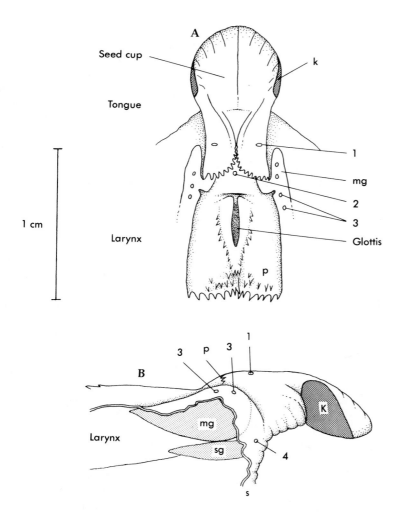

FIGURE 11-7

Tongue of the parrot *Psittacula roseata*. **A,** Dorsal view. **B,** Right lateral view, the mucosa of the oral cavity cut away to reveal glands. *k,* Horny shield; *mg,* mandibular gland; *sg,* sublingual gland. *1* to *4,* Openings of the following glands: *1,* paired lingual; *2,* medial lingual; *3,* mandibular; *4,* sublingual. In **B,** *s* is the site of the mandibular symphysis (Courtesy D.G. Homberger.)

When these are being emptied by compression of their walls, the water is strained through frayed sheets of baleen, and organisms strained out of the water accumulate on the deeply furrowed tongue and are then swallowed. Since newborn whales have only rudiments of baleen, there is room in the oral cavity for the mobile tongue to be manipulated for suckling for about 6 months before it becomes immobilized. The tongues of most other mammals are protrusible, although tied to the floor of the oral cavity by a ligament, the **frenulum linguae.** In human beings, if the frenulum inhibits movements of the tip of the tongue, making the individual tongue tied, it can be snipped at its anterior edge. The tongues of turtles, crocodilians, and some birds are also unprotrusible. Garter snakes and a few others have no tongue.

The mucosa of the tongue contains receptors not only for taste but for other stimuli as well, including, in amniotes, stereognosis (perception of the form, weight, etc., of a solid body by feeling, handling, or lifting it). These encapsulated nerve endings enable insectivores to locate food by feeling, and a woodpecker knows whether it has impaled a grub or jammed into an unyielding object. Seed-eating birds use stereognostic information in manipulating a seed that is being husked in the seed cup (Fig. 11-7, A).

When tongues are not used in procuring food or water, they manipulate fluids and solids in the oral cavity; and the tongues of most tetrapods participate in swallowing. Long tongues are used by overheated mammals as a site for cooling the blood by evaporation of saliva while panting. Lizards clean their spectacles with them. Spiny papillae on the tongues' surface are used by carnivores for rasping bones and by many mammals for grooming. The latter use accounts for hairballs that often form in the stomach of domestic cats. And, of course, human speech as we know it would not be possible without a tongue.

Oral Glands

All tetrapods have multicellular oral glands that, in one species or another, secrete watery or viscous fluid containing mucus, ptyalin (a starch-digesting enzyme), toxins, or other substances. Saliva is a mixture of secretions and is most evident in mammals. Moisture is essential for taste buds to function, viscous secretions keep the tongue sticky, and toxins are defensive or tranquilize prey.

Oral glands are usually named according to location. Labial glands open into the oral vestibule at the base of the lips, molar glands lie near the molar tooth, infraorbital glands are in the orbit, palatal glands open onto the palate, and sublingual and submandibular glands open via papillae under the tongue. Intermaxillary (internasal) glands lie near the premaxillae. In frogs the latter consist of up to 25 small glands, each with its own duct that delivers a sticky secretion on the palate. The large poison gland of venomous snakes is a palatal gland. Its duct opens at the base of a maxillary tooth, and the venom exudes into a groove or tube in the fang (tooth). In *Heloderma*, the only poisonous lizard, sublingual glands secrete the toxin. The parotid gland of mammals is the largest oral gland, and ptyalin is always one of its products. Not all mammalian salivary glands secrete ptyalin, however. The parotid duct crosses the masseter muscle under the skin of the cheek and opens into the oral vestibule opposite one of the upper molar teeth (Fig. 11-8). The poison glands of reptiles resemble the parotid gland histologically. The openings of some of the oral glands of birds are shown in Figs. 11-4 and 11-7.

Aquatic vertebrates have adequate moisture in their oral cavity for taste buds to function, and production of digestive enzymes would be a waste of energy since any secretion would be diluted and washed away. Male cat-fishes are an exception. During breeding seasons they carry fertilized eggs in temporary epithelial folds, or crypts, in the mouth, and goblet cells in the crypts produce a copious secretion. The brood pouches and goblet cells atrophy after the eggs hatch.

Teeth

Bony dermal structures that prevented escape of live food from the oro-pharyngeal cavity or that were used for crushing shellfish, biting flesh, or rasping vegetation varied widely among early jawed fishes. Some placo-derms had short jaws covered with thin plates of dermal bone that no doubt participated in food procurement or retention. Some had long jaws covered with one or two *long* heavy bony **dental plates** with sharply notched blade-like cutting edges, and the plates were braced against heavy cheek plates. In others, the bony plates bore long prongs of dentin that were affixed to the plates. Some had small plates that bore clusters of broad-based conical or star-shaped denticles, and the plates were shed at intervals. In still oth-ers the plates were apparently not shed. In placoderms all *these denticle-bearing bony plates invested the underlying endoskeleton* of the jaws. In acan-thodians each denticle, or tooth, usually broad based, some conical, some slender and curved, some with a dozen spiny cusps arranged in whirls, was attached *individually* to the endoskeleton instead of being borne on plates, and the jaws often were flanked by small dermal scales. Some acan-thodians were toothless or had lower teeth only. Peyers' comprehensive treatise on comparative odontology (Selected Readings) is illustrated with over 50 pages of photographs.

Until the phylogenetic relationships between acanthodians, placoderms, and later cartilaginous and bony fishes can be ascertained, only the broad-est generalizations regarding the origin of teeth can be made: *they are deriv-atives of dermal armor.* Evidence comes not only from paleontology but from the comparative histology of armor and teeth and from the observation that placoid scales, even today, show a gradual transition from scales to teeth as they approach the cutting edges of the jaws (Fig. 11-9).

Teeth are composed of a core of **dentin** surmounted by a crown of enam-el (Fig. 11-10). At the base of a developing tooth and forming the core as the tooth grows is a live **dermal papilla** that supports nutritive vessels and nerves and provides, at its periphery, odontoblasts that secrete dentin. An epidermal **enamel organ** composed of ameloblasts secretes enamel (Figs.

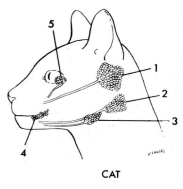

CAT

SNAKE

FIGURE 11-8

Oral glands in a reptile and mammal. *1,* Parotid; *2,* sub-mandibular; *3,* sublingual; *4,* molar; *5,* infraorbital; *6,* poi-son gland of rattlesnake; *7,* maxillary tooth with groove for transfer of toxin; *8,* tongue.

FIGURE 11-9

Schematic representation of the continuity of, *1*, placoid scales, *2*, teeth, and, *3*, stomodeal denticles (not to scale).

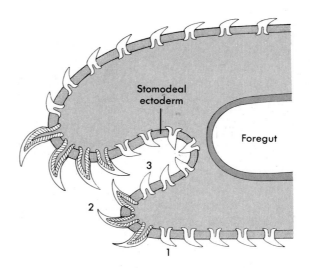

11-10 and 11-11). When the tooth is fully formed, the papilla remains as **pulp** in the root canal. An enamel organ is present but functionless in armadillos and a few other vertebrates, hence their teeth have no enamel. Around the root in bony vertebrates is a thin layer of **cementum,** a variety of bone. Collagenous fibers anchor the cementum in the bone of the jaw. In nonsocketed teeth the relationship of cementum into the jawbone is more intimate.

Agnathans, sturgeons, numerous teleosts (sea horses, among others), some toads, sirens (urodeles), turtles, and modern birds are toothless. However, one species of terns develops an embryonic set that does not erupt, and at least one genus of turtles has enamel organs. Toothless mammals develop a set that either does not erupt or, after erupting, is soon lost.

Teeth vary among vertebrates in number, distribution within the oral cavity, position with reference to the summit of the jaw, degree of permanence, and shape. They are numerous and widely distributed in the oral cavity and pharynx of living fishes. They develop on jaws, palatal bones, and even on the pharyngeal skeleton. For example, the blue sucker has 35 to 40 teeth on the last gill arch. In early tetrapods, too, teeth were widely distributed on the palate; and even today most amphibians and many reptiles have teeth on the vomer, palatine, and pterygoid bones and occasionally on the parasphenoid. They are confined to the jaws in crocodilians, toothed birds, and mammals, and they are least numerous among mammals. Teeth, therefore, like dermal armor, have tended toward a more restricted distribution with the passage of time.

Jaw teeth may be attached to the outer surface or to the summit of the jawbone, as in many teleosts **(acrodont dentition);** they may be attached to the inner side of the jawbone, as in anurans, urodeles, and many lizards **(pleurodont dentition);** or they may occupy bony sockets, or alveoli **(thecodont dentition,** Fig. 11-12). Socketed teeth are found in many fishes, in

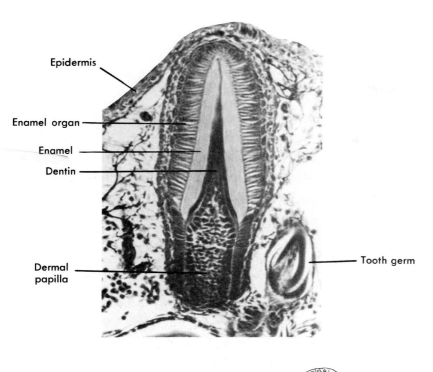

Epidermis

Enamel organ

Enamel

Dentin

Dermal papilla

Tooth germ

FIGURE 11-10

Unerupted tooth of a garfish. The odontoblasts are seen lined up at the periphery of the dermal papilla. Tooth germ is a replacement for a tooth that will be shed. The relationship of odontoblasts to dentin is diagrammed in Fig. 6-6.

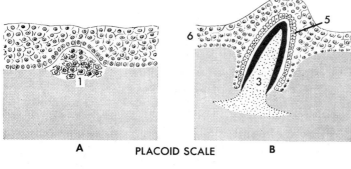

A   PLACOID SCALE   B

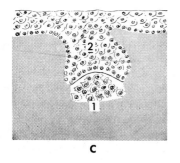

C

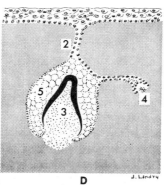

D

TOOTH

MORPHOGENESIS

FIGURE 11-11

Development of placoid scale and mammalian tooth. Dermis, except dermal papilla, is *gray*; enamel is *red*. *1*, Dermal papilla; *2*, dental ridge, an ingrowth of the epidermis; *3*, dentin, produced by the dermal papilla; *4*, germ of replacement tooth; *5*, enamel organ; *6*, epidermis.

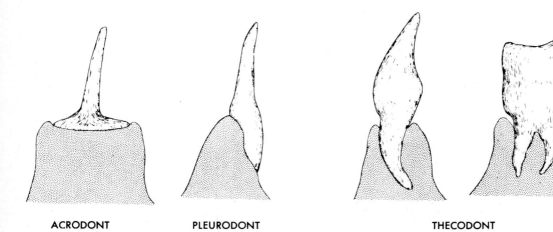

ACRODONT   **PLEURODONT**   **THECODONT**

FIGURE 11-12

Variations in the relationship of teeth to jaws. Acrodont teeth are attached either at the outer surface of the jaw or, as shown, at the summit. Pleurodont teeth are attached to the inner surface of the jaw. Thecodont teeth occupy alveoli. The bone of the jaw is seen in cross section.

crocodilians, extinct toothed birds, and mammals. The sockets are deepest in mammals.

Most vertebrates through reptiles have a succession of teeth, and the number of replacements during a lifetime is indefinite but numerous **(polyphyodont dentition)**. It has been estimated that an elderly crocodile may have replaced its front tooth 50 times. They and other submammalian bony vertebrates that have been studied replace teeth in waves that sweep along the jaws eliminating and replacing every other tooth. Thus in one wave, in tetrapods at least, even-numbered teeth are lost and in the next wave odd-numbered ones are lost. Meanwhile, tooth germs for the next wave of eruptions are forming. Whether the loss and replacement waves sweep from back to front or reverse is not agreed on at present. There is evidence that they sweep in different directions in different species. The waves ensure a balanced distribution of teeth along the jaws throughout life. In sharks, tooth germs form in the dermis on the oropharyngeal cavity side of the jaws and while growing migrate onto the cutting edge of a jaw (that is, from position 3 to position 2 in Fig. 11-9) as the tooth that is being replaced moves beyond the edge before falling away. The cause of the migration is not known.

Only in mammals is there a definite number of teeth in a species, with rare exceptions. Most mammals develop two sets, deciduous, or milk teeth, and permanent teeth **(diphyodont dentition);** and there is a definite sequence in which the teeth erupt. For example, numbering the permanent set in human beings 1 to 8 from front to rear, the sequence of eruption is 6, 1, 2, 4, 5, 3, 7, 8. Eruption of number 8, the last molar, is delayed in higher primates, and this "wisdom tooth" is sometimes imperfect, unerupted, or missing. A first set provides the constantly changing infant jaw with small temporary teeth adequate for an infant's diet until the jaws are more stabilized structurally and have elongated sufficiently to accommodate large teeth for macerating coarse foods.

A few mammals develop only a first set **(monophyodont dentition).** In a platypus the deciduous set is replaced by horny teeth. In toothless whales the first set, although formed within the jawbone, may not erupt, and if they do they are usually shed. The freshwater manatee from the Amazon river and the Australian Rock Wallaby do not have "sets," teeth being replaced throughout life by the forward migration of new teeth formed at the rear of the jaws. In the manatee, migration is at the rate of one or two millimeters a month. Thin, bony sockets separate the roots of successive teeth, and the bony septa are resorbed under pressure from the migrating teeth. The grasses eaten by the manatee contain abrasives that appear to be necessary for the teeth to move forward, since babies fed experimentally on milk show no forward migration of teeth. Proboscidians have a slow but constant succession of molar teeth that move forward from the rear.

MORPHOLOGIC VARIANTS IN FISHES. Most sharks are fish eaters and have numerous rows of jaw teeth that are either flat, sharp, notched triangles that are used for cutting, or they are single or multipointed tusks that

FIGURE 11-13

Jaws of the Port Jackson shark, *Heterodontus,* showing transition of placoid scales to teeth. The rounded teeth are used for crushing molluscs. (Courtesy Ward's Natural Science Establishment, Inc., Rochester, N.Y.)

curve toward the pharynx and hold a struggling prey until it can be swallowed whole. Each tooth has a broad basal plate of dentin embedded in the dermis. A minority eat shellfish and, although the teeth at the entrance have curved caudally directed spines, the rest form batteries of rounded denticles used for crushing shells (Fig. 11-13). Tiny stomodeal denticles line the pharynx in some sharks, and near the jaws these have transitional shapes between denticles and teeth.

The dental armor of holocephalans and modern lungfishes is reminiscent of that of earlier jawed fishes, consisting of a few large plates of enamel-covered dentin that bear rows of various sized rounded moundlike denticles that usually become sharp spines at the entrance to the oropharyngeal cavity. *Chimaera,* a holocephalan that eats molluscs, has on each side of the upper jaw one large anterior and one small posterior tooth plate that together cover the entire upper jaw. There is a single large plate on each side below. In modern lungfishes the plates are restricted to the palate and medial aspects of the lower jaw.

The jaw teeth of actinopterygians, amphibians, and most reptiles are simple pointed cones attached to one or more membrane bones. Small teeth may be interspersed among large ones, and those in front are sometimes larger and curved slightly to the rear. Specialized shapes sometimes appear on one jaw or the other. Gars, for instance, have a few fanglike teeth shaped at their ends like arrows; and the fangs of venomous snakes, borne on the maxillae, are curved or bladelike, grooved on the rear surface, or tubular, for injecting venom. When all teeth are similar the dentition is described as **homodont.**

MORPHOLOGIC VARIANTS IN MAMMALS. All but a very few mammals exhibit **heterodont dentition;** that is, the teeth vary morphologically from front to rear, being identified as **incisors, canines, premolars,** and **molars,** the latter two being "cheek teeth." Even those mammals with homodont dentition—modern cetaceans and sirenians—had Cretaceous mammalian ancestors whose dentition was heterodont, a condition inherited from synapsid reptiles. (Fig. 3-23 shows a therapsid reptile with a prominent canine tooth.) Therefore, homodonty in modern members of those orders is secondary. Some aquatic carnivores have a tendency toward homodonty, and this condition, too, is secondary, since these mammals are descendants of land-dwelling carnivores.

Incisors, located anteriorly, have one horizontal cutting edge and a single root. They are best developed in herbivorous mammals, which use them for holding, cropping, or gnawing. Those of rodents and lagomorphs have enamel on the anterior surface only, the back surface being dentin. Since enamel, being harder, wears down more slowly than dentin, sharp chisel-like teeth are produced from gnawing, an activity that is mandatory for rodents in order to keep the incisors from growing too long. Incisors may be totally absent, as in sloths, or lacking on the upper jaw, as in bovines (Fig. 11-14, ox). Elephant tusks are modified incisor teeth (Fig. 11-14, mastodon).

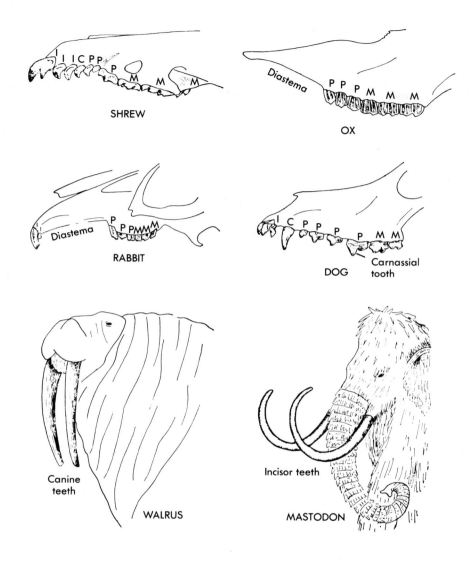

FIGURE 11-14

Mammalian upper teeth, showing generalized pattern (shrew), and specializations in a herbivore (ox), in a gnawing animal (rabbit), and in a carnivore (dog). Extreme specialization of canines is seen in the walrus and of incisors in the mastodon. *I,* Incisor; *C,* canine; *P,* premolar; *M,* molar.

Canines lie immediately behind the incisors (Fig. 11-14, dog). In generalized mammals, incisors and canines scarcely differ in appearance (Fig. 11-14, shrew). In carnivores the canines are spearlike and used for piercing flesh (Fig. 11-15, *H*). They are also the tusks of the walrus (Fig. 11-14). Canine teeth are absent in rodents and lagomorphs and so there is a toothless interval, or **diastema,** between the last incisor and the first cheek tooth (Figs. 11-14, rabbit, and 11-15, *F*). Canine teeth attained their greatest length on the upper jaw of the now extinct saber-toothed cats, in which they extended as much as 20 cm below the lower jaw with the mouth closed.

The molariform teeth—premolars and molars—of carnivores and herbivores are very different, those of carnivores being specialized for shearing

FIGURE 11-15

Some morphological varieties of mammalian teeth. **A,** Early mammalian teeth. **B** to **D,** Cheek teeth from a dog, horse, and elephant, respectively. **E,** Complete permanent human dentition. **F,** Upper jaw dentition of a mouse. The molars are bunodont. **G,** Lower molar from a young tapir, whose diet is leaves and twigs. **H,** Lower left flesh-tearing canine from a jaguar. **I,** One of a series of similar lower right molars from a crabeater seal. These teeth are used for straining plankton.

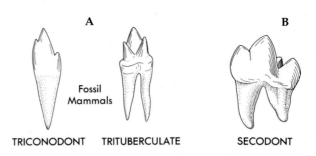

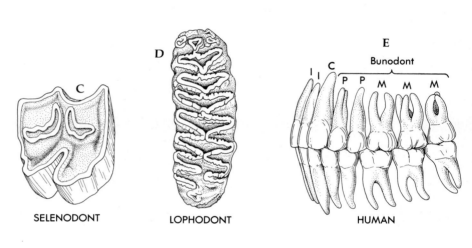

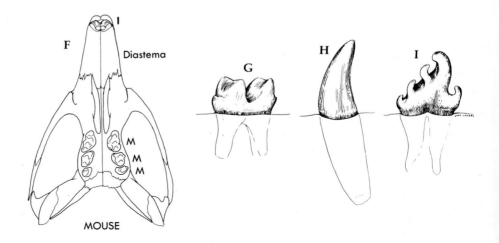

flesh and those of herbivores for grinding. There is little difference between premolars and molars in any case. The latter are characterized chiefly by the fact that they are not replaced by a second set in mammals with diphyodont dentition. They are, in reality, late arrivals of the first set.

The portion of a cheek tooth above the gum line—the crown—in carnivores is laterally compressed, and the sharp edges of the cusps of the upper and lower set fit between one another to produce a shearing effect when the jaws are closed. The last upper premolar and the first lower molar, known as **carnassial teeth,** are exaggerated instances of this condition (Fig. 11-14, dog). The cheek teeth caudal to the carnassials tend to be less developed. Teeth such as those just described are **secodont,** and are characteristic of carnivores.

The cheek teeth of many ungulates are specialized for grinding vegetation or chewing a cud and, in the more primitive ungulates, for crushing seeds, nuts, and fruits. The crowns are broad, and the enamel is disposed in vertical crescent-shaped columns separated by dentin (Fig. 11-15, C). Chewing causes the softer dentin to wear down more quickly than the enamel, thereby providing sharp, crescentic, rasping ridges with a wide variety of configurations that macerate the food during the complex side-to-side and forward-backward movements of the artiodactyl jaw. These are **selenodont** teeth. The condition is exaggerated in proboscidians, the enamel and dentin being intricately folded and disposed as transverse ridges (lophs) on enormous dental plateaus (Fig. 11-15, D). These **lophodont** teeth reach a foot or more in length and a third of a foot in width in the largest elephants.

The cheek teeth of some mammals lack sharp ridges and pointed cusps and have, instead, low rounded or moundlike cusps completely covered with enamel so that no dentin reaches the surface. These **bunodont** teeth wear down evenly (Fig. 11-15, E). They are found in rhinos, some hogs, some primitive ruminants, some rodents such as the white-footed mouse, and in man.

Rodents, the largest mammalian order and with the largest variety of diets, exhibit the largest variety of teeth, some of them low crowned with long roots **(brachyodont),** as in squirrels, some high crowned and with short roots **(hypsodont),** as in wood rats. The cheek teeth of man are brachyodont, and those of horses are hypsodont. Tusks are hypsodont. Among unusual teeth are those of the crabeater seal (Fig. 11-15, I).

Cheek teeth of early placentals were **triconodont,** having three conelike prominences arranged in a straight line. Later, these cones became arranged in a triangle, resulting in **trituberculate** secodont teeth (Fig. 11-15, A). The formation of crests connecting the tubercles in various configurations resulted in selenodont and lophodont teeth. Much of our knowledge of the evolution of mammals through geologic time is based on detailed studies of fossil molariform teeth; and much of the success of early eutherians can be attributed to dental adaptations.

The first Cretaceous eutherians had three incisors, one canine, four premolars, and three molars on each side of each jaw, a total of 44 teeth. This may be expressed by the formula $\dfrac{3-1-4-3}{3-1-4-3}$. Adults of some of the generalized mammals of today still exhibit this formula. Formulas for a few selected mammals of varied dietary habits are:

Cretaceous eutherian		$\dfrac{3-1-4-3}{3-1-4-3}$
Insectivores	Solenodon $\dfrac{3-1-3-3}{3-1-3-3}$	Mole $\dfrac{3-1-4-3}{3-1-4-3}$
Marsupials	American opossum $\dfrac{5-1-3-4}{4-1-3-4}$	Numbat $\dfrac{4-1-3-5}{3-1-3-6}$
Primates	Tarsius $\dfrac{2-1-3-3}{1-1-3-3}$	Catarrhines $\dfrac{2-1-2-3}{2-1-2-3}$
Carnivores	Canines $\dfrac{3-1-4-2}{3-1-4-3}$	Felids $\dfrac{3-1-3-1}{3-1-2-1}$
Lagomorphs	Rabbits $\dfrac{2-0-3-3}{1-0-2-3}$	Pika $\dfrac{2-0-3-2}{1-0-2-3}$
Rodents	Hamster $\dfrac{1-0-0-3}{1-0-0-3}$	Squirrel $\dfrac{1-0-2-3}{1-0-1-3}$
All bovines		$\dfrac{0-0-3-3}{3-1-3-3}$

Teeth, along with the tongue and hyoid, constitute a functional triad that procures, manipulates, and, in mammals, masticates foodstuffs at the entrance to the digestive tract, then starts a bolus of food on its way to digestive sites.

HORNY EPIDERMAL TEETH. Horny teeth sometimes take the place of bony ones. Agnathans have horny teeth in the buccal funnel and on the tongue that are used for rasping. Anuran tadpoles have several rows of horny teeth on temporary lips perched on poorly developed jaws; they are used for rasping algae and other vegetation, which is the tadpole's diet. At metamorphosis the horny teeth of anurans are shed and replaced by bony ones that develop on membrane bones overlying the embryonic upper and lower jaw cartilages. Before hatching, turtles, crocodilians, *Sphenodon*, birds, and monotremes have a temporary horny egg tooth that is used for cracking the shell; and after a baby platypus has lost its first set of bony teeth horny teeth replace them throughout life. The horny beaks of turtles and birds often have serrations that perform some of the functions of teeth, although ancestral birds did not need them. They had bony teeth.

PHARYNX

The pharynx is the part of the digestive tract that had pharyngeal pouches in the embryo. The pharynx of fishes, which is a respiratory organ, is described in Chapter 12. The embryonic pharynx is described in Chapter 1.

In adult vertebrates lacking gill slits, that is, metamorphosed amphibians and amniotes, the pharynx is the part of the foregut immediately preceding the esophagus. The most constant features of the tetrapod pharynx are the **glottis** (a slit leading into the larynx), the openings of **auditory tubes,** and the opening into the esophagus (Fig. 11-3). In mammals a cartilaginous flap, the **epiglottis,** overlies the glottis. In swallowing, the larynx is drawn forward (upward in humans) against the epiglottis and prevents foreign substances from entering the pathway to the lungs. In many species below mammals a flap of mucosa performs this function.

The pharynx of adult mammals consists of a **nasal pharynx (nasopharynx)** above the soft palate and an **oral pharynx (oropharynx)** that begins at the caudal border of the soft palate and extends to the glottis and esophageal opening. The nasal passageways, created by formation of a secondary palate, empty into the nasal pharynx anteriorly, and the two auditory tubes, derived from the first pair of embryonic pharyngeal pouches, open into its lateral walls. The nasal pharynx is completely closed off from the oral pharynx during swallowing, when extrinsic muscles draw the free caudal border of the soft palate tight against the roof of the pharynx.

The oral cavity in mammals leads to the oral pharynx. The site of continuity is a narrow passageway called the **isthmus faucium.** The isthmus is bounded dorsally by the caudal border of the soft palate and laterally by the **pillars of the fauces,** which are two muscular folds on each side of the throat that arch downward from the lateral edge of the soft palate to the side of the tongue **(glossopalatine arch)** and pharynx **(glossopharyngeal arch).** The arches can be seen in your own pharynx by looking in a mirror while saying "Ah!" Between the two pillars on each side is a **palatine tonsil,** which develops in the wall of the embryonic second pharyngeal pouch. A remnant of the pouch often remains as a pocketlike crypt alongside the palatine tonsil. The palatine tonsils are part of a ring of lymphoidal (adenoidal) masses that encircle the isthmus. Other masses are the **pharyngeal tonsils,** often called adenoids, in the nasal pharynx and **lingual tonsils** at the root of the tongue.

A **laryngeal pharynx** is present in some mammals, including humans. It is a caudal extension of the oral pharynx behind the larynx and leading to the esophagus. It exists only in species in which the esophageal opening is farther caudad than the glottis. In monkeys and humans a fleshy **uvula** hangs from the caudal border of the soft palate into the oral pharynx.

In some teleosts a pair of elongated muscular tubes, **suprabranchial organs,** evaginate from the roof of the pharynx on each side near the esophagus, extend cephalad above the membranous roof of the pharynx behind the skull, and then turn caudad to terminate as blind sacs. Elongated gill rakers from the last two gill arches form funnel-shaped baskets that extend into the entrances of the tubes, and each tube is surrounded by a cartilaginous capsule to which the striated muscle in its walls is attached. The epithelium at the blind ends has many goblet cells, and the sacs contain quantities of plankton, sometimes compressed into a bolus. It may be that one function of these in gill-breathing fishes is to trap plankton from incoming water and concentrate it into mucified masses, which are then forced out and swallowed. In at least one air-breathing teleost the cavity is filled with air and the epithelial lining is highly vascular and serves as an accessory respiratory membrane.

ESOPHAGUS

The esophagus is a distensible muscular tube, shortest in neckless vertebrates, connecting pharynx and stomach. Striated muscle at the cephalic end of a long esophagus is gradually replaced farther down by smooth muscle, although it may continue onto the stomach wall, especially in cud-chewing mammals that regurgitate their food for leisurely chewing.

One of the few specializations of the esophagus is the **crop** of some birds (Fig. 11-16). The crop is a paired or unpaired membranous diverticulum of the esophagus occurring primarily in grain eaters and used for initial storage of food. An enzyme for preliminary digestion may also be secreted. In

FIGURE 11-16

Crop, esophagus, and stomach (gizzard) of a grain-eating bird.

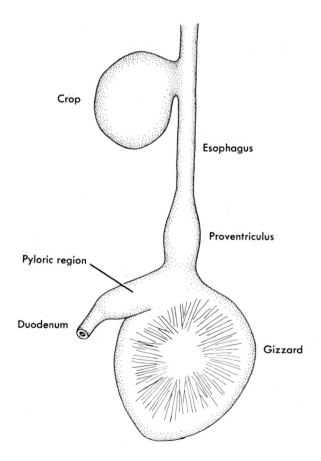

Crop

Esophagus

Proventriculus

Pyloric region

Duodenum

Gizzard

male and female doves a glandular part of the lining of the crop undergoes fatty degeneration under the stimulation of prolactin. The cells are then shed and regurgitated along with partially digested food as "pigeon's milk," which is fed to nestlings. A specialization of the esophagus of sanguinivorous bats is described in Chapter 3.

STOMACH

The stomach is a muscular chamber or series of chambers at the end of the esophagus. It serves as a storage and macerating site for ingested solids and secretes digestive enzymes that partially liquefy food before injection into the small intestine. Cyclostomes scarcely have a stomach, and a boundary between esophagus and stomach is indefinite or lacking below birds. Even in birds and mammals, the mucosa of part or all of the stomach may resemble that of the esophagus (Fig. 11-17). The stomach terminates at the **pyloric sphincter.**

The stomach is straight when it first develops in the embryo (Fig. 11-24) and may remain so throughout life in lower vertebrates. More often, flexures develop producing a J- or U-shaped stomach (Figs. 11-18 and 11-19). As a result, the stomach may exhibit a concave border **(lesser curvature)** and a convex border **(greater curvature,** Fig. 11-7, carnivore). The stomach also undergoes torsion in higher vertebrates so that part of it lies crosswise in the trunk. As flexion and torsion become pronounced during development of mammalian stomachs, the dorsal mesentery of the stomach **(mesogaster)** becomes twisted and finally suspended from the greater curvature, which was originally the dorsal border of the stomach. The part of the dorsal mesentery attached to the greater curvature is then called the **greater omentum.** Because the mesentery became twisted, it encloses a **lesser peritoneal cavity** continuous with the main peritoneal cavity via an **epiploic foramen.**

The stomach of some vertebrates, especially of fishes, exhibits one or more ceca (Fig. 11-18). In birds the stomach is often divided into **proventriculus** and **gizzard** (Fig. 11-16). The proventriculus, or glandular stomach, secretes a digestive enzyme, and the gizzard converts the food into a mash.

FIGURE 11-17

Distribution of esophageal-like epithelium (gray) and of typical gastric glands (triangles) in the stomachs of selected mammals. *d*, Duodenum; *e*, esophagus. Broken line at far right outlines region that, in *Homo*, is called the fundus.

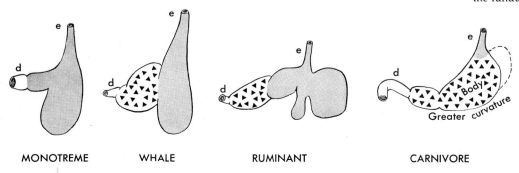

MONOTREME   WHALE   RUMINANT   CARNIVORE

The gizzard is lined with a horny membrane and often contains pebbles. The proventriculus and gizzard are best developed in birds that eat seeds and grain and least developed in carnivorous birds. Crocodilians also have gizzardlike stomachs (Fig. 11-19).

Mammalian stomachs are sometimes divided into several chambers. This is especially true in ruminants (Fig. 11-20). Grasses or grain is chewed briefly and then swallowed, after which it passes into the **rumen,** where preliminary digestion, especially by bacterial action, occurs. From the rumen, food moves into the **reticulum,** the lining of which is honeycombed (reticulated) by ridges and deep pits. In the reticulum, food is formed into a cud, which is regurgitated at will to be leisurely rechewed. After thorough chewing of the cud, the food is again swallowed. This time it passes into the **omasum,** where salivary enzymatic action continues. Finally it enters the **abomasum.** This segment exhibits the usual varieties of gastric glands, and the lining, along with that of the omasum, exhibits longitudinal ridges **(rugae)** found also in other vertebrate stomachs. The rumen, reticulum, and perhaps the omasum could equally well be considered specialized parts of the esophagus analogous to the crop of birds.

FIGURE 11-18

Digestive tracts of four teleosts. **A,** *Fundulus.* **B,** *Cyprinodon.* **C,** *Elops.* **D,** *Trichiurus.* *e,* Esophagus; *gc,* cecumlike stomach; *p,* pylorus; *pc,* pyloric ceca; *s,* stomach.

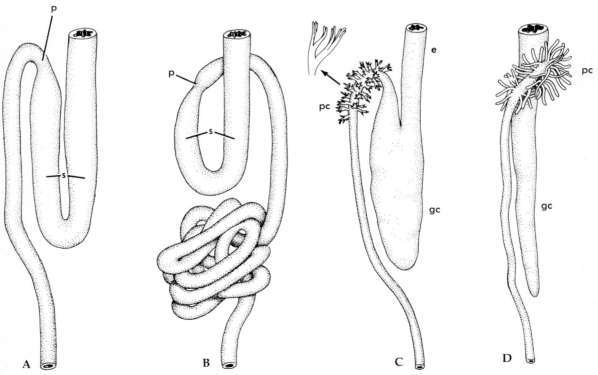

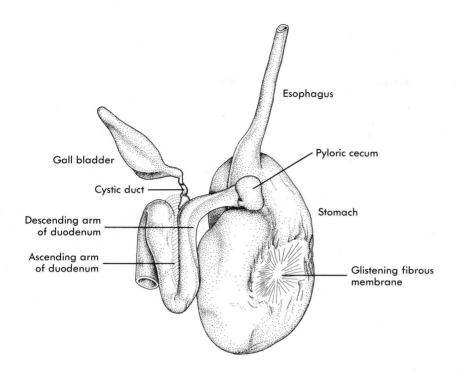

Esophagus

Pyloric cecum

Gall bladder

Cystic duct

Stomach

Descending arm
of duodenum

Ascending arm
of duodenum

Glistening fibrous
membrane

FIGURE 11-19

Gizzardlike stomach and associated structures of a caiman, ventral view. The gallbladder has been lifted cephalad from its normal position between the stomach and descending arm of the duodenum. The stomach has thick muscular walls, except in the midsection where there is a glistening fibrous membrane.

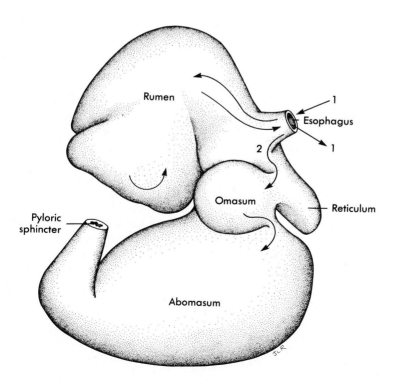

Rumen

Esophagus

1

1

2

Omasum

Reticulum

Pyloric
sphincter

Abomasum

FIGURE 11-20

Stomach of a ruminant (calf). Arrows indicate, *1*, route of food when initially swallowed; *2*, route after cud is swallowed. Cud goes directly to the omasum.

FIGURE 11-21

A, Cross section of mammalian small intestine. Except for villi, which are present but shorter in lizards and birds, the four layers listed at bottom of figure are found throughout the digestive tract in all vertebrates.

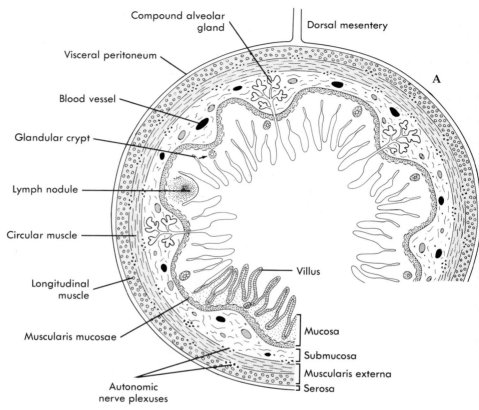

Compound alveolar gland

Dorsal mesentery

Visceral peritoneum

Blood vessel

Glandular crypt

Lymph nodule

Circular muscle

Longitudinal muscle

Muscularis mucosae

Villus

Mucosa

Submucosa

Muscularis externa

Serosa

Autonomic nerve plexuses

A

INTESTINE

The intestine commences at the pyloric sphincter and ends at the cephalic end of the cloaca or, if there is no cloaca, at an anus. It is relatively straight in fishes and in tetrapods with elongated bodies. It is tortuous in other tetrapods, which increases the area for aborption. A spiral valve, or **typhlosole,** does this for many fishes (Fig. 11-1, shark).

The intestine is, of necessity, muscular and its lining is glandular (Fig. 11-21). The first section, or **duodenum,** receives ducts from the accessory digestive organs—liver, gallbladder, and pancreas. Beyond the duodenum the small intestine of lizards, birds, and mammals is lined with fingerlike or leaflike **villi** (low in birds, tall in mammals) that increase the absorptive surface. They are supplied with dead-end lymphatics known as **lacteals** (Fig. 11-21, *B*) which collect absorbed lipids and, by contracting and relaxing forcefully, pump the milky fluid, or **chyle,** to larger lymphatics that bypass the liver en route to the heart. In mammals the small intestine is divided into **jejunum** and **ileum** on the basis of the shape of the villi, the nature of the mucosal lining, and the number of **lymph nodules (Peyer's patches),** which is greater in the jejunum. The small intestine is the chief site of digestion and absorption. Emptying of the small intestine is regulated by an **ileocolic sphincter** in tetrapods (Fig. 11-22).

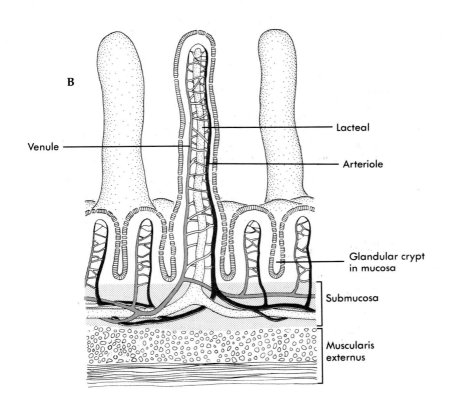

FIGURE 11-21, cont'd

B, Blood supply and lymph drainage of a villus, longitudinal view.

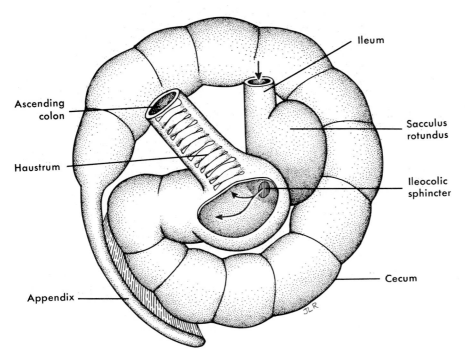

FIGURE 11-22

Ileocolic junction, cecum, and vermiform appendix of rabbit. Arrows indicate available pathways.

The intestine of fishes is simpler than that of tetrapods, and is not divisible into small and large intestines. In amphibians the large intestine is straight and short. In mammals, and in some reptiles and birds, it is divisible into **colon** and **rectum.** The colon commences at the iliocolic sphincter, is variously coiled, and, in human beings, it ends in a **sigmoid (S-shaped) flexure.** The rectum is the straight terminal portion of the large intestine in the pelvic cavity. The chief function of the large intestine is recovery of any water in the residue of digested foodstuffs.

CECA

Ceca are blind diverticula that occur anywhere from esophagus to colon. A digestive cecum in its simplest form is seen in an amphioxus (Fig. 2-5, *B*). The crop sac of birds is an esophageal cecum.

In modern fishes pyloric and duodenal ceca are common, especially in species lacking spiral valves (Fig. 11-18). Up to 200 have been counted in mackerel. Beyond the duodenum ceca are rare in fishes and amphibians. Ileocolic ceca are common in amniotes; usually there are two in birds (Fig. 11-1, turtle, chicken, pig, and man). In animals that feed on cellulose ileocolic ceca house cellulose-digesting bacteria, may be coiled, and may exceed the large intestine in capacity. In insectivorous or carnivorous mammals they are short or absent. The cecum terminates in a **vermiform appendix** in anthropoids, rodents, rabbits, and many other mammals (Fig. 11-22).

Ceca farther along on the colon are not rare. *Hyrax,* for instance, has a large bicornuate cecum on the descending colon (Fig. 11-23). The liver, gallbladder, and pancreas commence ontogenetically as ceca, and the rectal gland of elasmobranchs is a cecum that secretes sodium chloride.

LIVER

The liver arises as a hollow diverticulum **(liver bud)** from the ventral wall of the future duodenum (Fig. 11-24, *A*). The bud, occasionally multiple, invades the ventral mesentery and then grows cephalad into the ventral mesentery of the stomach. The growing tip of the bud gives rise to numerous sprouts that become the lobes of the liver and the gallbladder. The anterior pole of the liver finally becomes anchored to the septum transversum by a coronary ligament. Each lobe of the adult liver is drained by a **hepatic duct** that flows into a **common bile duct.** The terminal segment of the common bile duct is embedded in the wall of the duodenum for a short distance, where it is called the **ampulla of Vater.**

Although most of the embryonic ventral mesentery disappears during development, the mesentery ventral to the duodenum and stomach that was invaded by the liver bud remains as the **hepatoduodenal ligament** (connecting duodenum and liver) and as the **gastrohepatic ligament** (connecting pyloric stomach and liver). These two ligamentous mesenteries constitute the **lesser omentum,** which serves as a bridge transmitting the common bile duct, hepatic artery, and hepatic portal vein. The embryonic

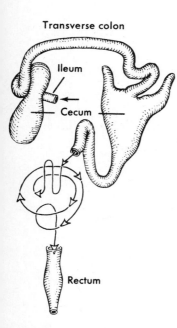

Transverse colon

Ileum

Cecum

Rectum

FIGURE 11-23

Large intestine of a mammal (*Hyrax*) with several ceca. Arrows indicate direction taken by missing segment.

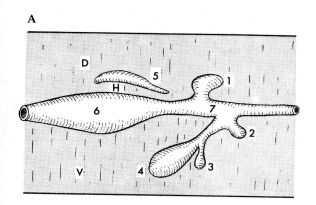

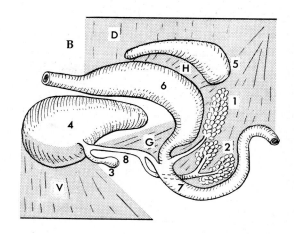

FIGURE 11-24

Development of liver, pancreas, spleen, stomach, and associated mesenteries. **A,** An early stage. **B,** A later stage. *1,* Dorsal pancreatic bud from duodenum; *2,* ventral pancreatic bud from common bile duct; *3,* gallbladder; *4,* liver bud in **A** and liver in **B;** *5,* spleen; *6,* stomach; *7,* duodenum; *8,* common bile duct. *D,* Dorsal mesentery; *G,* lesser omentum; *H,* gastrosplenic ligament; *V,* ventral mesentery, which remains as the falciform ligament in **B.**

mesentery ventral to the liver remains in adults as the **falciform ligament.**

Blood from the stomach, intestines, and pancreas is shunted via a system of portal veins to the capillaries of the liver where glucose in excess of immediate tissue needs is removed from the blood stream, converted to glucose-6-phosphate, and then stored as glycogen. Amino acids are removed from liver capillaries for deamination, the by-products of which are ammonia, uric acid, and urea. These are carried away by hepatic veins and excreted by the kidneys. The liver also manufactures bile, which emulsifies ingested lipids in the intestine in preparation for digestion and absorption, and it manufactures several blood proteins, including fibrinogen and prothrombin, which are essential for clotting. The fetal liver is an important source of embryonic red blood cells, and the adult liver breaks down hemoglobin freed from aged red cells. The breakdown results in red and green bile pigments (bilirubin and biliverdin) that become part of the bile, which is stored in the gallbladder. This list of functions is not exhaustive. The shape of the liver conforms to the space available in the coelom. In animals with long trunks the liver is elongate; in those with short trunks it is short and broad.

GALLBLADDER

One sprout of the liver bud expands to become the gallbladder and its **cystic duct.** Because of the embryonic origin of the gallbladder, the cystic duct empties into the common bile duct. Bile enters and leaves the gallbladder via the cystic duct. A gallbladder develops in most vertebrates, including hagfishes; however, no gallbladder develops in lampreys, many birds, rats, perissodactyls, or whales. The gallbladder serves primarily for storage of bile. Animals lacking one have little fat in their diet. Human beings can live without a gallbladder but they must avoid lipids in their diet.

EXOCRINE PANCREAS

Like the adrenal complex, the pancreas consists of two histologically distinct and functionally independent components—an exocrine portion consisting of **acini** (terminal alveoli of a compound gland) that secrete digestive enzymes and are drained by pancreatic ductules, and an endocrine component, the pancreatic **islands of Langerhans** that secrete hormones into the blood stream (Fig. 17-13). In cyclostomes and a few other vertebrates these components are not part of the same organ.

The pancreas varies from a diffuse to a compact organ, often consisting of several lobes, and situated in the ventral mesentery of the stomach and duodenum. It is especially compact in elasmobranchs, diffuse in most teleosts. When diffuse, the tissue is distributed along the blood vessels in the mesenteries, occasionally accompanying them into the substance of the liver. In cyclostomes there is no grossly observable pancreas, the exocrine and endocrine components are separated, and some of the exocrine cells remain in the intestinal epithelium.

The pancreas arises typically as one or two ventral pancreatic buds from the liver bud and as a single dorsal bud from the foregut immediately caudal to the stomach (Fig. 11-24, A). The ventral buds invade the ventral mesentery of the duodenum and stomach and form a **ventral lobe, or body** of the pancreas (Fig. 11-24, B, 2). The dorsal bud becomes the **dorsal lobe** or **tail** of the pancreas. There are variants of this pattern. In sharks the entire pancreas develops from the dorsal bud, and in most mammals it develops from one ventral and one dorsal bud. It is likely that three buds is a primitive condition. No pancreatic bud forms in cyclostomes.

Vertebrates may have as many pancreatic ducts as there were embryonic pancreatic buds, but more often one or more of the ducts loses its connection with the bile duct or gut and the pancreas is drained by the remaining duct or ducts. For example, in sheep the duct of the dorsal lobe loses its connection with the gut and the pancreas drains into the common bile duct. In pigs and oxen, the duct of the ventral lobe loses its connection with the common bile duct and the pancreas drains directly into the duodenum. In still other mammals, both ducts remain. When one is larger, as in cats and human beings, the other is an **accessory pancreatic duct.**

CLOACA

The cloaca is a chamber at the end of the digestive tract. It receives the intestine and urinary and genital ducts, and opens to the exterior via the vent. It is shallow or nonexistent in adult lampreys, chimaeras, living female coelacanths, ray-finned fishes, and mammals above monotremes, having either failed to keep pace in growth with the rest of the animal or having become partitioned into two or three separate passages, as in placental mammals (Fig. 14-41, C and D). When there is no adult cloaca, the intestine opens directly to the exterior via an anus. The cloaca is discussed in Chapter 14 and illustrated as follows: amphibians, Figs. 14-8 and 14-16; reptiles, Figs. 14-25 through 14-27 and Figs. 14-34 and 14-35; birds, Figs. 14-28 and 14-36; monotremes, Fig. 14-37.

Chapter Summary

1. Filter feeding is the oldest method of feeding. It is still employed by larval lampreys and with modifications by some teleosts, sharks, and cetaceans.

2. The embryonic digestive tract consists of foregut, midgut, and hindgut. Stomodeal and proctodeal invaginations establish an entrance and exit for the tract.

3. The mouth is the anterior opening to the digestive tract. Major subdivisions of the adult tract are oropharyngeal or oral cavity, pharynx, esophagus, stomach, and intestine. Chief accessory organs are tongue, teeth, oral glands, pancreas, liver, and gallbladder.

4. Nasal passageways open into the oropharyngeal cavity in lobe-finned fishes and into the oral cavity in tetrapods with primary palates. Multicellular oral glands open onto the roof, walls, and floor but are scarce in fishes.

5. Fishes and perennibranchiate amphibians have a primary tongue overlying the hyoid skeleton. In higher amphibians a glandular field contributes to the tongue, and in amniotes paired lateral lingual swellings contribute. An entoglossal bone is in the tongue of lizards and birds, a pair of paraglossals in many birds.

6. Teeth are vestiges of dermal armor. They consist of dentin formed by odontoblasts in dermal papillae, enamel formed by ameloblasts of enamel organs, and cementum. In lower vertebrates they are more numerous, more widely distributed in the oral cavity, more frequently replaced, and more alike throughout the oral cavity. A few vertebrates in every class are toothless.

7. Dentition may be monophyodont, diphyodont, or polyphyodont; acrodont, pleurodont, or thecodont; homodont or heterodont; secodont, selenodont, lophodont, or bunodont; brachyodont or hypsodont. Triassic cheek teeth were triconodont.

8. The pharynx is the part of the foregut that has pharyngeal pouches in embryos and gill slits in adult fishes and perennibranchiate amphibians. In adult mammals it may be divided into nasal, oral, and, sometimes, laryngeal pharynges. Openings lead from or to the spiracle, gill pouches, suprabranchial organs, oral cavity, nasal passageways, auditory tubes, larynx, esophagus, and vocal sacs depending on the species. Lymphoid tissue encircles it in mammals.

9. The esophagus connects the pharynx with the stomach. The crop sac is a diverticulum of the esophagus in birds.

10. The stomach is a muscular enlargement at the end of the esophagus. It is compartmentalized in birds (proventriculus and gizzard) and cud-chewing ungulates (rumen, reticulum, omasum, and abomasum). It terminates at the pyloric sphincter.

11. The intestine is digestive and absorptive. It is relatively straight in fishes, tortuous in tetrapods. Spiral valves, coils, ceca, and villi increase the absorptive area. The segments of the intestine in higher tetrapods are duodenum, jejunum, ileum, colon, and rectum.

12. The liver, gallbladder, and pancreas arise as evaginations of the foregut. There are usually one liver bud and two or three pancreatic buds.

13. A cloaca is characteristic of most adult vertebrates. It opens to the exterior via a vent. In placental mammals the embryonic cloaca becomes divided into two or three passageways, one of which is a rectum that opens to the exterior via an anus.

SELECTED READINGS

Alexander, R.M.: Mechanics of feeding action of various teleost fishes, Journal of Zoology, London **162**:145, 1970.

Butler, P.M., and Joysey, K.A., editors: Development, function, and evolution of teeth. New York, 1978, Academic Press.

Denison, R.H.: Feeding mechanisms of Agnatha and early gnathostomes, American Zoologist **1**:177, 1961.

Domning, D.P.: Marching teeth of the manatee, Natural History **92**:5 (May), 1983.

Gans, C., and Gorniak, G.C.: How does the toad flip its tongue? Test of two hypotheses, Science **216**:1335, 1982.

Gorniak, G.C.: Trends in actions of mammalian masticatory muscles, American Zoologist **25**:331, 1985.

Herring, S.W.: The ontogeny of mastication, American Zoologist **25**:339, 1985.

Kardong, K.V.: Evolutionary patterns in advanced snakes. The evolution of fangs. American Zoologist **20**:269, 1980.

Lauder, G.V.: Patterns of evolution in the feeding mechanisms of actinopterygian fishes, American Zoologist **22**:275, 1982.

Liem, K.F.: Adaptive significance of intra- and interspecific differences in the feeding repertoires of cichlid fishes, American Zoologist **20**:295, 1980.

Lombard, R.E., and Wake, D.B.: Tongue evolution in the lungless salamanders Plethodontidae. II. Function and evolutionary diversity, Journal of Morphology **153**:39, 1978.

Moss, M.: Enamel and bone in shark teeth; with a note on fibrous enamel in fishes, Acta Anatomica **77**:161, 1970.

Nickel, R., and others: The viscera of the domestic mammals, New York, 1973, Springer-Verlag.

Oguri, M.: Rectal glands of marine and fresh-water sharks; comparative histology, Science **144**:1151, 1964.

Osborn, J.W.: The evolution of dentitions, American Scientist **61**:548, 1973.

Peyer, B.: Comparative odontology, Chicago, 1968, The University of Chicago Press.

Pivorunas, A.: The feeding mechanisms of baleen whales, American Scientist **67**:432, 1979.

Savitzky, A.H.: Hinged teeth in snakes: An adaptation for swallowing hard-bodied prey, Science **212**:346, 1981.

Zeigler, A.C.: A theory of the evolution of therian dental formulas and replacement patterns, Quarterly Review of Biology **46**:226, 1971.

Symposium in American Zoologist

Adaptive radiation within a highly specialized system: The diversity of feeding mechanisms of snakes, **23**:337, 1983.

RESPIRATORY SYSTEM

12

In this chapter we look at some of the methods and mechanisms that embryonic, larval, and adult vertebrates have developed for obtaining oxygen and eliminating carbon dioxide in aquatic and terrestrial environments. We will find that many fishes are aerial respirators, and that lungs may be older than tetrapod limbs. Along the way we will see how a crocodilian or a whale can breathe with a mouth full of water, why a nursing kitten does not have to interrupt swallowing during inhalation, and how the position of the human larynx affects human speech.

The process of obtaining oxygen from the environment and eliminating carbon dioxide is external respiration. It is accomplished via respiratory membranes that, except in embryos, are part of some organ. Organs that are essential for external respiration constitute, collectively, the respiratory system.

External respiration precedes internal respiration, which is usually defined as the exchange of oxygen and carbon dioxide between capillary blood and tissue fluids. Because carbon dioxide quickly inhibits cellular activity, the continual elimination of this gas from the vicinity of the cell, and then from the organism, is essential. The role of the circulatory system in the total process of respiration is therefore vital.

External respiration is carried on through respiratory membranes. Except in very early embryos these must be highly vascular, the epithelium must be thin, the surface must be moist, and it must be in contact with the environment, as in external gills; or else the environment must be brought into contact with the respiratory surface, as in lungs.

The chief organs of external respiration in adult vertebrates are external and internal gills, the oropharyngeal mucosa, swim bladders or lungs, and the skin. Less common adult respiratory devices include bushy or filamentous outgrowths of the pectoral fins, as in the male *Lepidosiren*, or of the posterior trunk region and thigh, as in the African hairy frog; the cloacal, rectal, or anal lining; and the lining of the esophagus, stomach, or even of the intestine. Embryos employ a variety of respiratory surfaces including extraembryonic membranes.

In both water and air, respiration through the skin is employed extensively by modern amphibians. It is also used by some fishes, especially those that lack scales and therefore have capillary beds close to the surface. Aquatic urodeles, lacking dermal armor and a significant stratum corneum, acquire as much as three fourths of their oxygen from the water through the skin. Tree frogs acquire only one fourth by that route and terrestrial species of *Rana* acquire one third. Regardless of the proportionate role of skin and lungs in oxygen uptake in amphibians, up to nearly 90 percent of the carbon dioxide is excreted through the skin. Cutaneous respiration is not significant in amniotes because the thick stratum corneum insulates the capillaries from the atmosphere.

GILLS

It was pointed out in Chapter 1 that internal gills of fishes arise as evaginations of the lateral pharyngeal wall—pharyngeal pouches—that meet invaginations of the ectoderm—ectodermal grooves—to form a branchial plate that eventually ruptures; and that this establishes a passageway between the pharynx and the exterior (Fig. 1-7). Gill filaments then develop in the walls of some of these passageways to give rise to gill pouches.

We will examine first the respiratory systems of agnathans. Whether, in doing so, we are observing a more primitive manifestation of vertebrate gill systems than is seen in any gnathostome, or whether the system has been grossly modified from an early vertebrate ancestor is not known, since the intact gill apparatus of ostracoderms has not been preserved in fossils thus far recovered.

Agnathans

The well vascularized gills of agnathans line gill pouches in the lateral walls of the pharynx. Hagfishes in the genus *Myxine* have five or six pair of pouches, rarely seven (Fig. 12-1). The various species of *Bdellostoma*, another hagfish genus, have from five to fifteen pairs. Afferent branchial ducts conduct respiratory water from the pharynx to the pouches, and efferent ducts lead from the pouches to the exterior. Each efferent duct usually has its own external aperture (Fig. 1-9, *Bdellostoma*), but in myxinoids the efferent ducts unite to open via a single common external slit on each side (Fig. 12-1).

In hagfishes, respiratory water enters the median naris and passes via a **nasopharyngeal duct** to a velar chamber at the anterior end of the pharynx (Fig. 12-1). The wall of the chamber is a pulsating muscle, or **velum,** which pumps water from the velar chamber into the pharynx, thereby creating a vacuum in the chamber. The vacuum draws additional water into the naris. Muscle in the walls of the gill pouches expels the water. The velum is supported by pharyngeal cartilages and pulsates 50 to 100 times a minute in alert animals and 25 to 35 times a minute in sleeping animals. On the left side only, a **pharyngocutaneous duct** connects the pharynx with the last efferent branchial duct, or, in some hagfishes, with the exterior. Periodically, debris or particles too large to enter the afferent branchial ducts are forcefully ejected through the pharyngocutaneous duct in a manner somewhat analogous to coughing. The pharyngocutaneous duct arises in the same manner as the gill pouches—from a pharyngeal slit. It appears to be a specialized last gill pouch.

Lampreys have seven pairs of voluminous gill pouches that are lined with gill lamellae and communicate directly with the pharynx via internal

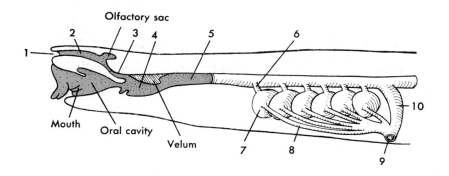

FIGURE 12-1

Respiratory system in the hagfish *Myxine glutinosa*, left lateral view. *1*, Naris; *2*, nasal duct; *3*, nasopharyngeal duct; *4*, velar chamber; *5*, pharynx; *6*, afferent branchial duct; *7*, gill pouch; *8*, efferent branchial duct; *9*, common external gill aperture (present on both sides); *10*, pharyngocutaneous (pharyngeal) duct (present on left side of animal only).

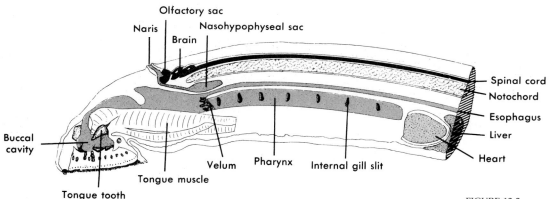

FIGURE 12-2

Cephalic end of the adult lamprey *Petromyzon*, sagittal section. A nasohypophyseal duct connects the naris with the nasohypophyseal sac. The pharynx ends blindly.

gill slits, and with the exterior via external slits (Figs. 12-2 and 1-9, *Petromyzon*). Respiratory water enters and exits via the external gill slits. This is necessary because the buccal funnel is usually attached to the flesh of the prey, and the nasal duct that commences at the naris ends at the nasohypophyseal sac (Fig. 12-2). Pulsations of the pharyngeal musculature, especially that in the walls of the gill pouches, draw respiratory water through the external gill slits and expel it by the same route. The external apertures are guarded by thin flaps of skin that serve as two-way valves.

The pharynx of lampreys becomes subdivided into a dorsal esophagus and a ventral pharynx at metamorphosis with the result that the adult pharynx ends blindly (compare larva and adult in Fig. 17-15). The connection between esophagus and pharyngeal tube is guarded by a velum which, in this case, is a valve that prevents blood rasped from a host from being diluted by pharyngeal respiratory water and wasted via the external gill slits.

The gills of agnathans and those of gnathostomes have little in common structurally. The fact that the visceral skeleton is not an integral part of the branchial apparatus of agnathans accounts for much of the difference.

Cartilaginous Fishes

Most elasmobranchs—sharks, skates, and rays—are pentanchid, having five pairs of gill pouches, each with an internal and external gill slit, and a pair of functional **spiracles** located dorsally immediately in front of the hyomandibular cartilages (Figs. 12-3 and 12-4, *A*). The shark *Hexanchus* is exceptional, having a spiracle and six gill pouches; *Heptanchus,* another shark, has a spiracle and seven, this being the largest number in any gnathostome. No gill surface develops in the posterior wall of the last pouch. The external gill slits in sharks are visible on the side of the head; those in Rajiformes are on the underside (Fig. 12-8). The slits are said to be "naked" because, unlike in other gnathostomes, there is no operculum. In embryos the spiracle and gill slits are the same size and aligned in a row, but the spiracle does not keep pace in growth with the other slits (Fig. 1-7). A miniature gill-like structure, or **pseudobranch,** consisting of a vascular rete (network) develops in its anterior wall. The spiracle has a one-way (intake)

FIGURE 12-3

Correspondence between the nine demibranchs and specific pharyngeal arches in *Squalus acanthias.* The hyoid is the second arch. The second holobranch (location indicated in *white* in top figure) has been excised and displayed in schematic cross section (bottom figure). In the central figure the five gill pouches are represented in *black* and the demibranchs are *crosshatched. 1-9,* demibranchs.

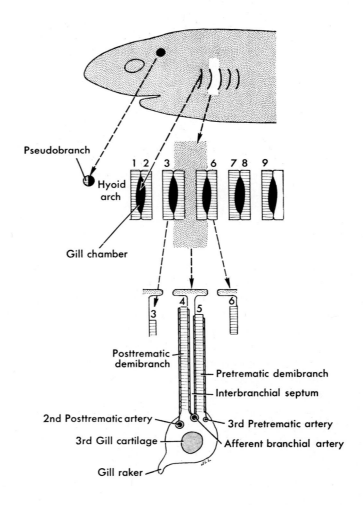

Pseudobranch

Hyoid arch

Gill chamber

Posttrematic demibranch

Pretrematic demibranch

Interbranchial septum

2nd Posttrematic artery

3rd Pretrematic artery

3rd Gill cartilage

Afferent branchial artery

Gill raker

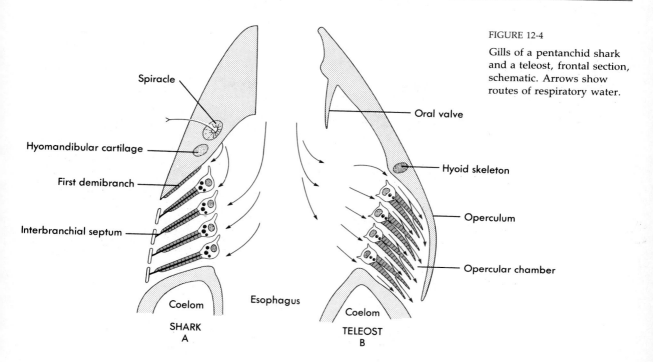

FIGURE 12-4

Gills of a pentanchid shark and a teleost, frontal section, schematic. Arrows show routes of respiratory water.

Spiracle

Oral valve

Hyomandibular cartilage

Hyoid skeleton

First demibranch

Operculum

Interbranchial septum

Opercular chamber

Coelom

Esophagus

Coelom

SHARK
A

TELEOST
B

valve and is the exclusive incurrent aperture for respiratory water in Raji-formes and for much of the respiratory water in all sharks except some fast-swimming voracious ones in which the spiracle becomes secondarily closed by a membrane of skin. Its dorsal position in rays minimizes the entrance of muddy debris-laden water into the pharynx of these bottom-dwelling species.

If a gill slit is probed, the instrument will enter a gill chamber. The anterior and posterior walls of the first four chambers exhibit a gill surface, or **demibranch.** The fifth and last chamber lacks a demibranch in its posterior wall. Ceratohyal cartilages support the demibranch in the anterior wall of the first chamber, and epibranchial and ceratobranchial cartilages (Fig. 8-31, *A* and *B*) of visceral arches 3-6 support the remaining eight demibranchs. The relationships of the nine demibranchs to specific visceral arches is diagrammed in Figs. 12-3 and 12-4, *A*. The demibranch in the anterior wall of a gill chamber is a **pretrematic demibranch.** The one in the posterior wall is a **posttrematic demibranch.** Separating the two demibranchs of a single gill arch in elasmobranchs is an **interbranchial septum** that is supported by many long, tapering, and in some species branching, cartilaginous **gill rays** that radiate from the gill cartilage into the interbranchial septum along its entire dorsoventral extent. The two demibranchs of a single gill arch, together with the associated interbranchial septum, cartilages, blood vessels, branchiomeric muscles, nerves, and connective tissue, constitute a **holobranch** (white area in upper drawing of Fig. 12-3). Stubby **gill rakers** projecting from the pharyngeal border of each gill cartilage guard the vertical slitlike entrances to the gill chambers, protecting the delicate gills from mechanical injury.

The functional surface of each demibranch consists of large numbers of transverse shelflike folds, or lamellae, of gill mucosa, the epithelium of which is very thin. The folds multiply the surface area for gaseous exchange. Underlying the lamellar epithelium is a rich capillary bed supplied by afferent branchial arterioles that receive blood from the ventral aorta via afferent branchial arteries (Fig. 13-18, *B*). Since the ventral aorta carries blood low in oxygen, the blood entering the capillary bed is low in oxygen. As oxygen-laden respiratory water flows steadily across the gill lamellae, it does so in a direction opposite that of the oxygen-poor blood flowing in the capillaries. As a result, the partial pressure of oxygen in the water always exceeds that in the blood even though the blood is becoming increasingly oxygenated. This **countercurrent flow** of blood and water maximizes the efficiency of gaseous exchange, enabling the blood to "make the most" of its opportunity to acquire oxygen. Other examples of the countercurrent principle will be found elsewhere in the text. From the capillaries oxygenated blood passes via efferent branchial arterioles to pre- or posttrematic arteries, then into efferent branchial arteries and the dorsal aorta (Fig. 13-18, *B*). From there it is distributed to all tissues of the body.

It has been mentioned that respiratory water enters the pharynx of most elasmobranchs via both the mouth and spiracle. Most of the water entering via the spiracle in sharks flows into the first two gill pouches, whereas that entering via the mouth enters the last three pouches. Water is sucked through the spiracle and the open mouth when the branchiomeric muscles described earlier constrict the external gill slits and expand the pharyngeal chamber, creating a vacuum within the chamber. This inspiration phase slows down and then ceases when the pharynx has become filled with water. The mouth then closes, the gill chambers are expanded by the action of levator and hypobranchial muscles, and the gill chambers fill with water. In the third phase of respiration—expiration—the water is forced out of the gill chambers via the external gill slits by constrictor muscles. In the interval between expiration and the next inspiration both the mouth and the gill slits are temporarily open. Experiments have shown that pressure is nearly always higher in the pharyngeal chamber than in the gill pouches. This assures a steady, uninterrupted flow of water over the gill lamellae, rather than tidal surges. There is a respiratory rhythm of about 35 per minute when the shark is at rest. Sharks swimming with their mouth open are utilizing the forward motion of the body to accumulate water in the pharynx, thereby reducing the cost of external respiration in terms of energy expended.

The holocephalan *Chimaera* has only four gill pouches (Fig. 12-5), the spiracle closes during early ontogeny, interbranchial septa are short and do not reach the skin, and a fleshy operculum extends caudad from the hyoid arch, hiding the gills and deflecting excurrent water caudad. In some of these traits *Chimaera* resembles teleosts.

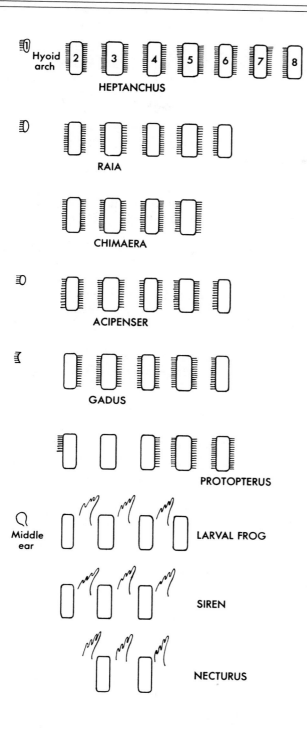

FIGURE 12-5

Open pharyngeal slits in selected aquatic vertebrates and distribution of gill surfaces (horizontal lines) in fishes. *Heptanchus* is a primitive shark. Note that *Gadus*, a cod, retains a vestigial opercular gill even though the spiracle is closed. *1* to *8*, Pharyngeal slits. The position of external gills is indicated in the amphibians.

Bony Fishes

The gill apparatus of cartilaginous and bony fishes exhibits the same basic pattern. A series of visceral arches support holobranchs, and a stream of respiratory water flows over the demibranchs en route from the pharyngeal cavity to the exterior. The chief difference lies in the length of the interbranchial septa and the presence of an **opercular chamber** and an **operculum** (Fig. 12-4, *B*). The interbranchial septa are very short, with the result that the demibranchs are unattached distally (Fig. 12-6). The operculum is a bony flap that commences at the hyoid arch and extends caudad, covering the gill chambers on each side. Extending from the ventral edge of each operculum is an accordianlike **branchiostegal membrane** supported, in modern actinopterygians, by numerous long bony **branchiostegal rays.** Shorter **gular bones,** of more ancient vintage, are found in some of the more primitive ray-finned fishes (Fig. 8-10, *A*). The two opercular membranes are united midventrally beneath the pharynx to form an opercular chamber that respiratory water pours into after passing over the gills and before being expelled via a cleft at the caudal end of the operculum. The cleft varies in size and is exceptionally small and round in eels. Infrequently in teleosts it opens by a midventral aperture.

Water is drawn into the pharynx via the mouth by lowering the pharyngeal floor with the mouth open and the operculum closed. Simultaneous expansion of the opercular chamber by unfolding of the branchiostegal membrane draws the incoming water across the gills and into the chamber.

FIGURE 12-6

Operculum and holobranch of a teleost. The holobranch, **B,** is seen in cross section in the plane X-X'. Arrows indicate direction of efferent water flow.

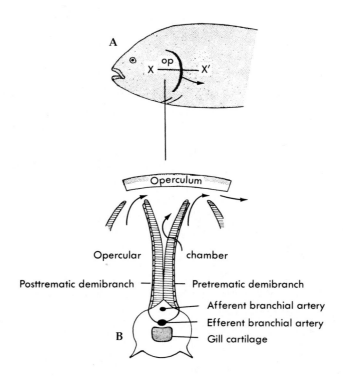

The mouth is then closed and the water is forced to the exterior via the opercular cleft by elevating the pharyngeal floor and compressing the opercular chamber. An **oral valve** immediately behind the mouth prevents escape of water by that route. Thus a suction pump and a pressure pump, operating rhythmically, keep the gills bathed in oxygenated water. A few teleosts, including mackerel and tuna, have a paucity of branchiomeric musculature and are obliged to swim with the mouth open in order to create a flow of water over the gills.

Most bony fishes other than dipnoans, chondrosteans, and garfishes have four holobranchs and five gill chambers, the demibranch in the anterior wall of the first gill chamber having been lost (Fig. 12-5, *Gadus*). Additional demibranchs have been lost in dipnoans (Fig. 12-5, *Protopterus*). In fact, the gills of *Protopterus* and *Lepidosiren* are so inefficient that these fishes suffocate when forcibly held under water. A spiracle is present in chondrosteans (Fig. 12-5, *Acipenser*), but it closes during embryonic life in other bony fishes.

Larval Gills

Larval gills are of three kinds—**external gills** that are outgrowths from the external surface of one or more gill arches; **filamentous extensions of internal gills** that project through gill slits to the exterior; and the **internal gills** of late anuran tadpoles, which are hidden behind the larval operculum.

External gills usually develop before gill slits open and before any opercular fold has started to develop. They can be waved about by branchial muscles and can often be retracted. They develop in the embryonic or larval stages of most dipnoans (*Ceratodus* is an exception), in all amphibians (including apodans), and in a few ganoid fishes such as sturgeons and *Polypterus*. The latter has only one pair (Fig. 12-7, *A*). Although external gills are larval structures, there are perennibranchiate urodeles (Table 3-1) and a perennibranchiate lungfish, *Protopterus* (Fig. 13-19, *B*). The latter has four pairs of larval gills and retains three pairs in a reduced state throughout life.

External gills develop in anuran tadpoles on pharyngeal arches III to V. Later, when pharyngeal pouches II to V rupture to the exterior, the pouch linings become folded to form a set of internal gills. Thereupon, a fleshy operculum grows backward from the hyoid arch over the gill region, enclosing external gills and internal gills in an opercular chamber that retains only a single excurrent pore, this being located at the posterior edge of the left operculum. Thereafter, the external gills gradually atrophy and the internal gills function until metamorphosis.

Elasmobranchs develop either in a yolk-laden egg case attached to underwater vegetation, or they develop in the maternal uterus. In the embryos of these fishes temporary filamentous external projections of internal gills are present during early developmental stages, projecting into the surrounding fluid (Fig. 12-7, *C* and *D*). Here they not only perform a respiratory function but, in viviparous species, they absorb nutrients from

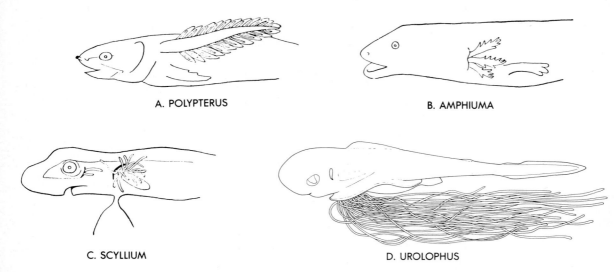

A. POLYPTERUS

B. AMPHIUMA

C. SCYLLIUM

D. UROLOPHUS

FIGURE 12-7

Larval gills. **A-D,** A bony fish, amphibian, shark, and ray, respectively. In *Amphiuma* the gills are resorbed before the larva escapes from the egg envelope. The spiracle in *Urolophus,* presently behind the eye, will ultimately be located dorsally. Pectoral fins characteristic of Rajiformes have just appeared as limb buds. (After Daniel.)

the fluid in the maternal uterus, supplementing the nourishment provided by the yolk sac. A few viviparous chondrostean and teleost larvae utilize filamentous external gills in a similar manner. Absorption of nutrients from uterine or other maternal fluids is known as **histotrophic nutrition.**

Excretory Role of Gills

Although usually thought of as respiratory organs, gills perform vital excretory functions. Marine fishes excrete the common marine salts chiefly via salt-secreting glands located on the lamellae of the gills. (Freshwater species secrete most of their excess chloride via the kidneys.) The gills of lampreys and those of marine fishes that migrate between salt water and fresh water excrete chloride when in salt water, and they absorb chloride in fresh water, thereby assisting in maintaining homeostasis under these stressful conditions. Fishes without exception also excrete their nitrogenous wastes largely through the gills rather than via kidneys, as in tetrapods. And, lastly, fishes that acquire oxygen from air in swim bladders release most of their carbon dioxide into the water flowing over the gills. Carbon dioxide dissolves more readily in water than in air, hence the partial pressure of CO_2 in the water passing over the gills remains much lower than would be the partial pressure of this gas if excreted into the air of the swim bladder. Its presence in sufficient quantities in the swim bladder would inhibit the further excretion of carbon dioxide; and this, in turn, would inhibit the uptake of oxygen by the hemoglobin in the red blood cells of the gill capillaries.

Air Breathing in Bony Fishes

Vertebrates arose in an aqueous environment, and water was the earliest source of oxygen. During the Devonian, however, if not earlier, air became a source of oxygen for many fishes. One compelling factor may have been that the water at that time was warm and swampy and, as such, low in dissolved oxygen. The atmosphere, on the contrary, contains twenty times the oxygen that oxygen-saturated water can hold. Little wonder, then, that many Devonian fishes obtained some or all of their oxygen from above the surface of the water. Today's air-breathing fishes include dipnoans, chondrosteans, holosteans, and many teleosts. They snatch bubbles of air from just above the water surface, and the air comes in contact with the oropharyngeal lining or, as we will see shortly, with the lining of swim bladders in those few fishes in which these serve as lungs. A few teleosts swallow the bubble and extract oxygen in the stomach or intestine. Although oxygen is acquired from the atmosphere, most of the excess carbon dioxide is eliminated through the gills into the water. As will be pointed out in the next paragraph, fishes do not employ nostrils in aerial respiration.

NARES AND NASAL CANALS

The nostrils or **external nares** of Chondrichthyes and Actinopterygii open directly into blind olfactory sacs that contain the sensory epithelia for smell. Each nostril is divided into a forward-directed incurrent aperture through which water is driven as a fish swims forward, and a laterally or ventrally directed aperture through which water exits after having washed over the olfactory epithelium. In Sarcopterygii (other than *Latimeria*) the nostrils are connected with the oropharyngeal cavity by **nasal canals** through which water is able to pass, so that, as R.M. Alexander states it, the incurrent aperture of each nostril is on the surface of the head and the excurrent aperture is in the oropharyngeal cavity. The openings into the oropharyngeal cavity are **internal nares** or **choanae.** In agnathans an unpaired nasal canal connects the nostril of Petromyzontiformes with the olfactory sac, and an extension of the nasal canal in Myxiniformes, the nasopharyngeal duct, continues to the velar chamber at the cephalic end of the pharynx. In none of these fishes, including, as far as is known, air-breathing lungfishes, are nostrils useful for respiration. They are part of a sensory system for monitoring one constituent of the environment—chemicals in solution in the surrounding water.

If the crossopterygian ancestors of labyrinthodonts were not using their nostrils for breathing—and whether they were is not known—labyrinthodonts were opportunists, since they commenced to use nostrils for drawing *air* into their oral cavity. With development in reptiles of a secondary palate, the choanae opened farther caudad; and, as we have seen earlier, the longer the secondary palate, the farther caudad the choanae are located. In mammals they open into the nasopharynx above the soft palate (Fig. 11-3, cat).

Nasal canals (nasal passageways) arise from paired **nasal pits** and **oronasal grooves,** the dorsolateral walls of which roll together to form a tube (Fig. 12-8, B). Although sharks and rays lack nasal canals, oronasal grooves form and remain throughout life (Fig. 12-8, A). In mammals the olfactory epithelium is restricted to the upper chambers of the nasal passageways, whereas

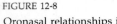

FIGURE 12-8

Oronasal relationships in, **A,** an adult skate and, **B,** a 6-week (12 mm) human fetus.

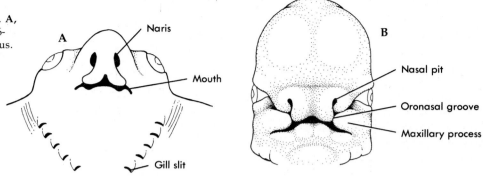

the ventral part has a ciliated glandular nasal epithelium. Hairs at the entrance to mammalian nasal canals trap insects and coarse particulate matter in the air; venous plexuses under the epithelium that covers the turbinal bones warm cold air; and air sinuses that open into the upper reaches of the nasal passageways of mammals serve as resonating chambers for vocalization (Fig. 12-12, *A*).

A fleshy, partially cartilaginous nose, or proboscis, develops in some mammals and carries the nostrils to positions characteristic of the species. (Compare the location of the nostrils in cats, humans, and elephants.) In whales, on the other hand, there is no nose, and the nostrils are situated dorsally, although in fetal whales they are farther forward. In some whales the two nostrils unite during ontogeny to become a median blowhole on top of the head.

SWIM BLADDERS AND THE ORIGIN OF LUNGS

Nearly every vertebrate from fish to man develops an unpaired evagination from the pharynx or esophagus that becomes one or a pair of sacs (swim bladders or lungs) filled with gases derived directly or indirectly from the atmosphere. The only adult vertebrates that do *not* have pneumatic sacs are cyclostomes, cartilaginous fishes, a few marine teleosts, some bottom dwellers such as flounders, and a few tailed amphibians, some of which have them as embryos. Imprints of paired sacs with ducts have even been found in a fossil placoderm. It can be said with considerable confidence that vertebrates that fail to develop pneumatic sacs have probably lost the genetic capability to do so.

After evaginating, pneumatic sacs may retain their unpaired connection with the foregut, or the duct may close. Pneumatic sacs hereafter will be called lungs in tetrapods and swim bladders in fishes, regardless of their function.

Swim bladders may be paired or unpaired. The pneumatic duct, when present in adults, connects to the esophagus ventrally in dipnoans and chondrosteans but has shifted toward the dorsal side in holosteans and teleosts (Fig. 12-9). Infrequently the pneumatic duct connects to the pharynx or stomach. Swim bladders are retroperitoneal, lie close to the kidneys, and bulge more or less into the coelom (Fig. 14-15). The walls contain elastic tissue and smooth muscle, and the lining is relatively smooth except in true lungfishes. Fishes are said to be **physostomous** when the duct is open, **physoclistous** when it is closed. Dipnoans, chondrosteans, holosteans, and a few teleosts are physostomous. Among the latter are catfish, carp, eels, herring, pickerel, and salmon. Only a few physostomous fishes use the swim bladder for respiration.

Swim bladders in teleosts serve chiefly as hydrostatic organs. The volume of gas in the bladder can be reflexly regulated at its source, thereby altering the specific gravity of the fish and increasing or decreasing its buoyancy. The gas in the hydrostatic swim bladder usually comes from the blood. It is actively transported into the chamber of the bladder from a network of small arteries (rete mirabile) in the lining of the bladder and named the **red gland.** The gland is supplied from the celiac artery, and its associated tortuous veins empty into the hepatic portal vein. The gas is reabsorbed in an area of modified epithelium near the caudal end of the bladder. In physostomes it may be bubbled through the mouth.

The gases in swim bladders differ among fishes. Some swim bladders contain almost pure (99%) nitrogen, some up to 87% oxygen, and all contain at least traces of four atmospheric gases—nitrogen, oxygen, carbon dioxide, and argon. In deep-water fishes, nitrogen may be transported from the blood into the bladder against a nitrogen pressure of as high as 10 atmospheres.

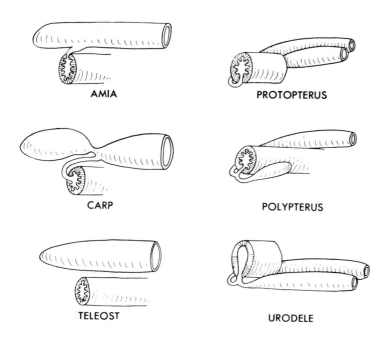

AMIA

PROTOPTERUS

CARP

POLYPTERUS

TELEOST

URODELE

FIGURE 12-9

Swim bladders and urodele lungs. In teleosts the embryonic pneumatic duct usually closes, and the swim bladder thereafter has no connection with the gut. The carp is a physostomous teleost with an anterior extension of the bladder connected with weberian ossicles.

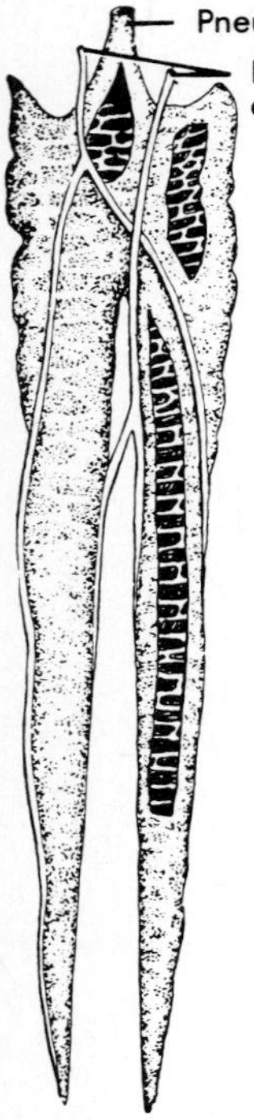

FIGURE 12-10

Swim bladders of *Protopterus*.

Swim bladders perform other functions in addition to their hydrostatic role. In one group of teleosts (Cypriniformes), a series of small bones, the weberian ossicles, connects the anterior end of the swim bladder and the sinus impar, a projection of the perilymph cavity (Fig. 16-7). Low-frequency vibrations of the gas within the swim bladder, evoked by waves of similar amplitude in the water, are transmitted by the ossicles to the inner ear. Therefore these fish can hear.

In a few fishes contractions of striated muscles attached to the swim bladder cause it to emit thumping sounds, or they force air back and forth between chambers separated by muscular sphincters to produce croaking and grunting sounds.

Swim bladders function as lungs in the chondrosteans *Polypterus* and *Calamoichthyes* and in the three living genera of true lungfishes (dipnoans), except that *Neoceratodus* uses its single swim bladder in this manner only when the oxygen content of the water is low. In these air-breathing fishes the swim bladder is lined with low septa and may exhibit thousands of tiny air sacs (Fig. 12-10). Like tetrapod lungs, they are supplied by arteries arising from the sixth embryonic aortic arch; and in dipnoans the venous return is to the left atrium, as in tetrapods.

An oropharyngeal pump forces air from the oropharyngeal cavity into the swim bladder. Expulsion of air from the bladder to the oropharynx is a result of the vacuum created by lowering the oropharyngeal floor while the mouth and nares are closed, elasticity of the bladder, contraction of smooth muscles in its walls, and pressure on the body wall exerted by the surrounding water. It is subsequently bubbled to the exterior through the mouth.

The striking similarities between swim bladders and lungs suggest that these are the same organs. In the Devonian when the fresh water was warm and periodically stagnant, and therefore low in dissolved oxygen, aerial respiration may have made the difference between extinction and survival. At that time, placoderms and most crossopterygians had pneumatic ducts. We can only speculate on which came first, respiratory or hydrostatic swim bladders, but these two conclusions seem well founded: they were functioning in aerial respiration long before vertebrates ventured onto land, and closure of the pneumatic duct in physoclistous fishes is probably a mutation from a more primitive, open duct condition.

LUNGS AND THEIR DUCTS

Tetrapod lungs arise as an unpaired evagination from the caudal floor of the pharynx (Fig. 1-6). The opening in the pharyngeal floor becomes a longitudinal slit, the **glottis.** The unpaired lung bud elongates only slightly before bifurcating to form bronchi and lungs (Fig. 12-21). The lung primordia push caudad underneath the foregut until they bulge into the coelom lateral to the heart. As they grow into the coelom, they carry along an investment of peritoneum, which becomes the visceral pleura. The part of the lung bud between glottis and lungs develops into larynx, trachea, and bronchi.

The Larynx and Vocalization

The larynx is a short air passageway between the glottis and the upper end of the trachea of tetrapods, the walls of which are supported by cartilaginous derivatives of the caudalmost pharyngeal arches (Table 8-4). The glottis and associated cartilages have become an instrument for vocal communication in species in which vocal cords develop. In urodeles the larynx is a primitive structure consisting of a single pair of lateral cartilages that surround the glottis and support it as their sole role. Most other tetrapods below mammals have two pair of laryngeal cartilages, **arytenoids** and **cricoids** (Fig. 12-11), and mammals have a third pair, the **thyroids.** (The cricoids of lizards and crocodilians are often called cricothyroids; but that they are partly homologous with the thyroid component of mammalian larynges has not been established.) Except in monotremes, the paired cricoids and thyroids of mammalian embryos tend to unite across the midventral line during early ontogeny (Figs. 8-38, cat and monotreme; 8-39; and 12-12, B). Additional small cartilages—cuneiforms, corniculates, procricoid, and others—develop in some mammals. The laryngeal cartilages are interconnected by ligaments and by intrinsic laryngeal muscles of branchiomeric origin. Extrinsic muscles of hypobranchial origin, particularly the sternothyroid and thyrohyoid of mammals, provide the larynx, as a unit, with a degree of mobility that is especially necessary during swallowing in adult human beings because, as will be explained later, the larynx in these primates is so low in the pharynx.

Folded within or stretched across the laryngeal chamber in anurans, some lizards, and most mammals are **vocal folds,** more frequently referred to as **vocal cords,** which endow these species with the ability to communicate vocally to one degree or another with other members of the species. The cords vary from mere fleshy folds surrounding the glottis and attached to the arytenoid and cricoid cartilages, as in anurans, to a pair of mammalian folds containing strong bandlike elastic ligaments that are stretched between the arytenoid and thyroid cartilage on each side of the sagittal plane. Air expelled from the lungs passes through the narrow slit, or glot-

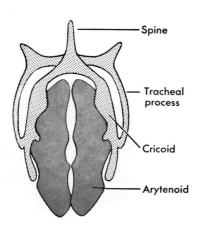

Spine

Tracheal
process

Cricoid

Arytenoid

FIGURE 12-11

Laryngeal skeleton of a frog. The glottis lies between the two arytenoid cartilages. The tracheal process is part of the cricoid cartilage.

FIGURE 12-12

Human pharynx, larynx, and associated structures. **A,** Sagittal view. **B,** Human larynx, frontal view. In **B,** the arytenoid cartilages and glottis are not seen. The former are dorsal to the thyroid cartilage, and the glottis is dorsal to the base of the epiglottis.

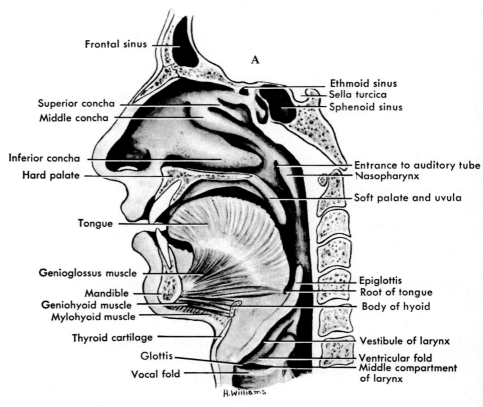

Frontal sinus

Superior concha
Middle concha

Inferior concha
Hard palate

Tongue

Genioglossus muscle
Mandible
Geniohyoid muscle
Mylohyoid muscle

Thyroid cartilage
Glottis
Vocal fold

A

Ethmoid sinus
Sella turcica
Sphenoid sinus

Entrance to auditory tube
Nasopharynx

Soft palate and uvula

Epiglottis
Root of tongue
Body of hyoid

Vestibule of larynx
Ventricular fold
Middle compartment of larynx

H. Williams

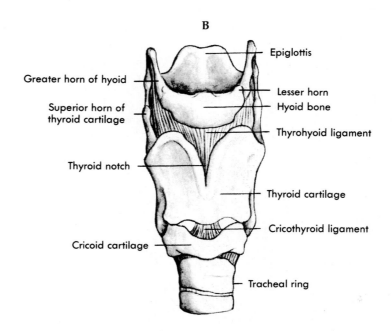

B

Greater horn of hyoid

Superior horn of thyroid cartilage

Thyroid notch

Cricoid cartilage

Epiglottis

Lesser horn
Hyoid bone

Thyrohyoid ligament

Thyroid cartilage

Cricothyroid ligament

Tracheal ring

tis, between the folds. When not in use for vocalization the cords are relaxed and exhaled air passes between them silently. When under tension the cords vibrate, giving rise to vocal sounds. During vocalization intrinsic muscles of the larynx alter the position of the thyroid and arytenoid cartilages with respect to one another, thereby regulating the tension on the cords. The pitch (frequency of vibration) of the human voice is a function of the amount of tension within the cords. Among mammals, hippos and a few others lack vocal cords and one breed of dogs, basenjis, have poorly developed ones and cannot bark. By contrast, the thyroid and associated hyoid bones of howler monkeys are enormous, causing a goiterlike bulge in the neck (Fig. 12-13). They surround a voluminous resonating chamber that enables the weird howl of this species to penetrate deep into the jungle.

Except in birds and mammals, vocalization is not a significant means of communication among tetrapods. Apodans and urodeles are mute as are most reptiles. Male anurans force air out of the lungs, over the vocal cords, and into the oropharyngeal cavity with the mouth and nares closed, and from there into the vocal sac or sacs, which deflate and refill with each croak (Fig. 11-3, *A*). The extent of participation of the vocal folds in anurans vis á vis the other components of the phenomenon, including the vocal sacs has not been fully explored. The vociferous croaking of male toads immediately after a rainstorm is a mating call. (It will be recalled that the eggs of oviparous females are laid in rainy weather, and that sperm are deposited over the eggs while they are being extruded.) Each species has its own breeding call for attracting females, but there may be several other calls in addition. Vocalization in birds is not a function of the larynx, which lacks vocal cords, but of a special avian voice box, the syrinx, to be discussed shortly.

Most mammals have "false" vocal cords which, in some species, produce vocal sounds, including purring in kittens. These are fleshy folds located at the entrance to the **laryngeal vestibule,** or **sinus of Morgagni,** an antechamber located just above the true vocal cords. The vestibule has become the enormous resonating chamber of howler monkeys referred to above.

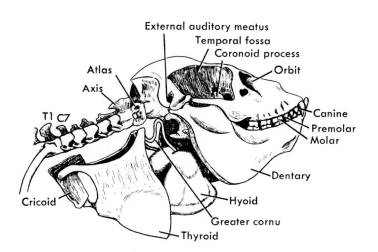

FIGURE 12-13

Hyoid bone and larynx of a howler monkey.

Anatomical adaptations prevent fluids from entering the glottis in aquatic tetrapods. Chief among these are valves at the entrance to the nares, whether of a frog or a whale. Crocodilians have a fleshy valve, the **palatine velum,** that consists of two large transverse folds suspended from the palate just anterior to the internal nares. When a matching fold at the base of the tongue is drawn upward against the velum, the glottis and pharynx are isolated from the oral cavity but have continuity with the nasal passageways. As a result, a crocodilian can be submerged with only two moundlike nostrils and the two elevated eyes protruding above the surface, thermoregulating, breathing while the oral cavity is full of water, and meanwhile keeping an eye on the environment above the water. When they submerge totally, valves in the nostrils prevent flooding of the nasal passageways and pharynx.

In most mammals, terrestrial and aquatic, the larynx is high in the oropharynx where it is able to lock into the nasopharynx during the act of swallowing liquids. This allows adult cats, monkeys, and many other mammals to simultaneously drink and breathe, and many mammalian babies to breathe while suckling, since a pair of pharyngeal recesses lateral to the larynx serve as bypasses to the esophagus for fluids, but not for solids. In marsupials, in which the mouth is unable to be withdrawn from the nipple, the arrangement prevents milk that is being pumped by the mother's mammary gland into the baby's esophagus from being sucked into the lungs as the baby breathes. In whales that have surfaced to breathe, it permits prolonged inhalation after water has been blown out of the nostrils, or blowholes, even though the cavernous mouth is running over with seawater (Fig. 12-14). In human babies the larynx is high in the neck until about 18 months of age, at which time it begins to descend in the neck, "altering the way in which a child breathes, swallows, and vocalizes" (Laitman). It remains high—does not descend—in monkeys or apes. Of course, a mammal cannot breathe through the mouth if the larynx is protruding into the nasopharynx, nor can air returning from the lungs be used for articulate human speech, which is the epitome of vocal communication in mammals. Human speech requires the lips, the tongue, and the increased space acquired between the base of the tongue and the glottis when the larynx descends. Thus the position of the human larynx contributes to the effectiveness of human speech.

Solids are prevented by an **epiglottis** from entering the larynx while swallowing. This is a fibrocartilaginous flap in the floor of the pharynx attached to the hyoid bone by an elastic ligament (Figs. 11-3, *B*, 12-12, and 12-14). The act of swallowing draws the larynx forward against the epiglottis, thereby blocking the entrance to the lower respiratory tract, that is, the part distal to the glottis.

The anterior wall of the larynx of some male lizards has a saccular evagination, the **gular pouch** or dewlap, just under the skin of the throat. When the pouch is inflated with air, it causes a ballooning of the throat; and when sunlight shines through its well-vascularized walls, the reddish pouch becomes a sex symbol. The pouch deflates noiselessly when the animal is disturbed. As mentioned in an earlier chapter, a process of the hyoid bone supports the inflated pouch.

Trachea, Syrinx, and Bronchi

The **trachea** is ordinarily as long as the neck. Therefore it is short in amphibians and long in amniotes. In birds and some turtles the part of the trachea within the neck is longer than the neck as an accommodation to the stretching, twisting, and looping that these necks are capable of. When a swan's neck is relaxed, or a turtle withdraws its neck into the shell, the trachea assumes an S-shape. In crocodilians, too, the trachea is longer than the neck in some species.

The walls of the trachea are prevented from collapsing under the negative pressure of inhalation by cartilaginous or bony rings or plates that form within the wall. The rings are usually incomplete dorsally, the ends being united by smooth muscle that can change the diameter of the tube in response to the need for a greater or lesser volume of tidal air. In crocodilians and birds, however, all tracheal rings are complete. Except in lower urodeles, the trachea bifurcates to become two primary **bronchi** that are similarly strengthened.

Birds have a special voice box, the **syrinx,** at the bifurcation of the trachea (Fig. 12-15). A **bronchotracheal syrinx** consists of a resonating chamber with walls strengthened by the last several tracheal rings and the first bronchial half rings. Mucosal folds project into the chamber, and a bony **pessulus** may form in a special semilunar membranous fold. In a **tracheal syrinx,** parts of the last several tracheal rings are missing, enabling the membranous wall of the trachea itself to vibrate. In a **bronchial syrinx** the membranous wall between two bronchial cartilages folds into the chamber when the cartilages are drawn together, and the fold vibrates with the flow of air.

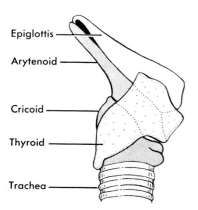

FIGURE 12-14

Laryngeal skeleton of a whale. The epiglottis and arytenoid can be drawn into the nasopharynx.

Epiglottis

Arytenoid

Cricoid

Thyroid

Trachea

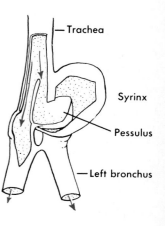

Trachea

Syrinx

Pessulus

Left bronchus

FIGURE 12-15

Asymmetrical bronchotracheal syrinx of a canvasback duck. Arrows indicate path of inhaled air.

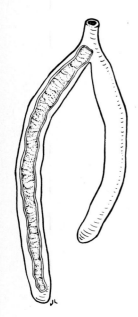

FIGURE 12-16

Lungs of *Necturus*.

Amphibian Lungs

The lungs of amphibians are simple sacs, long in urodeles (Fig. 12-16) and bulbous in anurans, conforming to the shape of the pleuroperitoneal cavity. The internal lining may be smooth throughout, there may be simple sacculations in the proximal part, or the entire lining may be pocketed. The left lung of caecilians is rudimentary, and the lungs of salamanders that inhabit swift mountain streams may be only a few millimeters long. Perhaps rudimentary lungs in the latter species made it possible for them to inhabit these streams, since buoyancy would be a disadvantage in swift currents. Plethodontids do not even form a lung *bud*.

The lungs of normally aquatic urodeles function mostly as hydrostatic organs, and respiration in species lacking gills takes place chiefly through the pharyngoesophageal lining and the skin. In water, which is the normal habitat of *Necturus*, only about 2 percent of the oxygen is acquired via lungs. Another perennibranchiate, *Siren*, a gill-breather like *Necturus* when in oxygenated water, frequently expresses air through the gill slits when obligated to breathe air. Urodeles using lungs either as swim bladders or as respiratory organs fill the lungs in the same manner as lungfishes: respiratory air enters the mouth, not the nostrils, and the muscular floor of the oropharynx pumps it to the lungs. Elasticity of the lungs probably plays a major role in emptying them.

The details of respiration in bullfrogs have been ascertained by Gans, DeJongh, and Farber. The mouth remains closed during respiration. In essence, they found that lowering the floor of the oropharyngeal cavity with the glottis closed draws air via the nostrils into the deep inferiocaudal reaches of the oropharynx where it is put on hold. Thereupon the glottis is opened and the air already in the lungs escapes to the exterior via the nostrils. Emptying of the lungs appears to be the result of the elasticity of the lung and the smooth muscle in its walls. In the third phase the fresh air stored in the floor of the oropharynx is forced into the lungs by raising the floor of the oropharynx with the nostrils again closed. In the fourth phase the reservoir of air on hold is replenished. Once this sequence has been completed, ventilation of the lungs is temporarily interrupted and, with the glottis closed, oscillation of the muscular floor of the oropharynx commences. The phenomenon can be observed on any frog by watching the rapid rhythmical pulsations of the floor of the oral cavity behind the lower jaw. The oscillations maintain a steady flow of air into and out of the oropharyngeal cavity, the lining of which acts as an accessory respiratory epithelium. The oscillations also probably flush out any residual air that escaped from the lungs during the exhalation phase of ventilation. Unlike other tetrapods, no *vacuum pump* is employed to fill the lungs of amphibians.

Reptilian Lungs

The lungs of *Sphenodon* (Fig. 12-17, *A*) and of snakes are simple sacs, although the caudal third of the lining in snakes is septate and contains residual air. In lizards (Fig. 12-17, *B*), crocodilians, and turtles the septa are so constructed that there are numerous large chambers. These lungs are spongy because of the numerous pockets of trapped air. Although asymmetry of the two lungs is usual in tetrapods, it is especially pronounced in legless reptiles, in which one lung is much shorter than the other and sometimes atrophied or, in a few snakes, absent altogether.

An enormous diverticulum of the left lung of puffing adders extends into the neck. Inflation of this air sac causes the neck to balloon. Similar **air sacs** extend among the abdominal and pelvic viscera in some lizards (Fig. 12-18). Dinosaurs and pterosaurs had them, and in these reptiles the sacs also invaded the centra of vertebrae, as they do in today's birds. Air sacs are used by reptiles to inflate the body as a scare tactic or for other defensive purposes; in birds, as will be seen shortly, they have become a vital part of the mechanics of respiration.

Most reptilian lungs, like those of amphibians, occupy the pleuroperitoneal cavity along with the other viscera. In turtles they lie against the inner surface of the carapace just caudal to the pectoral girdle on each side, and all but their caudal ends are retroperitoneal. The lungs of crocodilians and a

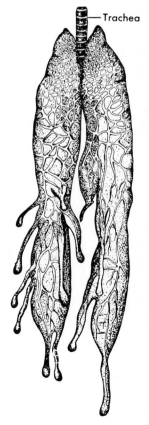

— Trachea

FIGURE 12-17

Lungs of, **A,** *Sphenodon* and, **B,** *Heloderma,* a large lizard.

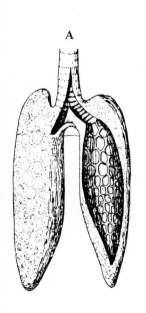

A

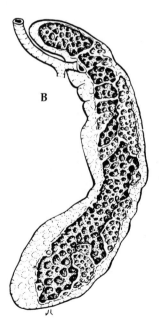

B

FIGURE 12-18

Lungs of a chameleon showing saccular diverticula (air sacs).

few squamates occupy separate subdivisions of the coelom, the paired **pleural cavities.** These are isolated from the rest of the coelom by a tendinous **oblique septum.** Although the septum plays no active role in respiration, it participates passively when the abdominal viscera that press against it are displaced cephalad or caudad by action of abdominal muscles.

Reptiles other than turtles inhale by employing intercostal muscles to rotate the ribs forward and outward, thereby increasing the volume of the pleuroperitoneal cavity. This action creates a vacuum around the lungs and other viscera, and atmospheric pressure forces environmental air into the nostrils and throughout the respiratory tract, causing the thin walled elastic lungs to balloon with oxygenated air. Return of the ribs to their resting position and, in some species, contraction of the abdominal wall musculature compresses the lungs and other viscera; and this effect, aided by the resilience of the lung walls, results in exhalation.

The process of respiration in turtles is complicated by the fact that their ribs cannot participate because they are fused with the carapace. Since the musculature of the abdominal wall has become vestigial, it, too, cannot serve as a vacuum pump. Movements of the nearby pectoral girdle therefore play the major role in ventilation of the lungs in these creatures by altering the volume of the pleural cavities.

Lungs and Their Ducts in Birds

Like crocodilian lungs, those of birds occupy paired pleural cavities separated from the rest of the coelom by a tendinous oblique septum. The septum in birds exhibits a modicum of striated muscle tissue.

Avian lungs are unique morphologically. Incoming air passes through the lungs nonstop and into spacious air sacs (Fig. 12-19, *A*). These extend among the viscera within the body cavity, lie among the flight muscles, and have long slender diverticula that extend into the interior of most bones including centra. Atmospheric pressure inflates the sacs when the ribs are rotated forward and upward and the sternum is actively depressed. The air sacs then act as bellows that force the air via **recurrent bronchi** back to the lungs. Within the lungs **secondary bronchi (dorsobronchi** and **ventrobronchi)** are interconnected by innumerable **parabronchi** (Fig. 12-19, *B*) with diverticula leading to air capillaries only a few thousandths of a millimeter in diameter and containing the respiratory epithelia (Fig. 12-20). The air capillaries associated with a single parabronchus are interconnected in three dimensions and air flows freely through the air capillary system and returns to the parabronchus from which it began. The air capillaries are suspended in a dense plexus of blood capillaries. After passing over the respiratory epithelia that line the air capillaries and returning to the parabronchus the air continues to the exterior. The *details* of airflow through the intrapulmonary duct system have not yet been ascertained. Air sacs and the open-ended duct system within the lungs result in complete replacement of the air in the lungs with every inflation-compression cycle of the bellows. Consequently, unlike in other tetrapods, there is no residual air in the lungs.

Air sacs are thin walled and distensible. Most birds have five or six pairs: (1) cervical sacs at the base of the neck; (2) interclavicular sacs dorsal to the furcula and sometimes united across the midline; (3) anterior thoracic sacs lateral to the heart; (4) posterior thoracic sacs within the oblique septum; (5) abdominal sacs among the abdominal viscera; and (6) axillary sacs, less common, lying between two layers of pectoral muscle. The sacs maintain a steady flow of air over the respiratory epithelia. During flight the bellows is operated by rhythmical contraction and relaxation of the flight muscles and the flapping of wings. When the bird is at rest the ribs and oblique septum operate the bellows. At this time the oxygen demand is lower, and the flow of air through the system is slower. The diverticula of the air sacs enter the bones via pneumatic foramina.

Air sacs are thermoregulatory, dissipating excess heat produced by the surrounding muscles during flight. The heat is transferred from the muscle directly into the air within the sacs and not via the bloodstream, air sacs having a relatively poor vascular supply. Thermoregulation in resting doves, at least, is also accomplished partly during respiration, heat being

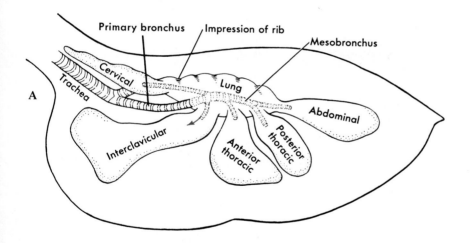

FIGURE 12-19

Lower respiratory tract of a bird. **A,** Air sacs. Air leaves the sacs via recurrent bronchi (not shown). **B,** Diagrammatic representation of a dorsobronchus, ventrobronchus, and several interconnecting parabronchi from which air capillaries evaginate.

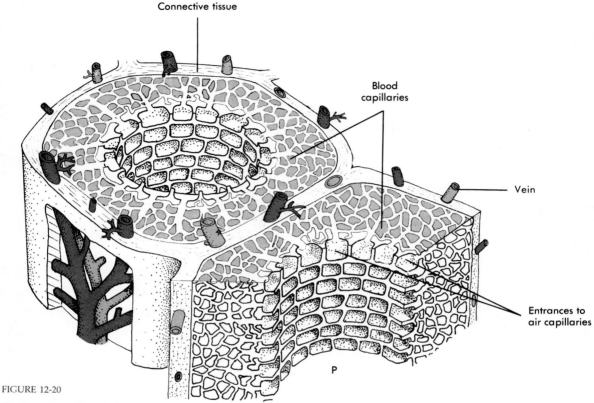

Connective tissue

Blood
capillaries

Vein

Entrances to
air capillaries

P

FIGURE 12-20

Parabronchus, *P,* and air
capillaries of a bird's lung.

transferred from the exceptionally vascular esophageal mucosa to the outgoing airstream as it passes through the adjacent trachea. A bird's chief
defense against overheating at rest is a reflexive increase in respiratory
rate.

As we have already seen, air sacs are not unique in birds. They are found
in some lizards and snakes, and they were present in dinosaurs and pterosaurs as revealed by pneumatic foramina in their bones. The extension of
air sacs into bones results in energy conservation during flight by decreasing the specific gravity of the bird, making it more buoyant in air. *Warm* air
in the sacs adds to their buoyant effect. Most ratites lack pneumatic bones
and so did *Archaeopteryx.*

Mammalian Lungs

The lungs of mammals are multichambered and usually divided into
lobes, with more lobes on the right (Fig. 12-21, 14-mm embryo). The lungs
of whales, sirenians, elephants, perissodactyls, and *Hyrax* lack lobes, and
in monotremes and rats, among others, only the right lung is lobed.

The left and right lungs occupy separate pleural cavities (Fig. 1-13).
These result from the formation of the mammalian diaphragm, the ontog-

eny, gross morphology, and innervation of which were discussed earlier. The parietal peritoneum of each pleural cavity lines the inner surface of the thoracic wall as the **parietal pleura** and covers the cephalic surface of the diaphragm as the **diaphragmatic pleura** (Fig. 12-21). At the root of each lung, where the primary bronchus and pulmonary vessels enter and leave, the parietal pleura is continuous with the **visceral pleura** on the surface of the lung. The space enclosed by these pleurae is the pleural cavity. It is only a potential cavity in life, since the parietal and visceral pleurae are in constant contact except for a thin layer of lubricatory lymph secreted by the pleural mesothelium.

Each primary bronchus penetrates a lung and divides into **secondary** and **tertiary bronchi,** which give rise to many **bronchioles.** These bronchioles branch into smaller and smaller tubes. The walls of the bronchi and larger bronchioles contain smooth muscle fibers, connective tissue, and irregular cartilaginous plates, and the lining is a ciliated pseudostratified columnar epithelium (an epithelium of tall ciliated columnar cells crowded very close together and the nuclei of which lie at different heights within

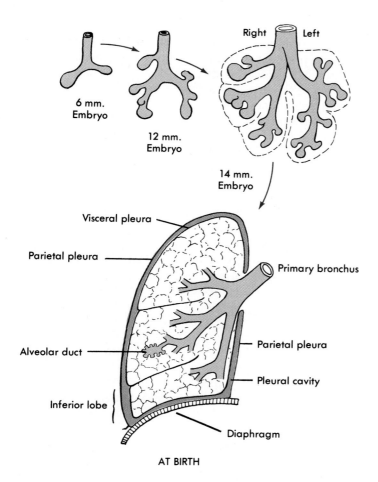

FIGURE 12-21

Development of mammalian lung. Embryo lengths are applicable approximately to both the fetal pig and man. The pleural cavity (lower figure) is in *dark red*.

the cell, giving the false impression in a cross section that the epithelium is multilayered). As the branches become smaller the cilia are lost, the epithelium becomes flatter, and first the cartilage and then the smooth muscle fibers disappear. Terminal bronchioles lead into delicate thin-walled **alveolar ducts,** the walls of which are evaginated to form clusters of **alveoli,** or respiratory pockets, estimated at 400 million in humans. Alveoli are lined with a simple squamous epithelium beneath which is a rich network of capillary beds. It is in the alveoli that gaseous exchange occurs.

Ventilation of the lungs in mammals is achieved primarily by the dome-shaped diaphragm functioning as a suction pump. It is anchored to the xiphoid process of the sternum ventrally, to the caudalmost ribs and their costal cartilages laterally, and to several lumbar vertebrae dorsally, and, when not under tension, the dome bulges cephalad into the thoracic cavity. Contraction of the diaphragmatic muscles, which insert on the central tendon, flattens the diaphragm, creating a potential vacuum between the parietal pleura on the inner surface of the rib wall and the visceral pleura on the surface of the lung. Consequently, atmospheric pressure forces environmental air to rush through the nostrils and into the lungs to fill the vacuum. Normal expiration is a passive phenomenon, as will be explained shortly.

Although diaphragmatic breathing is the major method of lung ventilation, mammals continue to employ the phylogenetically older method of costal breathing in an accessory capacity. Movable joints between the capitula of the ribs and the articular facets or demifacets on the thoracic vertebrae enable the ribs of mammals, as they do those of reptiles, to be rotated upward and outward, elevating the sternum and participating in the creation of the vacuum necessary for inspiration. Rib rotation will be painfully experienced by anyone unlucky enough to have a fractured rib! The ribs are moved by the combined action of intercostal and supracostal muscles (Table 10-2).

Expiration in mammals that are not panting is largely a result of the resilience of the thoracic walls, which assists the ribs to return to the resting position, relaxation of the diaphragmatic musculature, which returns the diaphragm to the domed condition, and the upward pressure exerted on the relaxing diaphragm by the abdominal viscera, which had been under compression. The abdominal wall also participates to a degree. Flattening of the diaphragm compresses the abdominal viscera downward so that the abdominal wall bulges out. When the pressure is relieved, the resilience of the abdominal wall causes it to press against the abdominal viscera, thereby potentiating the upward pressure they are exerting against the relaxing diaphragm. Thus a swift-acting *vacuum pump* (suction pump) and a slower *pressure pump* maintain appropriate partial pressure levels of oxygen in the alveoli of mammalian lungs. During forceful expiration the abdominal wall is employed to forcefully compress the abdominal viscera.

Chapter Summary

1. External respiration is the exchange of respiratory gases between organism and environment. It takes place via highly vascular membranes with thin moist epithelia.

2. Internal respiration is the exchange of gases between capillary blood and the tissues.

3. The chief adult organs of respiration are pharyngeal gills, oropharyngeal mucosa, swim bladders or lungs, and skin.

4. Cutaneous respiration is the chief method of respiration in aquatic urodeles and some scaleless fishes.

5. Internal gills develop in the walls of gill chambers. They consist of pretrematic and posttrematic demibranchs attached to pharyngeal arches. Respiratory water usually enters the mouth, but it enters the spiracle in some elasmobranchs, the naris in hagfishes, and external gill slits in lampreys and some fishes.

6. Elasmobranchs have naked gill slits. An operculum covers the gill chambers in chimaeras, bony fishes, and larval anurans. A branchiostegal membrane forms part of the wall of the opercular chamber in bony fishes.

7. The first pharyngeal slit persists in adults as an open spiracle in elasmobranchs and chondrosteans. In some species it houses a pseudobranch.

8. The hyoid demibranch tends to disappear in teleosts, and the number of demibranchs is reduced still further in lungfishes.

9. Larval gills may be external or internal. They are external in lungfishes, amphibians, and a few ganoids. Anuran larvae also develop internal gills. Filamentous internal gills project to the exterior in larval elasmobranchs, viviparous chondrosteans, and some teleosts.

10. In addition to respiration, gills function in salt homeostasis and excretion of nitrogenous waste.

11. External nares lead to blind olfactory sacs in jawed cartilaginous and ray-finned fishes. In lobe fins and tetrapods they are connected via nasal canals with the oropharyngeal cavity or pharynx. They are used for respiration in tetrapods only. Hagfishes have a nasopharyngeal duct that carries respiratory water. The nasohypophyseal duct of lampreys ends blindly.

12. Internal nares open anteriorly into the oropharyngeal cavity of dipno-ans and amphibians, farther caudad when there is a secondary palate. Despite internal nares, air-breathing fishes and aquatic amphibians take air into the mouth by gulping it.

13. Pressure and suction pumps create a flow of respiratory water or air into and out of the vertebrate body. The chief pumps are the oropha-ryngeal walls and floor of fishes and amphibians, the muscle-operated opercular chamber of bony fishes, intercostal muscles and ribs in tet-rapods, and the muscular diaphragm of mammals.

14. Pneumatic sacs arise from the foregut in nearly all vertebrates. They are called swim bladders in fishes, lungs in tetrapods.

15. Swim bladders and the lungs of aquatic urodeles are chiefly hydrostatic organs. Swim bladders are used for respiration in dipnoans, African lungfishes, and a few other physostomous fishes.

16. Swim bladders may serve also for sound transmission, sound produc-tion, and depth perception.

17. The glottis is the entrance to the larynx. It is protected against the entrance of water, milk in suckling mammals, and food particles by fleshy valves, by its ability to lock into the nasopharynx in many mam-mals, and by an epiglottis in most mammals.

18. The larynx is supported by a single pair of lateral cartilages in urodeles, arytenoid and cricoid cartilages in anurans, reptiles, and birds. A thy-roid cartilage is added in mammals. Other smaller cartilages may devel-op.

19. Laryngeal vocal folds are chiefly mammalian but are found also in anurans and some lizards. Birds use the syrinx for vocalization.

20. Tracheal walls are supported by bony plates, rings, or half-rings. The trachea bifurcates into two primary bronchi. At the base of the trachea in most birds is a syrinx.

21. Lungs arise as midventral evaginations of the pharynx. They occupy the pleuroperitoneal cavity in amphibians and most reptiles, pleural cavities in crocodilians, a few squamates, birds, and mammals. Oblique septa separate the pleural from the peritoneal cavities in crocodilians, some squamates, and birds. A muscular diaphragm that serves as a suction pump separates them in mammals.

22. Lungs are simple sacs in amphibians and most snakes, septate in other reptiles, spongy in birds and mammals. Saccular diverticula extend among the viscera in some reptiles and invade the bones in birds. In limbless tetrapods one lung is often rudimentary.

23. In birds inhaled air flows nonstop to air sacs and then via recurrent bronchi to secondary bronchi, parabronchi, and air capillaries before being exhaled. In addition to serving as bellows, avian air sacs are thermoregulatory and buoyant.

24. Expiration of air is largely passive in tetrapods, resulting from the resilience of the lungs, body wall, and viscera upon relaxation of the pump muscles. During flight in birds the pumping action of the flight muscles forces air out of the air sacs, through the lung duct system, and to the exterior.

SELECTED READINGS

Alexander, R.M.: The chordates, Cambridge, England, 1975, Cambridge University Press.

Atz, J.W.: Narial breathing in fishes and the evolution of internal nares, Quarterly Review of Biology 27:367, 1952.

Bartmar, G.: The vertebrate nose: remarks on its structure, functional adaptation and evolution, Evolution 23:131, 1969.

Gans, C., DeJongh, H.J., and Farber, J.: Bullfrog (Rana catesbeiana) ventilation. How does the frog breathe? Science 163:1223, 1969.

Hughes, G.M., and Morgan, M.: The structure of fish gills in relation to their respiratory function, Biological Review 48:419, 1973.

Laitman, J.T.: The anatomy of human speech, Natural History, August 1984, p. 20.

McMahon, B.R.: A functional analysis of the aquatic and aerial respiratory movements of an African lungfish Protopterus aethiopicus, with reference to the evolution of the lung-ventilation mechanism in vertebrates, Journal of Experimental Biology 51:407, 1969.

Randall, D.J., and others: The evolution of airbreathing in vertebrates, New York, 1981, Cambridge University Press.

Wood, S.C., and Lenfant, C.J.M.: Respiration: mechanics, control, and gas exchange. In Gans, C., editor: Biology of the reptilia, vol. 5, New York, 1976, Academic Press.

13

CIRCULATORY SYSTEM

In this chapter we'll look at the heart, blood vessels, and lymph channels of vertebrates from fishes to human beings. We'll find that in embryonic development, structure, and function they conform to a basic pattern, and we'll see how the pattern with respect to the blood circulatory system was modified step-by-step during the evolution of modern species. Then we will examine the abrupt changes that occur when a chick or a mammalian fetus shifts from life in amniotic fluid to breathing air. Along the way we'll look at a few more examples of countercurrents, some marvelous vascular retes, and the pacemaker of the vertebrate heart.

The circulatory system of vertebrates consists of the heart, arteries, veins or venous sinuses, capillaries or sinusoids, and blood **(blood vascular system)** and of lymph channels and lymph **(lymphatic system).** The blood carries oxygen from respiratory organs; nutrients from extraembryonic membranes, digestive tract, and storage sites; hormones and other substances associated with homeostasis and immunity to disease; and waste products of metabolism to the excretory organs. Blood also conducts heat to and from the skin and other surfaces where heat is exchanged, thereby regulating and equalizing internal temperatures. Lymph channels collect interstitial tissue fluids not taken up by the bloodstream and emulsified fats absorbed in the small intestine. Lymph vessels terminate in venous channels.

Arteries carry blood away from the heart. They have muscular and elastic walls (Figs. 13-1, *A*, and 13-2) capable of distention with each intrusion of blood. (Feel your pulse!) The smallest arteries, 0.3 mm or less in diameter, are **arterioles.** They dilate and constrict reflexly and thereby assist in regulating blood pressure. They terminate in blood capillaries. **Veins** commence in capillaries (other than the respiratory capillaries of the gills) and carry blood toward the heart. They have proportionately less muscle and elastic tissue and more fibrous tissue than arteries and are therefore capable of less distention or constriction. The smallest are **venules.** They begin in capillaries. The **heart** is a pump with highly muscular walls.

Capillaries (Fig. 13-3) generally consist of endothelium alone, although certain capillaries are accompanied by a delicate investment of mesenchyme and a scattering of smooth muscle fibers. Their lumen is just large enough to accommodate red blood cells in single file. In fact, the cells often must "squeeze through," becoming temporarily deformed until they "pop out" into the smallest venule. At the site where a capillary emerges from an arteriole, a short section of the capillary wall has smooth muscle fibers that, given an adequate neural or hormonal stimulus, close off the entrance to the capillary bed at that site. Relaxation of the constrictor muscle readmits blood to the capillary. Blanching or blushing of human skin is a result of the constrictor and dilatory effects of these **precapillary sphincters.** The smallest arterioles and venules are connected directly by short vascular **capillary shunts** that assure uninterrupted circulation between the arterial and venous sides of the capillary beds when other capillaries are constricted.

A **portal system** is a system of veins terminating in a capillary bed (Fig. 13-4). In most vertebrates blood from the capillaries of the tail passes via a **renal portal system** to capillaries of the kidneys before continuing to the heart. Blood from the digestive tract, pancreas, and spleen passes via a **hepatic portal system** to the capillaries of the liver before continuing to the heart. Blood containing pituitary regulating hormones from the hypothal-

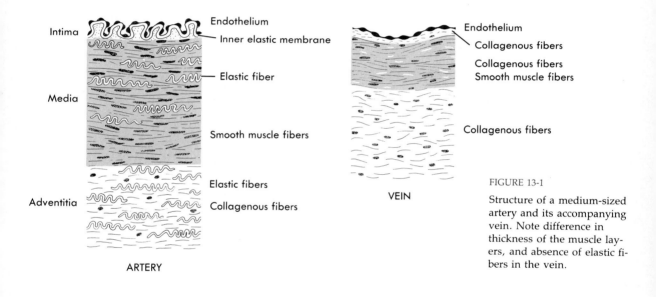

Intima — Endothelium — Inner elastic membrane

Media — Elastic fiber — Smooth muscle fibers

Adventitia — Elastic fibers — Collagenous fibers

ARTERY

Endothelium — Collagenous fibers — Collagenous fibers — Smooth muscle fibers — Collagenous fibers

VEIN

FIGURE 13-1

Structure of a medium-sized artery and its accompanying vein. Note difference in thickness of the muscle layers, and absence of elastic fibers in the vein.

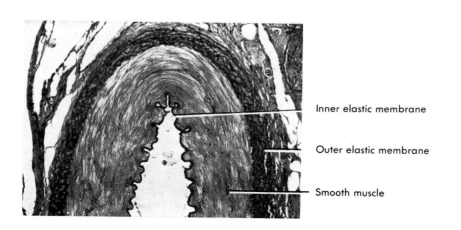

Inner elastic membrane

Outer elastic membrane

Smooth muscle

FIGURE 13-2

Microphotograph of a medium-sized artery. (From Bevelander, G., and Ramaley, J.A.: Essentials of histology, ed. 8, St. Louis, 1979, The C.V. Mosby Co.)

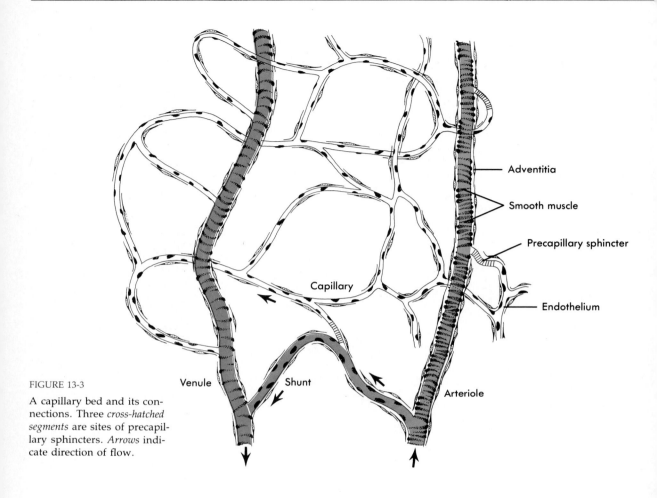

FIGURE 13-3

A capillary bed and its connections. Three *cross-hatched segments* are sites of precapillary sphincters. *Arrows* indicate direction of flow.

FIGURE 13-4

Chief portal systems of vertebrates. The renal portal system is lacking in typical mammals, and the hypophyseal portal system has not been demonstrated in most bony fishes. Channels indicated by broken lines are not part of a portal system.

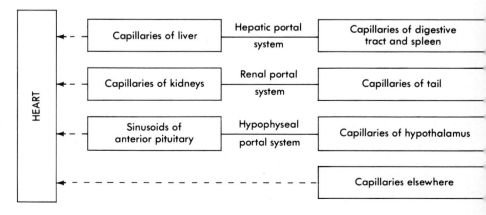

amus passes via a **hypophyseal portal system** to the adenohypophysis before continuing to the heart.

BLOOD

Blood is one of the tissues comprising the circulatory system. In human beings about 55% of blood is **plasma,** a viscous fluid that is 90% water and 10% dissolved solids. The solids consist of blood proteins (serum globulin, serum albumin, and fibrinogen), substances required by living cells (glucose, fat and fatlike substances, amino acids, and ions), essential products synthesized by living cells (enzymes, antibodies, hormones), and the waste products of living cells. **Serum** is essentially plasma from which fibrinogen has been removed.

Suspended in the plasma, and carried along in it as it flows, are **formed elements** consisting of red blood corpuscles (erythrocytes), white blood corpuscles (leukocytes), and platelets (thrombocytes). The formed elements can be separated from the plasma by centrifugation. With appropriate staining the following cell types can be identified (Fig. 13-5):

Erythrocytes
Leukocytes
    Granulocytes (polymorphonuclear leukocytes)
        Eosinophils
        Basophils
        Neutrophils
    Agranular leukocytes
        Lymphocytes
        Monocytes
Thrombocytes (platelets)

FIGURE 13-5

Cellular constituents of blood. **A,** Centrifuged whole blood. Plasma is a light yellow liquid; the formed elements constitute a dark sludge. **B,** Red and white blood cells and platelets. *1,* Erythrocytes; *2,* thrombocytes; *3 to 5,* neutrophilic, basophilic, and eosinophilic granulocytes; *6,* lymphocytes; *7,* monocytes.

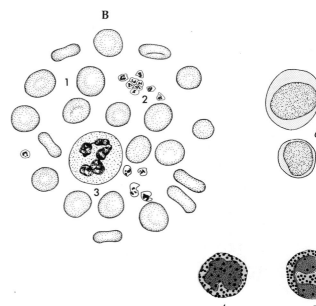

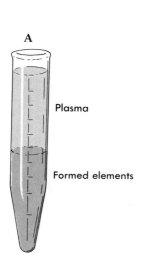

A

Plasma

Formed elements

B

The erythrocytes of most vertebrates are oval nucleated cells, but those of mammals are biconcave circular disks just large enough to squeeze through capillaries in a distorted shape, and without nuclei, having lost them before leaving the erythropoietic tissue where they develop. The circulating red cells of camels are an exception among mammals, resembling the oval nucleated cells of lower vertebrates. Erythrocytes may vary considerably in size among the species of any single major taxon. The most important constituent of red blood cells is hemoglobin, a compound protein composed of a simple protein, globin, and an iron-containing oxygenatable pigment, heme, with which available oxygen immediately binds loosely to form oxyhemoglobin (Hb O_2). Iron, therefore, is absolutely indispensible for life. The oxygen is freed from the hemoglobin in any capillary the surrounding tissue of which has a partial pressure of oxygen lower than that within the bloodstream. The chief erythropoietic tissues before hatching or birth are the liver, red bone marrow (when present), and the spleen. After hatching or birth the red cells are manufactured in red bone marrow, found in the center of flat bones and in the epiphyses of long bones of tetrapods. The average life of an erythrocyte in a human being is less than a month, and inasmuch as there are about 25 trillion erythrocytes in the human body, it is obvious that the erythropoietic centers are very active!

The white corpuscles are far less numerous than either red cells or platelets. Granulocytes and monocytes pass through capillary walls and enter the tissue spaces, where as phagocytes they serve as scavengers of broken down tissues. They have additional functions. Lymphocytes, the source of antibodies, are abundant in lymph nodes and other lymphatic masses that are discussed later. Some authorities classify monocytes as large lymphocytes. Platelets participate with fibrinogen in the clotting of blood. Whereas the blood comprises 5% to 10% of the body weight of amniotes, it represents only about 1.5% to 3% of the body weight of fishes.

Most of the preceding discussion is based on mammalian blood. Blood physiology differs among the vertebrate classes, and appropriate textbooks of physiology should be consulted for comparative purposes.

THE HEART AND ITS EVOLUTION

The vertebrate heart is chiefly a muscular pump that occupies the pericardial cavity. The walls consist of an **endocardium, myocardium,** and **epicardium,** which correspond, respectively, to the intima, media, and adventitia of the arteries. However, as described in Chapter 10, the myocardium is a special type of striated muscle, rather than smooth muscle as in arteries. It is especially thick in the ventricular wall. Lying on the epicardium is the **visceral pericardium,** the equivalent of the visceral peritoneum of the main coelom. The parietal and visceral pericardia are continuous with one another, being reflected over the blood vessels that enter and leave. The space between these is the **pericardial cavity.**

The tissues of the heart are supplied with arterial blood by **coronary arteries** and drained by **coronary veins.** The heart muscle pulsates in response to specific electrolytes in the blood that perfuses it. The rhythmicity of the beat, which will be discussed shortly, is imposed by the autonomic nervous system except in hagfishes, in which the heart has no nerve supply.

SINGLE- AND DOUBLE-CIRCUIT HEARTS. In fishes, blood passes from the heart to the gills and from there directly to all parts of the body, after which it returns to the heart (Fig. 13-6, fishes). Thus blood makes a single circuit during which it is pumped, oxygenated, distributed, and returned to the pump. No blood escapes oxygenation, and none fails to enter a capillary bed where oxygen can be released for use by the tissues.

In species that breathe with lungs rather than gills a **pulmonary circuit** carries oxygen-poor blood from the heart to the lungs and brings back oxygenated blood, whereupon a **systemic circuit** carries the oxygenated blood to all parts of the body and returns oxygen-depleted blood to the heart (Fig. 13-6, amniotes). Evolution of two-circuit hearts was accompanied by successive structural changes, discussed in the sections that follow.

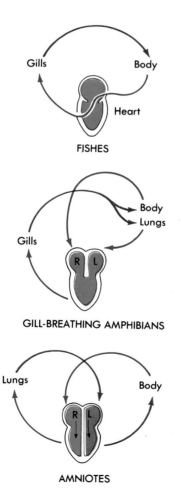

FIGURE 13-6

Normal circulatory routes in fishes, gill-breathing amphibians, and amniotes. *Red,* oxygenated blood. In the amphibian circuits shown, the animal is in well-oxygenated water and the gills are the functional respiratory organs. When the animal is forced to breathe air, much of the unoxygenated blood bypasses the gills, is oxygenated in the lungs, and the oxygen content of the pulmonary veins rises.

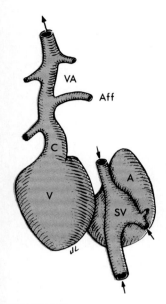

The Hearts of Gill-Breathing Fishes

The hearts of fishes other than dipnoans have four chambers in a series,
a **sinus venosus, atrium, ventricle,** and **conus arteriosus,** and blood flows
through these chambers in that sequence. Such a heart is seen in its simplest form in hagfishes (Fig. 13-7).*

The heart of sharks is typical of jawed fishes in general (Fig. 13-8). The
sinus venosus has thin walls, little muscle, and much fibrous tissue. Its
caudal wall is anchored to the anterior face of the septum transversum. The
sinus has some contractility, but is chiefly a collecting chamber for venous
blood that is returning from all parts of the body. It is filled by suction each
time the ventricle contracts and relaxes. Blood from the sinus gushes
through the sinoatrial aperture into the atrium between two one-way
valves as soon as the atrium begins to relax after emptying.

The atrium is a large thin-walled muscular sac that is a sort of staging
area for blood that is about to enter the ventricle to be propelled toward the
gills. From the atrium, blood pours into the relaxing ventricle through an
atrioventricular aperture that is guarded by a pair of one-way valves. These
prevent ventricular blood from being pumped back into the atrium when
the ventricle contracts.

The ventricle has very thick muscular walls and is the actual pumping
portion of the heart. The anterior end is prolonged as a muscular tube of
small diameter, the conus arteriosus, which extends to the extreme cephalic
end of the pericardial cavity, at which point it is continuous with the ventral
aorta. The conus is composed chiefly of cardiac muscle and elastic connec-

*Designation of fish hearts as two-chambered recognizes only atria and ventricles as cham-
bers. Thus fishes are sometimes said to have two-chambered hearts. The terminology is not
used in this text because it is not definitive.

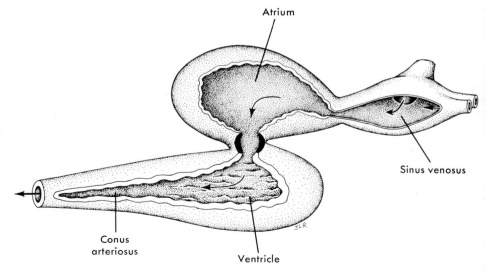

tive tissue. A series of semilunar valves facing forward within the conus prevents backflow of blood into the ventricle. Because of its elasticity it balloons with each delivery of ventricular blood, and then slowly constricts, thereby maintaining a steady arterial pressure in the ventral aorta so that the flow of blood through the gill capillaries, like the flow of countercurrent respiratory water over the gill filaments, is steady despite the rhythmicity of the ventricular beat. The conus of teleosts, unlike that of sharks, is short and has only one set of valves. The function of maintaining a steady flow of blood through the gills of teleosts is assumed by the **bulbus arteriosus,** a muscular expansion of the *ventral aorta.* Most perennibranchiate amphibians have a similar structure (Fig. 13-10).

The Hearts of Lungfishes and Amphibians

Modifications in the hearts of lungfishes and amphibians are correlated with aerial respiration by means of swim bladders or lungs. They enable oxygenated blood returning from the lungs to be separated from deoxygenated blood returning from elsewhere. One modification is the establish-

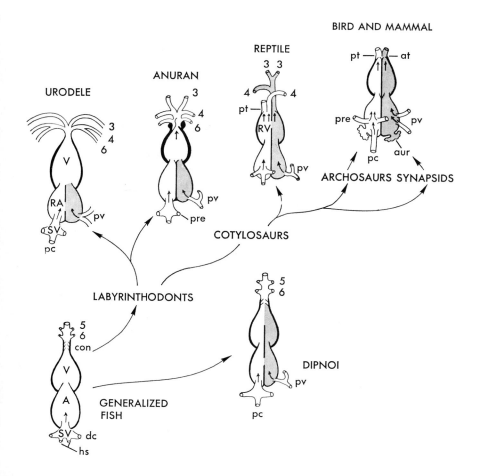

FIGURE 13-9

Modifications of the atria and ventricles that result in increased separation of oxygenated and deoxygenated blood. The parts of the heart shown are *A,* atrium; *RA,* right atrium; *V,* ventricle; *RV,* right ventricle; *SV,* sinus venosus; *con,* conus arteriosus; *aur,* auricle of mammalian heart. *3 to 6,* Third to sixth aortic arches. Other vessels are *at,* aortic trunk; *dc,* common cardinal vein; *hs,* hepatic sinus; *pc,* postcava; *pre,* precava (common cardinal vein); *pv,* pulmonary veins; *pt,* pulmonary trunk. Gray chambers contain chiefly, or only, oxygenated blood.

FIGURE 13-10

Heart and associated vessels of *Necturus*. Ventral view (left) and dorsal view (right). *Aff*, Common channel leading to second and third afferent branchial arteries; *B*, bulbus arteriosus; *C*, common cardinal vein; *H*, hepatic sinus; *L*, left atrium receiving pulmonary vein; *P*, pulmonary vein; *R*, right atrium; *SV*, sinus venosus; *V*, ventricle. A short conus arteriosus connects the ventricle with the bulbus. *Arrows* indicate direction of blood flow. Colors represent arteries *(red)* and veins *(blue)* but not necessarily oxygen content.

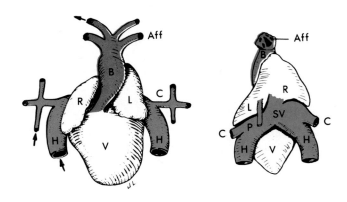

FIGURE 13-11

Conus arteriosus and afferent branchial arteries *(1 to 5)* of the lungfish *Protopterus*. The spiral valve distributes oxygen-rich blood *(red)* to the first three afferent branchial arteries and oxygen-poor blood *(blue)* to the last two, which support internal gills. An interventricular septum is present but not illustrated. The fourth and fifth afferent branchial arteries in this illustration are the fifth and sixth embryonic aortic arches illustrated in Fig. 13-19, *B*.

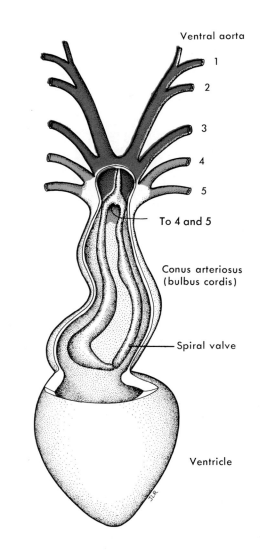

Ventral aorta

1

2

3

4

5

To 4 and 5

Conus arteriosus (bulbus cordis)

Spiral valve

Ventricle

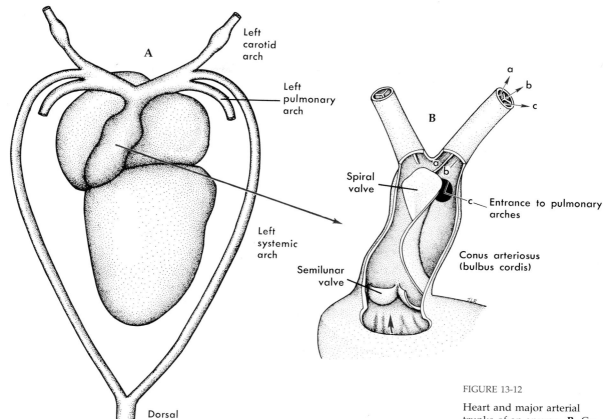

A

Left carotid arch

Left pulmonary arch

Left systemic arch

Semilunar valve

Dorsal aorta

B

a
b
c

Spiral valve

a
b

c — Entrance to pulmonary arches

Conus arteriosus (bulbus cordis)

FIGURE 13-12

Heart and major arterial trunks of an anuran. **B,** Conus arteriosus (bulbus cordis) opened to show passageways to left carotid arch (bristle a-a), left systemic arch (bristle b-b), and left pulmonary (pulmocutaneous) arch (bristle c-c). The ventral aorta is practically nonexistent in adults.

ment of a partial or complete partition within the atrium, so that there is a right and a left atrium (Figs. 13-9, dipnoi, urodele, anuran; and 13-10). The partition is complete in anurans and some urodeles. The pulmonary veins empty into the left atrium; therefore the blood in this chamber is oxygen rich. The sinus venosus empties into the right atrium; hence the blood in this chamber is low in oxygen. In lungless urodeles the atrium remains totally undivided.

A second modification is the formation of a partial interventricular septum (chiefly in lungfishes but also in *Siren,* a urodele) or of ventricular trabeculae (in amphibians). Trabeculae are shelves or ridges projecting from the ventricular wall into the chamber and running mostly cephalocaudad. Interventricular septa and ventricular trabeculae perform identical functions: they maintain separation of oxygenated and unoxygenated blood that began in the left and right atria.

A third modification is formation of a spiral valve in the conus arteriosus **(bulbus cordis)** in many dipnoans and in anurans. The valve directs oxygenated and deoxygenated blood into appropriate channels. In the lungfish *Protopterus* (Fig. 13-11) it shunts blood low in oxygen into the aortic arches that lead to internal gills. In anurans (Fig. 3-12), with every ventricular contraction it blocks and unblocks the common entrance to the left and right pulmonary arches, shunting unoxygenated blood (first blood out of the ventricle) to the pulmonary circuits, then oxygenated blood (next out) to the systemic circuits, as will be described later.

A fourth modification shortened the ventral aorta in dipnoans and anurans, so that it becomes practically nonexistent as embryonic development progresses. As a result, oxygenated and deoxygenated blood that has been kept separate in the heart by septa, trabeculae, and spiral valves moves *directly* from the heart into appropriate vessels (Figs. 13-11 and 13-12).

Urodeles retain a prominent ventral aorta. It exhibits a swollen muscular segment, the bulbus arteriosus, which maintains a steady, rather than pulsating, flow of blood through the gills (Fig. 13-10).

Many of the changes in dipnoans and amphibians—septa and trabeculae, for example—presage conditions in amniotes.

The Hearts of Amniotes

Amniote hearts have two atria, two ventricles, and except in adult birds and mammals, a sinus venosus (Figs. 13-13 and 13-23). In crocodilians the sinus venosus is partially incorporated into the wall of the right atrium. Birds and mammals have a sinus venosus during early development, but it fails to keep pace with the growth of the right atrium into which it empties, and is finally incorporated into the wall of that chamber. Thereafter the vessels that emptied into the sinus venosus empty directly into the right atrium (Fig. 13-14). The embryonic location of the sinus venosus is marked

FIGURE 13-13

Heart and associated vessels of *Sphenodon*. **A,** Ventral view. **B,** Dorsal view. *L,* Left atrium; *la,* left aortic trunk; *lcc,* left common carotid artery; *LP,* left precava; *P,* pulmonary trunk; *PO,* postcava; *PV,* pulmonary veins entering right atrium; *R,* right atrium; *ra,* right aortic trunk; *rcc,* right common carotid artery; *RP,* right precava; *SV,* sinus venosus. The ventricle is divided internally into two chambers. *Red* vessels contain oxygenated blood.

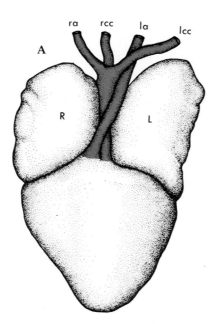
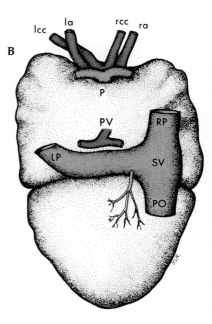

in adults by the **sinoatrial node** of neuromuscular tissue, which, as will be explained shortly, plays a role in regulation of the heartbeat.

The right and left atria of adult amniotes are completely separated by an interatrial septum. Nevertheless, they are confluent during embryonic development via an **interatrial foramen (foramen ovale),** which closes about the time of hatching or at birth. The site of the obliterated foramen ovale is marked in adult hearts by a depression, the **fossa ovalis,** in the medial wall of the right atrium. The right atrium receives blood from the sinus venosus (reptiles) or blood that previously emptied into the sinus venosus (bird and mammals). It also receives blood from the coronary veins. The left atrium receives blood from the pulmonary veins.

In mammals each atrium has an earlike flap, or **auricle,** containing a blind, saclike chamber (Fig. 13-37). Any functional advantage of the mammalian auricle has yet to be demonstrated.

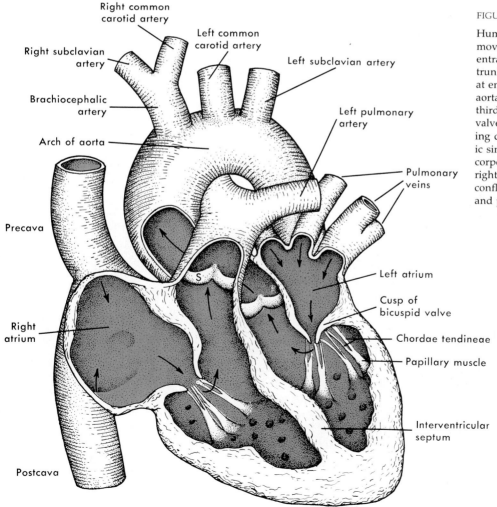

Right common
carotid artery

Left common
carotid artery

Right subclavian
artery

Left subclavian artery

Brachiocephalic
artery

Left pulmonary
artery

Arch of aorta

Pulmonary
veins

Precava

Left atrium

Cusp of
bicuspid valve

Right
atrium

Chordae tendineae

Papillary muscle

Interventricular
septum

Postcava

FIGURE 13-14

Human heart, auricles removed. *S,* Semilunar valve at entrance to pulmonary trunk. The semilunar valve at entrance to the ascending aorta is also shown. The third cusp of the tricuspid valve has been removed during dissection. The embryonic sinus venosus has been incorporated in the wall of the right atrium at the site of confluence of the precava and postcava.

The two ventricles are completely separated in crocodilians, birds, and mammals. In other amniotes the interventricular septum is incomplete. A cavum venosum and cavum pulmonale, characteristic of the interventricular region of turtles and squamates is discussed on p. 441. The internal walls of the ventricles frequently exhibit interanastomosing ridges and columns of muscle (**trabeculae carneae).** The **os cordis,** a bone in the interventricular septum of deer and bovines, was mentioned in Chapter 6.

Valves guard the passage from the atria into the ventricles. The valves are fibrous flaps (muscular on the right in crocodilians and birds) connected in mammals and some lower amniotes by tendinous cords (**chordae tendineae)** to **papillary muscles** projecting from the ventricular walls (Fig. 13-14). During relaxation of the ventricle (diastole), blood from the atria falls freely past the flaps (cusps) into the ventricles. During ventricular contraction (systole), the flaps are forced upward into the atrioventricular passageway, thereby preventing reflux of blood into the atria. Both valves have one or two flaps in reptiles and birds. In most mammals the left valve has two flaps (**bicuspid,** or **mitral, valve),** and the right has three (**tricuspid valve).**

Guarding the exits of the pulmonary and aortic trunks from the ventricles are **semilunar valves** that prevent backflow into the ventricles as they relax (Fig. 13-14).

INNERVATION OF THE AMNIOTE HEART. Within the sinus venosus of fishes and amphibians neural ganglia, innervated by the autonomic nervous system, impose a rhythmicity on the heartbeat in those classes. The **sinoatrial node,** the "pacemaker" of the amniote heart, is a comparable, although more specialized, functional site in amniotes. The node is innervated by postganglionic fibers of the autonomic nervous system (inhibitory fibers via the vagal nerve, excitatory fibers via cardiac nerves from the sympathetic nervous system). Impulses from the node are propagated by special elongated electroconductive muscle cells (**Purkinje fibers)** to the musculature of the atrium and to an **atrioventricular node.** From the atrioventricular node an **atrioventricular bundle** of Purkinje fibers propagates the regulatory impulses via the interventricular septum to the myocardium of the ventricle. A battery-operated pacemaker is used to regulate the human heartbeat when necessary.

Morphogenesis of the Heart

The heart of all vertebrates commences as a single, almost straight, pulsating tube that receives incoming blood at the caudal end and empties into the embryonic ventral aorta anteriorly (Fig. 13-15, *A* and *D*). As development progresses, the tube, whether of a shark or a human being, bends to the animal's right, then twists into an S shape, so that the atrial region, previously at the caudal end, is carried dorsad and cephalad until it lies where it is found in adult fishes (Fig. 13-8). (In studying illustrations keep in mind that, when seen from *ventral* view, the tube will be bending to the observer's *left.*) The twisting and bending is probably correlated with the

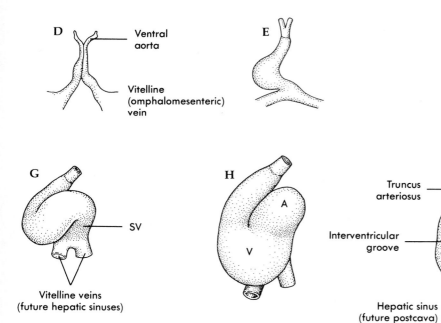

Dorsal view

FROG

CHICK

Ventral views

FIGURE 13-15

Development of an anuran and avian heart. *A*, Atrium; *BA*, bulbus arteriosus; *SV*, sinus venosus; *V*, ventricle. Oriented for optimal viewing of bending and twisting. For a view of the early chick heart in situ see Fig. 15-6.

confinement of the rapidly growing heart in a less expansive pericardial cavity.

In amphibians and amniotes the twisting is carried further, so that the atrial region finally lies cephalad to the ventricular region (Fig. 13-15). Following this the atrial chamber expands to form bilateral pouches, and a median dorsal fold within the chamber grows ventrad, separating the pouches into left and right atria. As a final major step in amniotes a longitudinal interventricular groove appears on the surface of the ventricle (Fig. 13-15, *I*), while internally an interventricular septum completes the division of the amniote heart into right and left sides.

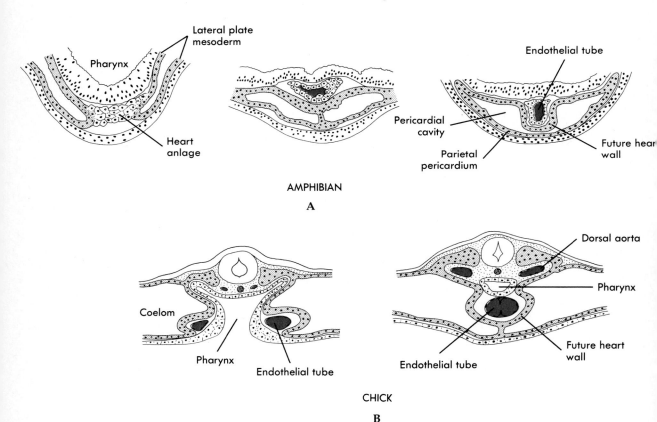

AMPHIBIAN

A

CHICK

B

FIGURE 13-16

Origin of heart from paired lateral plate anlagen. In amphibians the heart forms from a median mass of mesenchyme that has a bilateral origin. In amniotes two endothelial tubes of splanchnic mesoderm are brought together to form the primordium of the heart. (**A** after Hertwig; **C** after His.)

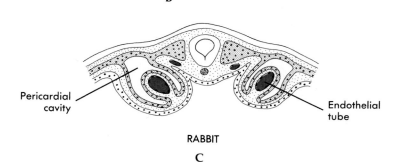

RABBIT

C

As hatching approaches in birds, the sinus venosus is incorporated almost completely into the wall of the right atrium. In mammals the sinus venosus fails early to keep up in growth with the rest of the heart, and its incorporation into the right atrial wall occurs sooner in organogenesis. Because of the imperative need for circulating oxygen and nutrients from their source to the embryonic tissues at the earliest possible moment, the heart is the first organ to function, and does so even before any nerves have reached it to impose a cardiac rhythm.

The initial straight tube that will become the heart of sharks and amphibians organizes from paired mesenchymal masses of lateral-plate somatic and splanchnic mesenchyme that aggregate beneath the pharynx to form a single tube (Fig. 13-16, *A*). In amniotes a pair of already organized endothelial tubes is brought together beneath the pharynx, they fuse, and a single tube results (Fig. 13-16, *B* and *C*). In either case, the heart is a bilateral contribution of lateral-plate mesoderm.

ARTERIAL CHANNELS AND THEIR MODIFICATIONS

Arterial channels supply most organs with oxygenated blood, although they carry deoxygenated blood to respiratory organs. In the basic pattern (Fig. 13-17) the major arterial channels consist of (1) a **ventral aorta (truncus**

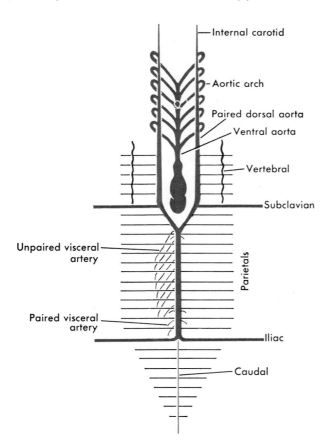

Internal carotid

Aortic arch

Paired dorsal aorta

Ventral aorta

Vertebral

Subclavian

Unpaired visceral artery

Paired visceral artery

Parietals

Iliac

Caudal

FIGURE 13-17

Basic pattern of the chief arterial channels of vertebrates.

arteriosus) emerging from the heart and passing forward beneath the pharynx (paired early in embryogenesis); (2) a **dorsal aorta,** paired above the pharynx and passing caudad above the digestive tract; and (3) six pairs of **aortic arches** connecting the ventral aorta with the dorsal aorta. Branches of these major channels supply all parts of the body. Modifications affect most prominently the aortic arches, which become adapted during embryonic development for respiration by gills or lungs.

Aortic Arches of Fishes

Adaptive modifications of the embryonic aortic arches for respiration by gills may be illustrated in developing sharks (Fig. 13-18). The ventral aorta in *Squalus* extends forward under the pharynx and connects with the developing aortic arches. The aortic arches in the mandibular arch are the first to develop. Shortly thereafter the other five pairs appear. Before the sixth pair is completed, the ventral segments of the first pair disappear, and the dorsal segments become the two efferent pseudobranchial arteries. The second

FIGURE 13-18

Changes in embryonic aortic arches I to VI of *Squalus* during development, lateral view. In **A,** buds (*white*) off the aortic arches are establishing pretrematic and posttrematic arteries and crosstrunks. *Broken lines* indicate sections of the aortic arches that become occluded, forcing blood into afferent branchial arterioles (not shown). In **B,** *I* has become the efferent pseudobranchial artery and *II* to *VI* have become efferent branchial arteries. *1, 3, 5, 7, 9,* Pretrematic arteries; *2, 4, 6, 8,* posttrematic arteries. *As,* Afferent pseudobranchial; *Hyp,* hypobranchial; *ic,* internal carotid; *S,* spiracle.

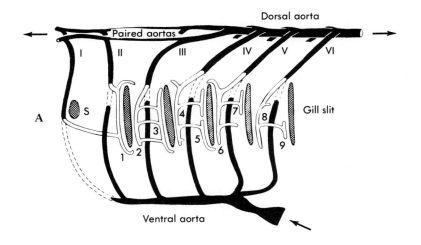

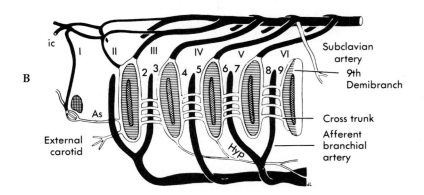

pair sprout buds that become the first pretrematic arteries. Other buds sprout from the third, fourth, fifth, and sixth aortic arches and give rise to posttrematic arteries. The posttrematic arteries then sprout crosstrunks, which grow caudad in the holobranch and by further budding establish the last four pretrematic arteries. Aortic arches II to VI soon become occluded at one site (broken lines in Fig. 13-18, *A*). The segments ventral to the occlusions become afferent branchial arteries. The dorsal segments become efferent branchial arteries. In the meantime, capillary beds are developing within the nine demibranchs. Afferent branchial arterioles connect the afferent branchial arteries with the capillaries. Efferent branchial arterioles return oxygenated blood from the capillaries to the pretrematic and posttrematic arteries.

As a result of these modifications of the embryonic aortic arches, blood entering an aortic arch from the ventral aorta of fishes must pass through gill capillaries before proceeding to the dorsal aorta. The aortic arches have thus been modified to serve the gills.

The same developmental changes convert the six pairs of embryonic aortic arches of bony fishes into afferent and efferent branchial arteries. The specific number converted determines the number of functional gills. In most teleosts the first and second aortic arches tend to disappear (Fig. 13-19, *A*). In *Protopterus* (Fig. 13-19, *B*) the third and fourth embryonic aortic arches do not become interrupted by gill capillaries.

In lungfishes a pulmonary artery sprouts off the left and right sixth aortic arch and vascularizes the swim bladders. This happens also in two ganoid fishes, *Amia* and *Polypterus*.* This is precisely how tetrapod lungs are vascularized!

Aortic Arches of Tetrapods

Embryonic tetrapods, like fishes, construct six pairs of embryonic aortic arches (Fig. 1-6). The first and second arches are transitory and not found in adults (Fig. 13-19, *C* to *H*). After arches I and II disappear, the third aortic arches and the paired dorsal aortae anterior to arch III are named internal carotid arteries. With the exception of a few tailed amphibians, tetrapods lose also the fifth aortic arches during embryonic life (Fig. 13-19, *E* to *H*). Pulmonary arteries sprout off the sixth arches to vascularize the lung buds (Figs. 1-6; 13-19, *C* to *H*; 13-20, *D*; and 13-21). Further modifications of the aortic arches and associated vessels will be discussed under amphibians, reptiles, birds, and mammals.

AMPHIBIANS. Most terrestrial urodeles retain four pairs of aortic arches (Fig. 13-19, *C*). Aquatic urodeles typically retain three because the fifth arches either drop out or unite with the fourth (Fig. 13-21). In aquatic urodeles that retain gills throughout life, afferent branchial arterioles pass to the gills, efferent branchial arterioles return oxygenated blood to the arch, and a short section of each arch becomes a gill bypass (Fig. 13-19, *D*).

*In most other actinopterygians the swim bladders are supplied from the dorsal aorta.

FIGURE 13-19

For legend see opposite page.

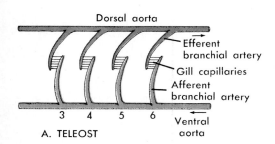

Dorsal aorta

Efferent branchial artery

Gill capillaries

Afferent branchial artery

Ventral aorta

3 4 5 6

A. TELEOST

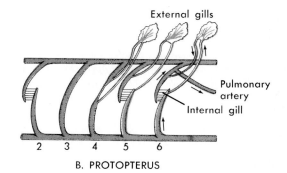

External gills

Pulmonary artery

Internal gill

2 3 4 5 6

B. PROTOPTERUS

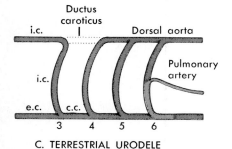

Ductus caroticus

i.c.

Dorsal aorta

Pulmonary artery

i.c.

e.c. c.c.

3 4 5 6

C. TERRESTRIAL URODELE

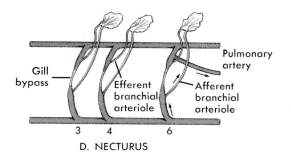

Pulmonary artery

Gill bypass

Efferent branchial arteriole

Afferent branchial arteriole

3 4 6

D. NECTURUS

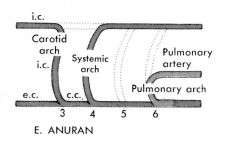

i.c.

Carotid arch

i.c.

Systemic arch

Pulmonary artery

Pulmonary arch

e.c. c.c.

3 4 5 6

E. ANURAN

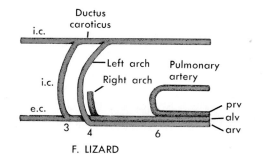

i.c.

Ductus caroticus

i.c.

Left arch

Right arch

Pulmonary artery

e.c. c.c.

prv
alv
arv

3 4 6

F. LIZARD

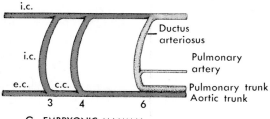

i.c.

i.c.

Ductus arteriosus

Pulmonary artery

Pulmonary trunk
Aortic trunk

e.c. c.c.

3 4 6

G. EMBRYONIC MAMMAL

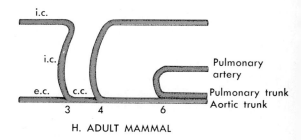

i.c.

i.c.

Pulmonary artery

Pulmonary trunk
Aortic trunk

e.c. c.c.

3 4 6

H. ADULT MAMMAL

FIGURE 13-19

Persistent left aortic arches in representative vertebrates. *2* to *6,* Second through sixth embryonic aortic arches. *Dotted lines* in **C** indicate vessel present in some species; in **E,** vessels that are functional in larvae. *Arrows* indicate direction of blood flow. *alv,* Aortic trunk from left ventricle; *arv,* aortic trunk from right ventricle; *cc,* common carotid artery, which is the paired segment of the embryonic ventral aorta; *ec,* external carotid; *ic,* internal carotid; *prv,* pulmonary trunk from right ventricle. In **B, C,** and **E** the ventral aorta carries venous blood *(blue)* during one phase of a single ventricular contraction, and arterial blood *(red)* during the next phase. In **B** the oxygen content of each vessel depends on the extent to which oxygen is being acquired via the gills as compared to the swim bladder. In **C** the sixth arch carries only oxygenated blood after the pulmonary artery is filled with unoxygenated blood. In **G** all blood in arches is mixed and colors designate predominant condition.

FIGURE 13-20

Embryonic modifications of the aortic arches (*1* to *6*) of a porcupine. (Modified from Struthers.)

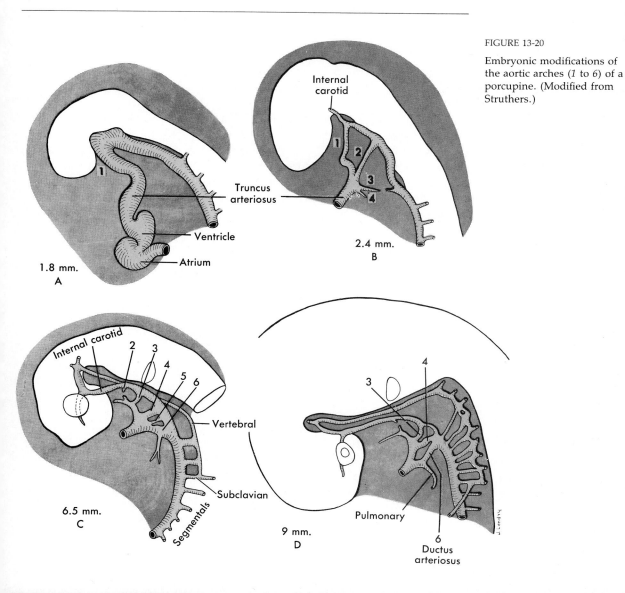

FIGURE 13-21

Efferent branchial arteries (*3, 4, 6*) of *Necturus* as seen from above the gills. Numerals designate dorsal segments of third, fourth, and sixth embryonic aortic arches. *Arrows* indicate direction of blood flow.

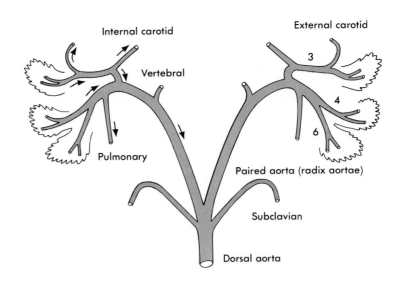

Bypasses carry little blood so long as the animal is using its gills; however, when the water is low enough in dissolved oxygen to cause the animal to gulp air, the gills shrink, and the bypasses carry more blood. Similarly, when resorption of the gills of *Siren* is brought about by thyroid hormone injections, the animal gulps air, and the bypasses carry all the blood entering the arches. On the ventral aorta of perennibranchiates the bulbus arteriosus (Fig. 13-10) maintains a steady nonpulsating arterial pressure in the gills.

Larval anurans (tadpoles) retain four aortic arches (III through VI) for awhile. Arch VI sprouts a pulmonary artery that vascularizes the lung bud. Arches III, IV, and V supply larval gills on the third, fourth, and fifth pharyngeal arches. Gill bypasses are prominent at first, then become temporarily occluded while the external gills are functioning.

With loss of gills at metamorphosis, three changes affect the aortic arches and associated vessels of anurans (Fig. 13-19, *E*): (1) Aortic arch V disappears. (2) The dorsal aorta between aortic arches III and IV **(ductus caroticus)** disappears. As a result, blood entering aortic arch III **(carotid arch)** can pass only to the head. (3) The segment **(ductus arteriosus)** of aortic arch VI dorsal to the pulmonary artery disappears. (This also occurs in a few urodeles. It persists in other urodeles and in apodans.) Blood entering arch VI **(pulmonary arch)** in anurans can now pass only to the lungs and skin. Aortic arch IV **(systemic arch)** on each side continues to the dorsal aorta to distribute blood to the rest of the body (Fig. 13-12, *A*).

Oxygenated blood from the left atrium and deoxygenated blood from the right are kept remarkably well separated as they pass through the ventricle in amphibians. In anurans this is accomplished by ventricular trabeculae, by the movement out of the ventricle of right atrial blood first, and by action of a spiral valve in the conus arteriosus (Fig. 13-12, *B*). At the start of ventricular systole the valve is flipped into a position that closes off the entrance to the systemic and carotid arches, thereby directing deoxygenated blood into the aperture leading to the pulmonary arteries. Then, as backpressure builds within the pulmonary arteries because of filling of the lung capillaries, the spiral valve flips into an alternate position, which directs oxygenated blood into the systemic and carotid arches. Late in ventricular systole some of the left atrial blood enters the pulmonary arches. In a marine toad studied by the injection of radioisotopes and using angiocardiography, the mixing was found to be only slightly greater.

Adult apodans retain three complete aortic arches (III, IV, VI), although the ductus arteriosus and ductus caroticus are reduced and carry little blood.

REPTILES. Modern reptiles exhibit three adult aortic arches: III, IV, and the base of VI (Fig. 13-19, *F*). Although the ductus arteriosus and ductus caroticus usually close before birth, both remain in primitive lizards, and in a few other reptiles one or the other may persist.

An innovation has been introduced in the ventral aorta of reptiles. Instead of developing a spiral valve to shunt fresh and deoxygenated blood into the proper arches, reptiles underwent a series of mutations that split the truncus arteriosus (unpaired segment of the ventral aorta) into three separate passages: two **aortic trunks** and a **pulmonary trunk** (Fig. 13-19, *F*).* The effects of these changes were as follows (Fig. 13-22, *B*): (1) The pulmonary trunk emerges from the right ventricle and leads to the left and right sixth aortic arches. Deoxygenated blood from the right atrium is therefore sent to the lungs. (2) One aortic trunk emerges from the left ventricle and carries oxygenated blood to the *right fourth* aortic arch and to the carotid arches. (3) The other aortic trunk leads out of *what appears to be* from external view the *right* ventricle and leads to the *left fourth* aortic arch. Studies of the oxygen content of this arch show that it, too, carries *oxygenated blood.*

Inasmuch as the left systemic arch appears to receive blood from the right side of the heart, how can it carry oxygenated blood? A series of studies using cinefluoroscopy have provided the answer. In turtles, lizards, and snakes, the interventricular septum is incomplete in the vicinity where the two aortic trunks leave the ventricle, and that region is converted into a separate pocket (**cavum venosum**) by trabeculae (Fig. 13-23). Oxygenated blood from the left ventricle is directed into this pocket, which leads to the two systemic arches. Therefore both left and right systemic arches receive oxygenated blood. Unoxygenated blood from the right atrium is directed by trabeculae toward the entrance to the pulmonary trunk, which is also located in a pocket, the **cavum pulmonale.** The arteries emerging from a turtle's heart are illustrated in Fig. 13-24, *A*.

---

*It is an attractive hypothesis that in early reptiles the truncus arteriosus was divided into two trunks that corresponded to the two ventricles. A pulmonary trunk from the right ventricle would have led to the sixth aortic arch, and an aortic trunk from the left ventricle would have led to the fourth and third aortic arches. This simple condition would then have been altered in three directions: toward the three trunks of modern reptiles, toward the condition in modern birds, and toward the condition in modern mammals.

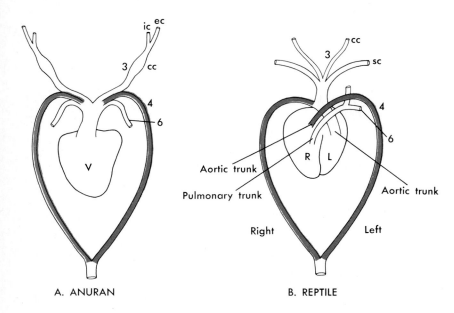

A. ANURAN

B. REPTILE

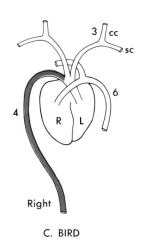

C. BIRD

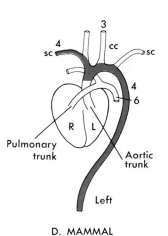

D. MAMMAL

FIGURE 13-22

Fate of the fourth aortic arches *(red)* in selected tetrapods, ventral view. *3*, Third arch (carotid); *4*, fourth arch (systemic); *6*, sixth arch (pulmonary). The relationships of the vessels have been adjusted to emphasize homologies. *cc*, Common carotid; *ec*, external carotid; *ic*, internal carotid; *sc*, subclavian artery; *L*, left ventricle. In **B**, both aortic trunks emerge from the cavum venosum, as shown in Fig. 13-23.

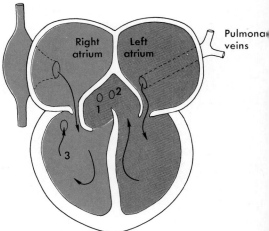

FIGURE 13-23

Circulation in a turtle's heart. *Red* indicates oxygenated blood. *1* and *2*, Entrances to left and right aortic trunks, respectively, in the cavum venosum; *3*, entrance to pulmonary trunk in cavum pulmonale of right ventricle. Ventral view.

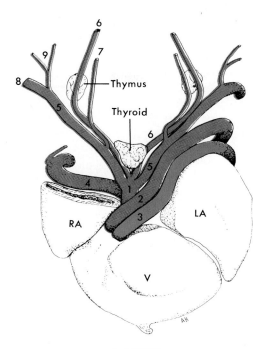

A. TURTLE

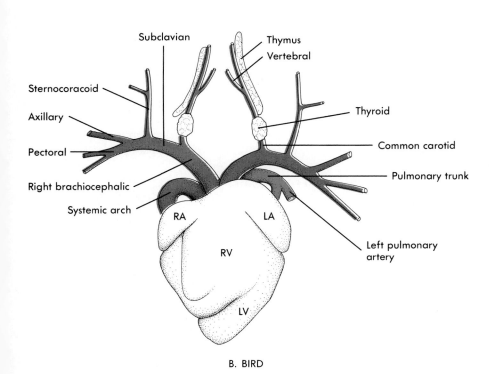

B. BIRD

FIGURE 13-24

Heart and aortic arches of turtle and bird, ventral view. *LA,* Left atrium; *LV,* left ventricle; *RA,* right atrium; *RV,* right ventricle; *V,* ventricular region of turtle. Numerical key for **A:** *1,* brachiocephalic artery, a branch of the aortic trunk emerging from left ventricle (see also Fig. 13-22, *B*); *2,* left aorta; *3,* pulmonary trunk; *4,* right aorta, a continuation of the aortic trunk that emerges from the left ventricle (see also Fig. 13-22, *B*); *5,* subclavian; *6,* common carotid; *7,* ventral cervical; *8,* axillary; *9,* arteries to pectoral and shoulder muscles. *Red,* oxygenated blood; *blue,* deoxygenated blood.

A shunting of blood from right to left ventricle, and therefore away from the lungs, occurs in aquatic turtles and snakes that remain submerged without breathing for long periods. (When insufficient oxygen is available for metabolism while submerged, aquatic turtles and snakes derive energy from glycolysis, an anaerobic process.) Shunting from right to left is also observed when radiant heat is applied to the body. This may be thermoregulatory, enabling blood warmed by basking in sunlight to avoid the lungs, where heat loss would occur during exhalation. This would be valuable in cold weather. Reverse shunting, left to right, causing some blood from the lungs to return to the lungs, takes place when resistance to blood flow in pulmonary vessels is lowered by dilation. The survival value of shunting is mostly conjectural. (Fetal mammals and unhatched birds also have a right-left shunt away from the lungs, the function of which is obvious, as described later.)

Crocodilians have no way to shunt blood between ventricles because the interventricular septum is complete. However, an opening, the **foramen of Panizza,** connects right and left systemic trunks at their base, and blood can be shunted between these vessels at that location. During normal respiration some of the well-oxygenated blood in the right arch can be shunted to the left.

Most snakes and limbless lizards lose their left sixth aortic arch, which is correlated with the rudimentary nature, or absence of the left lung; and in snakes the left third arch also disappears. The right third arch (right carotid artery) in these species has a bilateral distribution. All that adult snakes have left of the original six pairs of aortic arches are the right third, left and right fourth, and the ventral part of the right sixth.

BIRDS AND MAMMALS. In birds and mammals, for the first time since the introduction of pulmonary respiration, circulatory routes have evolved in which there is no opportunity for mixing of oxygenated and unoxygenated blood. This has been achieved by closing the interventricular foramen and dividing the ventral aorta into two trunks. The pulmonary trunk (Fig. 13-22, *C* and *D*) emerges from the right ventricle and leads only to the sixth aortic arches and lungs. The aortic trunk emerges from the left ventricle and leads to the third and fourth aortic arches. *The left fourth aortic arch disappears in birds, and most of the right fourth disappears in mammals.* Note, therefore, that the systemic arch of birds turns to the right (Figs. 13-22, *C*, and 13-24, *B*). The part of the right fourth aortic arch that remains in mammals becomes the proximal part of the right subclavian artery (Fig. 13-22, *D*).

In birds and mammals, therefore, six aortic arches develop in the embryo, and the first, second, fifth, and left fourth (in birds) or most of the right fourth (in mammals) disappear. The ductus caroticus also disappears. The ductus arteriosus functions until hatching or birth to shunt unoxygenated blood away from the lungs and into the dorsal aorta (Fig. 13-37), which has branches leading to the allantois (the embryonic respiratory organ). Circulation in fetal mammals is described later.

As a result of these modifications, all blood returning to the right side of the heart passes to the lungs. From there it returns to the left side of the heart to be recirculated (Fig. 13-6).

The left fourth aortic arch (systemic arch) of mammals is referred to simply as "the" aortic arch by mammalian anatomists. The common carotid and external carotid arteries were part of the paired embryonic ventral aortae, and the internal carotids form from the third aortic arches and paired dorsal aortae (Fig. 13-19, *H*). A few individual and species differences in the vessels arising from "the" aortic arch in mammals are illustrated in Fig. 13-25.

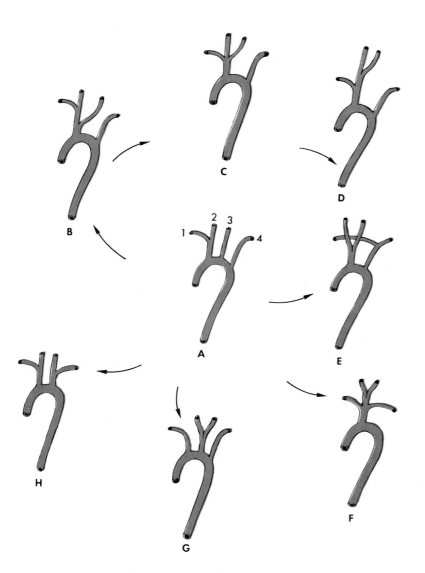

FIGURE 13-25

Selected individual and species differences in the relationships of the common carotid and subclavian arteries in selected mammals. **A,** Basic pattern, seen also in man,* porcupine,* rabbit,* pig; **B,** cat,* dog,* pig,* man, and rabbit; **C** and **D,** rabbit, domestic cat, with **D** occurring with lower frequency; **E,** anomalous right subclavian artery in cat, man, and rat; **F,** many perissodactyls; **G,** walrus; **H,** man, porcupine, and rabbit. An asterisk indicates this to be a predominant condition in the populations examined. If there is no asterisk, the condition is common but not predominant. *1,* Right subclavian; *2,* right common carotid; *3,* left common carotid; *4,* left subclavian. The anomalous condition shown in **E** has been induced experimentally in rats by irradiation of the fetus. Because all conditions illustrated are developmental modifications of a single embryonic pattern, any of the variants could, and probably do, occur in every mammalian species.

Aortic Arches and von Baer's Law

Whenever a respiratory surface develops on a pharyngeal derivative, buds off the nearest aortic arch usually vascularize that surface. When a paryngeal arch develops an external or internal gill, the aortic arch in that pharyngeal arch vascularizes that gill. If the pharyngeal floor evaginates to form a lung bud, a sprout from the nearest aortic arch vascularizes the bud.

Vascularization of pharyngeal derivatives for respiration has phylogenetic roots that extend back to the first chordates. In filter feeders, the external environment enters the pharynx, water devoid of foodstuffs passes between two pharyngeal arches, and oxygen is acquired by buds that sprout from the two aortic arches. The development of six aortic arches in all vertebrate embryos and the systematic modification or elimination of first one vessel and then another in successively higher vertebrates is an example of von Baer's law. This zoological concept is discussed in Chapter 1.

Dorsal Aorta

The dorsal aorta in the head and pharyngeal region is paired in embryos, and frequently in adults, although sometimes masquerading under other names such as internal carotid (in which blood flows to the brain) and ductus caroticus.

The dorsal aorta of the trunk is unpaired. It gives off a segmental series of paired somatic branches to the body wall and appendages, and a series of paired and unpaired visceral branches. It continues into the tail as the caudal artery.

SOMATIC BRANCHES. The **subclavian arteries** are enlarged segmentals (Figs. 13-17 and 13-20). They arise in embryos as branches of the paired or unpaired dorsal aortas or from the third aortic arch (some birds) or fourth aortic arch (some mammals) close to the aorta, but these embryonic relationships become obscured during subsequent development. In tetrapods the subclavian artery traverses the axilla as the **axillary,** parallels the humerus as the **brachial,** and divides in the forearm into **ulnar** and **radial** arteries. In many amniotes one branch on each side, the **vertebral** artery, passes cephalad (in the vertebroarterial canal when the canal is present) to contribute blood to the **circle of Willis** (Fig. 15-26). The vertebral is not well developed in birds and some reptiles.

A series of segmental arteries arise from the aorta along the length of the trunk. These give off short dorsally directed **vertebromuscular** branches to the epaxial muscle, skin, and vertebral column, and long **parietal** branches that encircle the body wall to the midventral line. Where there are long ribs in amniotes the parietals are called **intercostal** arteries. **Lumbar** and **sacral** arteries are parietals in those regions. Segmentals in the neck come off the vertebral arteries.

The **iliacs** are segmental arteries that supply the fins or limbs. In tetrapods the iliac becomes the **femoral,** where it follows the femur, **popliteal** in the knee, and **tibial** in the shank.

Branches of some of the anterior and posterior segmentals, especially of the subclavian and iliac, pass longitudinally in the body wall to anastomose (unite end-to-end). Anastomoses ensure that if one of the anastomosing vessels supplying a region becomes occluded, the vessel approaching from the opposite direction will fill the affected arterial tree beyond the occlusion. Anastomoses are common throughout the body.

VISCERAL BRANCHES. A series of **unpaired visceral branches** (splanchnic vessels) pass via dorsal mesenteries to the unpaired viscera, chiefly digestive organs, suspended in the coelom. The number of such vessels is largest in generalized species such as *Necturus*. As few as three unpaired trunks—frequently celiac, superior mesenteric, and inferior mesenteric—may occur in higher vertebrates. Anastomoses between two successive visceral branches occur along the entire length of the gut. Among anastomosing visceral branches in mammals are a superior pancreaticoduodenal branch of the celiac, which anastomoses with an inferior pancreaticoduodenal branch of the superior mesenteric; a middle colic branch of the superior mesenteric, which anastomoses with a left colic branch of the inferior mesenteric; and a superior rectal branch of the inferior mesenteric, which anastomoses with a middle rectal branch of the internal iliac. Anastomoses are also common on the greater and lesser curvatures of the stomach.

Paired visceral branches of the aorta include arteries to the urinary bladder, reproductive tract, gonads, kidneys, and adrenals. A series of gonadal and renal arteries occur in lower vertebrates, several pairs in reptiles and birds, and usually a single pair in mammals.

■   ■   ■

The early embryonic dorsal aorta of amniotes ends at the level of the hind limbs by bifurcating into right and left **allantoic (umbilical) arteries** that carry blood to the allantois (Fig. 13-36). Internal iliacs sprout off the umbilical arteries as development progresses, and the umbilicals finally become branches of the external and internal iliacs. Species-specific differential growth of the external iliac, internal iliac, and umbilical arteries results in species variations in the relations of these vessels (Fig. 13-26).

Coronary Arteries

The walls of all arteries and veins except the smallest are supplied with **vasa vasorum** ("vessels of the vessels"). The heart is no exception, and here the vessels are called coronary arteries and veins. In elasmobranchs the coronary arteries arise from **hypobranchial arteries** that receive aerated blood from several arterial loops around the gill chambers (Fig. 13-18). In frogs they arise from the carotid arch. In reptiles and birds they arise from the aortic trunk leading to the right fourth arch, or from the brachiocephalic. In mammals they arise from sinuslike dilations at the base of the ascending aorta just beyond the semilunar valves. In a few vertebrates, including urodeles, the coronary supply consists of many small arteries.

FIGURE 13-26

Contrasting termination of the dorsal aorta in two mammalian orders. There is no inguinal ligament in cats. The sources of paired vessels are frequently asymmetrical (compare the left and right superior epigastrics in the rabbit illustrated here). *co*, Common iliac; *da*, dorsal aorta; *df*, deep femoral; *ex*, external iliac; *fe*, femoral; *ie*, inferior epigastric; *ig*, inferior gluteal; *il*, iliolumbar; *in*, internal iliac; *mb*, muscular branch; *mr*, middle rectal; *ms*, median sacral; *sg*, superior gluteal; *su*, superficial epigastric; *um*, umbilical; *vs*, vesicular (absent in rabbits).

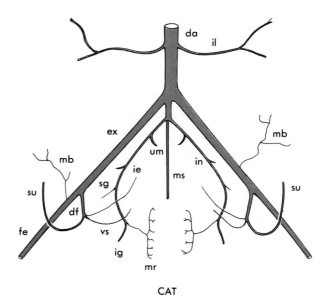

CAT

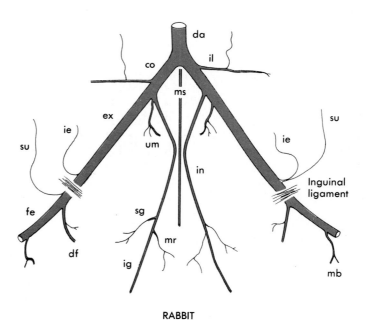

RABBIT

Retia Mirabilia

Certain arteries along their course become highly tortuous, then straighten out again. Such structures are retia mirabilia (singular, rete mirabile), or "wonderful networks." Retia are found in the head on the carotid arteries of a variety of vertebrates. They may modulate the blood pressure within the brain or other organs of the head. The pseudobranch of *Squalus acanthias* is a rete. It probably regulates the blood pressure in the eyeball. Whales have extensive retia consisting of generous-sized arteries in the thorax in a protected position beneath the transverse processes of vertebrae and within the bony vertebral canal beside the spinal cord. These retia are all confluent and are supplied by segmental arteries and drained by vertebral arteries that are en route to the brain. When the whale dives, the thoracic and abdominal viscera are compressed and blood is forced out of the visceral organs into the retia, which are protected from compression by their bony surroundings. These retia constitute a reservoir of blood that was oxygenated just before the whale dived. The oxygen is used by the brain during the dive, which may last for 2 hours.

Often the tortuous artery is associated with an equally tortuous vein in such manner that artery and vein lie side by side, with the blood flowing in opposite directions, another example of a countercurrent. In birds that wade in icy waters, countercurrents in retia above the thigh result in transfer of heat from the artery entering the leg to the vein emerging. This conserves body energy in the form of heat, and at the same time warms returning blood to body temperature. Polar bears and arctic seals have retia that serve the same function. On the other hand, retia do not occur in arctic animals in which it would be a disadvantage; for example, in those that are so well insulated by feathers or hair that the metabolic heat generated by exercise needs to be eliminated rather than retained. Mammals with testes in scrotal sacs have a vascular rete, the **pampiniform plexus,** in each inguinal canal. Heat is transferred from spermatic artery to spermatic vein, assuring that the temperature within the scrotal sac is lower than body temperature, a necessity for the viability of sperm in those species.

In the lining of the swim bladder of some fishes, conspicuous retia, the **red glands,** maintain high gas pressures within the bladder. The pressure forces some of the gas in the bladder to enter the veins of the bladder, into which lactic acid is also being secreted by the bladder cells. Lactic acid alters the pH of the venous blood, releasing oxygen from hemoglobin and carbon dioxide from bicarbonates, and these gases are returned to the bladder by arteries that receive them by countercurrent exchange in the rete. Thus the swim bladder builds pressure and volume as needed.

VENOUS CHANNELS AND THEIR MODIFICATIONS

A generalized venous system consists of the following major streams: cardinals (anterior, posterior, and common), renal portal, lateral abdominal, hepatic portal, hepatic sinuses, and coronary veins (Fig. 13-27, *A*). The hepatic portal system is derived from the embryonic subintestinal and the distal portion of embryonic vitellines, and the hepatic sinuses are derived

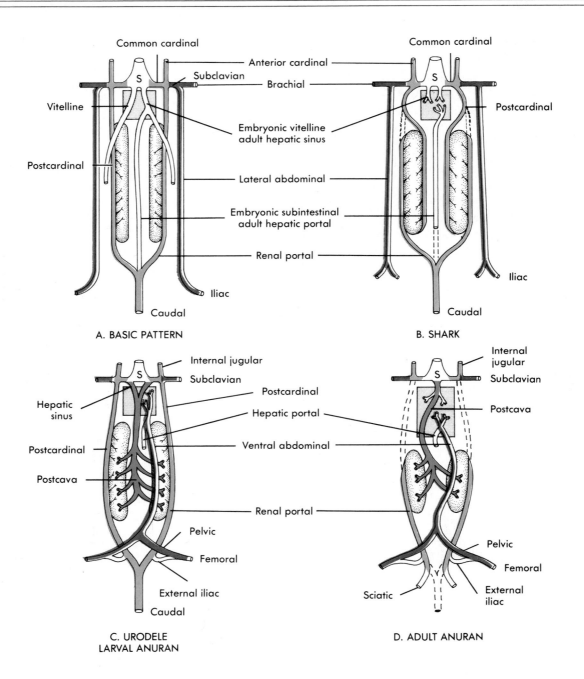

FIGURE 13-27

Modifications of the basic venous channels in sharks and amphibians. *Broken lines* indicate lost segments. *Light blue,* cardinal and caudal (renal portal) streams; *dark blue,* postcava, receiving renal veins; *red,* abdominal stream. *S,* Sinus venosus. The kidneys are *stippled,* and the liver is *gray.* The venous channels medial to the kidneys in **B, C,** and **D** develop from an embryonic subcardinal plexus.

from the vitellines between liver and heart. Two additional streams develop in lungfishes and tetrapods: a pulmonary stream from the lungs and a postcava from the kidneys. These eight channels and their tributaries drain the entire body—head, trunk, tail, appendages. As development progresses they are slowly modified by deletion of some vessels and addition of others. Modifications are few in lower vertebrates, more numerous in higher vertebrates.

The Basic Pattern: Sharks

An adult shark is an ideal living swimming blueprint of the basic venous channels of vertebrates. Inasmuch as very few changes occur in the embryonic channels during later development, knowledge of the venous channels of sharks and how they develop is a good introduction to the venous channels of other vertebrates.

CARDINAL STREAMS. The sinus venosus receives all blood returning to the heart of sharks. Most of this blood, except that from the digestive organs, enters the sinus by a pair of **common cardinal veins** (Fig. 13-27, B). These use the transverse septum as a bridge from the lateral body walls to the heart. They appear early in development, and remain essentially unchanged thereafter.

Blood from the head other than the lower jaw is collected by a pair of **anterior cardinal** (precardinal) **veins** lying dorsal to the gills. The anterior cardinals pass caudad and empty into the common cardinals. The embryonic anterior cardinals, like common cardinals, remain essentially unchanged throughout life.

The earliest *embryonic* **posterior cardinal** (postcardinal) **veins** are continuous with the caudal vein (Fig. 13-27, A). These embryonic postcardinals pass cephalad lateral to the developing kidneys and receive a series of renal veins from them en route. They then empty into the common cardinals. Their cephalic ends in adult sharks expand to become **posterior cardinal sinuses.**

While these embryonic postcardinals are functioning, a new pair of postcardinal veins is forming *between* the kidneys from a subcardinal plexus. (A similar plexus in turtles is illustrated in Fig. 13-33). These new veins become confluent with the old postcardinals at the anterior end of the kidneys, and they, too, drain the kidneys. As more and more blood from the kidneys flows into the new veins the older postcardinals are lost anterior to the kidneys (Fig. 13-27, B). Thereafter the name *posterior cardinal* is applied to the newer vessels. In adults they drain chiefly the kidneys, body wall, and gonads.

RENAL PORTAL STREAM. At an early stage in development in sharks some of the blood from the caudal vein continues forward beneath the gut as a subintestinal vein that drains the digestive tract (Fig. 13-27, A). Later the connection of the caudal vein with the subintestinal is lost (Fig. 13-27,

B). When the old posterior cardinals are lost anterior to the kidneys, all blood from the tail thereafter enters the capillaries surrounding the kidney tubules **(peritubular capillaries).** The result is a **renal portal system.**

LATERAL ABDOMINAL STREAM. Commencing at the pelvic fin from which it receives an **iliac vein** and passing forward in the lateral body wall on each side is a **lateral abdominal vein** (Fig. 13-27, *B*). At the level of the pectoral fin it receives a **brachial vein,** after which the vessel turns abruptly toward the heart to enter the common cardinal vein. The part of the abdominal stream between brachial and common cardinal is the **subclavian vein.** In addition to collecting blood from the paired fins, the abdominal stream also receives a **cloacal vein,** a metameric series of **parietal veins** from the lateral body wall, and minor tributaries. This basic channel remains unmodified during subsequent development.

HEPATIC PORTAL STREAM AND HEPATIC SINUSES. Among the first vessels to appear in any vertebrate embryo are paired **vitelline,** or **omphalomesenteric, veins** from the yolk sack, whether functional or not, to the heart (Figs. 4-10 and 15-6). One of the vitelline veins is soon joined by the embryonic **subintestinal vein** that drains the digestive tract (Fig. 13-27, *A*). As the developing liver enlarges it encompasses the vitelline veins, causing them to be broken into many sinusoidal channels. Caudal to the liver one vitelline vein disappears, and the other, with the subintestinal vein, becomes the **hepatic portal system** that drains the digestive organs of the coelom, and the spleen. Between liver and sinus venosus the two vitelline veins become **hepatic sinuses** (Fig. 13-27, *B*).

Other Fishes

The venous channels of other fishes are much like those of sharks. Cyclostomes have no renal portal veins and no left common cardinals, although two common cardinals develop in embryos. Abdominals are lacking in most ray-finned fishes, and the pelvic fins are drained by the postcardinals. In dipnoans the pelvic fins are drained by an unpaired ventral abdominal vein that ends in the sinus venosus, and the right postcardinal is missing. Blood from the swim bladders of all ray-finned fishes empties into the hepatic or common cardinal veins; in dipnoans it empties into the left atrium. Coronary veins in all fishes empty into the sinus venosus.

Tetrapods

The early embryonic venous channels of tetrapods are basically the same as those of embryonic sharks. We have just seen how, by adding a vessel here and dropping one there during development, the basic pattern of a shark embryo is converted into the veins of an adult shark. We will now see how the same embryonic pattern is converted into the veins of adult tetrapods.

CARDINAL VEINS AND THE PRECAVAE. Embryonic tetrapods have postcardinals, precardinals, and common cardinals. In urodeles the embryonic postcardinals persist between the caudal vein and common cardinals throughout life (Fig. 13-27, C). In this respect a necturus is less modified than a shark. In anurans, most reptiles, and birds the segment of the embryonic postcardinals anterior to the kidneys disappears as development progresses, and in anurans the connection with the caudal vein is lost. (Figs. 13-27, D, 13-28, and 13-29). In lizards, crocodilians, and mammals the anterior segment of the right postcardinal persists under the name **azygos,** and part of the left persists as the **hemiazygos** (Figs. 13-30 to 13-32).

Common cardinal veins in tetrapods are better known as **precavae,** and anterior cardinals are called **internal jugular veins.** Although most mammals retain both the left and right precavae, some, including cats and humans, lose the left precava during embryonic life (Fig. 13-30). In this case a transverse vessel, the **left brachiocephalic,** shunts blood from the left side of the head and left arm to the right precava. A remnant of the left precava remains as a **coronary sinus.** The persisting right precava in mammals is known also as the **superior vena cava.**

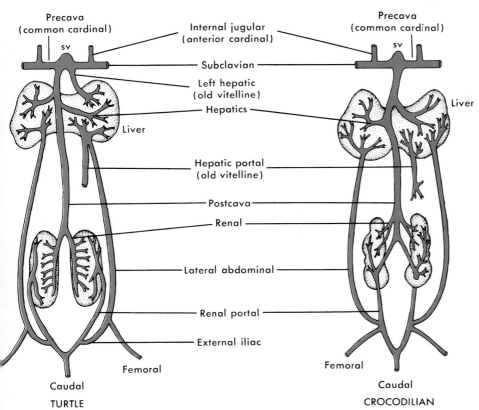

Precava
(common cardinal)
sv
Liver
Caudal
Femoral
TURTLE

Internal jugular
(anterior cardinal)
Subclavian
Left hepatic
(old vitelline)
Hepatics
Hepatic portal
(old vitelline)
Postcava
Renal
Lateral abdominal
Renal portal
External iliac

Precava
(common cardinal)
sv
Liver
Caudal
Femoral
CROCODILIAN

FIGURE 13-28

Systemic veins of two reptiles. Only vessels of the basic pattern illustrated in Fig. 13-27, A, are shown. A strong branch of the renal portal vein of crocodilians continues directly to the postcava without ending in the kidney capillaries. *sv,* Sinus venosus.

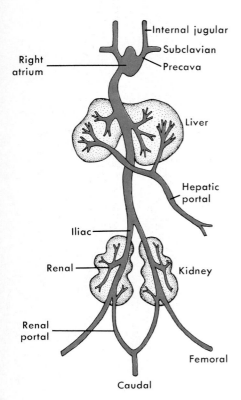

FIGURE 13-29

Major systemic veins of a bird.

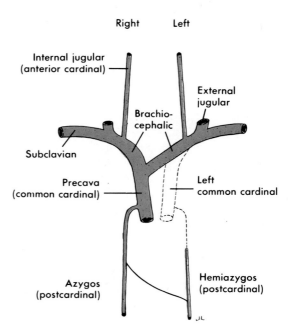

FIGURE 13-30

Basic anterior venous channels of cat and man, ventral view. *Broken lines* indicate vessels obliterated during ontogeny. These sometimes remain as anomalies in adult mammals, including cats, pigs, and humans. Internal and external jugulars are sometimes confluent. Compare channels with those of a rabbit illustrated in Fig. 13-32.

FIGURE 13-31

The azygos (on the animal's right) and the hemiazygos (on the animal's left) in a rhesus monkey, ventral view. This condition is one of many variants in this species. Similar variants are seen in humans. The azygos is shown receiving transverse connections from the hemiazygos and flowing into the precava.

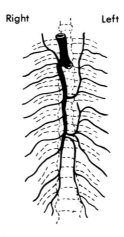

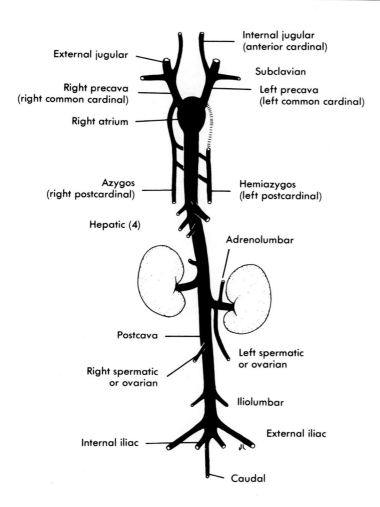

External jugular

Internal jugular
(anterior cardinal)

Right precava
(right common cardinal)

Subclavian

Left precava
(left common cardinal)

Right atrium

Azygos
(right postcardinal)

Hemiazygos
(left postcardinal)

Hepatic (4)

Adrenolumbar

Postcava

Left spermatic
or ovarian

Right spermatic
or ovarian

Iliolumbar

External iliac

Internal iliac

Caudal

FIGURE 13-32

Major systemic venous channels of a rabbit, ventral view. *Broken line* indicates obliterated segment of left posterior cardinal vein. Entrance of one spermatic or ovarian vein into a renal vein is uncommon in rabbits, common in cats.

THE POSTCAVA. The postcava arises during embryonic life in a subcardinal venous plexus that receives renal veins from the kidneys (Fig. 13-33). One subcardinal channel (usually the right) predominates, grows into the mesentery in which the liver is developing, and becomes confluent with the hepatic sinuses. This vessel is the postcava. The enlarging liver envelops it but does not break it into capillaries. Thus the postcava becomes an expressway from kidneys to heart via the hepatic sinuses (Fig. 13-27, *C*). In most tetrapods the two hepatic sinuses ultimately fuse to form a median vessel that becomes part of the postcava (Fig. 13-27, *D*). In mammals the postcava is also called the **inferior vena cava.**

In crocodilians, some of the blood from the hind limbs that enters a renal portal vein bypasses the kidney capillaries and flows directly into the postcava (Fig. 13-28, crocodilian). In birds most, if not all, and in mammals all blood from the hind limbs flows directly into the postcava (Figs. 13-29 and 13-32).

FIGURE 13-33

Embryonic subcardinal venous plexus and origin of the postcava (*dark blue*) in the turtle *Chrysemys*, ventral view. The mesonephros is indicated by *broken lines*. *PC*, Postcava; *PO*, left postcardinal; *RP*, renal portal vein; *SC*, subclavian vein.

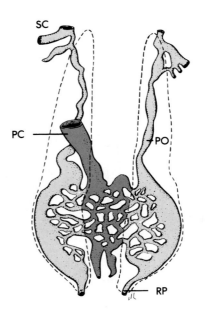

With establishment of a postcava, blood that previously passed from the kidneys to the heart via postcardinal veins now uses the postcava, and the postcardinals are reduced or disappear.

ABDOMINAL STREAM. In early tetrapod embryos, paired lateral abdominal veins commence in the body wall at the level of the future hind limbs, pass cephalad in the lateral body wall, receive veins from the developing forelimbs, and terminate in the common cardinal veins or sinus venosus, as in sharks (Fig. 13-27, *A*). As development progresses in tetrapods this stream alters its course.

In amphibians the two embryonic lateral abdominal veins, which lie close together on either side of the midventral line, unite in the midventral body wall as far forward as the level of the embryonic falciform ligament that connects the ventral body wall with the developing liver. Blood in this unpaired segment, now known as the **ventral abdominal vein,** establishes connections with small venous channels within the falciform ligament, one of these channels enlarges, and soon all the blood in the ventral abdominal vein is crossing the falciform ligament and emptying into the liver (Figs. 13-27, *C* and *D*, 13-28, and 13-29). Thereafter the abandoned segments of the lateral abdominal veins anterior to the liver disappear. The result of these embryonic changes is the establishment of a portal stream between the capillaries of the incipient hind limbs and those of the liver. The rerouting of the abdominal stream dissociates it from the drainage of the anterior limbs (Fig. 13-27, *D*).

In reptiles the two lateral abdominals do not unite (Fig. 13-28). Otherwise, they undergo the same modification as in amphibians, using the falciform ligament as a bridge across the coelomic cavity to the liver, and losing their connection with the common cardinal veins. They acquire temporary tributaries, the two **allantoic veins,** as they pass along the ventral body wall. The allantoic connection disappears when the extraembryonic portion of the allantois is lost at hatching. In birds none of the embryonic abdominal stream remains in adults (Fig. 13-29).

In mammals the segment of the **umbilical veins** in the ventral body wall beginning at the navel and continuing in the falciform ligament to the liver is all that is left of the abdominal stream (Fig. 13-36, 2). The segment of the umbilical veins within the umbilical cord is the reptilian allantoic vein with a new name. No connection develops with the drainage of the hind limbs. The abdominal (umbilical) stream in mammals has no function other than to drain the placenta; and when the umbilical cord is severed at birth blood no longer flows through the umbilical vein between the navel and the liver, and the umbilical vein becomes converted into the **round ligament of the liver.** It is seen in the dissecting room as a fibrous cord in the free border of the falciform ligament extending between the umbilicus and the liver. As a result of these mutations the very ancient abdominal stream, which at one time drained the posterior appendages, body wall, and anterior appendages, has evolved into a vessel draining only the mammalian placenta and confined to embryos.

The umbilical vein in embryonic mammals carries a heavy load of blood and erodes a broad channel, the **ductus venosus,** directly through the substance of the liver and into the postcava (Fig. 13-36, 5). After birth the channel becomes a ligament, the **ligamentum venosum.**

RENAL PORTAL SYSTEM. The renal portal system of amphibians has acquired a tributary, the **external,** or **transverse, iliac vein** (not homologous with the iliac veins of amniotes), which carries some blood from the hindlimbs to the renal portal vein (Fig. 13-27, C and D). This channel provides an alternate route from the hind limbs to the heart. The connection has persisted in reptiles (Fig. 13-28). It may have been one of the factors responsible for the ultimate loss of the abdominal stream in mammals.

Snakes have no hind limbs, so the renal portal system is seen in its primitive relationships (Fig. 13-34); but in crocodilians, as has been mentioned, some blood passing from the hind limbs to the renal portal stream is able to bypass the kidney capillaries, going straight through the kidneys into the postcava; and in birds this has become the chief pathway. In mammals above monotremes the renal portal system disappears as an adult structure. However, transitory vascular connections from the posterior cardinals into the early embryonic kidney of placental mammals may represent an abortive renal portal system.

From the foregoing it can be seen that during phylogeny the posterior appendages (fins first, limbs later) have been drained by a series of vessels (Fig. 13-35): first the abdominal stream, then the renal portal, and finally the postcava. The venous realignment necessitated by the displacement of the caudal end of the nephrogenic mesoderm during formation of the adult mammalian kidney, discussed in Chapter 14, may have been one factor in the ultimate loss of the renal portal system in mammals.

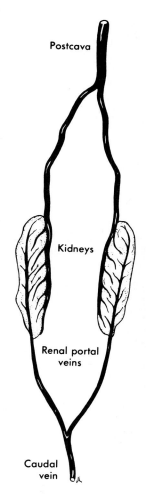

FIGURE 13-34

Renal portal system and venous drainage of the kidneys in a snake. The postcava commences at the confluence of right and left efferent renal veins.

FIGURE 13-35

Venous routes from paired fins and limbs, diagrammatic. Vessels represented by thin lines carry relatively less blood. *Blue* in a vessel indicates a portal stream. (*The abdominal stream in reptiles is paired.)

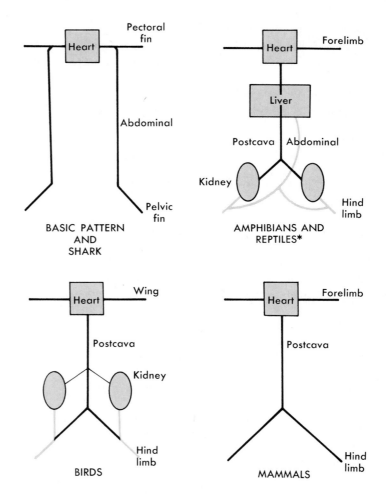

HEPATIC PORTAL SYSTEM. The hepatic portal system is similar in all vertebrates. It drains chiefly the stomach, pancreas, intestine, and spleen and terminates in the capillaries of the liver. Its origin in sharks from the embryonic vitelline and subintestinal veins has been described earlier. It arises in a similar manner in tetrapods. The abdominal stream (allantoic in birds, umbilical in mammals) becomes a tributary of the hepatic portal system, commencing with amphibians. Veins from the swim bladders are usually tributaries of the hepatic portal system in bony fishes.

CORONARY VEINS. Many amphibians seem to lack a definitive coronary system. In frogs, one coronary vessel **(vena bulbi anterior)** enters the left precava, and another **(vena bulbi posterior)** empties into the ventral abdominal vein near the liver. In reptiles, birds, and mammals a few coronary veins empty into the right atrium; most empty into the coronary sinus. The coronary sinus lies on the surface of the heart in the **coronary sulcus,** a groove between the left atrium and ventricle.

CIRCULATION IN THE MAMMALIAN FETUS, AND CHANGES AT BIRTH

In the mammalian fetus blood passes from the caudal end of the dorsal aorta into the umbilical arteries (Fig. 13-36). These extend out the umbilical cord to the placenta. From the placenta oxygenated blood returns to the fetus via an umbilical vein that traverses the falciform ligament to enter the liver. Some of this blood enters liver capillaries, but most of it continues nonstop via the ductus venosus, into the postcava, and finally into the right atrium. From the right atrium most of it passes via an **interatrial foramen** (**foramen ovale** of the heart) into the left atrium. The foramen is guarded by a one-way flaplike valve. The rest of the aerated blood, along with blood returning to the right atrium from the head, enters the right ventricle and is pumped into the pulmonary trunk. Because the ductus arteriosus is open and functional (Figs. 13-36 and 13-37), most of the blood in the pulmonary trunk is shunted into the dorsal aorta. This is an advantage, because blood in the ductus arteriosus is mostly unaerated and some of it will pass down the dorsal aorta to enter the umbilical arteries leading to the fetal respiratory membranes.

Blood coming from the lungs, which is unaerated and in small quantities, enters the left atrium and, along with the blood coming into the left atrium via the interatrial foramen, passes into the left ventricle. This blood

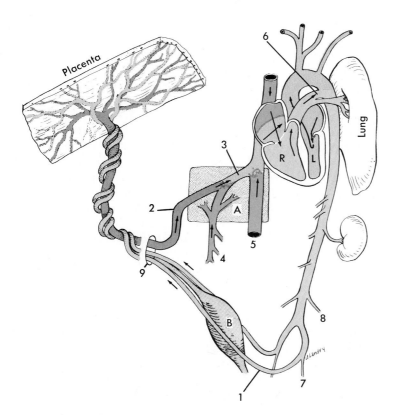

FIGURE 13-36

Circulation in the mammalian fetus. *1,* Umbilical artery; *2,* umbilical vein; *3,* ductus venosus; *4,* hepatic portal vein; *5,* inferior vena cava; *6,* ductus arteriosus; *7,* internal iliac, *8,* external iliac growing into hind limb bud; *9,* umbilicus; *A,* liver; *B,* base of the allantois, which is developing into a urinary bladder. After birth, the constricted distal part of the allantois *(gray)* becomes the urachus. *L,* Left ventricle; *R,* right ventricle. Much of the blood returning to the right atrium from the placenta passes through a foramen ovale (not illustrated) into the left atrium to be distributed via the left ventricle to the head and anterior limbs. *Darker red* indicates blood rich in oxygen; *darker blue,* low in oxygen; *light red,* mixed, considerable oxygen; *light blue,* mixed.

FIGURE 13-37

Heart, associated arteries, and ductus arteriosus *(arrow)* of fetal pig. The left auricle has been removed to show the pulmonary arteries. Blood in the dorsal aorta is mixed.

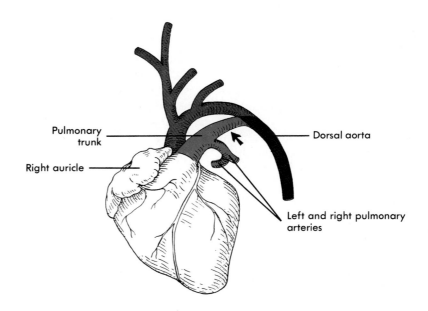

Pulmonary trunk

Right auricle

Dorsal aorta

Left and right pulmonary arteries

is then pumped into the ascending aorta. From this account it can be seen that the blood in the fetus is either venous (that is, lacking oxygen) or mixed, except in the umbilical vein. An essentially identical allantoic circulation occurs in unhatched chicks. The embryonic bird, however, depends on the vitelline (yolk sac) circulation for nourishment.

At birth major circulatory changes adapt the organism for pulmonary respiration:

1. The ductus arteriosus closes as a result of nerve impulses passing to its muscular wall. These impulses are initiated reflexly when the lungs are filled with air with the first gasp after delivery. In birds this is usually the day before hatching, when the imprisoned chick pecks a hole in its extraembryonic membranes and starts breathing the air entrapped between these membranes and the shell. When the chick inside the shell starts to peep, it already has air in its lungs! Shortly thereafter, all blood entering the pulmonary trunk goes to the lungs, and the ductus arteriosus of birds and mammals becomes converted into an **arterial ligament (ligamentum arteriosum).**

2. The flaplike interatrial valve is pressed against the interatrial foramen by the sudden increase in pressure in the left atrium that results from the greatly increased volume of blood entering from the lungs. This valve prevents the unoxygenated blood in the right atrium from entering the left atrium, which now contains only oxygenated blood from the lungs. Within a few days the foramen ovale is permanently sealed and only a scar, the **fossa ovalis,** remains.

3. At birth the umbilical arteries and vein are severed at the umbilicus. Thereafter, no blood passes through the umbilical arteries beyond the urinary bladder, which it continues to supply. From bladder to navel the

umbilical arteries become converted into **lateral umbilical ligaments** in the free border of the ventral mesentery of the bladder.

4. Blood no longer flows through the umbilical vein, and this vessel becomes converted into the **round ligament of the liver.** At the same time, the ductus venosus is converted into the **ligamentum venosum.** (This occurs halfway through gestation in whales.) As a result of these changes the fetal mammal (and bird, too) is changed from an allantoic-respiring organism to one capable of breathing air.

Failure of the interatrial foramen to close or of the ductus arteriosus to fully constrict results in cyanosis (blueness of the skin) of the newborn, because blood continues to be shunted away from the lungs, resulting in insufficient oxygen in the blood.

LYMPHATIC SYSTEM

Lymph and lymph channels are found in all vertebrates. A lymphatic system consists of thin-walled **lymph channels, lymph** (a fluid in transit), **lymph hearts** as far up the phylogenetic scale as embryonic birds, **lymphoid masses** in the lining of the digestive tract and elsewhere (Fig. 11-21, *A*) and **lymph nodes** in birds and mammals (Figs. 13-38 and 13-39). In contrast to blood, lymph moves in one direction only: toward the heart. Lymph channels are seldom dissected in the anatomy laboratory because of their delicate and collapsible nature.

Either lymph spaces or discrete lymph vessels **(lymphatics)** penetrate most of the soft tissues of the vertebrate body other than the skeleton, liver, and nervous system. They commence as blind-ended lymph capillaries that collect interstitial fluids. Once inside the lymph capillaries, the fluid is colorless or pale yellow, and is called lymph. This fluid passes from one endothelial-lined channel to the next, and finally empties into a vein. Valves at these exits prevent the influx of venous blood into the lymph channels. A lymphatic network consisting of long, narrow, discrete tubular vessels with a modicum of smooth muscle in the walls is found only in birds and mammals. A series of valves line these tubes and help counteract the effect of gravity, especially in the limbs.

Lymphatics in intestinal villi of higher vertebrates collect fats absorbed from the small intestine after a meal (Fig. 11-21, *B*). If the meal was particularly fatty, the lymph in these vessels has a milky appearance. For this reason these specific lymphatics are called **lacteals.** The lymph within them is **chyle.** Certain lymphatics in cyclostomes, cartilaginous fishes, and even humans contain some red blood cells. The fluid in these vessels is **hemolymph.**

Lymph channels that drain the body wall, limbs, and tail of lower vertebrates empty into nearby veins at the base of the tail, in the trunk, or in the neck. Those draining viscera are often paired in lower vertebrates, but in mammals a single **thoracic duct** commences in a large abdominal lymph sinus, the **cisterna chyli,** and empties into the brachiocephalic or left subclavian vein, or into the external or internal jugular vein (Fig. 13-38). The

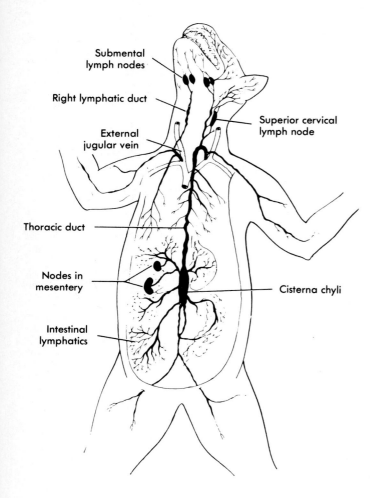

Submental
lymph nodes

Right lymphatic duct

External
jugular vein

Superior cervical
lymph node

Thoracic duct

Nodes in
mesentery

Intestinal
lymphatics

Cisterna chyli

FIGURE 13-38

A few of the superficial and
deep lymphatics of a mam-
mal.

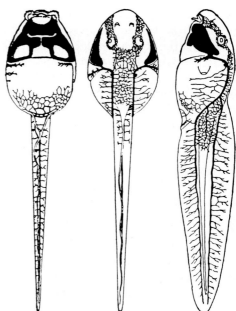

FIGURE 13-39

Superficial lymph channels
of a frog tadpole. (From
Kingsley, after Hoyer and
Udziela.)

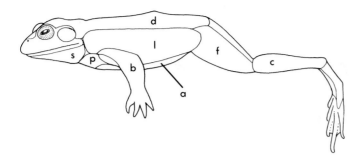

FIGURE 13-40

Subcutaneous lymph sacs of a frog. The outlines of the sacs are the sites of septa that extend between the skin and the musculature. *a*, Abdominal; *b*, lateral brachial; *c*, crural; *d*, dorsal; *f*, femoral; *l*, lateral; *p*, pectoral; *s*, submaxillary. (Modified from Gaupp.)

mammalian thoracic duct also receives lymphatics from the left side of the head and neck and from the left forelimb. One or more additional major lymphatics drain the right side of the body anteriorly.

The lymphatic system of anurans is remarkable because of the numerous wide lymph spaces tucked among the organs, into which lymph collects before draining into a nearby vein. These take the place of definitive lymphatics. They are especially well developed just under the skin of most of the body (Fig. 13-40). Because of these, the skin of a frog is easily removed from the underlying muscles; and anyone who has dissected a frog preserved in fluid has seen how puffy the skin becomes when the sacs are distended with fluids. The subcutaneous lymph spaces are separated from one another by connective tissue septa that attach the skin to the underlying muscles at these locations only.

The flow of lymph results from a number of factors, including lymph hearts situated at advantageous locations along lymph pathways in fishes, amphibians, and reptiles. These are pulsating sinusoidal swellings on lymph pathways with thin walls containing striated muscle fibers of unknown embryonic origin. They have been described from teleost fishes, but those of amphibians have been more thoroughly studied. Frogs have two pairs of lymph hearts, urodeles have as many as 16 pairs, and caecilians as many as 100 pairs. The anterior pair of lymph hearts in frogs, located just behind the transverse processes of the third vertebra, pump lymph into the vertebral veins, from which it is carried to the internal jugulars. A posterior lymph heart on each side of the urostyle empties into an external iliac vein. Amphibians, especially aquatic and semiaquatic amphibians, have more tissue fluids to manipulate than do other vertebrates, and as a result their lymph hearts move a proportionately larger volume of fluid hourly than the hearts of most other vertebrates. Semilunar valves at the exit of the hearts prevent backflow.

Lymph hearts are not present in birds after hatching, although embryonic birds have one on each side of the last sacral vertebra. None has been described in mammals. In these vertebrates lymph flow is maintained by activity of the skeletal muscles as they contract and relax, by movements of the viscera, and by rhythmical changes in intrathoracic pressure that result from breathing. Some of these factors are also operative in lower vertebrates.

Lymph nodes are masses of hemopoietic tissue interposed along the course of lymph channels of birds and mammals. They form when connective tissue capsules condense about lymphatic plexuses associated with strands of mesenchymal tissue. Nodes may be no larger than a pinhead, or they may be several centimeters in diameter. They are the "swollen glands" that can be palpated in the neck, axilla, and groin of humans when there is inflammation in the areas drained. Lymph enters a node via several afferent lymphatics, filters through the node, and leaves via a single large efferent lymphatic. The endothelium of their sinusoidal passageways includes many phagocytes that ingest bacteria and other foreign particles, and enmeshed in their reticulum are large numbers of small lymphocytes. The nodes are therefore the second line of defense against bacterial infections acquired through the skin, the first line being granulocytes that have emigrated from blood capillaries in the dermis.

Miscellaneous other lymphoid masses develop at one or another locations in the vertebrate body, particularly in the walls of the digestive tract. The largest of these is the **spleen,** which is absent in cyclostomes. Others include the **thymus** (not found thus far in Myxiniformes), **Peyer's patches,** which are indiscrete nodules in the mucosa and submucosa of the small intestine of amniotes, the unpaired **bursa of Fabricius** of very young birds, and mammalian **tonsils** and **adenoids.**

Chapter Summary

1. The circulatory system includes the blood vascular and lymphatic systems.

2. The blood vascular system consists of whole blood, the heart, arteries, capillaries, and veins. Whole blood consists of plasma, erythrocytes, leukocytes, and thrombocytes.

3. The lymphatic system consists of lymph, lymph channels, lymph hearts (absent in birds and mammals), lymph nodes in birds and mammals, and miscellaneous lymphoid masses. Lymph is transported from tissue spaces, and chyle from intestinal villi to major venous channels.

4. A sinus venosus occurs in fishes, amphibians, and reptiles. It is incorporated into the right atrial wall of adult birds and mammals. It receives venous blood from all parts of the body in fishes. It is the site of origin of cardiac rhythms.

5. A single atrium receives blood from the sinus venosus in most fishes. In lung-breathing vertebrates the atrium is partitioned into two chambers by a septum, which is incomplete in lungfishes and some urodeles, complete in other tetrapods. The right atrium receives the sinus venosus or the largest systemic veins, and in amniotes, coronary veins. The left atrium receives pulmonary veins.

6. A single ventricle occurs in fishes and amphibians. Lungfishes have an incomplete ventricular septum partially dividing the ventricle. Amniotes have two ventricles separated by an incomplete septum in most reptiles, a complete septum in crocodilians, birds, and mammals.

7. A conus arteriosus is the terminal chamber of the heart in fishes. In dipnoans and amphibians it is short, and is also called bulbus cordis. It is absent in adult amniotes.

8. A ventral aorta leads cephalad from the heart to the aortic arches. The initial unpaired segment is sometimes called truncus arteriosus. In teleosts and perennibranchiate urodeles the ventral aorta exhibits a swelling, the bulbus arteriosus.

9. Oxygenated and unoxygenated blood are kept separate in the hearts of dipnoans and tetrapods by partial or complete interatrial and interventricular septa, trabeculae, and spiral valves.

10. In most reptiles the truncus arteriosus is split longitudinally into two aortic trunks leading from the cavum venosum to the third and fourth arches, and a pulmonary trunk leading from the right ventricle to arch VI.

11. In crocodilians, birds, and mammals two trunks emerge from the heart: an aortic trunk (ascending aorta) supplying the carotid and systemic arches, and a pulmonary trunk supplying the pulmonary arteries.

12. An aortic arch is a blood vessel connecting ventral and dorsal aortae and located in a visceral arch. Typically, six pairs of aortic arches develop in each vertebrate embryo. During ontogeny the aortic arches are reduced in number, the highest vertebrates retaining the fewest. In fishes the aortic arches become interrupted by gill capillaries. In fishes and amphibians with external gills, detours from the aortic arches carry blood to the external gills. In lungfishes, two ganoids, and tetrapods the sixth aortic arch sprouts pulmonary arteries.

13. Aortic arch I usually disappears, in part at least. Elasmobranchs retain parts of aortic arches II to VI. Modern fishes and many terrestrial urodeles retain arches III to VI. Aquatic urodeles and amniotes retain parts or all of arches III, IV, and VI.

14. The dorsal aorta between aortic arches III and IV (ductus caroticus) disappears in some amphibians, most reptiles, and all birds and mammals. The dorsal segment of aortic arch VI (ductus arteriosus) disappears in anurans, most reptiles, and birds and mammals. Birds also lose the left fourth aortic arch, and mammals lose much of the right fourth. As a result of loss of the ductus caroticus, blood entering arch III (carotid) must continue to the head. As a result of loss of the ductus arteriosus, blood entering arch VI must pass to the lungs.

15. Birds and mammals lose arches I, II, V, the left or most of the right side of IV, the dorsal segment of VI on both sides, and the ductus caroticus.

16. As a result of modifications in the heart and ventral aorta and deletions in the aortic arches, a system of vessels appropriate for respiration via gills is translated phylogenetically and ontogenetically into one suitable for lung respiration.

17. There is little mixing of oxygenated and unoxygenated blood in any adult vertebrate. Mixing occurs in fetal birds and mammals.

18. Dorsal aortae in the trunk have segmental somatic branches, unpaired visceral branches to digestive organs in the coelom, and paired visceral branches to the urogenital system and adrenal glands.

19. Major primitive venous channels are anterior, posterior, and common cardinals, ventral abdominals, renal portals, hepatic portals, hepatic sinuses, and coronary veins. Dipnoans add postcavae, and tetrapods develop pulmonary veins. Subintestinals and vitellines are embryonic precursors of hepatic portals.

20. Anterior cardinals (internal jugulars) drain the head and empty into the common cardinals.

21. Postcardinals are absent in anurans, reptiles, and birds. They persist in mammals under new names (azygos, hemiazygos).

22. Abdominal veins drain pectoral and pelvic fins except in ray-finned fishes. They lose their connection with the forelimbs in tetrapods and with the hind limbs in birds and mammals. In mammals they remain as umbilical veins.

23. The renal portal system drains only the tail in fishes. It acquires a connection from the hind limbs in amphibians. In crocodilians and birds this connection partly bypasses the kidneys and goes directly to the postcava. There is no renal portal system in cyclostomes or adult mammals above monotremes.

24. The postcava becomes increasingly prominent in higher vertebrates. Commencing in dipnoans and amphibians as an alternate route to the heart from the kidneys, it finally drains hind limbs, most of the trunk, and tail.

25. Remnants of embryonic vascular channels in adult mammals include the round ligament of the liver (remnant of left umbilical vein), ligamentum venosum (remnant of ductus venosus), ligamentum arteriosum (remnant of left ductus arteriosus), lateral umbilical ligaments (remnants of paired umbilical arteries from urinary bladder to umbilicus), and fossa ovalis (occluded interatrial foramen ovale between the atria of the fetus).

26. Gill-breathing fishes have a single circulation as follows: heart, gills, body, heart. Birds and mammals have a double circulation: left side of heart, body (except lungs), right side of heart, lungs, left side of heart. Dipnoans, amphibians, and reptiles have a functionally double circulation with essentially no mixing of blood when environmental oxygen is adequate.

SELECTED READINGS

Berg, T., and Steen, J.B.: The mechanism of oxygen concentration in the swim bladder of the eel, Journal of Physiology **195:**631, 1968.

Brodal, A., and Fänge, R., editors: The biology of *Myxine*, Oslo, 1963, Norway Universitetsforlaget.

Heatwole, H.: Adaptations of marine snakes, American Scientist **66**(5):594, 1978.

Johansen, K., and Hanson, D.: Functional anatomy of the hearts of lungfishes and amphibians, American Zoologist **8:**191, 1968.

Kampmeier, O.F.: Evolution and comparative morphology of the lymphatic system, Springfield, Ill., 1969, Charles C Thomas, Publisher.

Lawson, R.: The anatomy of the heart of *Hypogeophis restratus* (Amphibia, Apoda) and its possible mode of action, Journal of Zoology **149:**320, 1966.

Nandy, K., and Blair, C.B.: Double superior venae cavae with completely paired azygos veins, Anatomical Record **151:**1, 1965.

Ottaviani, G., and Tazzi, A.: The lymphatic system. In Gans, C., and Parsons, T.S., editors: Biology of the reptilia, vol. 6, New York, 1977, Academic Press.

Ruszynák, I., Földi, M., and Szabó, G.: Lymphatics and lymph circulation: physiology and pathology, ed. 2, New York, 1967, Pergamon Press.

Satchell, G.H.: Circulation in fishes, Cambridge, England, 1971, Cambridge University Press.

Simons, J.R.: The heart of the tuatara, *Sphenodon punctatus,* Journal of Zoology **146:**451, 1965.

Struthers, P.H.: The aortic arches and their derivatives in the embryo porcupine (*Erethizon dorsatus*), Journal of Morphology and Physiology **50:**361, 1930.

White, F.N.: Circulation. In Gans, C., and Dawson, W.R., editors: Biology of the reptilia, vol. 5, New York, 1976, Academic Press.

Yoffey, J.M., and Courtice, F.C.: Lymphatics, lymph, and lymphoid tissue, Cambridge, Massachusetts, 1956, Harvard University Press.

Zweifach, B.W.: The microcirculation of the blood. In Vertebrate structures and functions: readings from Scientific American, with Introduction by N.K. Wessells, San Francisco, 1955-1974, W.H. Freeman and Co., Publishers.

UROGENITAL SYSTEM

14

In this chapter we will find that the urinary and reproductive organs of all vertebrates, male and female alike, develop in accordance with a basic architectural pattern, and we'll learn how the pattern is modified phylogenetically and ontogenetically. We will look at how marine and desert animals conserve water, and how freshwater vertebrates get rid of it; and we will briefly examine extrarenal routes for elimination of excess salts. At the end we will see how a simple embryonic chamber, the cloaca, is partitioned in male and most female mammals to produce one passageway for urine and reproductive products and a second for the digestive tract; and how, in female rodents and primates a third passageway forms solely for the reproductive tract.

Although the function of kidneys is different from that of gonads, the urinary and genital systems of gnathostomes are so intimately related, both developmentally and structurally, that neither can be discussed without reference to the other. For this reason, urinary and genital organs are discussed in a single chapter. The term *urogenital* is a combination of the Greek combining form *ouro-*, pertaining to urine, and an English derivative of the Latin word *genitalis*, which pertains to reproduction.

KIDNEYS AND THEIR DUCTS

Vertebrate life began in the water, and the early stages of the evolution of kidneys took place in that medium. Their function was osmoregulatory. They maintained the appropriate osmotic concentration of the blood by eliminating excess water, if any, by preventing the escape of water when necessary, and by regulating the excretion of certain salts. In the latter role they had the assistance of the gills. These same roles are performed by the kidneys of fishes today; and additional roles, including excretion of nitrogenous wastes, have been acquired by terrestrial vertebrates.

Basic Plan and the Archinephros

Vertebrate kidneys, or nephroi, are all built in accordance with a basic structural pattern consisting of (1) **glomeruli,** usually incorporated in renal corpuscles (Fig. 14-1, *A, 5*); (2) **tubules,** surrounded by peritubular capillaries (Fig. 14-1, *A* and *B*); and (3) **a pair of longitudinal ducts** (Fig. 14-1). Variations in the details from fish to humans are principally in the number and arrangement of glomeruli and in the relative length of the tubules.

Glomeruli are tufts of arterial capillaries where water, ions, and certain other substances are removed from the bloodstream. In all vertebrates they

FIGURE 14-1

Basic structure of vertebrate kidney. **A,** Functional renal unit, or nephron. *1,* Afferent glomerular arteriole; *2,* efferent glomerular arteriole; *3,* vessel from renal portal system; *4,* tributary of renal vein; *5,* Renal corpuscle. Bowman's capsule has been opened to show the glomerulus. **B,** Primary, secondary, and tertiary tubules in a single body segment. **C,** The increased number of tubules per segment disrupts the metamerism of the kidney.

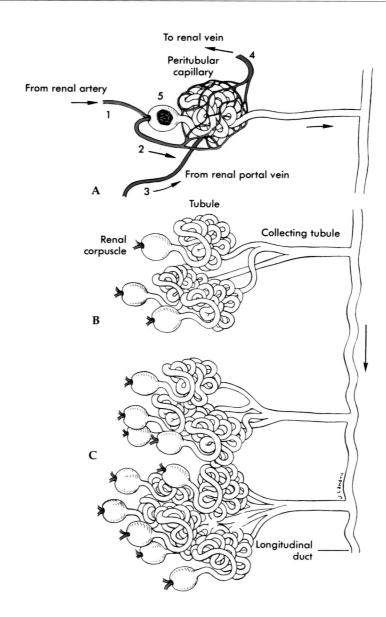

FIGURE 14-2

A, External glomerulus suspended in coelom. **B,** Internal glomerulus surrounded by Bowman's capsule to form a renal corpuscle.

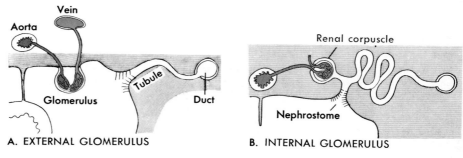

A. EXTERNAL GLOMERULUS

B. INTERNAL GLOMERULUS

are the chief site for removal of water. In some lower vertebrates they are large enough to be seen with the naked eye or a hand lens; in others they are microscopic. The most primitive glomeruli are suspended in the coelom and hence are called **external glomeruli** (Fig. 14-2, *A*), as opposed to **internal glomeruli** (Fig. 14-2, *B*), which are encapsulated by part of a kidney tubule to form a renal corpuscle. External glomeruli in today's vertebrates are confined to embryos and larvae. Supplying a glomerulus is an **afferent glomerular arteriole,** and emerging from it is an **efferent glomerular arteriole** that leads to a peritubular capillary bed (Fig. 14-1, *A*). Venules drain the peritubular capillaries and lead to renal veins. Only cyclostomes lack peritubular capillaries, but as compensation the kidney ducts have an unusually rich capillary bed that takes their place functionally.

Kidney tubules are microscopic passageways that collect glomerular filtrate, or when associated with an external glomerulus, coelomic fluid, and conduct it to a longitudinal duct. They increase in complexity among the vertebrate classes, being short and straight in cyclostomes, longest and most tortuous in mammals. Typical tubules commence as **Bowman's capsule,** a modified portion of the tubule wall surrounding a glomerulus. The capsule and glomerulus together constitute a **renal corpuscle.** A corpuscle, tubule, and the associated peritubular capillaries constitute a **nephron,** the functional unit of gnathostome kidneys.

Glomeruli function much like a simple filter, which under the force of blood pressure allows water and a variety of small molecules—smaller than blood proteins—to be expressed through its walls and into the lumen of the tubule. Much of this glomerular filtrate, which includes among other substances, glucose, certain salts,* and water, may not be excess; and components of the filtrate are selectively reclaimed during its passage through specific segments of the tubule (Fig. 14-3). As a result the filtrate becomes altered in passage. All of the glucose, for example, is ordinarily reclaimed. It contains a miniscule amount of the energy of the universe, it is not easily acquired, and except in pathological conditions, is never in excess, because once acquired it is quickly stored in the liver. Water and certain salts may or may not be in excess, depending on the environment. When in excess, water is allowed to continue throughout the length of the kidney tubules and to enter the kidney duct without being intercepted. Under conditions that tend to lead to dehydration, water is reclaimed under the influence of an antidiuretic hormone from the posterior lobe of the pituitary gland. The elimination or recapture of certain salts is also hormone controlled. The result of these exchanges between filtrate and the lining cells of the tubule is a final product, urine, which varies in constituents and in the concentration of ions from species to species, and may vary in a single individual from hour to hour, if not from minute to minute.

The more anterior tubules in a few adult fishes and in many embryos and larvae have a ciliated funnel, or **nephrostome,** that opens into the coelom (Fig. 14-2, *B*). Nephrostomes may be vestiges of a hypothetical ancestral

---

*Tubules in most fishes excrete largely Mg^{++}, Ca^{++}, SO_4^-, and phosphate. Most of the Na^+ and Cl^- is excreted extrarenally.

kidney, or **archinephros,** in which all glomeruli were external, all tubules short and segmental, and which extended the length of the coelom (Fig. 14-4). The glomeruli would have secreted into the coelomic fluid, which would have been swept by cilia into the nephrostome, just as in some annelids and in the protonephridia of an amphioxus. The nearest approach to such a kidney is seen in larval hagfishes and apodans, in which it is sometimes referred to as a **holonephros.**

Kidney tubules differentiate from a ribbon of embryonic nephrogenic (intermediate) mesoderm that lies beside the mesodermal somites and

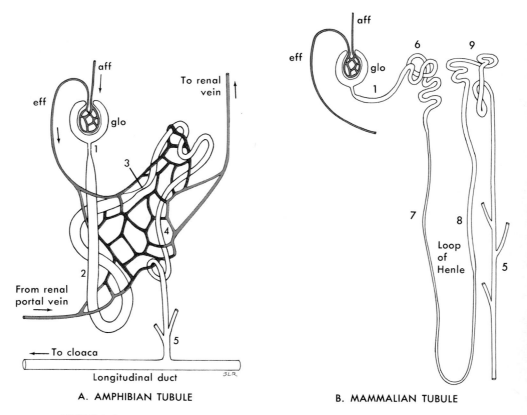

A. AMPHIBIAN TUBULE   **B. MAMMALIAN TUBULE**

FIGURE 14-3

Kidney tubules. **A,** Mesonephric tubule of an aquatic urodele showing associated vessels. **B,** Mammalian tubule, peritubular capillaries removed to reveal the segments. Note smaller diameter of segment 7. The segments are *1,* neck; *2,* proximal segment; *3,* intermediate segment; *4,* distal segment; *5,* collecting tubule; *6,* proximal convoluted tubule; *7,* descending arm of loop of Henle; *8,* ascending arm; *9,* distal convoluted tubule. The mammalian tubule is characterized by the loop of Henle. *aff,* Afferent glomerular arteriole; *eff,* efferent glomerular arteriole; *glo,* renal corpuscle consisting of a glomerulus (*red*) and Bowman's capsule. Dark vessels in **A** indicate mixing of arterial and venous blood in peritubular capillaries. The mammalian tubule is more than twice as long as that of the amphibian; and in placental mammals, the peritubular capillaries receive no portal blood.

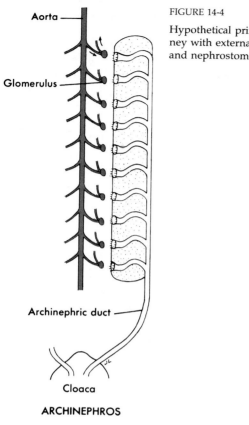

Aorta

Glomerulus

Archinephric duct

Cloaca

ARCHINEPHROS

FIGURE 14-4

Hypothetical primitive kidney with external glomeruli and nephrostomes.

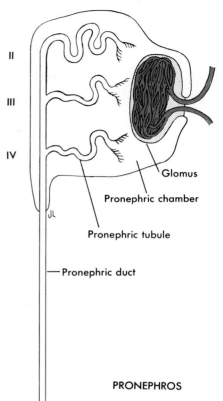

II

III

IV

Glomus

Pronephric chamber

Pronephric tubule

Pronephric duct

PRONEPHROS

FIGURE 14-5

Pronephric kidney of a 15 mm larval frog. *II, III,* and *IV,* Location of the second, third, and fourth somites. The glomus is three fused external glomeruli. The next tubule to form will be a mesonephric tubule at the level of somite VII.

extends the length of the embryonic trunk from immediately behind the head to the cloaca (Figs. 1-12 and 15-6). The earliest tubules appear at the anterior end of the ribbon, and additional tubules are added as the embryonic trunk elongates. The paired longitudinal ducts commence development as caudally directed extensions of the first tubules, and each duct grows caudad until it opens into the cloaca (Fig. 14-5). The ducts participate in the induction of additional tubules as the body elongates.

Kidneys and the Environment

Vertebrates that are submerged in fresh water inevitably acquire water by absorbing it through the gills, oropharyngeal membranes, or skin and by unavoidably swallowing it with their food. Salts, on the other hand, are scarce in fresh water, the chief source being food. Therefore one osmoregulatory necessity for freshwater organisms is to *excrete water and conserve salt*. Animals that are submerged in salt water face a different osmotic problem. Instead of accumulating too much water, they are in danger of accumulating too much salt. Survival in salt water depends on *conserving water and excreting salt*. Whether kidneys evolved in fresh water or salt water is not settled. Advocates of a freshwater origin point to the lack of ostracoderm fossils in early marine deposits. Proponents of the saltwater origin are convinced ostracoderm fossils exist in marine deposits and that the reason they have not been discovered is that species were not abundant when these sediments were being formed. Stahl presents both sides of the debate, meanwhile pointing out that negative evidence is not conclusive. Regardless of which view is correct, the archinephros exhibited by these early species was such that, when modified, it was able to maintain later vertebrates in salt water, in fresh water, on land, and even in a desert. By varying the size and number of glomeruli in species in different habitats, water can be excreted abundantly or sparingly; and by varying the length of specific segments of kidney tubules, either salt or water, as necessary, can be recovered from the filtrate, or salt can be excreted abundantly. In general, glomeruli are larger in freshwater fishes and aquatic amphibians, smaller in marine fishes and tetrapods, especially tetrapods living in arid environments. (Elasmobranchs are an exception. Freshwater and marine sharks have very large renal corpuscles.)

Some marine teleosts have renal corpuscles that are poorly vascularized, cystic, or vestigial. In some others the distal segments of their tubules are abbreviated or lost. Loss of glomeruli results in increased water retention, and shortening of the distal segment of the tubules eliminates a major site for resorption of salts and therefore increased salt excretion, both of which have survival value in salt water. There are, however, freshwater teleosts with the same mutations. In these, water excretion is principally tubular, and they seem to compensate for any excessive salt loss in urine by active uptake of salt via their gills. There is reason to believe that these are former saltwater species that adapted to a freshwater habitat. In fishes that migrate between salt and fresh water, cyclically secreted hormones regulate the kidney tubules as well as extrarenal sites of absorption and excretion of water and salts. In tetrapods, dehydration is the principal stimulus for release of hormones that reclaim water from glomerular filtrate.

Thus far nothing has been said about excretion of nitrogenous wastes, because in no fish is the kidney of primary importance in excretion of nitrogenous compounds. These are eliminated predominately through gills and sometimes via the skin in fishes. In tetrapods the kidneys dispose of nitrogenous wastes. Nitrogen is excreted chiefly as ammonia in freshwater teleosts and aquatic amphibians (ammonotelic animals), because ammonia is highly soluble in water, and water is plentiful. It is also excreted as ammonia by most marine fishes other than elasmobranchs. The solvent in these species is provided by drinking sea water and rapidly eliminating the salt. Nitrogen is excreted as urea by elasmobranchs and mammals (ureotelic animals) and as uric acid in a semisolid urine when water is scarce, as in terrestrial reptiles and birds (uricotelic animals).

Pronephros

The earliest embryonic tubules arise from the anterior end of the nephrogenic mesoderm. Because they are the first to appear and are anteriorly located, they are called **pronephric tubules.** They are segmentally arranged, one opposite each of the more anterior somites (Fig. 14-5).

Each pronephric tubule arises as a solid bud of cells that later organizes a lumen and, except in birds and mammals, a nephrostome. Associated with each pronephric tubule may be a glomerulus. The number of pronephric tubules is never large—three in larval frogs; seven in human embryos, opposite somites VII to XIII; and about 12 in chicks, commencing at somite V. The tubules lengthen and become coiled. The region of the nephrogenic mesoderm having these metameric tubules is the **pronephros,** and its duct is the **pronephric duct.**

Pronephric tubules are temporary. They function only until the tubules farther back are able to work. This is at the end of the larval stage in amphibians and at an equivalent developmental stage in fishes. At that time the glomeruli lose their connection with the dorsal aorta and commence to regress. The tubules regress more slowly, and traces may remain in adult fishes. However, *the pronephric duct does not regress.* It continues to drain the tubules farther back. Although a pronephros always develops in amniotes, it is doubtful that it ever functions as an excretory organ in these vertebrates.

Mesonephros

Under the partial stimulus of the pronephric duct acting as an inductor, additional tubules develop sequentially in the nephrogenic mesoderm behind the pronephric region. The new tubules establish connections with the existing pronephric duct. For at least several segments these tubules, too, may be segmentally disposed, exhibit the same convolutions as those anterior to them, and often have open nephrostomes. In fact, there is seldom justification for drawing at any specific point a boundary between the embryonic pronephros and the rest of the embryonic kidney. There is usually a gradual transition from tubules characteristic of the pronephric region to those found farther back.

In the transitional area secondary and tertiary tubules develop as buds from the initial (primary) tubule in each segment (Fig. 14-1, *B*). As these additional tubules enlarge and encroach on one another, the metamerism of the developing kidney is at first obscured, and then lost altogether. Another feature of the transitional region is the development of internal glomeruli and of tubules that are longer and more convoluted and lack nephrostomes. However, many embryonic fishes and amphibians develop nephrostomes for a long distance back, and some fishes retain a few throughout life.

With the disappearance of pronephric tubules *the pronephric duct is thereafter called the **mesonephric duct*** and the new kidney region it drains is the

FIGURE 14-6

Fate of the nephrogenic mesoderm in representative vertebrates. The pronephric duct of the embryo persists in adult anamniotes to drain the adult kidney.

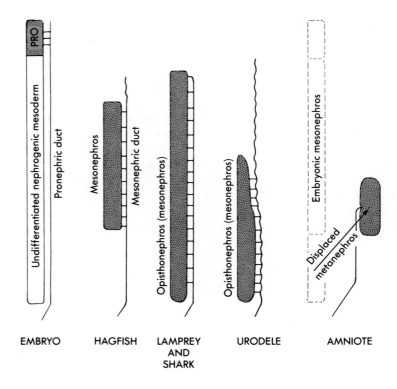

EMBRYO   **HAGFISH**   **LAMPREY AND SHARK**   **URODELE**   **AMNIOTE**

mesonephros (Fig. 14-6). It is the adult kidney of cyclostomes, jawed fishes, and amphibians, and the functional embryonic kidney of amniotes (Fig. 14-9). When the mesonephros serves as an adult kidney it is sometimes called an **opisthonephros** because it extends all the way to the cloaca.

The mesonephros of the adult hagfish *Myxine* is an expression of the embryonic holonephros. The glomerular part occupies a 10 cm segment of the nephrogenic mesoderm, commencing some distance behind the regressed pronephros and terminating some distance anterior to the cloaca. This segment consists of 30 to 35 large renal corpuscles up to 1.5 mm in diameter, strictly segmental, and connected to the mesonephric duct by very short tubules. There are no peritubular capillaries and no detectable renal portal system. Between the regressed pronephros and the glomerular segment are a variable number of corpuscles that lack glomeruli or have lost their connection with the longitudinal duct. Caudal to the glomerular segment are additional aglomerular corpuscles. The total number of corpuscles, typical and atypical, is about 70, and they are strictly segmental.

The adult kidneys of fishes and amphibians commence varying distances behind the pronephric region, depending on how many anterior embryonic tubules disappear. In lampreys, sharks, and caecilians the kidneys begin far forward and extend the length of the coelom (Fig. 14-6, lamprey and shark). In other fishes and amphibians a longer series of transitional tubules usually disappears (Fig. 14-6, urodele). In male fishes and amphibians the anteriormost tubules of the mesonephros usually have no connection with glomeruli and, instead, conduct sperm from the testis to the mesonephric duct. This part of the male mesonephros is the **sexual,** or **epididymal, kidney,** and the coiled part of the mesonephric duct that drains it is the **epididymis** (Figs. 14-7 and 14-8). The corresponding part of the mesonephroi of females is often nonfunctional.

The mesonephric ducts in some species of male sharks and some male salamanders are preempted for sperm transport and carry very little urine. In sharks an accessory urinary duct carries more or less of the urine depending on the species (Fig. 14-23, *B*). Kidney drainage in two contrasting families of urodeles is illustrated in Fig. 14-8.

The mesonephros is the functional kidney of amniotes during part of embryonic or fetal life (Fig. 14-9). In embryonic chicks it reaches its peak on the eleventh day of incubation—halfway through embryonic life. In placental mammals it peaks earlier—at 9 weeks of gestation in humans. The first mesonephric tubules of human embryos appear after 4 weeks of

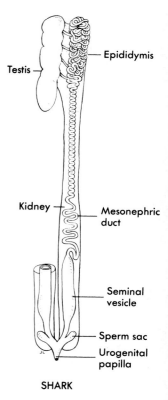

SHARK

FIGURE 14-7

Urogenital system of a male shark. An accessory urinary duct, not shown, drains the more posterior kidney tubules. Within the urogenital papilla there is a urogenital sinus.

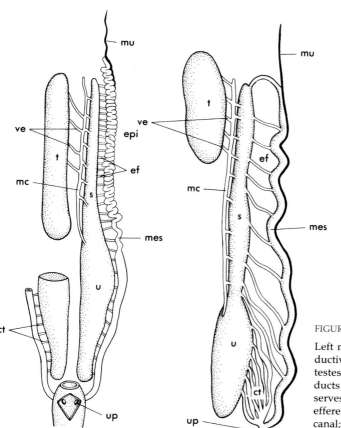

A. NECTURUS

B. AMBYSTOMA

FIGURE 14-8

Left mesonephroi (opisthonephroi) and associated reproductive structures of two male urodeles, ventral view. The testes have been displaced mediad and the mesonephric ducts laterad. The sexual portion of the mesonephros serves solely for sperm transport. *ct,* Collecting tubules; *ef,* efferent epididymal ducts; *epi,* epididymis; *mc,* marginal canal; *mes,* mesonephric duct; *mu,* rudimentary muellerian duct; *s,* sexual kidney; *t,* testis; *u,* uriniferous kidney; *up,* urogenital papilla opening into cloaca; *ve,* vasa efferentia.

FIGURE 14-9

Sagittal section of a 20 mm pig embryo. The mesonephros extends about half the length of the body. (From Phillips, J.B.: Development of vertebrate anatomy, St. Louis, 1975, The C.V. Mosby Co.)

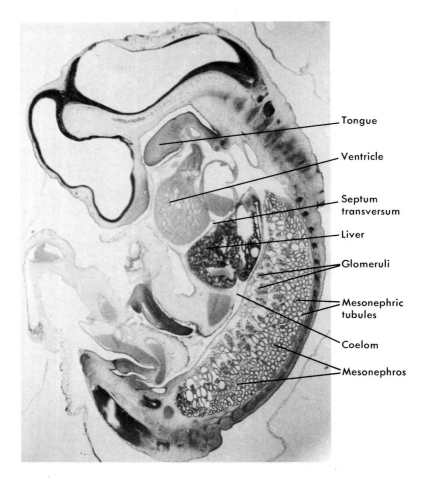

Tongue

Ventricle

Septum transversum

Liver

Glomeruli

Mesonephric tubules

Coelom

Mesonephros

embryonic life. A wave of differentiation sweeps along the nephrogenic mesoderm, but before the last mesonephric tubules have formed the earliest ones have already regressed (Fig. 14-10). As a result the human mesonephros at its peak consists of about 30 functioning renal corpuscles, although as many as 80 have been known to form. During the time the mesonephros is functioning in amniotes a new kidney, the metanephros, is organizing. When the metanephros is able to function, the mesonephros disappears except for remnants. However, it functions as late as the first hibernation in some lizards and the first molt in some snakes; and it is still functioning in monotremes and marsupials at hatching or birth.

MESONEPHRIC REMNANTS IN ADULT AMNIOTES. Vestiges of the mesonephros remain in adult mammals as two groups of blind tubules, the **epoophoron** and **paroophoron,** in the dorsal mesentery of the ovary (Fig. 14-11), and the **paradidymis** and **appendix of the epididymis** in the dorsal mesentery of the testis, both near the epididymis (Fig. 14-24). The mesonephric ducts remain as sperm ducts in all male amniotes. In females,

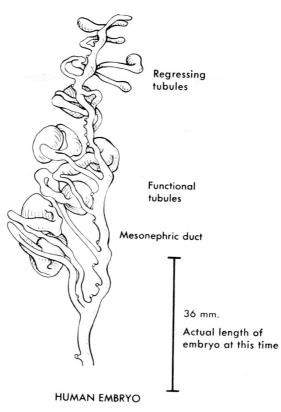

FIGURE 14-10

Right functional mesonephros of a 36 mm human embryo. (After Altschule.)

Regressing tubules

Functional tubules

Mesonephric duct

36 mm.

Actual length of embryo at this time

HUMAN EMBRYO

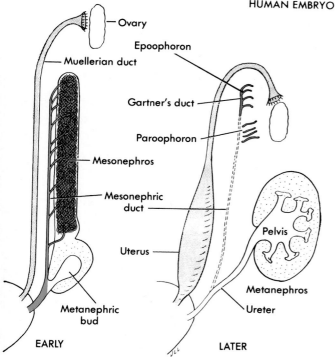

— Ovary

Epoophoron

— Muellerian duct

Gartner's duct —

Paroophoron —

— Mesonephros

— Mesonephric duct

Pelvis

Uterus —

Metanephros

Ureter

Metanephric bud

EARLY   **LATER**

FIGURE 14-11

Developmental changes in the urogenital system of a female amniote. In the early stage *(left)* the mesonephric kidney and duct are present *(red),* and the metanephric bud has formed at the caudal end of the nephrogenic mesoderm. A muellerian duct primordium is present. In the later stage *(right)* the mesonephric kidney and duct have regressed except for remnants *(red),* and the muellerian duct has differentiated to form a female reproductive tract.

mesonephric duct remnants include a short, blind **Gartner's duct (ductus deferens femininus)** in the mesentery of each mammalian oviduct (Fig. 14-11), and blind vestiges near the ovaries of some lower amniotes. **Vasa efferentia ovarii,** homologous with male vasa efferentia, are found in the ovaries of most female mammals.

Metanephros

The metanephros organizes from the caudal end of the nephrogenic mesoderm, which becomes displaced cephalad and lateral during development (Figs. 14-6, amniote, and 14-11). This is the same mesoderm that gives rise to the caudal part of the mesonephros of fishes and amphibians. The number of tubules that form from this caudal section in amniotes is extremely large—up to an estimated 4.5 million. Differentiation of the metanephric kidney begins when a metanephric bud sprouts from the caudal end of the mesonephric duct (Fig. 14-11). Surrounding the bud is nephrogenic mesoderm. The bud grows cephalad, carrying the metanephric blastema along with it. Eventually it gives rise to the metanephric duct, or **ureter,** and the **pelvis of the kidney** (Fig. 14-11, later). From the pelvis many fingerlike outgrowths invade the surrounding kidney blastema and become collecting tubules (Fig. 14-3, *B, 5*). Meanwhile, S-shaped tubules are organizing within the blastema. One end of each tubule grows toward and encapsulates a glomerulus to form a renal corpuscle; the other end grows toward, and finally empties into, a collecting tubule.

FIGURE 14-12

Mammalian kidney, frontal section. The renal vein and its tributaries have been removed. Glomeruli are confined to the cortex *(red).* Loops of Henle and common collecting tubules form the bulk of the medulla. Some mammalian kidneys have only one pyramid. *P,* pelvis.

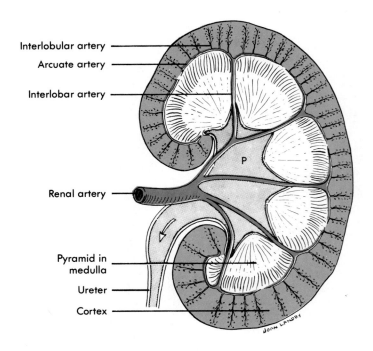

Interlobular artery

Arcuate artery

Interlobar artery

Renal artery

Pyramid in medulla

Ureter

Cortex

P

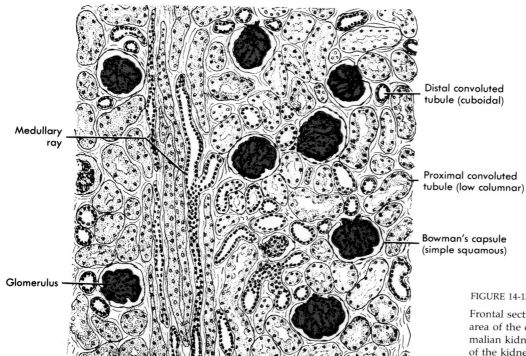

Distal convoluted
tubule (cuboidal)

Medullary
ray

Proximal convoluted
tubule (low columnar)

Bowman's capsule
(simple squamous)

Glomerulus

FIGURE 14-13

Frontal section of a small
area of the cortex of a mammalian kidney. The surface
of the kidney is near the top.
Medullary rays consist of collecting tubules and loops of
Henle that extend downward
into the medulla. (From Bevelander, G., and Ramaley,
J.A.: Essentials of histology,
ed. 8, St. Louis, 1979, The
C.V. Mosby Co.)

The tubules of mammalian kidneys have a long, thin, U-shaped **loop of Henle** (Fig. 14-3, *B*) interposed between proximal and distal convolutions. As the loops of the Henle elongate, they grow away from the surface of the kidney where the glomeruli are located and toward the renal pelvis. The kidney therefore has a **cortex,** containing renal corpuscles, and a **medulla,** consisting of hundreds of thousands of loops of Henle and common collecting tubules (Figs. 14-12 and 14-13). The loops and collecting tubules are aggregated into a number of conical **pyramids,** depending on the species, and give the medulla a striated appearance in frontal section (Fig. 14-12). Each pyramid—as many as 20 in some mammalian kidneys—tapers to a blunt tip, the **renal papilla,** that projects into a funnel-shaped outpocketing **(calyx)** of the pelvis. Each collecting tubule drains a small number of metanephric tubules (7 to 10 in humans), and empties into the pelvis at the tip of a papilla.

Blood in the straight capillaries **(vasa recta)** that parallel the two arms of each loop of Henle flows in a direction opposite that of the glomerular filtrate, another instance of countercurrent flow and its functional ramifications. The filtrate enters the proximal convoluted tubule at a given concentration of ions, depending on the species and the physiological state of the organism. In the descending arm the solute concentration normally

FIGURE 14-14

Lobulated metanephric kidneys.

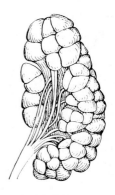

LIZARD   NEWBORN HUMAN   ADULT OTTER

increases progressively toward the bottom of the loop as water moves osmotically from the filtrate into the epithelial cells of the tubule and then into ascending vasa recta. In the ascending arm, under nondehydration conditions, the epithelial cells of the tubule are impermeable to water and the filtrate in that arm becomes increasingly hypotonic to the tissue fluid as a result of *active transport* of sodium out of the filtrate under the influence of the sodium-regulating hormone aldosterone. Consequently, a dilute urine of relatively high volume (depending on the expendibility of water) is excreted by a terrestrial mammal. When dehydration threatens, antidiuretic hormone enters the circulation; the epithelial cells of the ascending arm, distal convoluted tubule, and collecting duct become highly permeable to water; water enters them from the glomerular filtrate as a result of a favorable osmotic gradient; and the glomerular filtrate beyond the base of the loop of Henle becomes increasingly hypertonic as a result of the extraction of water. Under these conditions the excreted urine is hypertonic, and the volume is small. The physics and chemistry of the process of urine formation from glomerular filtrate is a complex phenomenon the details of which are beyond the scope of this text.

Reptilian, avian, and some mammalian kidneys are lobulated, each lobe consisting of clusters of many tubules (Fig. 14-14). The kidneys of snakes, legless lizards, and apodans are elongated, conforming to the slender body. Those of birds lie snugly against the contours of the sacrum and ilium. In most mammals the kidneys are smooth and bean shaped, and the arteries, veins, nerves, and ureters enter and leave at a median notch, the **hilum.**

Because of their embryonic origin as buds off the mesonephric ducts, the ureters at first terminate in these ducts, and they do so throughout life in male sphenodons and lizards (Fig. 14-25). In other reptiles and in birds and monotremes they ultimately open into the cloaca. In placental mammals the ureters open into the urinary bladder because of differential growth of components of the embryonic cloaca, to be explained later (Figs. 14-29 and 14-30).

Extrarenal Salt Excretion in Vertebrates

Vertebrates that live in an environment laden with salt, or that inhabit arid environments and cannot afford much body water to carry off accumulated salts, have extrarenal structures for salt excretion. Marine fishes have chloride-secreting glands on the gills, and elasmobranchs have rectal glands that perform this function. (The rectal glands of bullsharks caught in fresh water are smaller than those caught in salt water, and show regressive changes.) Marine reptiles and birds that scoop fish out of salt water have nasal glands that excrete salt. So do terrestrial lizards and snakes that live in arid habitats. These same lizards and snakes have atrophied glomeruli, which also conserves water.

The salt-secreting nasal glands of lizards are located outside the olfactory capsule and empty into the nasal canals via small ducts. Whitish incrustations of sodium chloride and potassium can be seen in the nasal canal or at the nostrils.

The nasal gland of marine birds is a large paired gland located above the orbit. It is drained by a long duct that opens close to a nostril. A groove extends from the opening to the tip of the beak. Within 15 minutes after these birds have drunk water containing sodium chloride and potassium, minute drops of fluid containing these salts trickle down the groove and drip or are shaken off the beak.

Sweat glands eliminate some salt in mammals, but salt loss by this route is incidental to secretion of water for its evaporative cooling effect. Whereas the excretion of electrolytes via most routes is regulated by hormones, their loss via sweat glands is unregulated. In fact, salt lost by this route usually must be replaced.

URINARY BLADDERS

Most tetrapods have a urinary bladder that serves as a storage site for urine before it is voided. Its chief survival value for terrestrial vertebrates seems to reside in the fact that it is a reservoir of water that may be needed later, hence should not be wasted. Infrequently, in one species or another, the fluid is put to another use. Antidiuretic hormone from the pituitary gland, the release of which is evoked when dehydration threatens, causes active water resorption from the bladder, after which the water is recycled. Most fishes have a small unpaired or bicornuate swelling at the end of the urinary pathway that is called a urinary bladder; however, these appear to be inconsequential as holding sites for urine. In some lower vertebrates the bladder may serve as a recovery site for certain essential ions that are scarce in the environment.

Fishes

The urinary bladder in most fishes, when present, is an insignificant terminal enlargement of the conjoined caudal ends of the urinary ducts. The urogenital sinus of male sharks and the urinary sinus of females, both

FIGURE 14-15

Caudal end of urogenital system of a male teleost (pike), left lateral view. The unpaired urinary bladder arises as a bud off the conjoined caudal ends of the two mesonephric ducts. Note absence of cloaca. (After Goodrich.)

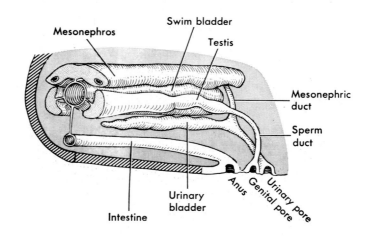

TELEOST

of which lie within a urogenital or urinary papilla, are the closest structures to a urinary bladder in sharks, but the sinuses are so small as to be useless as a *reservoir* for urine (Figs. 14-7 and 14-32). Homologous sinuses in Rajiformes, sometimes called urinary bladders, are only slightly larger, and the same thing may be said of the swellings at the conjoined ends of the urinary ducts of lower actinopterygians. The latter have been called **tubal bladders** in recognition of their location. They are frequently larger in females than in males. Dipnoans have a small diverticulum of the dorsal wall of the cloaca that is called a urinary bladder.

The most conspicuous structures bearing the name urinary bladder in fishes are vesicles that arise ontogenetically as evaginations from the caudal end of the conjoined embryonic mesonephric ducts, as in pikes, which are primitive teleosts (Fig. 14-15). Urine must back up into these, as it does into amphibian bladders.

There is no *apparent* survival value in the presence of a urinary bladder in most fishes for water conservation, inasmuch as freshwater fishes are immersed in water and, as has been pointed out, many marine fishes have the capacity to extract fresh water from sea water by drinking sea water and quickly excreting the salts. Cyclostomes have no structure that can be called a urinary bladder.

Tetrapods

The urinary bladders of amphibians through mammals arise during organogenesis as an evagination of the embryonic cloaca, and they empty into the cloaca in adults, except in placental mammals (Figs. 14-16, 14-25, 14-26, and 14-31). They are lacking in crocodilians, snakes, some lizards, and in birds other than ostriches. Their absence in birds reduces the energy requirements for flight.

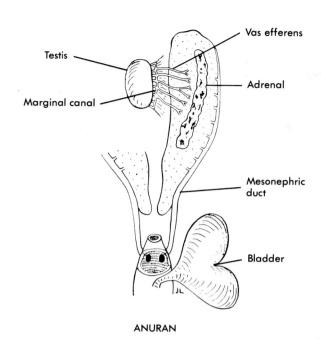

Testis

Marginal canal

Vas efferens

Adrenal

Mesonephric duct

Bladder

ANURAN

FIGURE 14-16

Urogenital system and adrenal of a male frog, ventral view.

In amniote embryos the cloacal evagination that gives rise to the bladder is prolonged beyond the ventral body wall as an extraembryonic membrane, the allantois (Fig. 4-11, *B*). Only the base of the allantois—the part proximal to the cloaca—contributes to the adult bladder (Fig. 13-36). After birth the part of the allantois within the coelom and distal to the bladder remains in mammals as a **urachus (middle umbilical ligament)** connecting the tip of the bladder with the umbilicus. The urachus, which is more prominent in some species (primates, for instance) than in others (cats, for instance) lies in the anterior border of the ventral mesentery of the bladder along with the obliterated umbilical arteries.

Anurans, turtles, and some lizards have unusually large bladders, and some freshwater turtles have **accessory urinary bladders** (Fig. 14-26). The latter are used by females to carry water for softening and moistening the soil when a nest for the eggs is being prepared; if they have an additional function it has not been demonstrated. Because the kidney ducts of amphibians, reptiles, and monotremes empty into the cloaca, urine in these tetrapods collects in the cloaca and then backs up or is forced into the bladder while the sphincter at the vent is closed. In placental mammals the kidney ducts empty directly into the bladder, and the bladder is drained by the urethra. More will be said about the urethra shortly.

The smooth muscle in the wall of the tetrapod bladder is disposed in a reverse pattern to that of the muscular coat of the digestive tract, there being an inner longitudinal and outer circular layer. The size of the lumen at all times is only sufficient to accommodate the quantity of contained urine. As an adaptation, the muscle fibers are interspersed with a larger than usual amount of loose connective tissue, which facilitates adjustments in the thickness of the bladder wall as the bladder fills or empties. Also, the cells of the epithelial lining possess, to an unusual degree, the ability to change their position by sliding over each other, thinning out as the bladder fills, and heaping as it empties. As a result, there is generally little unfilled space in the bladder.

GONADS

The embryonic gonads arise as a pair of elevated **gonadal (genital) ridges.** These are thickenings in the coelomic epithelium just medial to the mesonephroi (Fig. 14-17). The ridges are longer than the resulting mature gonad, which suggests that at one time gonads may have extended the length of the pleuroperitoneal cavity, as they do in living cyclostomes. Although the gonadal ridges are paired, a few adult vertebrates have a single testis or ovary because of fusion of the two ridges across the midline (lampreys, a few teleosts), or because one of the juvenile gonads fails to differentiate (hagfishes, some viviparous elasmobranchs, some female crocodilians, some lizards, most female birds). A few mammals also, among which are the platypus and some bats, have only one ovary. As the

FIGURE 14-17

Opossum embryo *(Didelphis)* 6½ days after cleavage, cross section. The gonadal primordia are shown in *black*. The umbilical veins carry oxygenated blood.

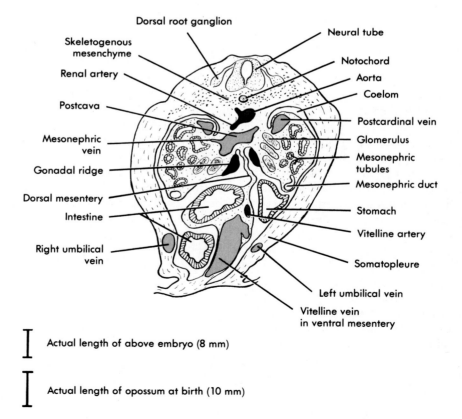

Dorsal root ganglion
Neural tube
Skeletogenous mesenchyme
Notochord
Renal artery
Aorta
Coelom
Postcava
Postcardinal vein
Mesonephric vein
Glomerulus
Mesonephric tubules
Gonadal ridge
Mesonephric duct
Dorsal mesentery
Stomach
Intestine
Vitelline artery
Right umbilical vein
Somatopleure
Left umbilical vein
Vitelline vein in ventral mesentery

Actual length of above embryo (8 mm)

Actual length of opossum at birth (10 mm)

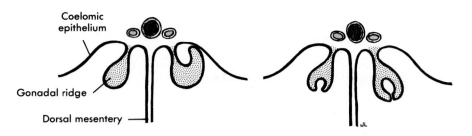

Coelomic
epithelium

Gonadal ridge

Dorsal mesentery

FIGURE 14-18

Two methods of entrapment of coelom to form a permanently hollow ovary in teleosts. The gonadal ridges are shown in cross section.

gonads approach sexual maturity they enlarge and usually acquire a dorsal mesentery, the **mesorchium** in males and **mesovarium** in females. Gonads are the source of gametes and of gonadal hormones.

Ovary

The ovary in some teleost fishes is a permanently hollow sac (Fig. 14-33). The condition results from entrapment of a small part of the coelomic cavity within the developing ovary (Fig. 14-18). Consequently the *ovarian cavity* is lined by germinal epithelium, which is the source of eggs. In some other teleosts the cavity within the ovary results from a secondary hollowing out of the interior of the ovary at each ovulation. In either case, the eggs, or in viviparous teleosts the young, are discharged into the ovarian cavity, which is continuous with the lumen of the oviduct (Fig. 14-33). The ovaries of most other fishes are compact (Fig. 14-32). The amphibian ovary is also a hollow sac, but the germinal epithelium is on the surface and eggs are shed into the coelom (Fig. 14-31). The ovaries of amniotes other than placental mammals are not compact either. They develop a large number of irregular fluid-filled cavities, or lacunae, which evidently provide nutritive support to the developing eggs in the overlying cortex. However, the mature, yolk-laden eggs are shed into the coelom (Figs. 14-34 and 14-36). At the end of each reproductive season the ovaries of most vertebrates below placental mammals regress to a state similar to that in juveniles.

The ovary of placental mammals is compact, the only cavitation being the **antrums** within the **graafian** (mature ovarian) **follicles** (Fig. 14-19, *A*). The ovary, like most others, is covered on its surface by a germinal epithelium from which arise oocytes, some of which become mature ova during the life of the individual. The wall of a graafian follicle at the ovarian surface thins out just prior to ovulation, and when the wall ruptures the ovum escapes into the coelomic fluid, surrounded by a corona of follicular cells. The escape is the process called ovulation. After the follicle ruptures, the cells that remain within it are altered histologically and physiologically under the stimulus of pituitary hormones, and organize a **corpus luteum,** which is one source of the progesterone needed for the maintenance of pregnancy. Prior to ovulation, the predominant secretory product of the cells was an estrogen. The role of these gonadal hormones is discussed in Chapter 17.

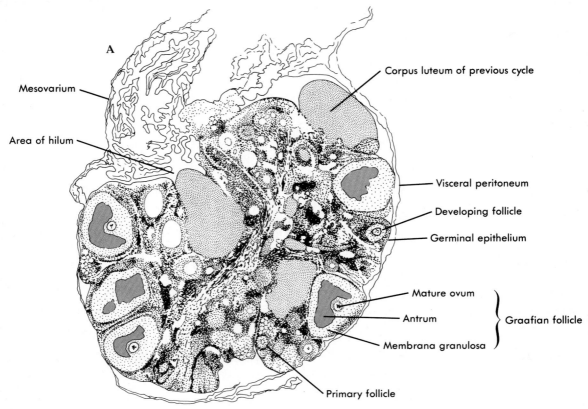

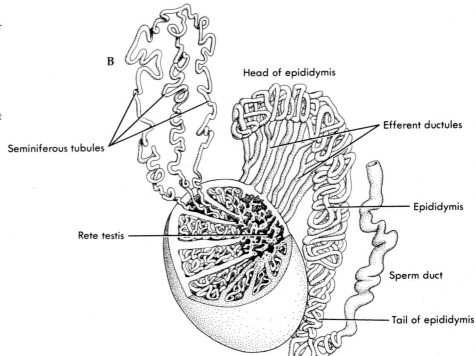

FIGURE 14-19

A, Section of golden hamster ovary at approach of estrus (heat) and ovulation, which occurs every fourth day unless pregnancy intervenes. The corpora lutea of two cycles overlap. **B,** Mammalian sperm channels. The efferent ductules (vasa efferentia) are modified mesonephric tubules. (**A** drawn from a histological section. **B** after George Corner.)

In many mammals a membranous fold of peritoneum develops close to the ovary and to the short oviduct and grows around these two structures, entrapping them and a small portion of the coelom in a cul-de-sac, the **ovarian bursa.** The bursa may be broadly open to the main coelom, as in cats and rabbits; it may communicate with the coelom by a mere slitlike passage, as in most carnivores and rats; or it may be closed completely, as in hamsters. The bursa increases the probability that all ovulated eggs will enter the oviduct rather than be lost in the coelomic fluid or implant in the coelom (ectopic pregnancy). Fetal rabbits are occasionally found implanted in the peritoneal lining of the ventral abdominal wall of pregnant rabbits.

Testis

Mature testes are usually smaller than ovaries because sperm, although more numerous by millions of times, are much smaller than eggs. The testes of placental mammals, however, are unique among vertebrates in that they are larger than ovaries because the eggs are microlecithal and only a few ripe ova are present at a time.

The basic components of vertebrate testes are essentially alike from fishes to human beings, consisting of a spermatogenic and a steroidogenic component. Spermatozoa are produced by the germinal epithelium, which except in lower fishes and tailed amphibians constitutes the lining of convoluted **seminiferous tubules** (Figs. 14-19, *B*, and 17-12). When mature, the spermatozoa separate from the germinal epithelium and, propelled by flagellum-like tails, traverse the tubule until they reach the **rete testis,** a network of very thin channels within the testis. The rete is drained by efferent ductules (vasa efferentia), which are modified mesonephric tubules (Fig. 14-24). Retia form in the gonadal ridges before differentiation in the direction of either a testis or an ovary, and rudiments persist in the mesovarium of most female vertebrates as blind tubules or solid cords, the **rete ovarii.**

In cyclostomes, most lower fishes, some teleosts, and urodeles sperm mature in numerous cystlike **seminiferous ampullae,** into which primordial germ cells migrate from a germinal epithelium located elsewhere within the testis or on its surface. At the end of the spawning season, after thousands or hundreds of thousands of spermatozoa have been shed from each cyst, the cysts collapse. These specific sites may or may not become a site of spermatogenesis during a subsequent reproduction cycle. The seminiferous ampullae of gnathostomes discharge into the rete testis. In cyclostomes the sperm are shed into the coelom.

The embryonic testis of anurans is subdivided into an anterior portion, **Bidder's organ,** which usually disappears before sexual maturity, and a more caudal portion, which becomes the adult testis. Bidder's organ persists in adult male toads (Fig. 14-20) and contains large undifferentiated cells resembling immature ova. If the testes are removed experimentally, Bidder's organs develop into functional ovaries, and the rudimentary female duct system (Fig. 14-20) enlarges under the influence of female hormones from the new ovaries.

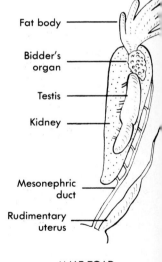

Fat body

Bidder's organ

Testis

Kidney

Mesonephric duct

Rudimentary uterus

MALE TOAD

FIGURE 14-20
Bidder's organ and the rudimentary female reproductive tract in a young male *Bufo,* ventral view. Only the left organs are illustrated. The mesonephric duct and rudimentary uterus empty into the cloaca.

Instances of sex reversal occur in nature in many submammalian vertebrate groups. Hens have been known to cease laying eggs, crow, and
develop other roosterlike characteristics. This comes about when the left
ovary atrophies and the right ovary, which is rudimentary, enlarges and
produces male hormones. (They apparently cannot produce sperm.)

During early development the gonads are indistinguishable as to sex,
and male and female ducts appear in every embryo. Under the influence of
a combination of sex chromosomes and pituitary and gonadal hormones
the indifferent early gonads develop into either testes or ovaries, and the
appropriate ducts, male or female, enlarge, whereas the other set remains
rudimentary or disappears. True hermaphroditism* (production of eggs
and sperm by the same individual) is common in cyclostomes and occasional in bony fishes, but is rare among other lower vertebrates and absent
among higher ones.

Translocation of Ovaries and Testes in Mammals

The caudal pole of each embryonic ovary and testis is connected by a
ligament to a shallow evagination of the coelom (**genital swelling,** Fig.
14-42, *B*), which becomes the **scrotal sac** in males, **labium majus** in females
(Fig. 14-21). In females the cephalic part of the ligament is named the **ovarian ligament,** and the caudal part is the **round ligament of the uterus** (Fig.
14-21, female). In males the ligament is the **gubernaculum.** Partly as a result

---

*Hermaphroditos was the son of Hermes and Aphrodite. While bathing in the mythical foun-
tain of Salmacis, he became united in one body with the nymph living in the fountain.

FIGURE 14-21

Caudal displacement of
mammalian gonads, ventral
view. The ovarian ligament
and round ligament of the
uterus collectively are homologous with the male gubernaculum. *Arrows* indicate the
route of translocation of the
right gonads. *C,* Scrotal recess of coelom; *Sd,* spermatic
duct arching over the ureter.

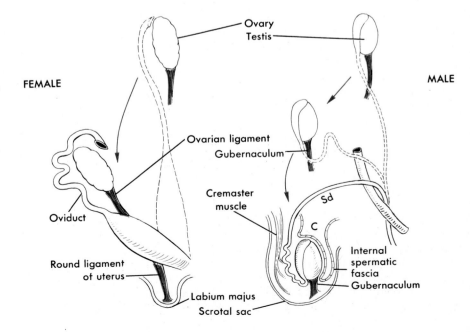

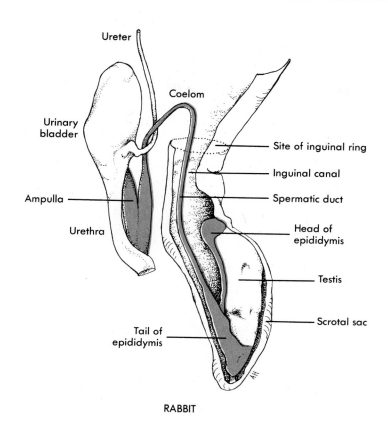

Ureter

Coelom

Urinary
bladder

Site of inguinal ring

Inguinal canal

Spermatic duct

Ampulla

Head of
epididymis

Urethra

Testis

Scrotal sac

Tail of
epididymis

RABBIT

FIGURE 14-22

Rabbit testis in scrotal cavity. The inguinal canals are broadly open to the main coelom at the inguinal ring; therefore the testes are retractable. Sperm pathway from testis to urethra is shown in *red*.

of shortening of the ligaments, partly because elongation of the ligaments does not keep pace with elongation of the trunk, and partly for unknown reasons,* the ovaries and testes are displaced caudad toward the labia or scrotal sacs. The ovaries are not displaced as far caudad as the testes.

The testes remain retroperitoneal and descend permanently into scrotal sacs in many mammals, including most marsupials, ungulates, carnivores, and higher primates. In others they are lowered into the sacs and retracted at will (rabbits, bats, a few rodents, some primitive primates, for example). The passage between the abdominal cavity and the scrotal cavity is the **inguinal canal** (Fig. 14-22). The opening of the canal into the abdominal cavity is surrounded by a fibrous **inguinal ring** (often the site of inguinal hernia). In species that retract the testes the canal remains broadly open. In species in which the testes are permanently confined to the scrotum the inguinal canal is only wide enough to accommodate the spermatic cord.

The **spermatic cord** contains the spermatic duct, arteries, veins, lymphatics, and nerves. These are all wrapped in a single sheath, the internal spermatic fascia, and all are dragged into the scrotum along with the testis. Scrotal sacs do not develop in monotremes, some insectivores, elephants,

---

*Removal of the gubernaculum does not always prevent descent of the testes.

FIGURE 14-23

The mesonephric duct *(black)* as a carrier of sperm and urine. **A,** Basic plan, carrying both sperm and urine. **B,** Carrying urine from the anterior end of the kidney only; chiefly a spermatic duct. **C,** Carrying sperm only.

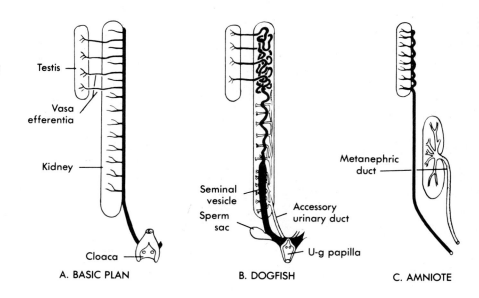

A. BASIC PLAN   B. DOGFISH   C. AMNIOTE

whales, and certain other mammals. In these the testes remain permanently in the abdomen.

In mammals whose testes are permanently in scrotal sacs the spermatic artery and vein lie side by side in tight coils within the inguinal canal. This rete of vessels is the **pampiniform plexus.** The arterial blood is at body temperature; the venous blood has been cooled in the scrotal sac. Heat is transferred in the plexus from arterial to venous blood so that blood reaching the testis is cool and blood returning to the internal circulation has been prewarmed. This countercurrent flow of warm and cool blood protects the sperm of these endothermic species from temperatures that would kill them, while conserving body heat. No such plexus occurs in birds, but avian sperm can withstand high temperatures.

MALE GENITAL DUCTS

Although the mesonephric ducts are part of the urinary system embryonically, they transport both sperm and urine in generalized urogenital systems (Fig. 14-23, *A*). Connections between the mesonephroi and the gonadal ridges (future testes) are established early in embryonic life (Fig. 14-24). Some of the more anterior mesonephric tubules—a few to 24 or more, depending on the species—grow across the developing mesorchium to connect with the tubules of the future rete testis. These modified mesonephric tubules in the mesorchium become **vasa efferentia** that carry sperm from the testis.

In some fishes and in urodeles there has been a tendency to form a new sperm duct, the longitudinal **marginal canal,** to receive the vasa efferentia (Figs. 14-8, *A* and *B*, and 14-23, *D* and *E*). This tendency may have been more common in Paleozoic ganoids, of which sturgeons and *Polypterus* are relics. The condition has reached a culmination in modern teleosts, in

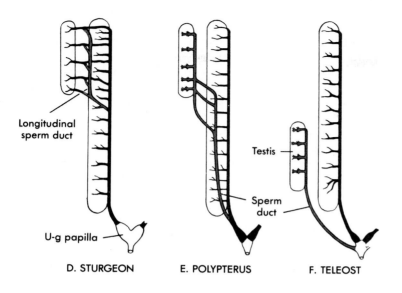

Longitudinal sperm duct

U-g papilla

D. STURGEON

Testis

Sperm duct

E. POLYPTERUS

F. TELEOST

FIGURE 14-23, cont'd

D to **F,** Tendency toward a separate sperm duct *(red)* in fishes, the mesonephric duct ultimately carrying only urine. Note the reverse of this condition in amniotes. *U-g papilla,* Urogenital papilla. For a variation in the termination of the teleost mesonephric duct see Fig. 14-15.

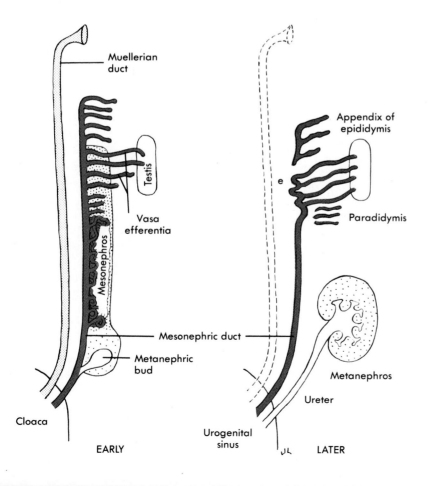

Muellerian duct

Testis

Vasa efferentia

Mesonephros

Mesonephric duct

Metanephric bud

Cloaca

EARLY

Appendix of epididymis

e

Paradidymis

Metanephros

Ureter

Urogenital sinus

LATER

FIGURE 14-24

Developmental changes in the urogenital system of a male amniote. In the earlier stage *(left)* some of the mesonephric tubules have invaded the testis to become vasa efferentia. In the later stage *(right)* the mesonephros has regressed except for remnants (appendix of epididymis, paradidymis), and the muellerian duct has regressed *(broken lines* at right). The mesonephric duct remains to carry sperm. Mesonephric duct and tubules are *red. e,* Epididymal portion of mesonephric duct.

which the mesonephric duct carries no sperm whatsoever (Fig. 14-23, *F*). When a duct carries only sperm, whether a mesonephric duct or a substitute, it is called a **vas deferens (ductus deferens).**

The spermatic ducts (male mesonephric ducts) empty into the cloaca in reptiles and birds (Figs. 14-25 to 14-28), and into a derivative of the cloaca in mammals. The anatomical relationships of the spermatic ducts in mammals are affected by (1) complete separation of the embryonic cloaca into a urogenital sinus and rectum (Fig. 14-41, *E*) and (2) caudal migration of the testes. As a result of subdivision of the cloaca, the spermatic ducts of mammals finally empty into the urogenital sinus, which is the male **urethra** (Fig. 14-29). As a result of caudal migration of the testes, the spermatic ducts become "caught" or "hung up" on the ureters, so that thereafter they loop

FIGURE 14-25

Urogenital organs of the male lizard *Anolis carolinensis,* ventral view. The kidneys are metanephric. The sperm duct is the persistent mesonephric duct. The hemipenes are seen in an everted (erected) position.

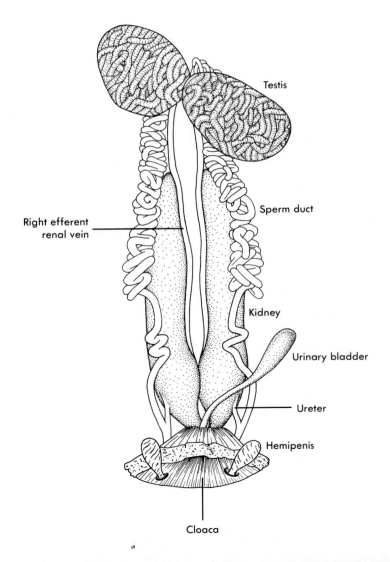

Testis

Sperm duct

Right efferent
renal vein

Kidney

Urinary bladder

Ureter

Hemipenis

Cloaca

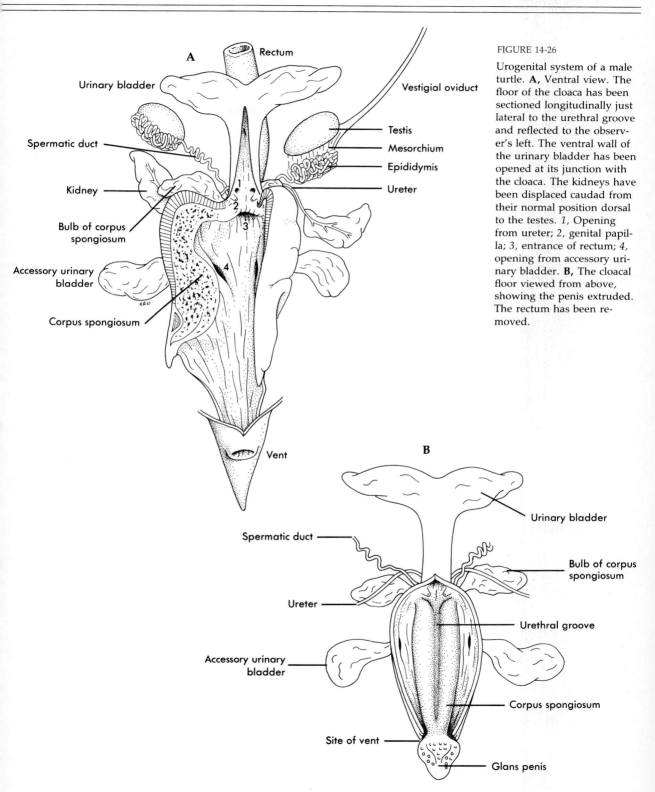

A

Rectum

Urinary bladder

Vestigial oviduct

Spermatic duct

Testis

Mesorchium

Epididymis

Ureter

Kidney

Bulb of corpus
spongiosum

1
2
3

Accessory urinary
bladder

4

Corpus spongiosum

Vent

B

Spermatic duct

Urinary bladder

Ureter

Bulb of corpus
spongiosum

Accessory urinary
bladder

Urethral groove

Site of vent

Corpus spongiosum

Glans penis

FIGURE 14-26

Urogenital system of a male turtle. **A,** Ventral view. The floor of the cloaca has been sectioned longitudinally just lateral to the urethral groove and reflected to the observer's left. The ventral wall of the urinary bladder has been opened at its junction with the cloaca. The kidneys have been displaced caudad from their normal position dorsal to the testes. *1,* Opening from ureter; *2,* genital papilla; *3,* entrance of rectum; *4,* opening from accessory urinary bladder. **B,** The cloacal floor viewed from above, showing the penis extruded. The rectum has been removed.

over the ureters en route from the testes to the urethra (Figs. 14-21 and 14-29). Just before their entrance into the urethra they frequently exhibit a swelling, the **ampulla of the ductus deferens** (Fig. 14-22).

Emptying into the urethra in the immediate vicinity of the entrance of the spermatic ducts are **prostate glands** and one or more other accessory sex glands that produce some of the constituents of semen (Fig. 14-30). The urethra in male mammals is frequently called **prostatic urethra** where prostate glands empty into it, **membranous urethra** from prostate glands to the base of the penis, and **spongy urethra** within the penis (Fig. 14-29).

Agnathans, male and female, lack genital ducts. Sperm and eggs are shed into the ciliated coelom, are propelled caudad by undulations of the body and by beating of the coelomic cilia, and exit via a pair of genital pores, to be described next. Whether this is a primitive route for shedding gametes in vertebrates is conjectural; living agnathans are a mosaic of primitive and specialized traits.

FIGURE 14-27

Urogenital system of a male alligator. **A,** Ventral view, ventral cloacal wall removed. The ureters open in the dorsal wall of the cloaca. **B,** Dorsal view of penis. The vas deferentia open onto the floor of the cloaca via genital papillae.

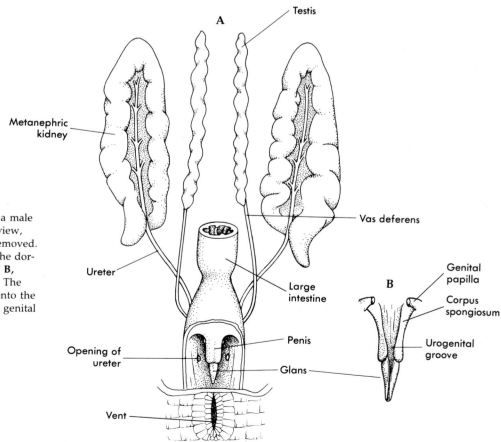

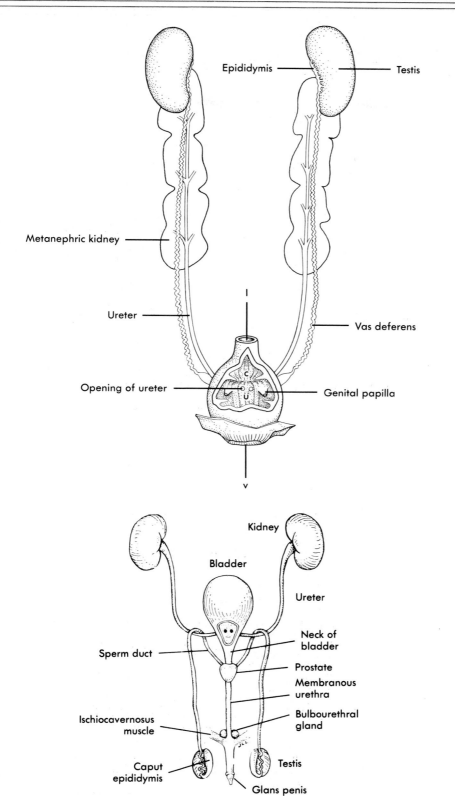

FIGURE 14-28

Urogenital system of a rooster, *c,* Coprodeum; *I,* large intestine: *u,* urodeum; *v,* vent.

FIGURE 14-29

Urogenital system of a male cat, ventral view.

FIGURE 14-30

Accessory sex glands of a male hamster, ventral view. The bladder and urethra have been opened to show entrances of ducts. *1*, Ureter; *2*, spermatic duct; *3*, prostatic urethra; *4*, coagulating gland; *5*, seminal vesicle; *6*, cranial prostate; *7*, caudal prostate; *8*, ampullary gland. A bulbourethral gland enters the urethra farther caudad.

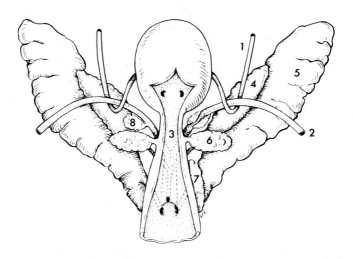

GENITAL PORES

Cyclostomes lack genital ducts, and sperm and eggs are shed into the coelom and exit via a pair of funnel-shaped **genital pores** in the caudal abdominal wall. These lead into a median papilla (genital in hagfishes, urogenital in lampreys) that opens to the exterior just behind the anus.

Similar pores lead from the coelom directly to the exterior in some elasmobranchs and a number of bony fishes; they have also been described in turtles and crocodilians. In none of these, however, do the pores convey gametes, because genital ducts are present. Whether they are homologous with the genital pores of agnathans is not known, and it seems unlikely. For this reason it is preferable to call them **abdominal pores** in gnathostomes. What role abdominal pores may perform in vertebrates having genital ducts is obscure. In some marine teleosts the pores are present only in females, and open only during breeding seasons. They may play a role in reproduction, or they may be functionless vestiges that retain a hereditary responsiveness to reproductive hormones.

INTROMITTENT ORGANS

When fertilization is internal, the male, with few exceptions, develops intromittent, or copulatory, organs for introducing sperm into the female reproductive tract. These are present in reptiles and mammals. They are also present in fishes with internal fertilization, the anuran family Ascaphidae, and a few birds. In most birds the cloaca of both sexes is eversible, facilitating sperm transfer.

The intromittent organs of elasmobranchs are grooved, fingerlike appendages of the pelvic fins known as **claspers** (Fig. 9-9). In basking sharks they transfer a sperm-filled spermatophore up to 3 cm in diameter. Embedded in the fin at the base of the clasper in some sharks is a muscular **siphon sac** that contributes copious quantities of an energy-rich mucopolysaccharide to the seminal fluid. In many teleosts the anal fin is modified for sperm transfer and is called a **gonopodium.** In *Ascaphus* the intromittent organ is a permanent tubular taillike extension of the cloaca.

Intromittent organs of amniotes are of two types: hemipenes and penis. Male snakes and lizards have a pair of **hemipenes,** which are pocketlike diverticula of the caudal wall of the cloaca that extend under the skin at the base of the tail (Fig. 14-25). Each is held in place by a retractor muscle. During copulation the muscle relaxes and the pocket turns inside out and protrudes through the vent in an erect position. Sperm passes along spiral grooves on its surface. Hemipenes are present, but much smaller, in females.

Male turtles, crocodilians, a few birds (swans, ducks, ostriches, a few others), and male mammals exhibit an unpaired erectile **penis.** In its simplest form, the penis is a thickening of the floor of the cloaca that consists chiefly of spongy erectile tissue, the **corpus spongiosum,** which bears a **urethral groove** on its dorsal surface and ends in a **glans penis** (Figs. 14-26 and 14-27). The urethral groove channels sperm and urine toward the vent. The erectile tissue consists of cavernous blood sinuses that when engorged cause the glans penis to be extruded through the cloacal aperture. The glans is richly supplied with sensory cutaneous endings that reflexly stimulate ejaculation. In mammals it is ensheathed by a fold of skin, the **prepuce,** except during erection. The penis of monotremes is reptilian in structure and in its location in the cloacal floor. In placental mammals the grooved embryonic corpus spongiosum rolls into a tube with the urethral groove, now the **spongy urethra,** folded inside, and it extends beyond the body at the pubic symphysis accompanied by two additional erectile masses, the **corpora cavernosa.**

The mammalian penis develops from a **genital tubercle** found both in male and female embryos (Fig. 14-42, *B*). It lies at the anterior end of two **genital swellings** that later become scrotal sacs in males or **labia majora** in higher female primates. In genetic males the tubercle becomes grooved, and then tubular, and elongates to form a penis. In females no tube usually develops, and the tubercle becomes the **clitoris.** The clitoris usually remains embedded in the floor of the urogenital sinus or vagina and is erectile. In female rodents, however, a tube does develop, and the clitoris becomes a penislike urinary papilla enclosing the urethra (Fig. 14-42, *C*). In female hyenas the urogenital sinus becomes enclosed within the clitoris, and the latter looks exactly like the male penis. It carries urine, copulation takes place through it, and young are delivered through it. It has never been shown that this is responsible for the maniacal laughterlike cry of the hyena.

FEMALE GENITAL DUCTS

The female reproductive tract typically consists of a pair of muscular tubes that commence at an **ostium** surrounded by a fimbriated **oviducal funnel,** or **infundibulum,** and empty into the cloaca (Fig. 14-31). The tubes differentiate from a pair of **muellerian ducts** present in the embryos of both sexes (Fig. 14-11). In mature females the ducts transport eggs, and certain

FIGURE 14-31

Urogenital system of female
necturus, ventral view.

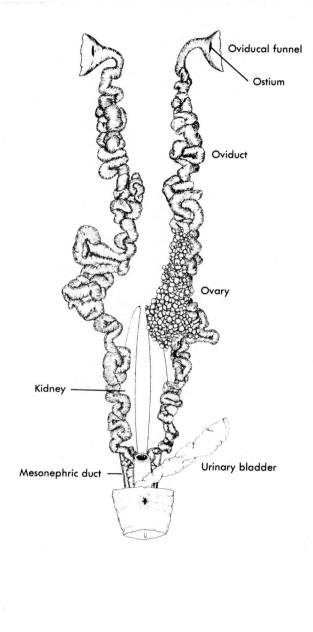

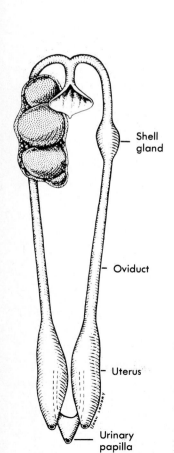

FIGURE 14-32

Reproductive system of female *Squalus,* ventral view.
The left ovary has been removed. The right ovary contains three mature eggs.
Within the urinary papilla
there is a urinary sinus.

segments are modified for specific functions, including, in one species or another, coating the eggs with protective or nutrient substances, holding eggs or living young and maintaining them in a viable state until the eggs are shed or young are delivered, expelling eggs or young, receiving the male intromittent organ, and storing and maintaining sperm in those numerous species in which the eggs are not mature at the time of mating.

Ostia appear to be phylogenetic derivatives of one or a few pronephric nephrostomes. That is how they arise in living elasmobranchs and amphibians; and in these same vertebrates muellerian ducts arise by longitudinal splitting of the pronephric ducts. This illustrates again the intimate phylogenetic and embryological relationship between genital and urinary organs. In most other vertebrates each muellerian duct arises as a longitudinal groove in the coelomic mesothelium paralleling the pronephric duct.* The groove subsequently becomes a tube, except at the anterior end, which remains open to become the ostium. At the caudal end it acquires an opening into the cloaca.

When fertilization is internal, sperm usually penetrate the eggs in the upper reaches of the oviduct, and eggs are propelled along the tract by cilia or peristaltic action of smooth muscles. Teleost eggs are often fertilized while still in an ovarian follicle. Female urodeles, however, have sperm storage pockets, **spermathecae,** in the dorsal wall of the cloaca, that receive and store spermatophore sperm and expel them onto mature eggs as they pass by later. Some lizards and snakes have spermathecae that are crypts of the oviducal lining located just above the shell gland, and domestic fowl have them at the uterovaginal junction. These amniotes lay a succession of eggs during the season, and spermathecae ensure the availability of viable sperm at the time of each ovulation. Sperm are sometimes stored in spermathecae for many months.

Fishes and Amphibians

In female elasmobranchs each muellerian duct gives rise to an oviduct with a **shell (nidimental) gland** and to a **uterus** that opens to the cloaca (Fig. 14-32). The cephalic half of the shell gland secretes albumen, and the caudal half secretes the shell. Two embryonic ostia unite to form a single adult ostium suspended in the falciform ligament, a condition not typical of vertebrates.

The oviducts of ray-finned fishes are peculiar. They are either short funnels at the caudal end of the coelom, or they are directly continuous with the ovarian cavity. In either case they lead to a genital pore located between the urinary aperture and the more anterior anus (Fig. 14-33). The genital pore is sometimes at the end of a genital papilla, and in teleosts the papilla is sometimes elongated to form a tubelike **ovipositor.** Their development is such as to make it doubtful that they are muellerian duct derivatives. Cyclo-

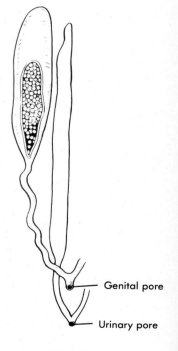

Genital pore

Urinary pore

TELEOST

FIGURE 14-33

Female reproductive system of a teleost. Ova are shed into the ovarian cavity. In some teleosts the ovary reaches almost to the genital pore.

---

*For this reason embryonic muellerian ducts are sometimes called paramesonephric ducts.

stomes have no oviducts; eggs exit the coelom via a pair of genital pores, as explained earlier.

The female tracts of lungfishes, urodeles, and caecilians are long and somewhat convoluted (Fig. 14-31). In anurans they are much more tortuous as an accommodation to the short wide trunk. The caudal end of amphibian ducts may become voluminous thin-walled **ovisacs** where eggs accumulate before being shed. When filled in anurans, ovisacs occupy the entire coelom and distend the abdomen. In ovoviviparous urodeles they serve as uteri. The two tracts open independently into the cloaca in lungfishes and most amphibians, but in toads they unite just anterior to the vent and exit to the cloaca through a common genital pore. The oviducal lining in amphibians is richly supplied with glands that secrete several jelly envelopes around each egg as it moves down the tube.

Reptiles, Birds, and Monotremes

The female tracts of reptiles, birds, and monotremes conform to the basic vertebrate pattern (Figs. 14-34 to 14-37). However, only one muellerian duct differentiates in crocodilians, some lizards, and most female birds (Fig. 14-36). In oviparous amniotes other than snakes and lizards, albumen glands line a segment of the oviduct, and all have a shell gland just anterior to the cloaca (Figs. 14-36 and 14-37). The shell remains leathery or becomes brittle in air, depending on the constituents of the secretion.

FIGURE 14-34

Urogenital system of female aquatic turtle, *Trionyx euphraticus,* ventral view. The left ovary has been removed. (Courtesy Mohamad S. Salih, University of Baghdad.)

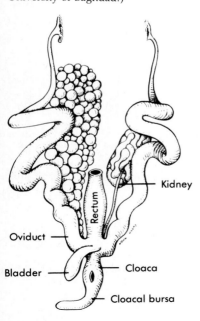

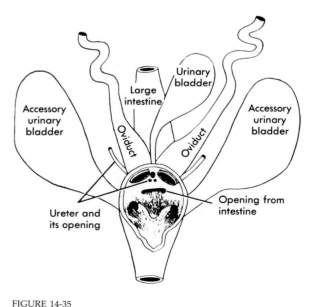

FIGURE 14-35

Cloaca and associated structures of a female terrestrial turtle, ventral view. Clitoris has been removed.

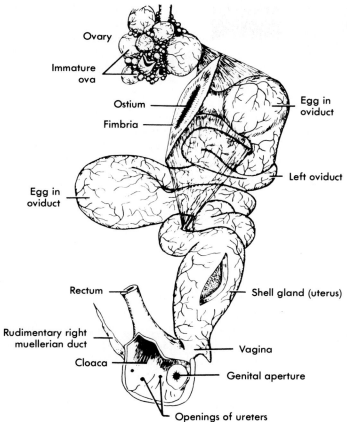

FIGURE 14-36

Reproductive tract of a hen. The presence of two eggs in the oviduct is unusual. The larger egg is traversing the magnum.

FIGURE 14-37

Genital tract and cloaca of female monotreme, ventral view. The right ovary is usually smaller than the left, and is sometimes rudimentary. The cephalic half of the cloaca is divided by a partition into urogenital sinus (receiving oviducts and ureters) and coprodeum.

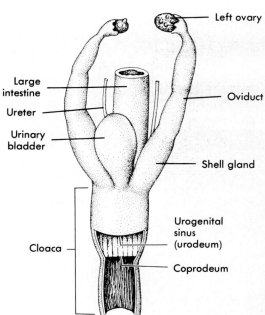

In birds the albumen-secreting region is the **magnum,** and the thick-walled shell gland is called, inappropriately, the **uterus** (Fig. 14-36). The short muscular terminal segment, the **vagina,** secretes mucus that seals the pores of the shell to water vapor but not to oxygen, thus retarding moisture loss from the egg after it has been laid. The vagina then expels the egg.

Placental Mammals

The muellerian ducts of placental mammals give rise to **oviducts, uteri,** and **vaginae.** Except in marsupials, muellerian ducts unite at their caudal ends. As a result the adult female tract above marsupials is paired anteriorly and unpaired posteriorly, terminating as an unpaired vagina. The oviducts, or **fallopian tubes,** as they are called in mammals, are relatively short, small in diameter, convoluted, and lined with cilia. They commence at an oviducal funnel bordered by a delicate membranous fringe, the **fimbria of the infundibulum.**

UTERI. In most marsupials there is no fusion of the embryonic muellerian ducts. Therefore the entire female tract is paired. They have a **duplex uterus** and paired vaginae (Fig. 14-38).

In other placental mammals there are varying degrees of fusion of the caudal ends of the muellerian ducts, which results most often in two **uterine horns,** a **uterine body,** and a single **vagina** (Fig. 14-39, rabbit). When there are two complete lumens with the body of the uterus, it is said to be **bipartite** (Fig. 14-40, hamster). When there is a single lumen within the body and there are two horns, the uterus is said to be **bicornuate** (Fig. 14-40, ungulates). There are species with uteri intermediate between the bipartite and bicornuate condition. When there are uterine horns, the blastocysts implant in the horns. In some mammals one horn is much larger,

FIGURE 14-38

Internal passageways of the reproductive tract of a female opossum. Compare with the external view in Fig. 14-39, marsupial.

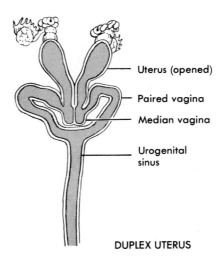

Uterus (opened)

Paired vagina

Median vagina

Urogenital sinus

DUPLEX UTERUS

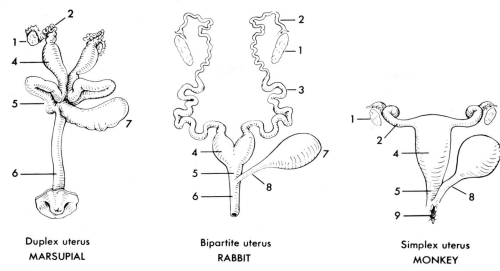

Duplex uterus
MARSUPIAL

Bipartite uterus
RABBIT

Simplex uterus
MONKEY

FIGURE 14-39

Reproductive tracts of three female mammals. *1*, Ovary; *2*, oviduct; *3*, horn of uterus; *4*, body of uterus; *5*, vagina; *6*, urogenital sinus; *7*, urinary bladder; *8*, urethra; *9*, vestibule of primate. The primate is a rhesus monkey. The marsupial is an opossum, shown also in Fig. 14-38.

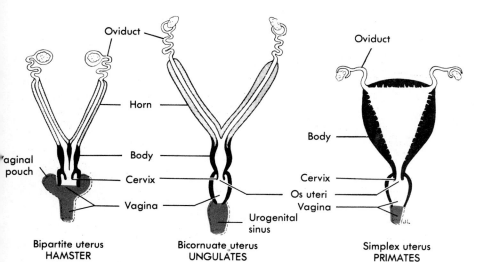

Bipartite uterus
HAMSTER

Bicornuate uterus
UNGULATES

Simplex uterus
PRIMATES

FIGURE 14-40

Uterine types among mammals. Blackened regions represent fused caudal ends of the muellerian ducts; *red* represents the cloaca or a derivative thereof. Note the two lumens in the body of the bipartite uterus.

and the blastocysts implant in that horn—the right in impalas—even though both ovaries produce eggs.

Apes, monkeys, humans, some bats, and armadillos have no uterine horns, and the oviducts open directly into the body of the **simplex** uterus (Figs. 14-39, monkey, and 14-40, primates). Except in ectopic pregnancies—pregnancies in which blastocysts implant in abnormal locations, such as the oviduct (tubal pregnancies) or coelom (abdominal pregnancies)—the usually single fetus or the twins, triplets, quadruplets, or quintuplets of these specific mammals all implant in the body of the uterus.

The body of all uteri narrows to form a **cervix** (neck), the lower end of which projects into the vagina as the **lips** of the cervix. The lips surround the opening **(os uteri)** leading from uterus into vagina. The cervix must dilate under the influence of hormones for the young to be delivered.

Sperm deposited in the vagina pass through the os uteri en route to the upper part of the oviducts, where only one sperm penetrates an egg. The uterine lining **(endometrium)** becomes highly vascular under the stimulus of hormones before implantation of a blastocyst. The thick, muscular layer of the uterine wall **(myometrium)** assists in ejection of the young at birth, provided it too has been hormonally prepared for this action.

VAGINAE. Above marsupials the vagina is the fused terminal portion of the muellerian ducts, and it usually opens into the urogenital sinus (Fig. 14-40, ungulates). In higher primates, however, the vagina extends almost to the exterior, opening into a shallow **vestibule** between the labia and caudal to the opening of the urethra (Fig. 14-39, monkey). In many rodents the vagina opens directly to the exterior immediately caudal to the urinary papilla (Fig. 14-42, C). The vaginal lining of mammals is cornified for reception of the penis.

The vagina of marsupials is unusual (Fig. 14-38). Just beyond the uteri the two muellerian ducts meet to form a **median vagina,** which may or may not be paired internally. Beyond the median vagina the two ducts continue as **paired (lateral) vaginae.** The pouchlike median vagina projects caudad and lies against the urogenital sinus, separated by a septum. At birth the fetus is usually forced through the septum directly into the urogenital sinus. The new passageway thus established may remain throughout life, which results in a **pseudovagina,** although it closes in opossums. As an adaptation to dual vaginae, the penis of male marsupials is forked at the tip.

Muellerian Duct Remnants in Adult Males

Although muellerian ducts do not fully mature in males, they often develop into prominent structures. In male elasmobranchs a pair of **rudimentary oviducts** encircle the anterior end of the liver and end in a rudimentary ostium in the falciform ligament. The **sperm sac** of sharks is a caudal remnant of the muellerian duct. A complete, although rudimentary, female tract is common in male amphibians (Figs. 14-8 and 14-20) and reptiles. In male anurans after removal of testicular hormones by orchidectomy the rudimentary muellerian ducts develop into functional oviducts and uteri. Remnants in male mammals include an **appendix testis**, a small cyst-like body on the cephalic end of the mammalian testis, and a **prostatic sinus (vagina masculina),** an unpaired sac near the prostate gland homologous with the female vagina.

Entrance of Ova into Oviduct

After seeing the very large size of a shark's egg, laboratory students often inquire how such a large egg can get into the ostium and down the relatively small oviduct. We know the answer with reference to the equally large egg of the chick or monotreme. Under the influence of hormones at the time of ovulation, the fringe **(fimbria)** of the oviducal funnel waves gently in an undulating movement. When it comes in contact with an egg, whether still in the ovary or separated from it, the fimbria clasps the egg, delicately at first and then more firmly, until the egg is engulfed by the funnel. At this time the egg is a shapeless mass of flowing yolk (like the yolk in a fresh chicken egg) contained in a nonrigid membrane. Muscular contraction of the funnel squirts the shapeless mass into the oviduct. There-upon, peristalsis of the wall of the oviduct moves the egg caudad. Cilia play a relatively unimportant role. In the case of the tiny eggs of mammals, however, the cilia are more important, although the fimbria also plays a role. In mammals the ovary is partially surrounded by the fimbria at all times, and this increases the probability that the egg will enter the oviduct. In mammals with an ovarian bursa the egg can go nowhere else.

THE CLOACA

The cloaca is the terminal segment of the hindgut that receives the large intestine and the urinary and genital ducts. It has become shallow or non-existent in adult lampreys, chimaeras, and ray-finned fishes; and in placen-tal mammals the embryonic cloaca is partitioned into several separate pas-sageways and no longer exists as an adult structure. With these exceptions, a cloaca is present in all vertebrates. It acquires an opening to the exterior when the cloacal membrane, which separates hindgut from proctodeum, ruptures (Fig. 1-1). The contribution of proctodeum to adult cloaca is minor, except in amphibians.

The cloaca of fishes and amphibians receives the large intestine and the mesonephric ducts, and in females the oviducts. In amphibians a urinary bladder opens from the ventral wall. The cloaca of reptiles, birds, and monotremes receives the same structures: large intestine, mesonephric ducts (but in males only, carrying sperm), oviducts in females, and urinary bladder unless absent. In addition, the ureters of reptiles, birds, and mono-tremes open into the cloaca, except in those few male reptiles in which the ureters retain their embryonic connection with the mesonephric duct (Fig. 14-25). The penis or clitoris, when present, is embedded in the cloacal floor, and a lymphoid pouch, the bursa of Fabricius, opens into the roof of the cloaca of young birds.

In reptiles, birds, and monotremes a horizontal partition, the **urorectal fold,** separates the cephalic portion of the cloaca into two chambers, a **coprodeum** that receives the large intestine, and a **urodeum** that receives the oviducts and ureters (Figs. 14-28 and 14-37). The terminal portion of the cloaca in birds is called the **proctodeum,** but it is not entirely homologous with the ectodermal structure of the same name in vertebrate embryos.

Fate of the Cloaca in Placental Mammals

We have seen how in monotremes a urorectal fold divides the cephalic end of the cloaca into a urodeum and coprodeum (Fig. 14-37). In placental mammals the urorectal fold grows caudad until it reaches the cloacal membrane separating the cloaca from the exterior. By this process the cloaca becomes completely divided into a **rectum** dorsally and a **urogenital sinus** ventrally (Fig. 14-41, *C* and *E*). Rupture of the cloacal membrane at two points provides an **anus** and a **urogenital aperture** (Fig. 14-42, *B*). The early embryonic urogenital sinus (Fig. 14-41, *B*) receives the mesonephric ducts, muellerian ducts (which are initially present in both sexes), and the future urinary bladder (allantois), like the urodeum of monotremes.

As development progresses in males, the muellerian ducts regress and the urogenital sinus elongates (compare Fig. 14-41, *B* and *E*). The urogenital sinus becomes continuous with the spongy urethra that has developed independently in the penis (compare Fig. 14-41, *E* and *F*). The urogenital sinus now consists of the prostatic and membranous urethra (Fig. 14-29). The ureters become reoriented to open into the bladder, whereas the mesonephric ducts (now spermatic ducts) continue to empty into the urogenital sinus (Figs. 14-29 and 14-41, *F*).

As development progresses in females, the mesonephric ducts regress, and the muellerian ducts unite at their caudal ends to form the body of the uterus and the vagina (Fig. 14-41, *C*). The part of the urogenital sinus between bladder and entrance of the vagina is the urethra (Fig. 14-41, *C*). As a result of these changes most adult female mammals have two caudal openings to the exterior, a urogenital aperture and an anus.

In most female primates (including humans) and in some rodents, an additional partition forms in the cloaca—this one in the urogenital sinus. It separates the urogenital sinus into a urethra and a vagina (Fig. 14-41, *D*). As a result, the embryonic cloaca in these species becomes subdivided into

FIGURE 14-41

Fate of the mammalian cloaca and allantois *(red),* muellerian ducts *(gray),* and mesonephric duct *(black).* **A** and **B,** Bisexual stages. Only the left muellerian and mesonephric ducts are shown. In **B** the cloaca is becoming subdivided by the urorectal fold into a urogenital sinus ventrally and a rectum dorsally. **C,** Typical adult female mammal. **D,** Female primate, a modification of the condition shown in **C.** In **C** and **D** the contributions of both the left and right muellerian ducts are shown. **E,** Developing male, showing intermediate stage in reorientation of mesonephric and metanephric ducts. **F,** Adult male.

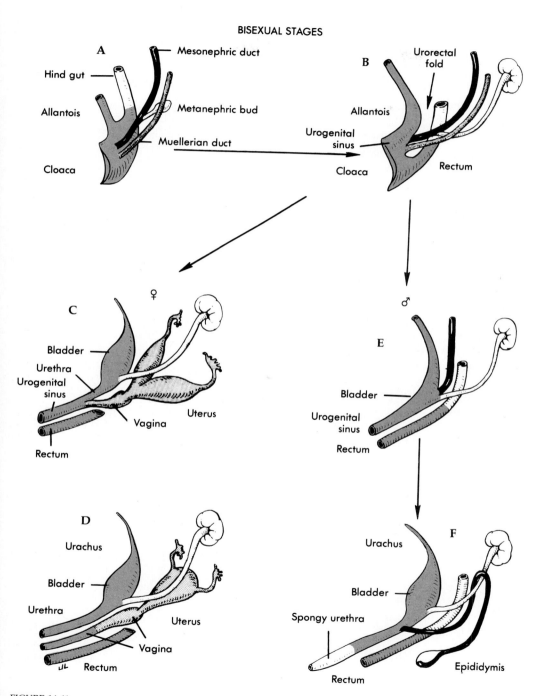

BISEXUAL STAGES

FIGURE 14-41

For legend see opposite page.

three passages: urethra, vagina, and rectum. Each passageway leads to the exterior via its own aperture (Fig. 14-42, C). In this regard the females of these species have evolved farther than the males. The vagina in these females has a dual origin. The cephalic part is derived from the fused muellerian ducts, and the terminal part is cloacal.*

Table 14-1 summarizes the fate in adult males and females of the chief sexually indifferent urogenital structures found in all mammalian embryos.

---

*In higher primates the urethra and vagina actually open into a shallow vestibule (Fig. 14-39) derived from the very distal end of the urogenital sinus.

TABLE 14-1   Some Homologous Urogenital Structures in Male and Female Mammals

Indifferent Structure	Mature Male	Mature Female
Mesonephric duct	Ductus deferens (vas deferens) Epididymis	Ductus deferens femininus* (Gartner's duct)
Mesonephric tubules	Vasa efferentia Appendix of epididymis* Paradidymis*	Vasa efferentia ovarii* Epoophoron* Paroophoron*
Muellerian duct	Appendix testis* Vagina masculina* (prostatic sinus)	Oviduct Uterus Vagina, cephalic to urogenital sinus
Gonadal ridge	Testis Rete testis	Ovary Rete ovarii*
Gubernaculum	Gubernaculum	Ovarian ligament Round ligament of uterus
Genital swellings	Scrotal sacs	Labia majora
Genital tubercle	Penis	Clitoris
Genital folds	Contribute to penis	Labia minora
Urogenital sinus	Urethra, prostatic and membranous portions	Urethra Urogenital sinus Lower vagina in rodents and primates

*Vestigial.

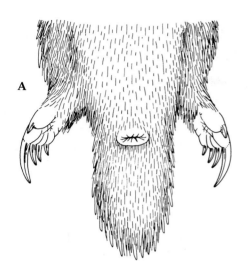

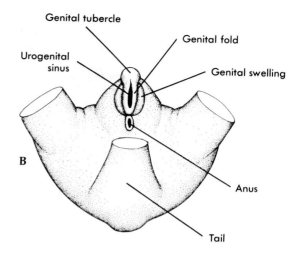

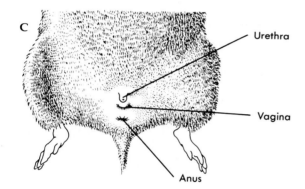

FIGURE 14-42

Apertures from mammalian cloaca or its adult derivatives. **A,** Cloacal aperture of *Echidna,* a monotreme. **B,** External genitalia of sexually indifferent stage of a 29 mm human embryo (about 12 weeks of age). The cloaca has been divided into urogenital sinus and rectum. The genital tubercle becomes penis or clitoris, the genital swellings become scrotal sacs or labia majora, and the urogenital sinus becomes the urethra in a male, is partitioned to form the vagina and urethra in certain females. **C,** Perineal region of female hamster. The urethra opens at the tip of a penislike urinary papilla.

Chapter Summary

1. Kidneys arise from a ribbon of nephrogenic mesoderm that extends the length of the trunk. A wave of differentiation sweeps along the ribbon, giving rise to convoluted tubules. The tubules become associated with glomeruli and open into a longitudinal duct that terminates in the cloaca when present.

2. Glomeruli are capillaries that filter water and other substances from the blood. The most primitive (external glomeruli) dangle into the coelom. The remaining glomeruli (internal glomeruli) are encapsulated by Bowman's capsule. Glomerulus and capsule constitute a renal corpuscle.

3. Kidney tubules are tiny convoluted ductules that collect glomerular filtrate, selectively reabsorb some substances, add others, and conduct the final filtrate to the longitudinal duct.

4. A renal corpuscle with its associated tubule is a nephron, the functional renal unit.

5. An archinephros is a hypothetical primitive kidney with external glomeruli, nephrostomes, and simple tubules arranged metamerically the length of the trunk. The holonephros of larval hagfishes and apodans resembles an archinephros.

6. The first embryonic tubules are segmental and often have nephrostomes in lower vertebrates. These tubules are temporary and constitute a pronephros.

7. A mesonephros organizes behind the pronephros. It is the functional kidney of adult fishes and amphibians and of embryonic amniotes. It consists of closely packed nephrons. It is also called opisthonephros in anamniotes.

8. The pronephric duct persists to drain the adult kidney of fishes and amphibians except in a few males, in which it is preempted for sperm transport.

9. Some of the anteriormost mesonephric tubules in male vertebrate embryos invade the testes and become vasa efferentia. As a result, the cephalic end of the adult kidney of male fishes and amphibians may be preempted for sperm transport, and is called a sexual (epididymal) kidney.

10. The metanephros is the adult amniote kidney. It organizes from the caudal end of the nephrogenic mesoderm, which is displaced craniad and laterad. Its duct (ureter) arises as a bud from the mesonephric duct and empties into the cloaca of reptiles and birds, urinary bladder of mammals.

11. The mammalian kidney tubule has a loop of Henle between proximal and distal convolutions. It lacks an afferent supply from the renal portal system except in monotremes. Renal corpuscles are in the cortex. Loops of Henle and common collecting tubules with associated vasa recta constitute the medulla.

12. Vertebrates excrete some salts extrarenally via gills, rectal glands in elasmobranchs, and nasal glands in marine reptiles, shore birds, and lizards and snakes in arid habitats.

13. Most vertebrates have urinary bladders. In fishes they are usually enlargements or evaginations of the mesonephric ducts (tubal bladders). In tetrapods they are evaginations of the floor of the cloaca, at least in the embryo. In dipnoans they are dorsal cloacal evaginations.

14. Gonads arise from paired gonadal ridges located close to the mesonephroi. Unpaired gonads result when paired ridges fuse or one fails to differentiate. Gonads tend to be displaced caudad in mammals.

15. The ovaries of most fishes are compact. Those of teleosts and amphibians are saccular. Those of reptiles, birds, and monotremes are lacunate. Placental mammalian ovaries are compact and often enclosed within an ovarian bursa.

16. Cyclostomes lack reproductive ducts, and the eggs and sperm exit from the coelom via genital pores. In most other vertebrates the mesonephric ducts carry sperm. Muellerian ducts convey eggs and the products of conception. The caudal ends of the muellerian ducts unite in female mammals above marsupials to form an unpaired uterine body and vagina.

17. Mammalian uteri are duplex, bipartite, bicornuate, or simplex, depending on the extent of fusion of the caudal ends of the muellerian ducts.

18. Male intromittent organs of fishes are modifications of pelvic or anal fins. Snakes and lizards have hemipenes. Turtles, crocodilians, some birds, and monotremes have an unpaired penis in the cloacal floor containing a corpus spongiosum. Mammals above monotremes have an external penis with a corpus spongiosum and two corpora cavernosa.

19. Female turtles, crocodilians, and mammals develop a homologue of the penis, a clitoris.

20. Most adult vertebrates below marsupials have a cloaca. In lampreys, chimaeras, male latimarians, and most ray-finned fishes the cloaca becomes shallow or nonexistent in adults.

21. In reptiles, birds, and monotremes the cephalic end of the cloaca is partitioned by a urorectal fold into a urodeum that receives the urinary and genital tracts, and a coprodeum that receives the large intestine. The caudal end of the cloaca is unpartitioned.

22. Above monotremes the cloaca is completely partitioned into urogenital sinus and rectum, and the two empty separately to the exterior.

23. In some rodents and most female mammals the urogenital sinus becomes subdivided into two passageways, so that there are three passageways to the exterior: urethra, vagina, and rectum.

24. Early gonads are indistinguishable as to sex, and duct systems for both sexes are present in embryos. Table 14-1 lists the chief sexually indifferent structures of the urogenital system of mammalian embryos and their fate in adult males and females.

SELECTED READINGS

Altschule, M.D.: The change(s) in the mesonephric tubules of human embryos ten to twelve weeks old, Anatomical Record **46**:81, 1930.

Breeder, C.M., and Rosen, D.E.: Modes of reproduction in fishes, Garden City, New York, 1966, Natural History Press.

Brodal, A., and Fänge, R., editors: The biology of *Myxine,* Oslo, 1963, Norway Universitetsforlaget.

Dunson, W.A.: Salt glands in reptiles. In Gans, C., and Dawson, W.R., editors: Biology of the reptilia, vol. 5, New York, 1976, Academic Press.

Forbes, T.R.: Studies on the reproductive system of the alligator. IV. Observations on the development of the gonad, the adrenal cortex and the Müllerian duct: Contributions in embryology, Carnegie Institution **28**:129, 1940.

Fox, H.: The urinogenital system of reptiles. In Gans, C., and Parsons, T.S., editors: Biology of the reptilia, vol. 6, New York, 1977, Academic Press.

Kent, G.C., Jr.: Reproductive systems of vertebrates. In Encyclopaedia Britannica, ed. 15, vol. 15, Chicago, 1974, Encyclopaedia Britannica, Inc., p. 707.

McCrady, E., Jr.: The development and fate of the urinogenital sinus in the opossum, *Didelphys virginiana,* Journal of Morphology **66**:131, 1940.

Moffat, D.B.: The mammalian kidney, New York, 1975, Cambridge University Press.

Mossman, H.W., and Duke, K.L.: Comparative morphology of the mammalian ovary, Madison, Wisconsin, 1973, University of Wisconsin Press.

Pang, P.K.T., Griffith, R.W., and Atz, J.W.: Osmoregulation in elasmobranchs, American Zoologist **17**:365, 1977.

Peaker, M., and Linzell, J.L.: Salt glands in birds and reptiles, Monographs of the Physiological Society, New York, 1975, Cambridge University Press.

Prosser, C.L., editor: Comparative animal physiology, ed. 3, vol. 2, Philadelphia, 1973, W.B. Saunders Co.

Schmidt-Nielsen, K.K.: Animal physiology, adaptation, and environment, ed. 3, Cambridge, England, 1983, Cambridge University Press.

Sharman, G.B.: Evolution of viviparity in mammals. In Austin, C.R., and Short, R.V., editors: Reproduction in mammals, book 6, Cambridge, England, 1976, Cambridge University Press, p. 32.

Stahl, B.J.: Vertebrate history: problems in evolution, New York, 1974, McGraw-Hill Book Co.

Taylor, D.H., and Guttman, S.I., editors: The reproductive biology of amphibians, New York, 1977, Plenum Press.

NERVOUS SYSTEM

In this chapter we will examine the nervous system and its parts. We will see how the components are assembled and what they do to ensure survival. Much of our attention will be focused on neurons, the living cells that perform the actual conducting, associative, and secretory activities of the system.

The vertebrate nervous system plays three basic roles. It acquaints the organism with its external environment and stimulates the organism to orient itself favorably in that environment; it participates in regulation of the internal environment; and it serves as a storage site for information. These functions are accomplished by the nerves, spinal cord, and brain in association with **receptors** (sense organs) and **effectors** (chiefly muscles and glands).

The organism exists in an external environment that is sometimes friendly, sometimes inimical, and seldom neutral. An environment is friendly if it contains food for nourishment, a mate for the propagation of the species, and shelter from enemies. An environment is unfriendly if it leads to the weakening of the organism or of the species.

The organism must constantly monitor the external environment in order that it may go deeper into a friendly area or withdraw from an unfriendly one. The information is supplied by afferent (sensory) nerves commencing in sense organs. The response (body movement) is initiated by nerve impulses over efferent (motor) nerves that stimulate the skeletal muscles of the body and thus cause the fish to swim or the tetrapod to crawl, run, or fly deeper into, or out of, an area. Information from the external environment is also employed in the regulation of internal secretions, such as seasonal release of reproductive hormones (Fig. 17-5).

The organism also has an internal environment that must be continually monitored and controlled. Afferent nerves from visceral receptors carry information in the form of nerve impulses to the central nervous system; efferent nerves carry impulses from the center to visceral effectors, chiefly smooth and cardiac muscles and glands.

Memory (information storage and recall) is a function of the nervous system. Without information storage no animal could modify its behavior in accordance with experience, and every situation would be faced as if it were the first time. In other words, there could be no *conditioned* responses. As experiences multiply, information accumulates and the penalty of past errors and the rewards of successes modify behavior accordingly. The brain seems to be the chief site of information storage.

The nervous system is subdivided for convenience into central and peripheral nervous systems. The **central nervous system** consists of the brain and spinal cord. The **peripheral nervous system** consists of cranial, spinal, and autonomic nerves, their branches, and certain autonomic ganglia and plexuses. The autonomic components innervate visceral effectors.

THE NEURON

To understand the anatomy of the nervous system, one must be acquainted with the neuron—the living nerve cell. The neuron is to the nervous system what a muscle cell is to the muscle system: it performs the specific function of the system. The neuron, rather than the nerve, transmits the nerve impulse. Neurons exhibit many shapes, but all have a **cell body** and one or more **processes** (Fig. 15-1). The longest process, distinguished cytologically by the absence of **Nissl material,** is the **axon,** or **nerve fiber** (Fig. 15-2). It transmits nerve impulses to a synapse or effector. Some nerve fibers extend short or long distances up or down the brain and spinal cord aggregated into functional groups called **fiber tracts** (*T* in Fig. 15-4, *A;* Fig. 15-10, low cervical). Nerve fibers in the peripheral nervous system are usually in nerves (Fig. 15-3). In fact, a **nerve** is one or more bundles of nerve fibers outside the central nervous system wrapped in a connective sheath **(epineurium)** and supplied by blood vessels, the **vasa nervorum.** The other processes of neurons are **dendrites** (Fig. 15-2). They are short extensions of the cell body that provide an increased surface for receipt of incoming impulses from axons. Like cell bodies, they display prominent Nissl mate-

FIGURE 15-1

Several morphological varieties of neurons. *A,* Motor neuron with cell body in the spinal cord or brain. The long process (axon) extends into the motor root of a spinal nerve and ends on a motor end plate as seen in Fig. 10-2. *B,* Dorsal root ganglion cell (sensory). The fiber terminates to the left in the spinal cord, where branches ascend or descend in the cord and also enter the dorsal horn of gray matter (Fig. 15-5, *B*). *C,* Sympathetic ganglion cell. *D* and *E,* Pyramidal and horizontal cells from the cerebral cortex. *F* and *G,* Purkinje and granular cells from the cerebellum. *H,* Group of embryonic dorsal root ganglion cells in transition from bipolar, *b,* to unipolar, *u.* *A, C, D, F,* and *G* are multipolar neurons, *E* is bipolar, *B* is unipolar. Arrows indicate direction of impulse.

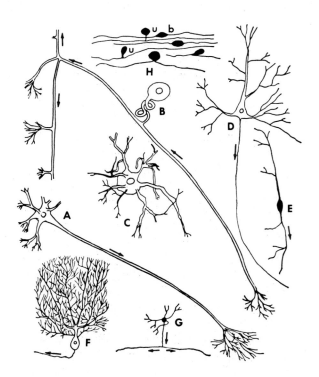

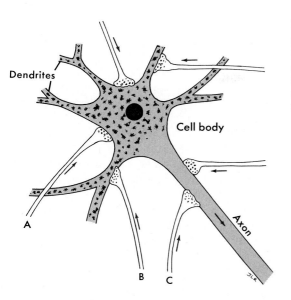

FIGURE 15-2

Synaptic endings on a motor neuron. *A,* Synapse between the synaptic knob (terminal button) of a telodendron and a cell body; *B,* between axon terminal and a dendrite; *C,* between axon terminal and another axon. Dark splotches in dendrites and cell body are Nissl material. A complete motor neuron is seen in Fig. 15-1, *A.*

FIGURE 15-3

Cross section of a small part of a nerve. The nerve fibers *(black dots)* are surrounded by fatty myelin sheaths of varying thicknesses.

rial, which constitutes sites of high protein synthesis. To observe how the neuron fits into the peripheral nervous system we will examine a sensory nerve, a motor nerve, and two mixed nerves.

A typical sensory nerve is diagrammed in Fig. 15-4, *A.* It commences in a sense organ (in this instance, the membranous labyrinth) and terminates in the brain. Like all nerves, it is made of nerve fibers. The cell bodies of sensory nerve fibers, with few exceptions, are found in a sensory ganglion on the pathway of the nerve. A **ganglion** is a group of cell bodies outside the central nervous system. A *sensory ganglion* contains sensory cell bodies. In lower vertebrates some sensory cell bodies are scattered along the nerve.

A typical motor nerve is diagrammed in Fig. 15-4, *B.* The cell bodies of most motor neurons are inside the central nervous system in a **motor nucleus.** Neurologically speaking, a **nucleus** is a group of cell bodies within the brain or cord. Motor nuclei contain the cell bodies of motor nerve fibers. The motor fibers of cranial nerve XII terminate in striated muscle. There are almost no purely motor nerves in vertebrates, since most nerves supplying somatic muscles have sensory fibers for proprioception from the muscle (Fig. 16-26).

Mixed nerves contain both sensory and motor fibers and are illustrated in Fig. 15-5. Their sensory cell bodies are in sensory ganglia, and, with the exception of some of the nerves of the autonomic nervous system, their motor cell bodies are in motor nuclei. Most vertebrate nerves are mixed.

The site where a nerve impulse is transferred from one neuron to another is a **synapse.** As an axon approaches a synapse, it sprays into a multitude of fine branches, or **telodendria,** each of which ends in a **synaptic knob (terminal button)** that is in contact with the cell body, dendrite, or axon of the next neuron (Fig. 15-2). Nerve impulses are propagated across the synapse by short-lived secretions, chiefly amines such as norepinephrine, acetylcholine, serotonin, melatonin, and others, which are released from the

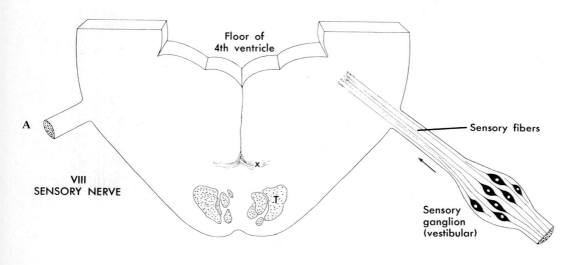

A

VIII
SENSORY NERVE

Floor of
4th ventricle

Sensory fibers

Sensory
ganglion
(vestibular)

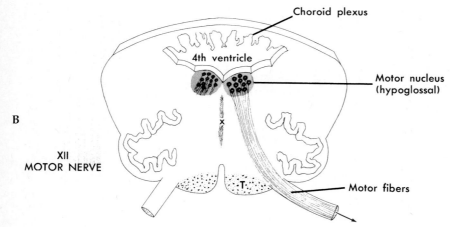

B

XII
MOTOR NERVE

Choroid plexus

4th ventricle

Motor nucleus
(hypoglossal)

Motor fibers

FIGURE 15-4

Typical locations of cell bodies *(black)* of sensory and motor fibers. **A,** Sensory nerve with sensory cell bodies in a sensory ganglion. Upon entering the brain the sensory fibers branch and pass toward synapses in several directions. **B,** Motor nerve (hypoglossal) with motor cell bodies in a motor nucleus in the brain. *T,* Descending fiber tract (corticospinal); *x,* decussating fibers. Arrows indicate direction of nerve impulses. Note the exceptional unipolar nature of the cell bodies in the vestibular ganglion.

synaptic knobs when a nerve impulse arrives. These amines are **neuro-transmitters.** Neurotransmitters are also released from axon terminals in contact with effectors, causing the effector (muscle, gland, pigment cell) to respond. The short life of neurotransmitters ensures against a protracted response from a single nerve impulse.

Most long axons within the central nervous system give off **collateral branches** along their pathway, thereby distributing a single impulse to a multiple of neurons. Collateral branches are seen on the descending branch of the sensory neuron illustrated in Fig. 15-1, *B.*

Some neurons **(neurosecretory neurons)** with cell bodies in the central nervous system secrete small polypeptides from their axon terminals. These polypeptides are **neurohormones.** Instead of terminating in synapses or at effectors, most neurosecretory fibers terminate at sinusoidal vascular channels into which they release their hormonal secretions (Fig. 17-1). They are discussed in Chapter 17.

FIGURE 15-5

Locations of cell bodies of mixed nerves. **A,** Mixed cranial nerve with sensory cell bodies in a sensory ganglion and motor cell bodies in a motor nucleus in the brain. Not all fiber components of this nerve are shown. **B,** Spinal nerve with sensory cell bodies in dorsal root ganglion and motor cell bodies in gray matter of ventral horn of cord. *C,* Central canal of cord; *D,* dorsal (posterior) horn of gray matter; *S,* somatic motor nucleus in ventral (anterior) horn of gray matter; *T,* descending fiber tract (corticospinal); *V,* visceral motor nucleus in lateral (visceral) horn of gray matter. Arrows indicate direction of nerve impulses.

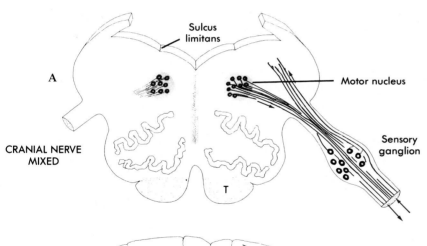

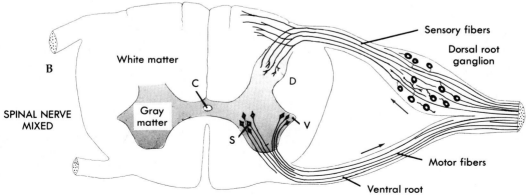

GROWTH AND DIFFERENTIATION OF THE NERVOUS SYSTEM

In order to gain additional insight into the architecture of the nervous system it is important to understand how the components of the nervous system develop. Neurulation has been discussed in Chapter 4. The next several sections describe maturation of the neural tube after neurulation and the embryonic origins of the neurons of the peripheral nervous system.

Neural Tube

An early embryonic neural tube is illustrated in Fig. 15-6. The cephalic end of the tube is the embryonic brain. The rest is future spinal cord. A cross section of the neural tube at this time exhibits three zones (Figs. 15-7, *B*, and 15-8, *A*): a **germinal layer** of actively mitotic cells, a **mantle layer** of cells proliferated from the germinal layer, and a **marginal layer.** Some of the mantle layer cells are **neuroblasts,** which sprout axons and dendrites to

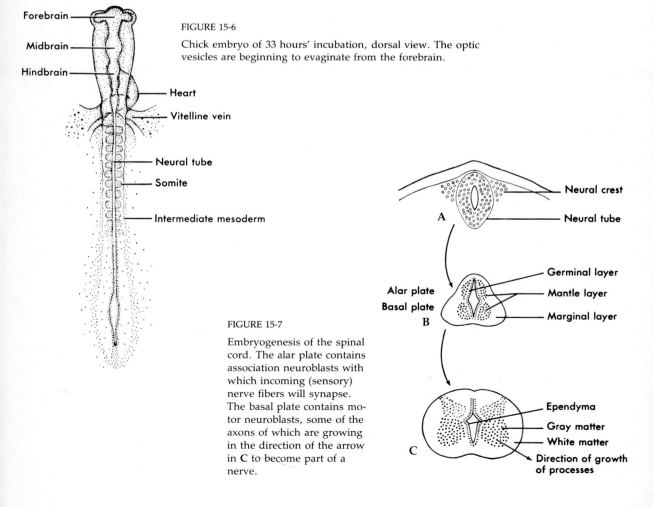

FIGURE 15-6

Chick embryo of 33 hours' incubation, dorsal view. The optic vesicles are beginning to evaginate from the forebrain.

Forebrain

Midbrain

Hindbrain

Heart

Vitelline vein

Neural tube

Somite

Intermediate mesoderm

Neural crest

Neural tube

Alar plate

Basal plate

Germinal layer

Mantle layer

Marginal layer

Ependyma

Gray matter

White matter

Direction of growth of processes

A

B

C

FIGURE 15-7

Embryogenesis of the spinal cord. The alar plate contains association neuroblasts with which incoming (sensory) nerve fibers will synapse. The basal plate contains motor neuroblasts, some of the axons of which are growing in the direction of the arrow in **C** to become part of a nerve.

become neurons. The rest are **spongioblasts** that give rise to neuroglia. As the neuroblasts differentiate, their axons grow into and add to the marginal layer, which therefore consists chiefly of nerve fibers. Because many axons become surrounded by a fatty myelin sheath, the marginal layer looks white when fresh and is called **white matter.** The cytoplasm of the cell bodies of the mantle layer causes this zone to look gray; hence the name **gray matter.**

Some of the nerve fibers that grow into the marginal layer turn upward or downward in the cord or brain to synapse with neurons elsewhere in the central nervous system. The earliest of these fibers lie close to the neurocoel and extend only one segment, a primitive condition. The later and longer ones become aggregated in long ascending and descending fiber tracts, each composed of functionally related fibers.

The embryonic cord, hindbrain, and midbrain consist of an **alar** and **basal plate** located, respectively, above and below a sulcus limitans (Figs. 15-7, B, and 15-34). The alar plate receives incoming (sensory) impulses, whereas the basal plate becomes motor in function.

When all the nerve cells have been formed that will ever be formed in the cord or brain, the cells of the germinal layer cease to divide. The undifferentiated cells that remain adjacent to the central canal **(neurocoel)** become ependymal cells (p. 523). The **ependyma** is the supportive lining of the central canal (Figs. 15-7, C, and 15-8, B, 5).

Development of Motor Components of Nerves

Many of the axons that sprout from neuroblasts in the basal plate grow out of, and away from, the neural tube (Fig. 15-7, C) to make contact with striated muscles. These become motor fibers of the cranial and spinal nerves. Because these fibers sprout from neuroblasts within the central nervous system, their cell bodies are within the adult brain or cord. Notice the location of motor cell bodies in Figs. 15-4, B, and 15-5.

Some of the fibers that sprout from neuroblasts in the basal plate grow out of the neural tube to make contact with neuroblasts in autonomic ganglia. These are **preganglionic fibers** of the autonomic nervous system (Fig. 15-11, PRE). The neuroblasts in autonomic ganglia (Fig. 15-11, sympathetic ganglion) sprout **postganglionic fibers** that grow toward, and innervate, smooth muscles and glands. Thus all motor neurons have their cell bodies in the cord or brain with one exception—postganglionic neurons of the autonomic system have their cell bodies in autonomic ganglia.

Most neuroblasts of autonomic ganglia are migrants from neural crests. A few, in amphibians at least, migrate outward from the basal plate of the neural tube.

Development of Sensory Components of Nerves

At the time the neural groove is closing to form a tube, a longitudinal ribbon of neurectoderm separates from the developing tube dorsolaterally on each side and soon segments to form a metameric series of **neural crests**

(p. 126), one pair in each body segment (Figs. 4-8, *B,* and 15-7, *A*). Some neural crest cells become neuroblasts that give rise to sensory neurons of spinal and cranial nerves. In doing so they pass through a bipolar stage (Fig. 15-1, *H*) in which one process grows into the alar plate of the cord or brain and the other grows into a sense organ. Thus there is established a neuronal connection between sense organ and central nervous system. Since each neural crest gives rise to a large number of sensory neurons, the result is a sensory ganglion on the nerve close to the central nervous system. Neurons whose cell bodies are in sensory ganglia are **first-order sensory neurons.** They conduct an impulse from a sense organ to the central nervous system. Inside the cord or brain they synapse with **second-order sensory neurons** (Fig. 15-11, 2) that conduct the impulse elsewhere within the central nervous system. They are sometimes called **association neurons.** Because most cell bodies of first-order sensory neurons arise from neural crests, we can make the following generalization: *the cell bodies of sensory neurons of cranial and spinal nerves are usually in sensory ganglia on the pathway of nerves.*

There are three major exceptions to the rule that the cell bodies of first-order sensory neurons are in ganglia on the pathway of nerves: (1) Olfactory nerve fibers sprout from neuroblasts in the embryonic olfactory epithelium. Their long processes grow into the nearest part of the brain, which is the olfactory bulb. Therefore the cell bodies of the olfactory nerves are in the olfactory epithelia (Figs. 15-27 and 16-20). (2) Neuroblasts that give rise to sensory fibers of the optic nerves are in the embryonic retinas (Fig. 16-6, *B,* optic cup), and their long processes grow brainward along the optic stalk. Therefore the cell bodies of optic nerve fibers are in the retina. Actually, the retina is part of the brain, since it arises as an evagination from the diencephalon and never separates from it. The term "optic nerve" is a misnomer. (3) The cell bodies of proprioceptive fibers in most, if not all, cranial nerves are not in ganglia. Neuroblasts within the alar plate of the embryonic midbrain sprout long processes that grow out to the muscles. Therefore the cell bodies of proprioceptive fibers in cranial nerves with the possible exception of the twelfth nerve of amniotes are in the midbrain in the **mesencephalic nucleus of the trigeminal nerve.**

Olfactory neurons and rods and cones (which are modified first-order sensory neurons) are the only neurons with cell bodies in sensory epithelia in vertebrates, and they are sometimes called **neurosensory cells.** These are the only sensory neurons in some diploblastic metazoans and appear to be phylogenetically older than sensory neurons with cell bodies in ganglia. The axons of some neurosensory cells in the tentacles of sea anemones (coelenterates) end in direct contact with *effector* cells, providing a fast firing one-neuron reflex arc. Such arcs are unknown in vertebrates. The phylogeny of neurons and their gradual organization into complex nervous systems have been discussed by Bullock and Horridge.

Most of the neuroblasts of the sensory ganglia of cranial nerves V, VII, VIII, IX, and X do not arise from neural crests but from epibranchial placodes. The contribution of neural crests to these ganglia varies with the species.

NEUROGLIA AND NEURILEMMA

Not all of the undifferentiated cells of the embryonic neural tube become neuroblasts. Nearly *half the bulk* of the brain and cord consists of a variety of interstitial cells known collectively as **neuroglia,** which provide mechanical and nutritive support for the neurons (Fig. 15-8). Glial cells arise from spongioblasts. They are smaller than most neurons, have nonnervous dendritic processes, and fill all the interstices in the brain and cord that are not occupied by neurons or blood vessels.

The oldest neuroglial elements are ciliated **ependymal cells** that line the neurocoel and have dendritic processes that extend radially to the margin of the neural tube. These are the only neuroglial elements of amphioxus and agnathans, and they are a stage in the ontogeny of the neuroglial cells of all higher vertebrates (Fig. 15-8, *A*). In an amphioxus, blood vessels do not enter the spinal cord, and the long dendritic processes of the ependymal cells evidently supply nutrients to the neurons. A progressive increase in the variety of glial cells is seen from cyclostomes to teleosts and from amphibians to amniotes, in which some ependymal cells lose their connection with the margins of the neural tube, become isolated among the nerve cell bodies and their processes, and become functionally specialized. Others form a **glial membrane** at the periphery of the cord and brain. Among glial cells are oligodendroglia, microglia, and astroglia (Fig. 15-8, *B*). The specific role of each variety is not yet fully known.

The dendritic processes of **oligodendroglia** wrap around axons in the central nervous system and elaborate myelin, a fatty insulator of electrical currents that speeds conduction of the nerve impulse in the axon (Fig. 15-3). **Microglia** are phagocytes that remove debris, including disintegration products of neurons after trauma or cell death. **Astroglia,** named

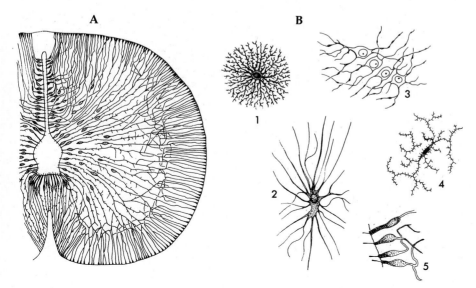

A   **B**

FIGURE 15-8

Neuroglia. **A,** Neural tube of a ten-week human fetus stained to show spongioblasts in the germinal, mantle, and marginal layers. Those that remain in the germinal layer at birth will be ependymal cells. After another two weeks differentiation of neuroglia will commence. **B,** Differentiated neuroglia. *1,* Protoplasmic astrocyte; *2,* fibrous astrocyte; *3,* oligodendroglia; *4,* microglia; *5,* ependymal cells. Not to scale. (**A,** after Cajal; **B,** after Rio Hortega.)

because their processes radiate like a star, have been implicated in the bioelectrical activity associated with the nerve impulse. In addition to serving in a number of metabolic roles, glial cells may play a yet-to-be-clarified role in information storage. Mechanical support seems only incidental to the other roles.

Glial cells are not found in nerves, but during ontogeny glialike **Schwann cells** migrate from the neural tube and neural crests into the nerves to form a living tube, the **neurilemma,** around each axon outside the central nervous system. Schwann cells, like oligodendroglia, elaborate myelin of varying thicknesses (Fig. 15-3). In general, heavily myelinated fibers conduct nerve impulses faster than lightly myelinated ones. The thickest sheaths are on fibers from encapsulated endings for touch, on proprioceptive fibers, and on motor fibers supplying striated muscle, which combination provides a very fast reflex motor response to danger signals arising on the body surface.

SPINAL CORD

The spinal cord occupies the vertebral canal surrounded by one or more meninges and a thick blanket of adipose tissue. Closely applied to the surface of the cord—and also of the brain—of most fishes is a rather delicate connective tissue membrane, the **meninx primitiva.** A similar meninx appears early in ontogeny in higher vertebrates; but during subsequent development in some teleosts and in amphibians, reptiles, and most birds the primitive meninx is replaced by an outer fibrous **dura mater** and an inner vascular **leptomeninx.** In mammals the leptomeninx ultimately differentiates into a weblike **arachnoid** membrane and a **pia mater.** The latter two meninges are united by connective tissue trabeculae that traverse a subarachnoid space occupied by a number of small vascular channels. The pia mater is intimately applied to the glial membrane. Thus mammalian cords and brains are surrounded by three meninges. Between the periosteum of the vertebrae and the dura mater is the blanket of peridural fat referred to above. It buffers the cord against mechanical trauma. It is not present in the skull.

The spinal cord commences at the foramen magnum, but in no vertebrate is there a specific landmark on the cord or brain that delimits the two. Instead, there is a gradual internal and external rearrangement of fiber tracts and nuclei on both sides of the foramen magnum which, in amniotes, is completed over the length of about one body segment. The transitional region is longer in lower vertebrates.

The adult cord extends to the caudal end of the vertebral column in vertebrates with abundant tail musculature. In other vertebrates the embryonic vertebral column elongates more rapidly than the spinal cord, with the result that at birth the cord is shorter than the column. In humans the spinal cord terminates at the third lumbar vertebra. In frogs it ends anterior to the urostyle. In a few bony fishes the cord is actually shorter than the brain. It is only an inch or so in length in one fish that is several feet long.

When the cord is as long as the vertebral column, each spinal nerve

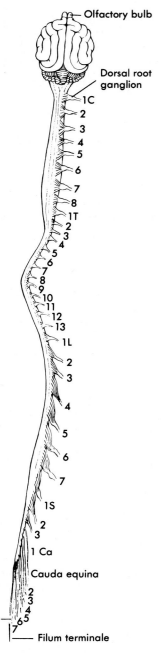

FIGURE 15-9

Brain and spinal cord of a cat. The epidural fat and dura mater have been removed. C, Cervical spinal nerve; T, thoracic; L, lumbar; S, sacral; Ca, caudal spinal nerves. Note origin of spinal nerves by multiple rootlets, and enlargements of the cord in the cervical and lumbar regions.

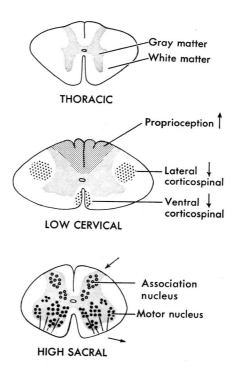

FIGURE 15-10

Human spinal cord in cross section at three levels, showing a few fiber tracts (arrows) at the cervical level and a few nuclei at the sacral level. The low cervical level is largest because it contains many cell bodies supplying the anterior limb, all fibers ascending to the brain from lower levels, and all fibers descending from the brain to lower levels. The corticospinal tracts carry voluntary motor impulses from the cerebral cortex. The motor horn in the thoracic region is small because there are no limb muscles to be supplied at this level. Association nuclei contain the cell bodies of second-order sensory neurons.

passes directly laterad to an intervertebral foramen through which it emerges from the vertebral canal. If, however, the column subsequently elongates more than the cord, the spinal nerves must then pass caudad within the vertebral canal to reach their foramina. As a result, the more caudal spinal nerves form a bundle of parallel nerves, the **cauda equina,** within the vertebral canal (Fig. 15-9). Nonnervous elements (ependyma and meninges) of the cord continue farther caudad as a delicate strand, the **filum terminale.**

The spinal cord exhibits cervical and lumbar enlargements at the level of the anterior and posterior appendages. The enlargements result from the large number of cell bodies and fibers innervating the appendage. When one pair of appendages is particularly muscular, such as the hind limbs of massive dinosaurs, the corresponding enlargement of the cord is especially pronounced. Conversely, the spinal cord of turtles is very slender in the trunk because the thoracic and abdominal musculature is greatly reduced. In many fishes the cord exhibits a neuroendocrine swelling, the **urophysis,** at its caudal end (Fig. 17-4).

The cord is flattened in cyclostomes but tends to be cylindrical or quadrilateral in higher vertebrates. In general, the neurocoel is relatively large in lower forms and constricted in higher ones.

A cross section of a typical cord reveals the nuclei arranged in a definite pattern surrounding the central canal, where they make up the gray matter (Fig. 15-10, high sacral). The nerve fibers occupy the periphery of the cord and, with neuroglia, constitute the white matter. The ascending and descending fibers are aggregated into fiber tracts that interconnect one level of the cord with another or with the brain. Fibers for touch constitute one tract, those for voluntary motor control another, and so forth (Fig. 15-10, low cervical). The fiber tracts of the cord are relatively few and simple in cyclostomes. They increase in number and complexity in amphibians because of the additional innervation of tetrapod limbs.

SPINAL NERVES
Roots and Ganglia

Spinal nerves, except in lampreys, have a dorsal and ventral root that unite just distal to the dorsal root ganglion (Fig. 15-11). Each root emerges from the cord as a series of rootlets which, in the dorsal root, unite proximal to the ganglion (Figs. 15-9, 15-32, and 15-37). Only in lampreys do the roots proceed to their destinations without uniting. As stated earlier, the dorsal root is predominantly sensory and the ventral root is purely motor. There is considerable evidence that in the earliest vertebrates (1) the dorsal and ventral roots did not unite but continued independently to their destinations; (2) the dorsal roots were mixed; (3) there were no dorsal root ganglia, bipolar sensory cell bodies having been scattered within the nerve along its course; and (4) when sensory cell bodies first aggregated in ganglia, they were bipolar. These conclusions are based partly on study of spinal nerves of lower chordates.

In an amphioxus only the dorsal root contains nerve fibers. These roots

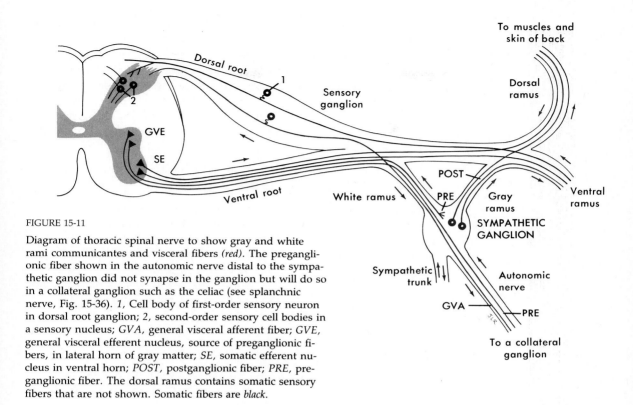

FIGURE 15-11

Diagram of thoracic spinal nerve to show gray and white rami communicantes and visceral fibers *(red)*. The pregangli-onic fiber shown in the autonomic nerve distal to the sympa-thetic ganglion did not synapse in the ganglion but will do so in a collateral ganglion such as the celiac (see splanchnic nerve, Fig. 15-36). *1,* Cell body of first-order sensory neuron in dorsal root ganglion; *2,* second-order sensory cell bodies in a sensory nucleus; *GVA,* general visceral afferent fiber; *GVE,* general visceral efferent nucleus, source of preganglionic fi-bers, in lateral horn of gray matter; *SE,* somatic efferent nu-cleus in ventral horn; *POST,* postganglionic fiber; *PRE,* pre-ganglionic fiber. The dorsal ramus contains somatic sensory fibers that are not shown. Somatic fibers are *black.*

arise from the cord at the level of a myoseptum and pass into the myosep-tum to be distributed to the skin (sensory) and viscera (visceral motor). The cell bodies for the visceral motor fibers are in the cord, and those for the sensory fibers are either in the cord or within the nerve, scattered along much of its length. Therefore there is no aggregation of sensory cell bodies at one location on a nerve, which means there are no sensory ganglia in an amphioxus. Also, the sensory cell bodies remain bipolar throughout life.

The ventral roots of amphioxus emerge from the cord between two myo-septa, do not unite with dorsal roots, and enter myomeres. Electron micros-copy indicates that their ventral roots are unique, containing not axons but bundles of very delicate long extensions of striated muscle fibers from the myomeres. These muscle filaments enter the cord through the ventral root. Inside the cord they are stimulated by nerve fibers that synapse with them. Thus the somatic muscles actually "come to the cord" for their stimuli. An analogous condition exists in echinoderms.

Dorsal and ventral roots alternate and remain independent in lampreys but unite in hagfishes except in the tail. Some of the cell bodies of sensory fibers of agnathans are aggregated in ganglia on the dorsal root, and most of these are bipolar, as in the embryo. Others are within the cord. Visceral motor fibers are in both roots, and the ventral root is entirely motor.

Above cyclostomes dorsal and ventral roots always unite. The dorsal root still contains numerous visceral motor fibers in many bony fishes, but in cartilaginous fishes and tetrapods most of the visceral motor fibers have been lost from the dorsal root. The cell bodies of first-order sensory neurons are in dorsal root ganglia. They are bipolar in cartilaginous fishes; bipolar, intermediate, and unipolar in bony fishes; chiefly unipolar in amphibians; and almost entirely unipolar in amniotes (Fig. 15-1, *B*). The ventral root is motor (somatic and visceral) with cell bodies within the cord.

Metamerism

In gnathostomes a spinal nerve arises from each segment of the cord except near the end of the tail. These nerves are metamerically distributed to the body wall and tail. At the level where a fin or limb bud forms, they supply the appendage (Figs. 10-15 and 15-12). The segmental distribution of spinal nerves in the somatopleure is best illustrated in fishes, since the metamerism of their body wall muscles is relatively undisturbed. Tadpoles have as many as 40 pairs of spinal nerves and lose all but 10 pairs when the tail is resorbed at metamorphosis.

Rami and Plexuses

Shortly after emerging from the vertebral canal, each typical spinal nerve divides into at least two branches (see 3 and 4 in Fig. 1-2). A **dorsal ramus** supplies the epaxial muscles and skin of the dorsum. A larger **ventral ramus** passes into the lateral body wall and supplies the hypaxial muscles and skin to the midventral raphe. In the thoracic and lumbar regions, **rami communicantes** connect these spinal nerves with the ganglia of the sympathetic trunk (Figs. 15-11 and 15-37, white ramus, gray ramus). They carry visceral fibers.

FIGURE 15-12

Innervation of the skin of the mammalian forelimb by successive spinal nerves. *C,* Cervical, and *T,* thoracic somites and area of cutaneous distribution of their associated nerves.

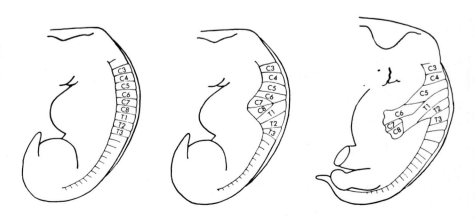

TABLE 15-1   Fiber Components of Typical Spinal Nerves

Components	Innervation
Sensory	
General somatic afferent fibers (GSA)	General cutaneous receptors (touch, pain, temperature, and pressure)
	Receptors on striated muscle, tendons, and bursae (proprioceptive)
General visceral afferent fibers (GVA)	Viscera, including general receptors in endoderm
Motor	
Somatic efferent fibers (SE)*	Myotomal muscle
General visceral efferent fibers (GVE)†	Smooth and cardiac muscle, and glands

*The fibers to myotomal muscle are designated simply as SE, rather than as GSE (general somatic efferent), because there are no special somatic efferent fibers. They include the extrinsic eyeball muscles.

†Autonomic fibers. Visceral fibers to skin are vasomotor (to arterioles), plumomotor or pilomotor (to erector muscles of feathers and hairs), or secretory (to skin glands), and also supply melanophores in lower vertebrates.

The ventral rami of successive spinal nerves often unite to form a plexus from which large nerve trunks arise (Fig. 10-15, mammal). The chief plexuses are the brachial and pelvic (lumbosacral in amniotes), which supply nerves to the anterior and posterior appendages. These plexuses are relatively simple in fishes but become increasingly complicated in tetrapods (compare shark and mammal, Fig. 10-15). Autonomic plexuses occur on visceral pathways.

Occipitospinal Nerves

In many fishes and amphibians one or more pairs of **occipitospinal nerves** arise between the vagal nerve and the first pair of spinal nerves (Fig. 15-30). They supply the hypobranchial musculature, including the tongue when present, and usually lack sensory roots. Embryonic frogs have an occipitospinal nerve immediately cephalad to the spinal nerve that supplies the tongue, but it becomes suppressed during later development. Cranial nerves XI and XII of amniotes lack sensory roots and appear to be derived in part from occipitospinal nerves.

Fiber Components of Spinal Nerves

The nerve fibers in a typical spinal nerve are of four functional varieties, two sensory and two motor (Table 15-1). Three of the varieties are referred to as **general** fibers (GSA, GVA, and GVE) to differentiate them from **special** types found only in cranial nerves.

BRAIN

The cephalic end of the embryonic neural tube in every vertebrate exhibits three primary brain vesicles—the future forebrain (**prosencephalon**), midbrain (**mesencephalon**), and hindbrain (**rhombencephalon**) (Fig. 15-6). The embryonic prosencephalon differs from the rest of the neural tube in that it is not divided into alar and basal plates. In adults it consists of two regions, **telencephalon** and **diencephalon.** The midbrain (mesencephalon) develops without further subdivision. The hindbrain differentiates into

FIGURE 15-13

Vertebrate brains. The posterior choroid plexuses have been removed to expose the fourth ventricle in necturus, frog, and snake. The prosencephalon, mesencephalon, and rhombencephalon are differentially colored. *1* to *9,* See perch for key; *10,* auditory lobe. Mammalian brains are shown in Fig. 15-22.

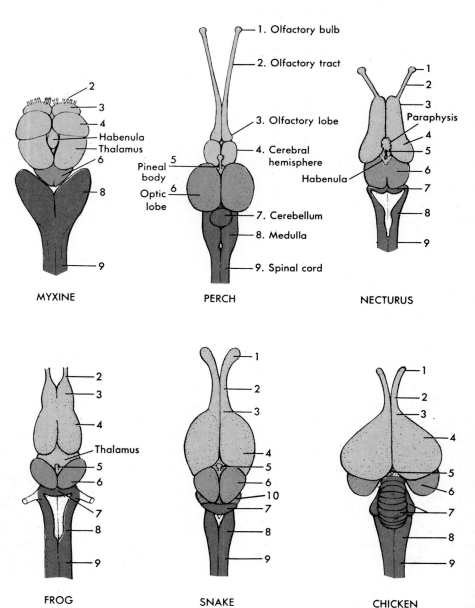

1. Olfactory bulb
2. Olfactory tract
3. Olfactory lobe
4. Cerebral hemisphere
5. Pineal body
6. Optic lobe
7. Cerebellum
8. Medulla
9. Spinal cord

Habenula
Thalamus

MYXINE

Paraphysis
Habenula

NECTURUS

PERCH

Thalamus

FROG

SNAKE

CHICKEN

TABLE 15-2   Major Subdivisions and Components of the Brain*

Component	Subdivision	
	Telencephalon	Rhinencephalon Cerebral hemispheres *Lateral ventricles*
Prosencephalon (forebrain)		
	Diencephalon	Epithalamus Thalamus Hypothalamus *Third ventricle*
Mesencephalon (midbrain)		Tectum Tegmentum *Cerebral aqueduct*
	Metencephalon	Cerebellum Tegmentum *Fourth ventricle*
Rhombencephalon (hindbrain)		
	Myelencephalon	Medulla oblongata *Fourth ventricle*

*For a listing of all parts discussed in this chapter, see p. 564.

metencephalon and **myelencephalon.** Differentiation involves thickening of the lateral walls and floor in some places, and dorsal, lateral, or ventral evagination in others until the adult brain has taken shape. Homologous parts in adults are readily demonstrable throughout the vertebrate series (Fig. 15-13). To display them from dorsal view in mammals requires removal of the overgrown cerebral hemispheres and cerebellum (Fig. 15-25). When these are removed, what remains is the **brain stem.** The brain, like the cord, is surrounded by meninges. The major subdivisions of the brain are listed in Table 15-2.

Metencephalon and Myelencephalon: The Hindbrain

The myelencephalon of the hindbrain is represented chiefly by the **medulla oblongata,** which merges imperceptibly with the spinal cord. The area of transition is characterized internally by gradual relocation of the fiber tracts, which constitute the white matter. As a result, whereas in the cord the gray matter is compact and centrally located in dorsal and ventral horns, in the brain it becomes dispersed into small and large nuclear masses of gray matter interspersed among fiber tracts (white matter).

The most conspicuous dorsal feature of the hindbrain is the **cerebellum,** a dorsal evagination of the metencephalon. It functions in the coordination of the skeletal muscles in response to input from the membranous labyrinth, lateral-line canals, and proprioceptors in muscles, joints, and tendons, and from reflex and voluntary motor centers in the forebrain. Its size is correlated with the complexity of the activities of the striated musculature. It is larger in fishes than in amphibians since swimming, which involves schooling, vertical movements, adjusting to water currents, and keeping the dorsal part of the body from tipping over, requires more synergistic muscle activity than dragging the belly along the ground or squatting on a lily pad. Lacking a large cerebellum, aquatic urodeles rely to a large extent on spinal cord reflexes and primitive nuclei in the hindbrain for muscle coordination when swimming. The cerebellum is largest in birds and mammals, which require a large computer-like neural center to coordinate the muscles of the head, neck, trunk, and appendages in such diverse activities as flying, running, climbing, balancing, or, in the case of human beings, playing a piano. The cell bodies of the cerebellum are in a cortex on the surface, a condition found elsewhere only in the cerebral hemispheres of higher amniotes. In cyclostomes the cerebellum is not well developed and does not cause a bulge on the brain.

Other topographical features of the hindbrain include various swellings that indicate underlying nuclei and elevated ridges or transverse bands that contain fiber tracts. These topographical markings are most prominent in mammals. However, one nucleus **(nucleus solitarius)** in the alar plate becomes enormous in fishes that have taste buds over the entire surface of the body. The resulting swelling, or **vagal lobe** (Fig. 15-14), is the termination of the many incoming sensory fibers for taste and contains the second-order sensory neurons whose fibers are projected to reflex and relay centers elsewhere in the brain. Among ventral ridges on the mammalian hindbrain are the **pyramids,** which contain the **corticospinal (pyramidal) tracts** that carry voluntary motor impulses from the cerebral cortex; and the **pons,** which consists of fibers crossing (decussating) from one side of the brain to the other (Fig. 15-15).

The cavity of the hindbrain is the fourth ventricle (Figs. 15-4, *B,* and 15-21). The cerebellum is part of its roof. The rest of the roof is a membranous **tela choroidea,** part of which hangs into the ventricle as the **choroid plexus of the fourth ventricle** (Fig. 15-17).

FIGURE 15-14

Brain of the buffalo fish *Carpiodes velifer*. Note unusual bulge (vagal lobe) on the alar plate of the medulla. Here terminate the many incoming taste fibers characteristic of bottom feeders.

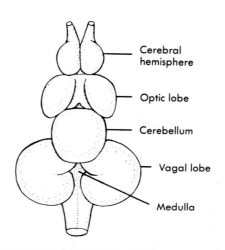

Cerebral hemisphere

Optic lobe

Cerebellum

Vagal lobe

Medulla

MODIFICATION FOR BOTTOM-FEEDING

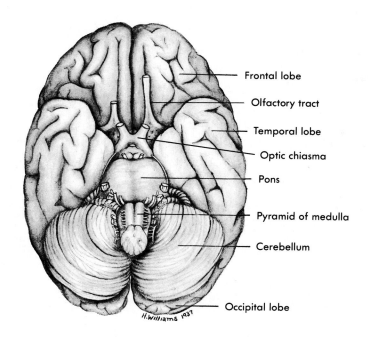

Frontal lobe

Olfactory tract

Temporal lobe

Optic chiasma

Pons

Pyramid of medulla

Cerebellum

Occipital lobe

H.Williams 1937

FIGURE 15-15

Brain of man, ventral view. The olfactory bulbs and pituitary have been cut away. The cut stump of the infundibular stalk is seen immediately caudal to the optic chiasma. (Adapted from Francis, C.C., and Martin, A.H.: Introduction to human anatomy, ed. 7, St. Louis, 1975, The C.V. Mosby Co.)

Mesencephalon: The Midbrain

The roof of the mesencephalon, or **tectum,** displays a pair of prominent **optic lobes** in all vertebrates. These bulging gray masses serve partly as optic reflex centers that receive fibers from the retina. They are especially large in birds, which have large eyes and rely on visual stimuli for much information about the environment. A pair of **auditory lobes** lies caudal to the optic lobes in the tectum commencing with reptiles, and the four bodies (optic lobes and auditory lobes) constitute the **corpora quadrigemina** (Fig. 15-25). Fishes have auditory nuclei in this location but they are not large enough to bulge from the surface. The auditory lobes receive input from the part of the membranous labyrinth that is sensitive to vibratory stimuli and from other sources. Phylogenetically, they enlarge along with the cochlea. A prominent nucleus in the alar plate (not in the tectum) of the midbrain is the **mesencephalic nucleus of the trigeminal nerve.** It contains the cell bodies of the proprioceptive fibers of most, if not all, cranial nerves that have them.

The part of the mesencephalon that lies in the floor of the neurocoel and includes the basal plate is greatly thickened by numerous motor nuclei and ascending and descending fiber tracts. This is the **tegmentum,** and it is continuous with the tegmentum of the hindbrain. In mammals some of these tracts become massive. One pair, the **cerebral peduncles,** is visible on the ventral surface of the midbrain where they emerge to the surface just caudal to the optic chiasma.

The ventricle of the midbrain is quite large in fishes and amphibians and extends dorsally into the optic lobes as the **optocoel.** In higher vertebrates the optic lobes are not hollow, and the midbrain ventricle is constricted to a narrow **cerebral aqueduct** (Fig. 15-23), also called the **aqueduct of Sylvius.**

Diencephalon

The diencephalon of vertebrates from fishes to mammals consists of three major components, **epithalamus, thalamus,** and **hypothalamus** (Fig. 15-16). Its neurocoel is the third ventricle. Intimately associated with the floor of the diencephalon is the **adenohypophysis,** known also as the anterior lobe of the pituitary. The major components of the diencephalon will be discussed in the sections that follow.

EPITHALAMUS. The epithalamus is the dorsalmost component of the diencephalon and, as such, is the roof of the third ventricle. It consists of a **pineal** or **parapineal organ** or both (Figs. 15-17 and 16-19), a thin but complex membrane that supports a **choroid plexus** (to be discussed shortly), and a pair of knoblike thickenings, the **habenulae** (Fig. 15-16). When a pineal and parapineal are both present, they are referred to collectively as the **epiphyseal complex.**

The pineal is a club-shaped or knoblike organ, sometimes threadlike or saccular, projecting above the diencephalon and, in amniotes, wedged between the caudal poles of the enlarged cerebral hemispheres (Figs. 15-13, 15-23, and 15-25). It is connected to the diencephalon by a stalk that sometimes contains an extension of the third ventricle. In lampreys, at least, the pineal is a photoreceptor. Otherwise, it functions as an endocrine organ

FIGURE 15-16

Diencephalon of a shark in cross section, generalized. Third ventricle is in *red.*

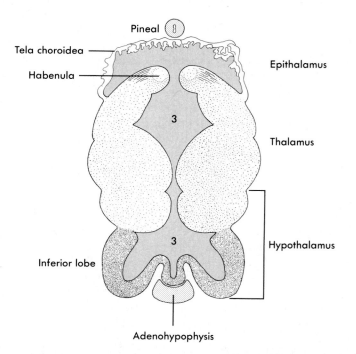

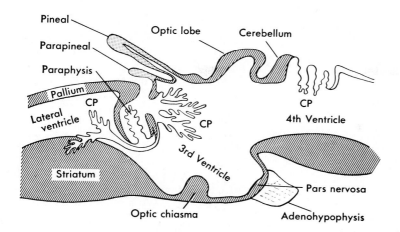

FIGURE 15-17

Diencephalon and adjacent areas of a larval anuran brain, sagittal section, anterior end to the left. *CP*, Choroid plexus of lateral, third, and fourth ventricles. The choroid plexus of the lateral ventricle enters the latter via an interventricular foramen. The slight swelling between the base of the pineal and optic lobe is the site of a transverse fiber tract, the habenular commissure that connects the two habenulae.

that is stimulated in part by light entering the lateral eyes, the impulses being relayed through cranial nuclei and cranial, spinal, and autonomic nerves. The pineal is vestigial or absent in hagfishes, one genus of electric rays, crocodilians, and some adult mammals (sirenians, porpoise, and others). It is relatively large in humans and sheep.

The parapineal was a constant feature of bony fishes of the Devonian and of ancestral amphibians and reptiles, serving as a median, or parietal, eye. It is now found only in a few bony fishes, larval anurans, *Sphenodon*, and some lizards in which it occupies a foramen of the skull just under a patch of translucent skin (Fig. 16-18). The pineal and parapineal at one time may have constituted a *pair* of dorsal photoreceptors. This is suggested by their embryogenesis and the relationships and morphology of the two organs in lampreys (Fig. 16-19, *A*). Both organs are present, the parapineal being connected by nerve tracts with the left side of the brain and the pineal with the right, and both contain photosensory cells. The sensory function of the pineal and parapineal organs is discussed in Chapter 16, and the endocrine function of the pineal is discussed in Chapter 17.

The habenulae are elevations caused by the presence of left and right habenular nuclei that preceded the evolution of the cerebral cortex. They receive fibers from olfactory centers, hypothalamus, and other nuclei of the forebrain and discharge over a main fiber tract and several accessory tracts to the thalamus and midbrain, being particularly involved in correlating reflex responses to olfactory stimuli. In Fig. 15-25 a paired fiber tract, the **stria medullaris,** is seen on the surface of the thalami passing toward the habenulae, which, in sheep, lie under cover of the pineal body. The stria medullaris is present in all vertebrates. The habenulae are largest in species that rely heavily on smell for locating food, such as sharks and bloodhounds, and they are inconspicuous in birds, which have a poorly developed olfactory sense, and in aquatic mammals, some of which, it may be recalled from Chapter 3, lack olfactory nerves. The habenulae of human beings are comparatively small. Those of hagfishes are united across the midline, but the underlying nuclei are paired (Fig. 15-13, *Myxine*).

THALAMUS. The thalamus is the largest subdivision of the diencephalon. It is a paired mass of many nuclei in the lateral walls of the third ventricle and comes to the surface dorsally just behind the cerebral hemispheres (Fig. 15-13, *Myxine* and frog). In amniotes it is hidden by the caudal poles of the hemispheres, which must be cut away to reveal it (Fig. 15-25). All sensory pathways ascending to the telencephalon from the spinal cord, hindbrain, or midbrain, synapse in a thalamic nucleus before continuing to the telencephalon. With development of a cerebral cortex in amniotes the thalamus becomes increasingly prominent, relaying more and more sensory information to the evolving cerebral cortex. As a consequence, the thalamic nuclei of mammals have become so numerous that the left and right thalami bulge into the third ventricle and meet at one site to form an oval bridge of gray matter, the **massa intermedia,** or false commissure (Figs. 15-21 and 15-23). The latter term reflects the fact that it does not consist of decussating fiber tracts as do other commissures.

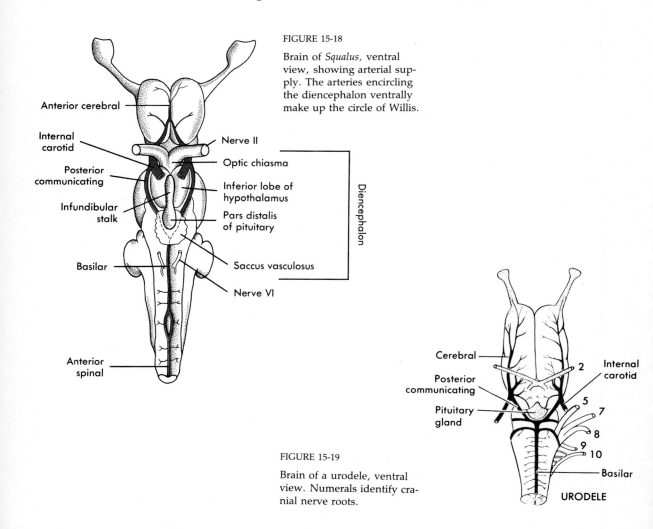

FIGURE 15-18

Brain of *Squalus,* ventral view, showing arterial supply. The arteries encircling the diencephalon ventrally make up the circle of Willis.

Anterior cerebral

Internal carotid

Posterior communicating

Infundibular stalk

Basilar

Anterior spinal

Nerve II

Optic chiasma

Inferior lobe of hypothalamus

Pars distalis of pituitary

Saccus vasculosus

Nerve VI

Diencephalon

FIGURE 15-19

Brain of a urodele, ventral view. Numerals identify cranial nerve roots.

Cerebral

Posterior communicating

Pituitary gland

2

Internal carotid

5

7

8

9

10

Basilar

URODELE

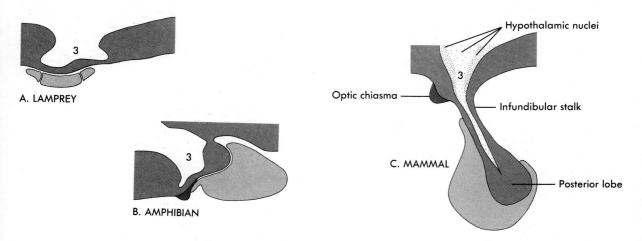

A. LAMPREY

B. AMPHIBIAN

C. MAMMAL

Hypothalamic nuclei

Optic chiasma

Infundibular stalk

Posterior lobe

FIGURE 15-20

Infundibular recess, 3, of third ventricle of lamprey, amphibian, and mammal. Adenohypophysis is *light gray.*

HYPOTHALAMUS. A ventral view of the diencephalon shows three prominent structures, the optic chiasma, hypothalamus, pituitary, and, in fishes only, a saccus vasculosus (Fig. 15-18). The **optic chiasma** is the cephalic boundary of the diencephalon ventrally. It is where the optic nerves reach the brain and where some or all of the nerve fibers cross to the opposite side before entering the brain to become optic tracts (Figs. 15-15, 15-19, and 15-29). The adenohypophysis of the pituitary is caudal to the optic chiasma and intimately associated with the diencephalic floor which, in sharks and higher vertebrates, evaginates to form an **infundibular (pituitary) stalk** (Fig. 15-20).

The **hypothalamus** is the floor and ventrolateral walls of the third ventricle (Figs. 15-16 and 15-20, *C*, hypothalmic nuclei). It is prominent on the ventral surface of the diencephalon in cartilaginous fishes because of the extraordinary size of the **inferior lobes,** which have connections with the cerebellum (Figs. 15-16 and 15-18). In all vertebrates the hypothalamus is an important neural center for homeostasis. Hypothalamic nuclei exert major reflex control over the autonomic nervous system, produce neurohormones that regulate the pituitary and gonads, and monitor the sodium chloride and glucose content of the blood. There is an appetite regulating center and, in endotherms, a temperature regulating center. In general, the anterior nuclei are associated with parasympathetic functions, the caudal nuclei with sympathetic ones. Passing around and, sometimes, through the nuclei are networks of nerve fibers and fiber tracts that communicate in a highly organized, genetically determined pattern with all parts of the brain; and entering and leaving the nuclei are afferent and efferent nerve fibers from some of these tracts.

The **saccus vasculosus,** or infundibular organ, is a highly vascularized, thin-walled ventral evagination of the diencephalic floor of elasmobranch and ray-finned fishes, located just behind the pituitary (Fig. 15-18). Within it is a fluid-filled recess of the third ventricle. Because it is a sense organ, a discussion of the saccus vasculosus will be found in Chapter 16.

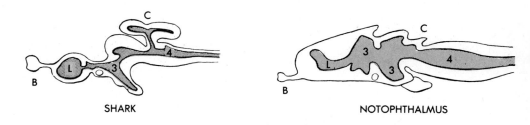

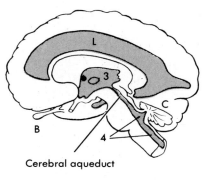

FIGURE 15-21

Sagittal brain sections show-ing the ventricles in *red. B,* Olfactory bulb; *C,* cerebel-lum; *L, 3,* and *4,* lateral, third, and fourth ventricles. The *black* foramen in the third ventricle in humans is the interventricular foramen. Immediately behind it is the massa intermedia.

THIRD VENTRICLE. The third ventricle is continuous caudad with the ventricle of the midbrain, and cephalad with the ventricle in each cerebral hemisphere via left and right **interventricular foramina** (Figs. 15-17 and 15-21). It becomes increasingly compressed laterally with expansion of the thalami toward the midline. An optic recess of the ventricle extends toward the optic chiasma, and an infundibular recess extends into the pituitary stalk (Fig. 15-20). Dorsally, a small evagination extends upward into the stalk of the pineal body (Figs. 15-17 and 16-19). The nonnervous roof of the ventricle **(tela choroidea)** remains thin, becomes highly vascularized, and hangs into the ventricle as the **choroid plexus of the third ventricle.**

Telencephalon

The telencephalon consists of **cerebral hemispheres** and **rhinencepha-lon.** In fishes, the olfactory lobe of the rhinencephalon is as prominent as the cerebral hemispheres, which reflects the importance of olfaction in their survival (Fig. 15-30). The hemispheres of reptiles and birds have increased in size (Fig. 15-13), and in mammals they have grown forward over the rhinencephalon, relegating it to an inconspicuous anteroventral location (Figs. 15-22 and 15-23). The increased size of the hemispheres mirrors their dominant role in mammalian behavior.

At the boundary between diencephalon and telencephalon immediately anterior to the epiphyseal complex, the thin roof of the ventricle evaginates upward in some members of every vertebrate class to form a wrinkled, thin-walled sac, the **paraphysis** (Fig. 15-17). It is present in lampreys, sharks, and urodeles but is seldom seen in the classroom because it is easily torn. It is confined to embryos in most amniotes other than *Sphenodon*. Little is known of its function. It resembles a choroid plexus, but its role seems to be different. The paraphysis is assigned arbitrarily to either the telencephalon or diencephalon since in some embryos its specific origin is not evident.

The neurocoel of the telencephalon is continuous with the third ventricle. Except in actinopterygians it consists of a pair of lateral ventricles, one in each cerebral hemisphere. The paired condition is a result of the ontog-

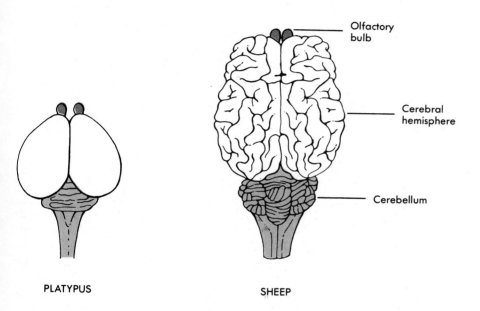

PLATYPUS

SHEEP

FIGURE 15-22

Brain of a primitive mammal (platypus) lacking cortical gyri, and brain of sheep. *1*, Olfactory bulb; *2*, cerebral hemisphere; *3*, cerebellum.

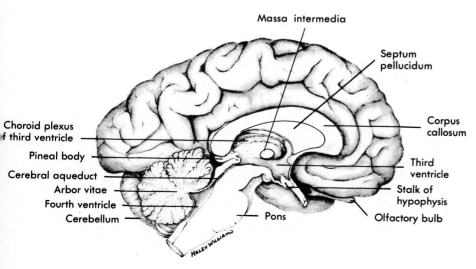

FIGURE 15-23

Human brain, left half, sagittal section. (Adapted from Francis, C.C., and Martin, A.H.: Introduction to human anatomy, ed. 7, St. Louis, 1975, The C.V. Mosby Co.)

FIGURE 15-24

Evolution of the cerebral hemispheres as seen in cross sections. Only the left hemisphere is shown in the lower figures. Reptiles have added a neostriatum to the old paleostriatum. Birds added a hyperstriatum. Note the striatal complex (now called basal ganglia) in mammals, and the addition of a cortex on the surface of the mammalian hemisphere. In bony fishes the telencephalic ventricle is unpaired.

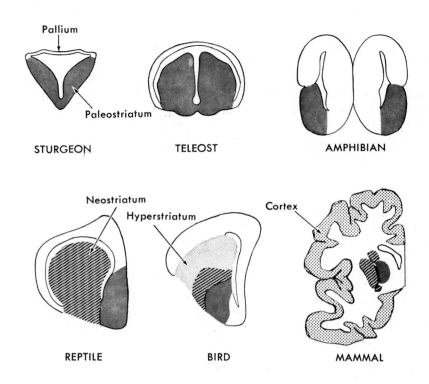

eny of the hemispheres. They commence as a pair of anteriorly directed evaginations, or **telencephalic vesicles,** of the early forebrain that arise and bulge forward immediately anterior to the optic vesicles. (The prior optic vesicle stage of development of the forebrain is seen in Fig. 15-6.) Within each evagination is an extension of the neurocoel. As a result, each cerebral hemisphere contains a lateral ventricle that is continuous with the third ventricle via an aperture which, in adults, is an interventricular foramen. In the midline between the two interventricular foramen of adults is a perpendicular **lamina terminalis** that demarcates the original cephalic boundary of the embryonic prosencephalon. The extent of development of the telencephalic vesicles once they have evaginated varies with the phylogenetic level of the taxon. The process of development of the telencephalon of actinopterygians is such that the telencephalic ventricle is median rather than paired (Fig. 15-24, sturgeon and teleost).

RHINENCEPHALON. The rhinencephalon differentiates at the anterior end of the telencephalon. It consists of **olfactory bulbs, tracts,** and **lobes** (Fig. 15-13, perch, and others). In mammals it is dwarfed by the huge cerebral hemispheres and often hidden by them when the brain is viewed from above (Figs. 15-22 and 15-23). The olfactory bulbs lie close to the olfactory epithelium separated by the olfactory capsule. Fiber tracts connect the rhinencephalon with the cerebral hemispheres and certain other parts of the brain. The rhinencephalon is rudimentary or regressed in many marine mammals, and is poorly developed in birds other than those that feed on carrion, vision having become the predominant source of environmental information in birds. An accessory olfactory bulb is present in tetrapods that have a large vomeronasal organ (p. 592).

CEREBRAL HEMISPHERES. When one thinks of cerebral hemispheres, what usually comes to mind are the huge cerebral hemispheres of mammals with their thick cortex of gray matter. However, cerebral cortex is a relatively recent acquisition. The hemispheres of fishes are primitive.

Each hemisphere of fishes consists chiefly of a **paleostriatum** (Fig. 15-24, sturgeon and teleost). The paleostriata are masses of motor nuclei whose cell bodies receive chiefly input from the rhinencephalon anteriorly and, to a lesser degree, from some of the sensory nuclei of the brain stem and cord, the latter being relayed via synapses in the thalamus. Efferent pathways from the striata form ancient descending tracts that end in motor nuclei of cranial and spinal nerves. Thus olfactory and other sensory stimuli result in reflexive, survival-oriented responses of the skeletal musculature of the head, trunk, and tail, chief among which are locomotion and feeding. The density and arrangement of the bundles of incoming and outgoing nerve fibers traversing the area are responsible for the striated appearance of this region when viewed with light microscopy. The paleostriatum, ventricle (unpaired in bony fishes), and **pallium,** which is the thin roof of the ventricle, constitute the chief mass of the cerebral hemispheres of fishes (Fig. 15-13, perch, and 15-30). The paleostriatum (corpus striatum) is the highest level to which sensory impulses of fishes are projected, and therefore the highest motor center of the brain in these vertebrates. There are considerable differences in the structure and fiber connections of the hemispheres of the major groups of fishes, and the functional significance of their components is, as yet, poorly understood.

The paleostriatum is still prominent in the hemispheres of amphibians, although additional telencephalic nuclei have developed, some of which establish additional connections for olfactory functions. Others receive an increased number of projections of sensory fibers from the thalamus. Amphibian hemispheres are still under the predominant control of olfactory stimuli, but they play a larger role in correlating reflex motor responses to a variety of external stimuli than do the cerebral hemispheres of fishes.

The cerebral hemispheres of reptiles are massive compared to those of amphibians. This is largely a result of the development of additional nuclei and associated fiber tracts and neuroglia that constitute a **neostriatum** in the ventrolateral wall of the hemispheres (Fig. 15-24, reptile). The hemispheres now bulge laterally, dorsally, and backward over the diencephalon so that most of the latter is no longer visible from dorsal view (Fig. 15-13, snake). The paleostriatum continues to receive olfactory input and to be a center for initiating motor activity. The neostriatum acquires sensory input from the thalamus and projects many fibers into the paleostriatum, thereby adding capacity for coordination of sensory input, the effect of which is reflected in the nature of skeletal muscle responses. As will be recalled, the skeletal musculature of reptiles, especially that of the head, neck, and limbs, has become much more complex than that of amphibians. Subsequent comparative neurological studies involving morphology, physiology, and behavior will alter some of the current perception of the functional significance of the striata of reptiles. In reptiles for the first time a trace of cortex appears on the pallium.

Bird hemispheres are essentially reptilian, but still another stratum of nuclei, the **hyperstriatum,** is superimposed (Fig. 15-24, bird). To the hyperstriatum come many sensory impulses which, after being relayed to the older striata, evoke stereotyped behavior such as nest building, incubation of eggs, and care of young. Presumably the hyperstriatum plays a role in avian memory, hence may be essential for migratory behavior and homing. It must be remembered, however, that lampreys and many fishes exhibit remarkable migratory feats with the brain structure that is available. The avian cortex is better developed than in reptiles, but almost complete experimental ablation was reported to have little observable effect.

In mammals the paleostriatum and neostriatum persist as part of a nuclear complex buried deep within the cerebral hemispheres in the wall of each lateral ventricle and known now as **basal ganglia** (Fig. 15-24, mammal). The striata continue to be the source of descending (motor) fiber tracts that end in synapse with motor cell bodies of somatic efferent nerve fibers of cranial and spinal nerves. The reflex motor function of the striata has been largely preempted in mammals by voluntary motor control exerted by the mammalian motor cortex. Pathological changes in these ancient nuclei in man result in certain motor afflictions such as the rhythmical tremor of the limbs seen in aged victims of paralysis agitans (Parkinson's disease).

The roof of the lateral ventricles has become expanded in all available directions, which enables it to accommodate the greatly expanded mammalian **cortex.** The latter consists largely of gray matter because of the tremendous number of cell bodies of neurons (13 billion, more or less, in the human cortex). The cortex is now the most conspicuous part of the brain. This additional cortex-covered roofing is **neopallium,** the primitive pallium **(archipallium)** having been relegated by differential growth to an inconspicuous anteroventrolateral location overlying the olfactory nuclei. In many mammals, though not all, the cerebral cortex is so voluminous that it becomes disposed in ridges **(gyri)** and grooves **(sulci)** (Figs. 15-22 and 15-23; in Fig. 15-22 compare the appearance of the cortex of the platypus, a primitive mammal, with that of the sheep). As a result of the balloonlike expansion of the neopallium, the striata, diencephalon, and midbrain have all become hidden from dorsal view (Fig. 15-22). Removal of the overgrown neopallium and of the cerebellum reveals one component of the primitive striata, that is, the **caudate nucleus** (one of the basal ganglia), and also the thalamus and epithalamic structures, the corpora quadrigemina of the pallium of the mesencephalon, and the floor of the fourth ventricle (Fig. 15-25). Viewing the mammalian brain from this aspect should convince the most dubious observer that the mammalian brain, including that of man, is built on the same architectural plan as the brain of the lowest vertebrate.

The mammalian cortex has at least four roles. (1) It is the highest center to which sensory impulses can be projected. These impulses give rise in the

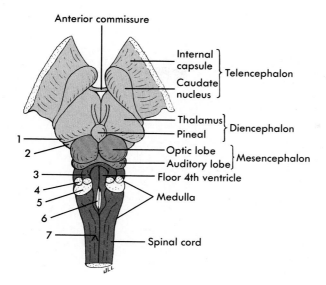

Anterior commissure

Internal capsule
Caudate nucleus
} Telencephalon

Thalamus
Pineal
} Diencephalon

Optic lobe
Auditory lobe
} Mesencephalon

Floor 4th ventricle

Medulla

Spinal cord

1
2
3
4
5
6
7

FIGURE 15-25

Brain stem of a sheep. The cerebral hemispheres and cerebellum have been cut away to reveal the primitive vertebrate structures. *1,* Location of lateral geniculate body of thalamus; *2,* medial geniculate body; *3* to *5,* anterior, middle, and posterior cerebellar peduncles, which carry fibers to and from the cerebellum (the peduncles had to be cut when the cerebellum was removed); *6,* hypoglossal trigone in the floor of the fourth ventricle marking the location of the hypoglossal nuclei; *7,* posterior funiculus containing ascending fibers for proprioception. The caudate nucleus is part of the striatum of the cerebral hemisphere. Optic and auditory lobes make up the corpora quadrigemina. The habenulae lie under cover of the pineal. A paired fiber tract, the stria medullaris, is seen passing toward the habenulae on the surface of the thalamus. Three shades of *red* indicate forebrain, midbrain, and hindbrain.

cortex to sensations of a discriminative (epicritic) nature, such as recognizing minor differences in the temperature, texture, or weights of objects held in the hand. This ability contributes to the esthetic enjoyment of sensory stimuli. (2) It appears to be one of the locations where past experiences are stored as memory. (3) It is a center where data, incoming or recalled, may be correlated, analyzed, and employed in making choices. (4) It is the center from which voluntary motor activity is initiated. The cortex is therefore the "thinking" part of the brain. The manner in which the cerebral cortex is employed in the solution of human problems will determine the future fate of civilization insofar as it is under the control of man.

For convenience in discussing cortical functions the mammalian neocortex has been subdivided topographically—and arbitrarily—into **frontal, parietal, occipital,** and **temporal lobes** (Fig. 16-24). Under the cortex in the roof of the ventricles except in monotremes and marsupials lies a broad transverse sheet of commissural (decussating) nerve fibers, the **corpus callosum.** It connects the cortices of the two hemispheres (Fig. 15-23). Separating the lateral ventricles is a thin, double-walled vertical partition, the **septum pellucidum.**

Blood Supply

No blood vessels penetrate the brain of an amphioxus. With increased metabolic demands in early vertebrates, a modicum of slender capillary loops penetrated the cord and brain ventrally, arising from a fine arterial network in the meninx primitiva. This is the condition in lampreys, in which the arterial network is supplied by the paired dorsal aortae under the name of **internal carotid arteries.** The latter are sometimes the only source of arterial blood to the brain below mammals. Secondary sources are the

FIGURE 15-26

Arterial supply to mammalian brain based on a domestic
cat. The anterior and posteri-
or communicating arteries
complete an arterial circle of
Willis around the base of the
diencephalon. The specific
pattern varies from species
to species and within populations of a single species.

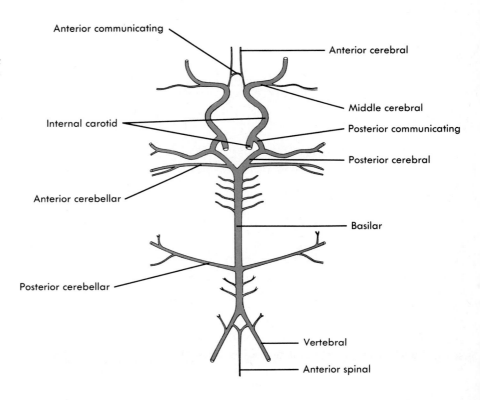

anterior spinal artery and, in some reptiles and in mammals, **vertebral
arteries** (Fig. 15-26). Anastomoses between the basilar and the internal
carotid arteries in sharks, a few lower amniotes, and in mammals result in
an arterial loop, the **circle of Willis,** which surrounds the diencephalon
ventrally (Figs. 15-18 and 15-26). In these vertebrates the internal carotid
supplies principally the forebrain, whereas the basilar supplies principally
the remainder. There is considerable variation in the specific patterns of
distribution of the circle of Willis, even within populations of a single species.

Brains are drained of venous blood by large cranial venous sinuses that
communicate with the internal jugular (anterior cardinal) veins. The
venous blood traverses a precaval (common cardinal) vein before emptying
into the heart. There is no lymphatic drainage from the cord or brain.

Choroid Plexuses and Cerebrospinal Fluid

The cavities of the brain and cord are filled with cerebrospinal fluid, a
clear watery fluid of low specific gravity that is secreted in part by the
choroid plexuses. As mentioned earlier, a choroid plexus consists of the
thin ependymal roof of the third and fourth ventricles and of the pia mater
or leptomeninx invaded by a rich vascular plexus. From the third ventricle
the plexus extends forward into the ventricles of the telencephalon (Fig.
15-17).

Cerebrospinal fluid moves sluggishly caudad into the central canal of the spinal cord partly by ciliary action. Occlusion of the interventricular foramina results in accumulation of fluid in the lateral ventricles. When this condition is present in newborn mammals the existence of fontanels in the skull enables the head to become greatly enlarged, a condition known as hydrocephalus. In vertebrates with an arachnoid meninx the fluid passes into subarachnoid spaces via a median and two lateral apertures in the dorsolateral walls of the fourth ventricle under cover of the cerebellum. From here it seeps along narrow perivascular spaces that accompany blood vessels entering and leaving the substance of the cord and brain, and it follows the rootlets of the spinal and cranial nerves as they emerge from the brain. It even seeps to the inner ear where it contributes to the perilymph.

Cerebrospinal fluid is removed in part in lower vertebrates by passing directly into the venous sinuses of the brain. In higher vertebrates much of it is removed by clusters of **arachnoid villi,** which are macroscopic tufts of the pia-arachnoid meninx that hang into the large middorsal venous sinuses and contain extensions of the subarachnoid spaces. Through these tufts cerebrospinal fluid from the subarachnoid spaces is secreted into the blood stream. Cerebrospinal fluid may assist in protecting the central nervous system from concussion, but more importantly it *selectively* exchanges metabolites with the tissues it bathes.

CRANIAL NERVES

The first ten cranial nerves of all vertebrates are distributed in accordance with a basic pattern. On a functional basis these nerves may be grouped as follows: purely or predominantly sensory nerves (I, II, and VIII), eyeball muscle nerves (III, IV, and VI), and branchiomeric nerves (V, VII, IX, and X). Amniotes have two additional cranial nerves (XI and XII) that are purely or predominantly motor.

The site where a cranial nerve emerges from the surface of the brain, or enters it, is its **superficial origin** as distinguished from the origins of its fibers, which usually come from a number of nuclei scattered throughout the brain stem. Unlike spinal nerves, cranial nerves with the exception of those supplying myotomal muscles (III, IV, VI, XII) do not have dorsal *and* ventral roots. Nerves III, VI, and XII have ventral roots only, and nerve IV emerges from the roof of the brain. Branchiomeric nerves and nerve XI have **lateral roots.** The three purely sensory nerves have superficial origins that will be described shortly.

Predominantly Sensory Cranial Nerves*

NERVE I (OLFACTORY). The cell bodies of the olfactory nerve fibers are located in the olfactory epithelium, and the fibers terminate in the olfactory

---

*Nerves II and VIII, although functionally sensory, contain a number of efferent fibers from the brain to the deep layer of the retina or to the vestibular hair cells. These fibers apparently influence the discharge of sensory impulses from the receptor.

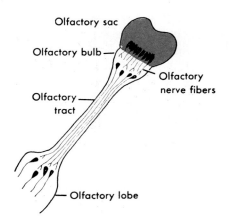

Olfactory sac

Olfactory bulb

Olfactory nerve fibers

Olfactory tract

Olfactory lobe

FIGURE 15-27

Olfactory sac (*red,* containing olfactory epithelium) and rhinencephalon (olfactory bulb, tract, lobe) of dogfish shark. Olfactory nerve fibers synapse in olfactory bulb with second-order neurons. The olfactory sac is not part of the brain.

FIGURE 15-28

Olfactory sacs (*red*) and rhinencephalon of two sharks. Olfactory nerve fibers between sac and bulb form a discrete nerve in *Scoliodon* but not in *Squalus.*

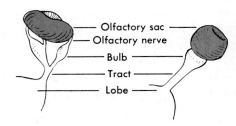

Olfactory sac
Olfactory nerve
Bulb
Tract
Lobe

A. SCOLIODON   B. SQUALUS

bulb (Fig. 15-27). In *Squalus* the olfactory epithelium lies so close to the bulb that an olfactory nerve cannot be distinguished as an anatomical entity. In *Scoliodon* (Fig. 15-28, *A*), another shark, and in some teleosts, the sac containing the olfactory epithelium is sufficiently removed from the olfactory bulb that an olfactory nerve is demonstrable. In most vertebrates, one or more short bundles of olfactory fibers **(filia olfactoria)** extend between the olfactory epithelium and the bulb and constitute collectively the olfactory nerve.

In mammals the olfactory epithelium is in the upper part of the nasal passage, separated from the olfactory bulb by the cribriform plate of the ethmoid bone (derived from olfactory capsule). The foramina in the ethmoid bone (Fig. 8-4) transmit the filia olfactoria. When the brain of a vertebrate is lifted from the cranial cavity, the olfactory nerve bundles are torn, and only stumps remain attached to the brain. The number of olfactory fibers is small in birds which, as explained earlier, have poorly developed rhinencephalons. It is also small in the platypus, which spends most of its time in creeks or rivers, or in a burrow, hence has few natural enemies on land. Sense organs similar to the corpuscles of Herbst (Fig. 16-23) located on their beaks substitute for olfaction as they rout for food in the water bottom. In many marine mammals the olfactory nerve is vestigial.

In tetrapods that have a vomeronasal organ, the olfactory nerve has a separate subdivision, the **vomeronasal nerve.** The vomeronasal organ is discussed in Chapter 16.

A **terminal nerve** lies close to the olfactory bulb and tract in representatives of all vertebrate classes except agnathans and birds. It arises from the ventral surface of the forebrain and consists of general sensory and, at least in some instances, vasomotor fibers that supply a restricted area of the nasal mucosa. Recently the terminal nerve of lower vertebrates has been implicated in the pheromonal triggering of sexual responses via fiber connections with hypothalamic nuclei associated with reproduction. Anatomic relationships of the nerve suggest that it may be a vestige of a hypothetical ancient branchiomeric nerve anterior to the trigeminal. The nerve conventionally has been without a numerical designation, or it has been called **"nerve 0."** It should not be confused with the vomeronasal nerve.

NERVE II (OPTIC). The cell bodies of the optic nerve fibers are in the retina. The nerve emerges from the rear of the eyeball and extends to the optic chiasma where the optic tracts begin. Except in mammals, all or nearly all the optic nerve fibers decussate in the chiasma to enter the opposite side of the brain. A few fibers do not decussate in anurans, lizards, snakes, and some birds. In primates whose eyes are directed forward only the fibers from the nasal side of the retina cross (Fig. 15-29). Other mammals are intermediate between the two conditions. Overlap of the visual field results in depth perception (binocular vision).

FIGURE 15-29

Decussation of optic nerve fibers in optic chiasma in vertebrates below mammals contrasted with higher primates. Below mammals only half of the arrow is projected onto the opposite retina. In higher primates the entire arrow is projected onto each retina, and from a slightly different angle of vision, resulting in depth perception.

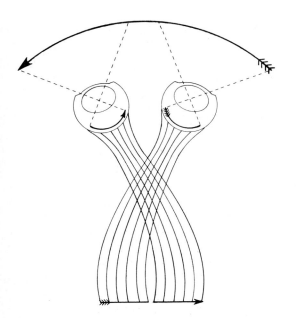
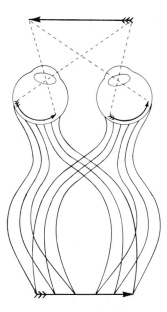

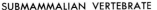
SUBMAMMALIAN VERTEBRATE   HIGHER PRIMATE

NERVE VIII (VESTIBULOACOUSTIC). The eighth nerve* in all vertebrates has an anterior and a posterior root that comes off the medulla very close to nerves V and VII. The branches of the **anterior root** in fishes innervate the ampullae on the anterior vertical and horizontal semicircular ducts, and the utriculus. The **posterior root** innervates the ampulla on the posterior vertical duct, the sacculus, and the lagena of the membranous labyrinth.

In tetrapods the lagena enlarges and becomes the cochlea for hearing. As a result, the posterior root enlarges and, with its branches, becomes the **cochlear nerve.** The anterior root is then called the **vestibular nerve.** The nerves have a **cochlear** or **vestibular ganglion** on their pathway. The cell bodies in these ganglia are exceptional for first order sensory neurons in that they remain in a primitive unipolar state. Since the cochlear ganglion spirals within the cochlea of mammals it is also called the **spiral ganglion.**

In some mammals (rat and mouse but not cat or bat) large unipolar sensory cell bodies are distributed within the eighth nerves along their entire length. These are cell bodies of second-order sensory neurons. They are stimulated by collateral (side) branches of incoming fibers; and because their axons end on other second-order neurons in the cochlear nuclei of the medulla they reinforce (augment) stimuli coming from the receptor.

Eyeball Muscle Nerves

NERVES III, IV, AND VI (OCULOMOTOR, TROCHLEAR, AND ABDUCENS). The third, fourth, and sixth nerves supply the superior oblique (IV), external rectus (VI), the four remaining extrinsic eyeball muscles (III), and certain other myotomal muscles of the eyes (Table 10-3). These eyeball muscle nerves resemble spinal nerves that have lost their dorsal roots. In addition to somatic motor fibers the nerves contain sensory fibers for proprioception from the muscles innervated.

Nerve III arises ventrally from the mesencephalon. Nerve IV is the only nerve arising dorsally from the brain (anterior roof of the fourth ventricle) and one of the few nerves with motor fibers that decussate before emerging. Nerve VI emerges ventrally at the anterior end of the hindbrain. Nerves IV and VI are the smallest of the cranial nerves, having fewest fibers. The cell bodies of all fibers in these nerves, both motor and proprioceptive, are in nuclei within the central nervous system. Therefore the nerves have no sensory ganglia.

Nerve III contains also visceral motor fibers that end in the ciliary ganglion of the autonomic nervous system (Fig. 15-36). From the ganglion, postganglionic fibers pass to the muscles of the iris diaphragm that constrict the pupil (sphincter pupillae muscles), and to the ciliary body of the eye, which governs either the position or thickness of the lens for visual accommodation, depending on the species.

---

*Conventionally called auditory or acoustic in tetrapods, although it has more than an auditory function.

Branchiomeric Nerves

One characteristic of vertebrates is the development of a series of embryonic pharyngeal arches. The fate of the skeleton and muscles of these arches has already been discussed. Whether the animal is to become fish or tetrapod, the muscles derived from the first arch are supplied by cranial nerve V; those from the second arch, by nerve VII; from the third arch, by nerve IX; and from succeeding arches, by nerve X. Since V, VII, IX, and X innervate branchiomeric muscles, they are branchiomeric nerves.

Branchiomeric nerves are mixed nerves. In addition to innervating branchiomeric muscles, they have other important motor and sensory functions. Some of their motor fibers are components of the autonomic nervous

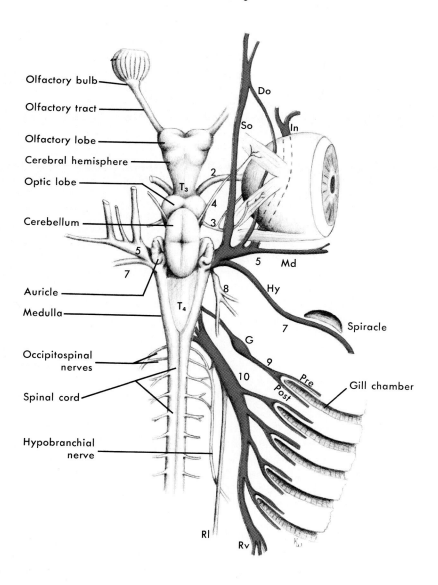

FIGURE 15-30

Brain and cranial nerves II to X of *Squalus acanthias,* dorsal view. Branchiomeric nerves are shown in red. *Do,* Deep ophthalmic; *G,* petrosal ganglion; *Hy,* hyomandibular; *In,* infraorbital; *Md,* mandibular; *Pre,* pretrematic; *Post,* posttrematic; *So,* superficial ophthalmic; *Rl,* ramus lateralis (lateral-line nerve) of vagus; *Rv,* ramus visceralis of vagus; T_3 and T_4, tela choroidea of third and fourth ventricles. The pineal body has been removed. The three eyeball muscles shown are, commencing anteriorly, superior oblique, superior rectus, and lateral rectus.

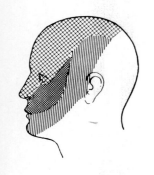

0 Spiracle

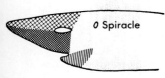

▨ Ophthalmic

▩ Maxillary (infraorbital)

▥ Mandibular

FIGURE 15-31

Cutaneous distribution of the trigeminal nerve of vertebrates.

system; their sensory fibers supply several varieties of sense organs; and each nerve includes proprioceptive fibers. The distribution of branchiomeric nerves in *Squalus* exemplifies the generalized pattern (Fig. 15-30). Alterations in tetrapods are chiefly the result of elimination of gills and other adaptations to terrestrial life.

NERVE V (TRIGEMINAL). The fifth nerve arises from the anterior end of the hindbrain and typically exhibits three divisions: **ophthalmic, maxillary,** and **mandibular.** In fishes the ophthalmic may be subdivided into superficial and deep ophthalmic nerves and the maxillary is also called infraorbital. All branches contain sensory fibers. Only the mandibular branch contains motor fibers.

Via its three divisions cranial nerve V is sensory to the ectoderm of the head, including the teeth, anterior part of the tongue, and nasal epithelium, for general cutaneous sensation (Fig. 15-31). The mandibular branch also contains proprioceptive fibers. The cell bodies of all sensory fibers except proprioceptive are found in the **trigeminal ganglion** unless, as in some lower vertebrates, the ophthalmic division has its own ganglion.

The mandibular nerve is motor to all muscles derived from the first pharyngeal arch. The predominant distribution is therefore to muscles of the jaws. It also operates the **tensor tympani** muscle attached to the malleus in mammals. This is not surprising since the malleus appears to be the displaced posterior tip of the lower jaw of synapsid reptiles. Table 10-5 gives the motor distribution of cranial nerve V.

NERVE VII (FACIAL). The seventh nerve arises from the anterior end of the hindbrain in close association with the fifth. In sharks the fifth and seventh nerves have common (trigeminofacial) roots.

Nerve VII is sensory to the neuromast organs on the head of fishes and aquatic amphibians. With adaptation to land, these sensory fibers were lost. Branches also supply taste buds in the pharynx at the level of the first and second arches, any taste buds on the external surface of fishes, and taste buds on the anterior part of the tongue in tetrapods. Nerve VII also contains general sensory fibers from the endoderm of the second arch and proprioceptive fibers from the muscles of the arch. The cell bodies of all sensory fibers, except proprioceptive, are in the **facial ganglion.**

The facial nerve is motor to muscles of the second arch (Table 10-5). These include the mimetic muscles of mammals. Since the stapes is a derivative of the hyoid arch, the stapedial muscle of mammals is also innervated by the seventh nerve. The seventh nerve in mammals also contains visceral motor fibers to the submandibular and sphenopalatine ganglia (Figs. 15-33 and 15-36). The submandibular ganglion innervates the submandibular and sublingual salivary glands. The sphenopalatine innervates the lacrimal gland and mucous membranes of the nose.

NERVE IX (GLOSSOPHARYNGEAL). The ninth nerve arises from the medulla. In sharks it has three major branches and *typifies the distribution of branchiomeric nerves.* These branches are **pretrematic** (sensory), **pharyngeal**

(sensory), and **posttrematic** (mixed).* The pretrematic branch supplies the anterior demibranch of the first gill chamber for general sensation (Fig. 15-30). The pharyngeal branch supplies taste buds and general visceral receptors in the pharyngeal mucosa at the level of the third visceral arch. The posttrematic branch is sensory to the demibranch in the posterior wall of the first gill chamber, motor to the muscles of the third visceral arch, and proprioceptive from those muscles. A small **lateral line branch** of IX (not illustrated) supplies a short segment of the lateral-line canal at the junction of head and trunk.

Preganglionic fibers of the autonomic nervous system are present in IX. They have not been fully explored in all vertebrates, but in mammals they innervate the otic ganglion, from which postganglionic fibers pass to the parotid salivary glands (Fig. 15-36).

With loss of gills and neuromast organs during adaptation to life on land, the ninth nerve lost many fibers. It continues to supply surviving taste buds on the third arch mucosa (on the posterior part of the tongue of mammals) and general receptors on the posterior part of the tongue and in the upper pharynx. Of the branchiomeric muscles of the third arch, only a stylopharyngeus remains in mammals.

The sensory cell bodies of nerve IX, except those for proprioception, are found in the **petrosal ganglion** of lower vertebrates, and in the **petrosal (superior)** and **inferior glossopharyngeal ganglia** of birds and mammals. The superior ganglion, chiefly somatic, is derived from a neural crest. The inferior ganglion, chiefly visceral, is derived largely from an ectodermal placode. In fishes the lateral-line branch has its own ganglion.

NERVE X (VAGUS). The vagus arises from a series of rootlets along the lateral aspect of the medulla. The branchiomeric components of the vagus in *Squalus* consist of a series of four trunks (more in elasmobranchs with more gill chambers), each exhibiting a pretrematic, posttrematic, and pharyngeal branch distributed as in nerve IX (Fig. 15-30). The pretrematic branches supply the anterior walls of the last four gill chambers. Posttrematic branches supply the posterior walls of these chambers and are motor to arches IV to VII. The vagus is therefore the chief respiratory nerve of fishes. The pharyngeal branches supply the pharyngeal epithelium for taste and general sensation.

In addition to branchiomeric components, the vagus in *Squalus* has two other major trunks. The **ramus lateralis** is sensory to the lateral-line canal all the way to the tip of the tail, and the **ramus visceralis** supplies efferent visceral fibers to the coelomic viscera via terminal ganglia of the autonomic nervous system of the trunk and conducts afferent visceral fibers from those viscera.

---

*The maxillary and mandibular branches of nerve V in fishes are functionally equivalent to pretrematic and posttrematic nerves farther caudad. The pretrematic branch of VII in fishes has joined the maxillary of V to form an infraorbital trunk, and the posttrematic branch of VII (behind the spiracle) is the hyomandibular nerve. The pharyngeal branch of VII in sharks is better known as the palatine.

During the process of adapting to land the vagus of tetrapods lost those functions associated solely with life in the water but retained the others. The prominent lateral-line branch disappeared. The sensory branches to gill chambers were lost. However, general receptors in the pharyngeal epithelium and taste buds in the vicinity of the glottis continue to be supplied by the vagus. Surviving also are the motor branches to any remaining branchiomeric muscles of the fourth and successive arches. These are chiefly the cricothyroid, cricoarytenoid, and thyroarytenoid muscles. These branches also include proprioceptive fibers. Since much of the distribution of the vagus has been lost in tetrapods, the ramus visceralis has become the major component of the nerve. It continues to supply afferent and efferent fibers to the heart and other coelomic viscera. Included in the ramus visceralis of amniotes are autonomic fibers contributed by the cranial roots of the spinal accessory nerve (Fig. 15-32).

The cell bodies of all sensory fibers of the vagal nerve, except those for proprioception, are found in one or more ganglia. In some elasmobranchs each of the four or more branches to the gills has its own **epibranchial ganglion,** and it is likely that these branches were at one time four separate cranial nerves. In birds and mammals nerve X, like IX, has have two sensory ganglia, a **superior,** chiefly somatic, and an **inferior,** chiefly visceral, derived respectively from neural crests and epibranchial placodes.

The innervation of the mucosa of the oral cavity and pharynx by nerves V, VII, IX, and X, *in that sequence,* in all vertebrates from fish to man, demonstrates the negligible effects of life on land on the sensory innervation of the moist pharyngeal endoderm.

FIGURE 15-32

Vagus, spinal accessory, and hypoglossal nerves of a typical mammal. The hypoglossal rootlets are in series with the ventral roots of the spinal nerves. Components in black are spinal nerve contributions to the accessory and hypoglossal nerves. C_1, Dorsal rootlets of the first cervical spinal nerve; C_6, ventral rootlets of the sixth cervical spinal nerve. Nerve XI contributes an internal ramus (*arrows*) to the vagus.

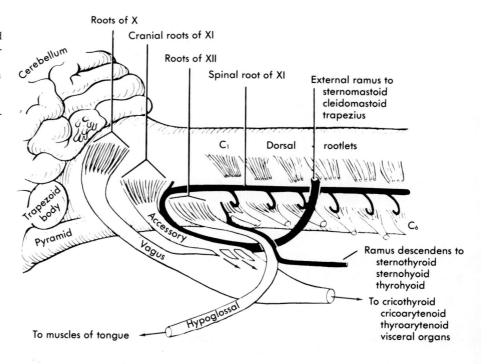

Accessory and Hypoglossal Nerves

NERVE XI (SPINAL ACCESSORY). The spinal accessory constitutes an eleventh cranial nerve in amniotes. It is purely motor. In reptiles and birds its rootlets emerge from the medulla immediately behind the vagus. In mammals, additional rootlets emerge from the spinal cord, as will be explained shortly.

The **medullary (cranial, bulbar) roots** in all vertebrates are composed of motor fibers that innervate some of the minor striated pharyngeal muscles of the pharynx and larynx such as those associated with the vela of the palate, and of preganglionic fibers of the autonomic nervous system which are distributed (via peripheral autonomic ganglia) to the same coelomic viscera supplied by the vagus. In fact, these cranial fibers, which are referred to collectively as the **internal ramus of the accessory nerve,** enter the vagus nerve before the latter leaves the cranial cavity, and are distributed by the vagus. In reptiles and birds the cranial root has a branch, the **external ramus,** that passes into the neck region to innervate the sternomastoid, cleidomastoid, and trapezius muscles, or their homologues. In mammals the external ramus of the accessory nerve has its superficial origin from rootlets that emerge from several segments of the cervical region of the spinal cord—five or six in humans, seven in horses, fewer in many mammals (Fig. 15-32). The nucleus of origin of these spinal fibers extends over an equal number of segments in the ventral horn of gray matter of the cord. These spinal rootlets in mammals unite to form a common trunk, the **spinal root of the accessory nerve,** which passes cephalad close to the cord and enters the cranial cavity via the foramen magnum. Within the cranial cavity the spinal root joins the cranial roots to form the spinal accessory nerve (Fig. 15-32). A short distance beyond this juncture the fibers of the medullary roots join the vagus as the internal ramus, as in lower amniotes, and the fibers from the spinal root become the external ramus, which in mammals leaves the cranial cavity via the jugular foramen along with nerves IX and X.

What is the phylogenetic history of this nerve of amniotes? This was a subject of discussion among early comparative anatomists for many years with no strong evidence favoring any single hypothesis. The components of the nerve are not new; only the way they are assembled is new. If we look at the location of the cell bodies of fibers in the external ramus in reptiles and birds, we find they are in neither the SE column that supplies myotomal muscles, nor in the SVE nucleus that supplies branchiomeric muscles (Fig. 15-34). Their specific location varies among reptiles and birds, but is close to the SVE nuclei that supply motor fibers to the branchiomeric muscles innervated by nerve X. In mammals they are in the anterior horn of gray matter of the spinal cord in what has sometimes been called the spinal accessory nucleus. It is close to, but not clearly a part of, the somatic efferent column. They have evidently migrated, or been displaced, from one nuclear column or the other. As pointed out in Chapter 10, the *muscles* supplied by this ramus are evidently derived from the caudalmost branchiomeric musculature of fishes, including the cucullaris. This leads to the conclusion that *nerve fibers* in the external ramus, like similar fibers in cranial nerves V, VII, IX, and X, are of branchiomeric origin. With reference to the medullary roots, it was pointed out earlier in this chapter that a series of occipitospinal nerves arises between the vagus and the first spinal nerves of fishes and some amphibians. These nerves are either missing in amniotes, or they have contributed to the spinal accessory. Thus the eleventh cranial nerve appears to have been derived from an ancestoral posteriormost component of the branchiomeric series, and from one or more occipitospinal nerves of lower vertebrates. It is highly likely that modern research techniques, when applied, will provide additional clues to the solution of this question.

NERVE XII (HYPOGLOSSAL). The twelfth nerve of amniotes is motor except for proprioceptive fibers. It arises from the hypoglossal nucleus in the medulla by a series of ventral rootlets and passes into the base of the tongue from which location it sends branches to the hyoglossus, styloglossus, genioglossus, and lingualis muscles (Figs. 10-6, 15-4, *B*, and 15-32). On emerging from the hypoglossal foramen in mammals the nerve receives fibers from one or more anterior cervical spinal nerves. Some of these spinal nerve fibers are distributed by the hypoglossal nerve to the geniohyoid muscle. The rest of the spinal fibers leave nerve XII to become the **ramus descendens of the twelfth nerve,** which joins a loose plexus of cervical spinal nerves supplying the hypobranchial muscles of the neck (sternohyoid, sternothyroid, thyrohyoid, or their homologues).

That the hypoglossal is a cranial rather than a spinal nerve is dictated solely by the location of the foramen magnum. *Actually,* it is a spinal nerve that became "locked up" in the braincase. Like spinal nerves, the hypoglossal nerve of some mammals develops an embryonic dorsal root and dorsal root ganglion **(Froriep's ganglion).** The root and ganglion later disappear.

The twelfth nerve is a derivative of the occipitospinal series of fishes and amphibians, as may be deduced from the following facts: (1) Occipitospinal nerves have been reduced in number above fishes, whereas the number of cranial nerves has increased. (2) The twelfth nerve, like many occipitospinal nerves, lacks a dorsal root. (3) The occipitospinal nerves of lower vertebrates supply hypobranchial muscles, and the tongue is hypobranchial muscle. (4) Whereas the tongue muscle is supplied by the last cranial nerve in amniotes, it is supplied by the first spinal nerve in amphibians.

Innervation of the Mammalian Tongue: An Anatomic Legacy

The innervation of the mammalian tongue illustrates how a single organ may be served by numerous nerves, depending on the ontogenetic and phylogenetic history of its parts (Fig. 15-33). The mucosa of the anterior part of the tongue is first arch endoderm and is therefore innervated by nerve V for general sensations. The taste buds on this part of the tongue are innervated by nerve VII, which supplies taste buds on the first and second arches in fishes. The mucosa on the posterior part of the tongue is innervated by nerve IX for both general sensation and taste because of the origin

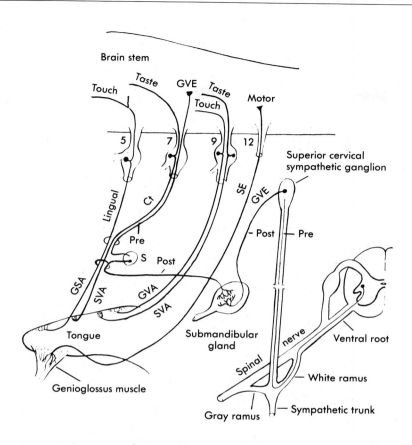

FIGURE 15-33

Innervation of the tongue and submandibular gland of a mammal (based on cat and man). *5, 7, 9,* and *12,* Cranial nerves; *Ct,* chorda tympani; *Pre* and *Post,* preganglionic and postganglionic fibers of the autonomic nervous system; *S,* submandibular ganglion of the autonomic system. A key to the fiber components (*GSA, SE,* and so forth) is given in Tables 15-1 and 15-3.

of this mucosa from the third visceral arch. The muscles of the tongue are myotomal and are innervated by nerve XII.

Although four cranial nerves innervate the tongue, only three nerves may be traced into it, since the taste fibers from nerve VII (in the chorda tympani branch) unite with the lingual branch of nerve V just before the latter reaches the tongue.

Fiber Components of Cranial Nerves

As mentioned earlier, fibers in spinal nerves may be classified in four functional categories (GSA, GVA, SE, and GVE), each supplying specific types of general receptors or effectors (Table 15-1). One or more of these components may be found also in most cranial nerves: SE in III, IV, VI, and XII; GVE (autonomic fibers) in III, VII, IX, X, and XI; GSA chiefly in V; and GVA in VII, IX, and X. In addition to the foregoing *general* fiber components, certain cranial nerves contain *special* components of which there are three types—SSA, SVA, and SVE (Table 15-3). Each of the foregoing varieties of nerve fibers commences or terminates in the brain or cord in a continuous or discontinuous column of nuclei preempted by that specific component (Fig. 15-34). In anamniotes there is less segregation of these cell bodies in clearly defined nuclei than in amniotes.

FIGURE 15-34

Cross section of medulla of an amniote showing location of certain nuclei. Sensory nuclei are in the alar plate; motor nuclei, in the basal plate. The two plates are delimited by the sulcus limitans. *SE,* Somatic motor fibers from cell bodies in somatic motor column. *VA,* Sensory nucleus for all GVA and SVA fibers except those for smell. The remaining nuclei are identified in Tables 15-1 and 15-3.

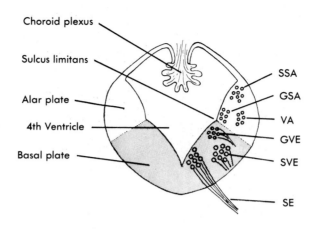

TABLE 15-3   Special Fiber Components of Cranial Nerves*

Components	Innervation and Nerve

Special somatic afferent fibers (SSA)	Special somatic receptors
	    Retina (II)
	    Membranous labyrinth (VIII)
	    Neuromast organs (VII, IX, and X)

Special visceral afferent fibers (SVA)	Special visceral receptors
	    Olfactory epithelium (I)
	    Taste buds (VII, IX, and X)

Special visceral efferent fibers (SVE)	Branchiomeric muscle (V, VII, IX, X, and XI)

*In addition to these special components, cranial nerves other than I, II, and VIII have one or more of the general components listed in Table 15-1.

AUTONOMIC NERVOUS SYSTEM

The autonomic nervous system is that part of the nervous system that innervates glands and smooth and cardiac muscle. It consists chiefly of autonomic nerves, plexuses, and ganglia (Figs. 15-35 to 15-37). It does not, however, constitute an anatomical entity; that is, it cannot be completely dissected away from the rest of the nervous system, since its components commence inside the central nervous system and emerge via cranial or spinal nerves. It is entirely a visceral motor system. However, sensory fibers from the viscera use autonomic pathways to reach cranial and spinal nerves.

Two motor neurons in series conduct the impulse from the brain or cord to a typical visceral effector. The cell body of the first neuron **(preganglionic neuron)** is in a visceral efferent nucleus in the central nervous system, and the preganglionic fiber terminates in an autonomic ganglion. The cell body of the second neuron **(postganglionic neuron)** is in an autonomic ganglion. Its fiber extends to the effector (see Fig. 15-11, pre and post, and nerve III in

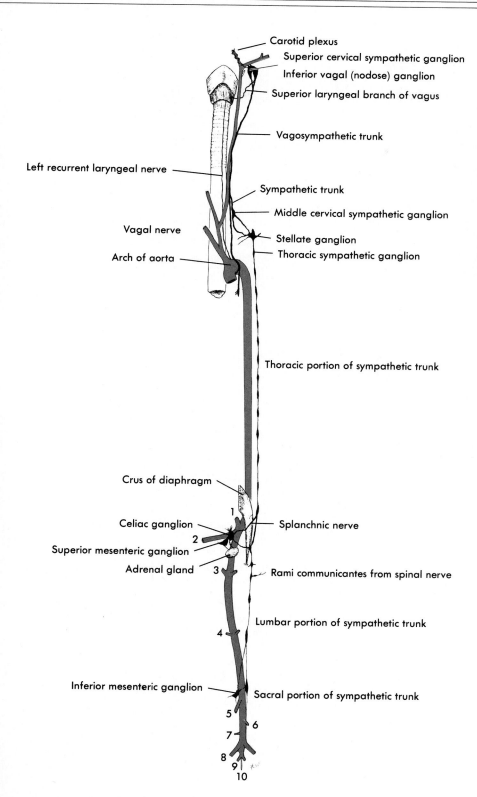

Carotid plexus
Superior cervical sympathetic ganglion
Inferior vagal (nodose) ganglion
Superior laryngeal branch of vagus

Vagosympathetic trunk

Left recurrent laryngeal nerve

Sympathetic trunk
Middle cervical sympathetic ganglion

Vagal nerve

Stellate ganglion
Thoracic sympathetic ganglion

Arch of aorta

Thoracic portion of sympathetic trunk

Crus of diaphragm

Celiac ganglion

Splanchnic nerve

Superior mesenteric ganglion

Adrenal gland

Rami communicantes from spinal nerve

Lumbar portion of sympathetic trunk

Inferior mesenteric ganglion

Sacral portion of sympathetic trunk

FIGURE 15-35

Left sympathetic trunk and associated structures of a cat. *1* to *10*, Major branches of abdominal aorta: *1*, celiac; *2*, superior mesenteric; *3*, renal; *4*, spermatic or ovarian; *5*, inferior mesenteric; *6* and *7*, left and right iliolumbars; *8*, external iliac; *9*, internal iliac; *10*, median sacral. The recurrent laryngeal nerve, a branch of the vagus, is not part of the autonomic system. In cats, as in some other mammals, the vagus nerve and sympathetic trunk in the neck are wrapped in a common sheath to form a vagosympathetic trunk.

FIGURE 15-36

For legend see opposite page.

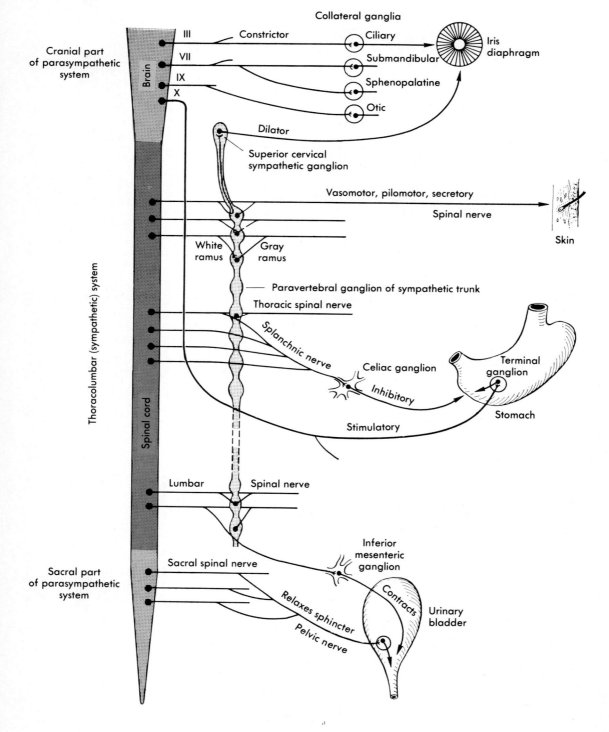

FIGURE 15-36

Representative components of the autonomic nervous system of a mammal. Innervation of iris diaphragm, skin, stomach, and urinary bladder. *Arrows* emphasize dual control exerted elsewhere than in the skin by craniosacral and thoracolumbar systems. Preganglionic fibers are those with a cell body *(black dot)* in the central nervous system. Postganglionic fibers are those with a cell body in a ganglion. Other spinal nerves in addition to those shown contain autonomic fibers. Sympathetic trunk is *gray.*

Fig. 15-36). Autonomic fibers have a very thin myelin sheath or one that is detectable only by electron micrography.

The autonomic nervous system is composed of (1) a **sympathetic system (thoracicolumbar system** in higher vertebrates) that emerges from the cord via most of the spinal nerves of the trunk; and (2) a **parasympathetic system (craniosacral system)** that emerges from the brain via cranial nerves III (except cyclostomes), VII, IX, X, and XI and, in tetrapods, from the cord via sacral spinal nerves (Fig. 15-36). Most visceral effectors except those in the skin are supplied by postganglionic fibers from both systems (Fig. 15-36, iris diaphragm, stomach, urinary bladder). The stimulatory effects of one system modulate the inhibitory effects of the other to bring about an appropriate response. The heart of cyclostomes has no innervation; in other fishes it is innervated by the vagus only; and in tetrapods it has a double innervation, the vagus being inhibitory, the sympathetic system excitatory. All major components of the sympathetic system of reptiles, birds, and mammals are already present in amphibians. Differences in higher tetrapods are more quantitative than qualitative.

Autonomic ganglia may be classified in three categories: **paravertebral, collateral,** and **terminal.** *Paravertebral ganglia* lie close to the vertebral column (Fig. 15-37). In teleosts and all tetrapods they are interconnected by longitudinal strands of autonomic fibers to form a ganglionated chain, the **sympathetic trunk** (Figs. 15-35, 15-36, and 15-37). There is usually one paravertebral ganglion for each spinal nerve except in the neck and sacral regions of tetrapods. There are fewer in the neck and none in the sacral region. Paravertebral ganglia of the trunk are connected to the nearest spinal nerve by a **white ramus communicans,** which conducts preganglionic fibers from the spinal nerve to the ganglion (Figs. 15-11, 15-36, and 15-37).* Postganglionic fibers with cell bodies in paravertebral ganglia supply viscera of the head, neck, coelom, and visceral effectors in the skin (vasomotor, plumomotor or pilomotor, secretory, chromatophores.) Postganglionic fibers going to the skin are returned to the spinal cord by a **gray ramus communicans** (Figs. 15-11 and 15-37). The skin of cyclostomes and elasmobranchs, however, appears to lack autonomic innervation, and there are no gray rami. White rami also conduct visceral afferent fibers *to* spinal nerves (Fig. 15-11, GVA).

---

*There are no white rami in the neck or sacral region.

Collateral ganglia are in the head (**ciliary, submandibular, sphenopala-tine, otic** in mammals) and in the abdomen at the base of the major branches of the aorta (**celiac, superior mesenteric, inferior mesenteric ganglia,** and others, Figs. 15-35 and 15-36). They are not part of a chain. Collateral ganglia in the trunk receive preganglionic *sympathetic* fibers from spinal nerves via autonomic nerves such as the splanchnic (Fig. 15-36) and supply postganglionic fibers to abdominal and pelvic viscera. The neural connections of collateral ganglia of the head of mammals are given in Table 15-4.

On emerging from collateral ganglia and from the sympathetic ganglia of the neck, bundles of postganglionic fibers often form autonomic plexuses rather than nerves. The plexuses adhere to the adventitia of nearby blood vessels and accompany these vessels to the organs they supply. For example, fibers from the superior cervical sympathetic ganglion, which is located close to the bifurcation of the external and internal carotid artery, accompany the external carotid artery and its branches to the iris diaphragm and to the salivary glands (**carotid plexus,** Fig. 15-35); and postganglionic fibers that commence in the coeliac ganglion accompany the celiac artery and its branches to the stomach, spleen, pancreas, liver, and gall bladder (**celiac plexus).** Autonomic fibers also form plexuses in certain other locations (Fig. 15-37).

Terminal ganglia, found only in the trunk, are embedded in the walls of the organs they innervate—heart, lungs, stomach, and urinary bladder, for instance—where they receive synaptic endings of preganglionic fibers of the *parasympathetic* system via the vagus or sacral spinal nerves and their visceral branches. The cell bodies in these ganglia send very short postganglionic fibers to the innervated tissue (Fig. 15-36, stomach, urinary bladder).

The neurotransmitter released at the endings of preganglionic fibers within all autonomic ganglia, and that released at postganglionic terminals of the parasympathetic system is chiefly acetylcholine. That synthesized by cell bodies in sympathetic ganglia and released by their postganglionic fibers at most effectors other than sweat glands is chiefly norepinephrine (noradrenaline) along with small amounts of epinephrine (adrenalin).

TABLE 15-4   Innervation and Peripheral Distribution of Autonomic Ganglia of the Head of Mammals

Ganglion	Receives Fibers from:	Projects Fibers to:
Ciliary	Oculomotor nerve	Ciliary body of eye
		Sphincter muscles of iris
Submandibular	Facial nerve	Submandibular gland
		Sublingual gland
Sphenopalatine	Facial nerve	Lacrimal gland
		Glands of nose and pharyngeal mucosa
Otic	Glossopharyngeal nerve	Parotid gland

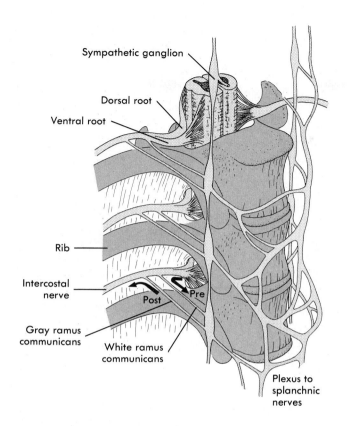

FIGURE 15-37

Paravertebral (sympathetic) ganglia and their interconnections in thoracic region of a cat, oblique ventral view. Arrows indicate direction of preganglionic *(pre)* and postganglionic *(post)* impulses. The postganglionic fibers in the intercostal nerves are vasomotor, pilomotor, and secretory to the somatopleure.

The cells of the adrenal medulla, like cell bodies in sympathetic ganglia, synthesize norepinephrine and epinephrine; they, too, are innervated by preganglionic fibers of the sympathetic system; and they, too, arise from neural crests. In fact, the adrenal medulla is, by origin, innervation, and function, equivalent to a ganglion of the sympathetic nervous system in which the cell bodies simply fail to sprout postganglionic processes. As a result, the secretions are carried away by the blood stream. Further discussion of these matters will be found in Chapter 17.

In all vertebrates the autonomic nervous system is primarily involuntary; that is, motor impulses are initiated reflexly by visceral afferent impulses. One is seldom aware of the stimulus or the response. Some voluntary control is possible by the use of biofeedback.

Chapter Summary

1. The neuron is the functional unit of the nervous system. It consists of a cell body and one or more processes. The long process is an axon, or nerve fiber.

2. A nerve is a bundle of nerve fibers outside the central nervous system. A tract is a bundle of nerve fibers within the central nervous system.

3. A nucleus is a group of similarly functioning cell bodies inside the central nervous system. Nuclei constitute the gray matter of the brain and cord.

4. A ganglion is a group of cell bodies outside the central nervous system. Ganglia are sensory except those of the autonomic nervous system, which are motor.

5. Neurotransmitters are amines that transmit nerve impulses across synapses. Neurosecretions are polypeptide hormones secreted by neurosecretory neurons.

6. The embryonic neural tube caudal to the forebrain consists of alar and basal plates that are sensory and motor, respectively. The alar plate gives rise to second-order sensory neurons. The basal plate gives rise to motor neurons other than postganglionic. The alar plate gives rise to second-order sensory neurons. Most first-order sensory neurons arise from neural crests or ectodermal placodes.

7. The cell bodies of motor neurons are inside the brain or cord with one exception: the cell bodies of postganglionic fibers of the autonomic nervous system are in autonomic ganglia.

8. Cell bodies of first-order sensory neurons in higher vertebrates are found in sensory ganglia on the nerves with three major exceptions: the cell bodies of olfactory nerve fibers are in the olfactory epithelium, the cell bodies of optic nerve fibers are in the retina, and those of proprioceptive fibers of cranial nerves are in the mesencephalon. Olfactory neurons and rods and cones are sometimes called neurosensory cells.

9. The spinal cord exhibits enlargements at the level of paired appendages and, in fishes, a urophysis. When shorter than the vertebral column, the cord terminates in a filum terminale surrounded by a cauda equina.

10. Neuroglia, found only in the cord and brain, consists of ependyma, oligodendroglia, microglia, and astroglia. In the periphery, Schwann cells constitute a neurilemma and elaborate myelin sheaths.

11. A meninx primitiva surrounds the brain and cord in fishes. A dura mater and leptomeninx are present in most vertebrates. In a few birds and in mammals the leptomeninx differentiates into a pia mater and arachnoid membrane.

12. Most spinal nerves exhibit sensory ganglia on their dorsal roots. The following cranial nerves have sensory ganglia: V (trigeminal), VII (facial), VIII (cochlear and vestibular), IX (petrosal; in mammals, inferior and superior glossopharyngeal), and X (in mammals, inferior and superior vagal).

13. Spinal nerves are metameric in origin and distribution. Most spinal nerves exhibit dorsal and ventral roots and dorsal, ventral, and communicating rami. Ventral rami often unite to form simple or complicated plexuses.

14. There is evidence that in the earliest vertebrates dorsal and ventral roots did not unite, dorsal roots were mixed and ventral roots were motor, and the cell bodies of the sensory fibers were not aggregated in ganglia. These conditions are found in lampreys, except that some sensory cell bodies aggregate on the dorsal root. Above cyclostomes, dorsal and ventral roots unite, ventral roots are motor, and dorsal roots are chiefly or wholly sensory.

15. Spinal nerves contain the following fiber components: GSA, SE, GVA, and GVE. Cranial nerves may contain one or more of the preceding and also one or more of the following: SSA, SVA, and SVE.

16. Occipitospinal nerves lacking sensory roots and supplying hypobranchial musculature arise between the vagus and first typical spinal nerves. They are more numerous in lower vertebrates and are represented in amniotes by part of nerve XI and by nerve XII.

17. Anamniotes have 10 pairs of cranial nerves, amniotes have 12. Nerves I, II, and VIII are purely sensory. Nerves III, IV, and VI supply the myotomal muscles of the eyeball. Nerves V, VII, IX, and X are branchiomeric.

18. Cranial nerve V is the chief nerve for cutaneous sensation on the surface of the head and the ectodermal part of the oral cavity. Nerves VII, IX, and X supply neuromast organs and taste buds.

19. Nerve XI has cranial and spinal roots and is derived from the caudal end of the branchiomeric series and from occipitospinal nerves. The internal ramus contains GVE fibers that are distributed with the vagus. The external ramus supplies the trapezius and sternocleidomastoid muscles.

20. Nerve XII represents one or more occipitospinal nerves and supplies the muscles of the tongue.

21. The terminal nerve is a vestige of an anterior branchiomeric nerve. It supplies a restricted region of the nasal mucosa with general sensory and vasomotor fibers.

22. The autonomic nervous system innervates smooth and cardiac muscles and glands. Preganglionic fibers of the craniosacral (parasympathetic) division emerge from the brain via cranial nerves III, VII, IX, X, and XI and from the sacral region of the cord via sacral spinal nerves. Preganglionic fibers of the thoracolumbar (sympathetic) division emerge from the cord via thoracic and lumbar spinal nerves.

23. Autonomic ganglia are paravertebral (sympathetic chain), collateral (in head or near abdominal aorta), and terminal (in trunk close to or within the organ innervated). They contain cell bodies of postganglionic neurons. They are of neural crest origin.

24. The autonomic ganglia of the head and their associated nerves are ciliary (III), sphenopalatine (VII), submandibular (VII), and otic (IX). These are parasympathetic ganglia.

25. Most viscera are supplied by sympathetic and parasympathetic fibers, but skin receives only sympathetic fibers.

26. Cerebrospinal fluid is secreted by choroid plexuses in the lateral, third, and fourth ventricles. The fluid fills the brain ventricles and central canal of the spinal cord and escapes to the meningeal spaces via foramina in the roof of the fourth ventricle.

27. The brain has three major subdivisions: prosencephalon (forebrain), mesencephalon (midbrain), and rhombencephalon (hindbrain). The more prominent components of these subdivisions are listed below.

PROSENCEPHALON (FOREBRAIN)

Telencephalon
  Rhinencephalon
    Olfactory bulbs
    Olfactory tracts
    Olfactory lobes
  Cerebral hemispheres
    Corpora striata (basal ganglia)
      Caudate nucleus
    Corpus callosum
    Neocortex on pallium
    Paraphysis
    Lamina terminalis
  Lateral ventricles

Diencephalon
  Epithalamus
    Habenulae
    Pineal
    Parapineal
  Thalamus
    Massa intermedia
  Hypothalamus
  Optic chiasma
  Infundibular stalk and pituitary
  Saccus vasculosus
  Third ventricle
  Circle of Willis

MESENCEPHALON (MIDBRAIN)

Optic lobes ⎫
Auditory lobes ⎬ Tectum (roof)
Tegmentum
  Cerebral peduncles
  Mesencephalic nucleus
Cerebral aqueduct

RHOMBENCEPHALON (HINDBRAIN)

Metencephalon
    Cerebellum
    Pons
    Fourth ventricle

Myelencephalon
    Medulla oblongata
    Vagal lobes
    Pyramidal tracts
    Fourth ventricle

SELECTED READINGS

Ariens Kappers, C.U., Huber, G.C., and Crosby, E.C.: The comparative anatomy of the nervous system of vertebrates, including man, 2 vols., New York, 1936, The Macmillan Co. (Republished by Hafner Publishing Co., 1960, New York).

Bullock, T.H., and Horridge, G.A.: Structure and function in the nervous systems of invertebrates, 2 vols., San Francisco, 1965, W.H. Freeman and Co., Publishers.

Demski, L.S., and Northcutt, R.G.: The terminal nerve: a new chemosensory system in vertebrates? Science **220:**435, 1983.

Flood, P.R.: A peculiar mode of muscular innervation in amphioxus. Light and electron microscopic studies of the so-called ventral roots, Journal of Comparative Neurology **126:**181, 1966.

Gans, C., Northcutt, R.G., and Ulinski, P., editors: Biology of the reptilia, vols. 9 and 10, New York, 1979, Academic Press.

Gillilan, L.A.: A comparative study of the extrinsic and intrinsic arterial blood supply to brains of submammalian vertebrates, Journal of Comparative Neurology **130:**175, 1967.

Hopkins, W.G., and Brown, M.C.: Development of nerve cells and their connections, New York, 1984, Cambridge University Press.

Kier, E.L.: The cerebral vesicles: a phylogenetic and ontogenetic study, St. Louis, 1977, The C.V. Mosby Co.

Morell, P., and Norton, W.T.: Myelin, Scientific American **242**(5):88, 1980.

Nelson, D.O., Heath, J.E., and Prosser, C.L.: Evolution of temperature regulatory mechanisms, American Zoologist **24:**791, 1984.

Nicol, J.A.C.: Autonomic nervous system in lower chordates, Biological Reviews **27:**1, 1952.

Nieuwenhuys, R.: An overview of the organization of the brain of actinopterygian fishes, American Zoologist **22:**239, 1982.

Norris, H.W., and Hughes, S.P.: The cranial, occipital, and anterior spinal nerves of the dogfish, Squalus acanthias, Journal of Comparative Neurology **31:**293, 1920.

Northcutt, R.G.: Evolution of the telencephalon in nonmammals, Annual Review of Neuroscience **4:**301, 1981.

Northcutt, R.G.: Evolution of the vertebrate central nervous system: patterns and processes, American Zoologist **24:**701, 1984.

Pearson, R., and Pearson, L.: The vertebrate brain, New York, 1976, Academic Press.

Pick, J.: The autonomic nervous system: morphological, comparative, clinical, and surgical aspects, Philadelphia, 1970, J.B. Lippincott Co.

Reiter, R.J., editor: The pineal gland, New York, 1984, Raven Press.

Rovainen, C.M.: Neurobiology of lampreys, Physiological Reviews **59:**1007, 1979.

Sarnat, H.B., and Netsky, M.G.: Evolution of the nervous system, New York, 1974, Oxford University Press.

Shuangshoti, S., and Netsky, M.G.: Choroid plexus and paraphysis in lower vertebrates, Journal of Morphology **120:**157, 1966.

Stevens, C.F.: The neuron, Scientific American **241**(3):54, 1979.

Symposia in American Zoologist

Recent advances in the biology of sharks. Section III. Central nervous system and sense organs, **17:**411, 1977.

Evolution of neural systems in the vertebrates: functional anatomical approaches, **24:**689, 1984.

16

SENSE ORGANS

In this chapter we will study the organs that enable vertebrates to monitor two environments, the one surrounding them and the one within. Since the former presents more challenges to moment-by-moment well-being, proportionately more time will be devoted to monitors of the ambient environment, the exteroceptors. Along the way, we will look at the sensory feedback mechanism that enables skeletal muscles to be prepared for any challenge.

Time has resulted in the evolution of a large variety of simple and complex sense organs or receptors for monitoring the external and internal environment. Sense organs are transducers of specific forms of kinetic energy. They change mechanical, electrical, thermal, chemical, or radiant energy into nerve impulses in sensory neurons.

Vertebrates arose with a full complement of essential sense organs and the necessary central nervous system pathways for processing the information: a mechanically stimulated lateral-line canal system, electroreceptors, a mechanoreceptive vestibular system that may have had some auditory competence, light receptors (lateral and median eyes) and their central pathways, an olfactory system, chemoreceptors other than olfactory, a general somatic sensory system, and a general visceral sensory system. Collectively, these systems provided all the information necessary for survival in the Ordovician Period. Because many of the receptors are, and have been, protected by skeletal capsules that fossilize readily, or because they leave indelible impressions in fossilized dermal armor, or because of telltale foramina, it can be said with confidence that all these systems were present in the oldest known ostracoderms. During subsequent vertebrate evolution they have undergone mutation—some more than others. Some, such as the lateral eyes, have been refined; others, such as the lateral-line system, were lost during adaptation to terrestrial life. Meanwhile, new systems such as thermal receptors in the infrared range have made an appearance and play a role in survival on land.

Somatic receptors provide information about the external environment and the individual's orientation in it. Their sensory fibers terminate in somatic sensory nuclei in the alar plate of the cord or brain (Fig. 15-34, SSA, GSA). As a result of receiving information via somatic receptors, animals make appropriate striated muscle responses that tend to ensure survival. For example, stimuli from an approaching potential enemy will initiate locomotion or a change in posture, and input from surrounding fishes in a school of fishes reflexly maintains the individual's orientation in the dynamic mass. The category of somatic receptors does not include chemical receptors such as those for what are termed "taste" and "smell."

Visceral receptors provide information about the environment within the animal, and olfactory and gustatory information. Their sensory fibers end in the alar plate of the cord or brain in visceral sensory nuclei (Fig. 15-34, VA), or in the olfactory bulb. As a result, the internal environment is reflexly maintained or adjusted as necessary. Since much of the stimulation for the olfactory epithelium and certain other chemical receptors originates in the environment, those receptors not only evoke visceral reflexes but, like somatic ones, they also evoke reorientation of the organism in space.

Although the dendritic endings of most sensory nerve fibers are stimulated directly, in some sense organs intermediary *nonnervous* **receptor cells** serve as transducers, and the receptor cells are innervated by terminals of sensory neurons. **Hair cells** are one variety of receptor cells (Fig. 16-1). They are elongated or bulblike epithelial cells with a variable number of short stereocilia and usually a single kinocilium projecting into the fluid that bathes the sensory epithelium. The cilia are sometimes embedded in a gelatinous mass, the **cupula.** Receptor cells with apical processes instead of cilia are found in taste buds (Fig. 16-22).

It is sometimes convenient to classify sense organs as general and special. **General receptors** are widely distributed on the surface and in the interior. **Special receptors** are receptors having a limited distribution that is usually confined to the head except in fishes and aquatic amphibians.

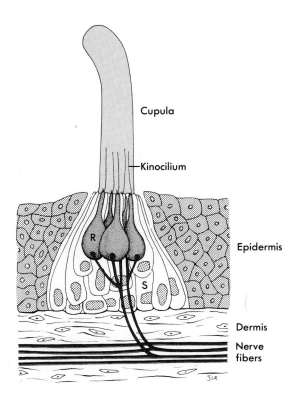

Cupula

Kinocilium

Epidermis

S

Dermis

Nerve fibers

FIGURE 16-1

A neuromast organ in the epidermis of a necturus. *R,* Hair cell (receptor cell) with a single kinocilium and three or four short delicate stereocilia (at base of kinocilium) extending into the cupula. Contacting its base is a heavily myelinated sensory nerve fiber. *S,* supporting cell.

SPECIAL SOMATIC RECEPTORS
Neuromast Organs

Neuromasts are receptors in the skin of fishes and aquatic amphibians that consist of **hair cells** with cilia that are embedded in a gelatinous cupula, **supporting (sustentacular) cells,** and **sensory nerve endings** (Fig. 16-1). The organs are disposed in groups of several or many, depending on the location and the species. They monitor principally mechanical stimuli in the surrounding water, although some also serve as electrical, thermal, or chemoreceptors. The simplest neuromast organs lie in shallow pits or grooves in the epidermis, where the cupula can project directly into the surrounding water. These **external neuromasts** are characteristic of cyclostomes, larval amphibians, and aquatic urodeles.

In jawed fishes, some neuromasts lie in fluid-filled pits that have sunk beneath the epidermis but retain an opening to the exterior. They are frequently referred to as pit organs (Fig. 16-2, *A*). **Ampullary neuromast organs** consist of neuromasts in subcutaneous ampullae that have long ducts, each leading to its own pore on the surface. Among these are the **ampullae of Lorenzini** of many fishes. *Amia* has as many as 3700 ampullae on the head alone. Some neuromasts of electric rays are in ductless **vesicles of Savi.** Pit organs, ampullae, and vesicles of various types are found chiefly, but not exclusively, on the head.

FIGURE 16-2

Neuromast organs in two fishes. **A,** Several varieties in *Squalus acanthias*. **B,** The unique elevated cephalic and lateral-line canals of *Chimaera*.

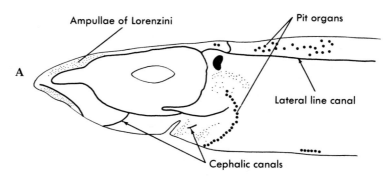

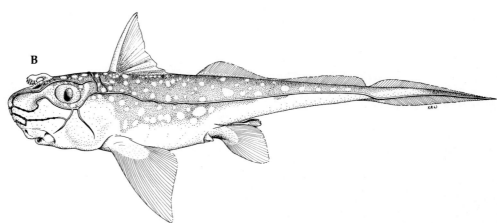

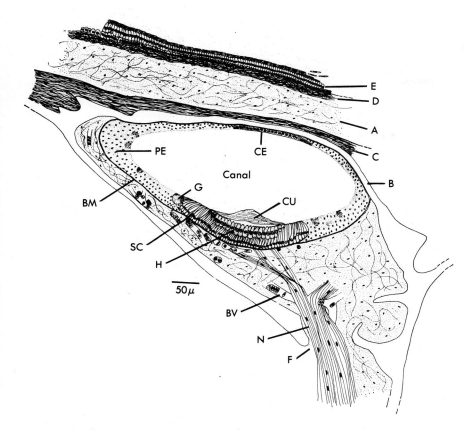

FIGURE 16-3

Lateral-line canal of a bony
fish in cross section at the
site of a neuromast organ.
The canal passes longitudinally embedded in bone under the skin. *A, C,* and *F,*
Dermis; *E,* epidermis. The
epithelial lining of the canal
rests on a basement mem-
brane *(BM),* and consists of a
cuboidal epithelium *(CE),* a
pseudostratified epithelium
(PE), goblet cells *(G),* and the
neuromast organ. The latter
is composed of sustentacular
cells *(SC),* hair cells *(H),* and
a cupula *(CU).* The sensory
nerve *(N)* penetrates the
bone via a foramen *(F).* A
blood vessel is seen ap-
proaching the receptor.
(From Branson and Moore).

The most prominent system of neuromast organs is the **lateral-line canal
system** of fishes and aquatic amphibians, and its cranial component, the
cephalic canal system (Fig. 16-2, *A* and *B*). In its simplest manifestation, it
consists of groups of external neuromasts located in cuplike depressions
along the length of shallow canal-like epidermal grooves on the body surface. Grooves on the surface of dermal armor of acanthodians and placoderms indicate that in early jawed fishes the system was a network of superficial canals covering the entire body. The network became reduced in later
fishes by the loss of many branches, and by interruption of the canals at one
or more locations. In sharks the canals have sunk into the skin of the head
and trunk but remain as open grooves on the tail. In chimaeras the canals
form elevated ridges on the surface where they are readily visible (Fig. 16-2,
B). When in enclosed canals, pores open to the surface at regular intervals.
In modern bony fishes the canals become even more isolated from the external environment, often being embedded in dermal bone underlying the
integument (Fig. 16-3). In some teleosts the canals are restricted to the head.

Free-living amphibian larvae, particularly anuran tadpoles, ordinarily
lose the lateral-line and cephalic canal systems at metamorphosis. Among
amphibian larvae that do not are those of the urodele *Notophthalmus.* When
these larvae metamorphose into red efts and migrate to land the neuromast
organs become buried under the proliferating stratum corneum. Later—

several years later in some localities—when the eft returns to water as a sexually mature newt, the stratum corneum is shed and the canal systems are again exposed. No traces of cephalic or lateral line canal systems are found in permanently terrestrial amphibians or in any amniote.

The major portion of the canal systems of fishes responds to mechanical stimuli such as compression waves of low frequency in the water, water currents, and, perhaps, hydrostatic pressure. Some portions respond to low electric potentials in certain species. Other regions possibly respond to thermal stimuli. Water currents include, for instance, those generated in flowing streams, those caused by locomotor movements of nearby fishes, and waves reflecting from stationary objects. Response to compression waves of low frequency was the original basis for the belief that fishes can hear. (As will be seen shortly, at least some fishes *can* hear.) But the role, if any, of the canal systems in audition is not yet certain. Input from mechanoreceptors would enable an organism to orient itself appropriately in flowing streams **(rheotaxis),** to participate in schooling, and to avoid potential enemies. Blind carnivorous fishes inhabiting dark caves locate some of their food with this input. The roles of the canal systems vary with the habitat of the species (swiftly flowing mountain streams, for example, as opposed to ocean depths), with feeding habits of the species (predators as opposed to vegetarians), with schooling proclivities, and with other environmental and behavioral differences.

Most fishes produce enough electric potential when their muscles contract to make their presence detectable by electroreceptive neuromasts at short range. And, as mentioned in an earlier chapter, some fishes have electric organs capable of generating large potentials. Environmental salinity gradients may also be a source of electrostimuli. Electroreception has been demonstrated in components of the lateral line system of teleosts and in the ampullae of Lorenzini of other fishes. Nerve fibers from electroreceptors terminate in central nuclei that are distinct from those receiving mechanoreceptive input. Since electroreceptive nuclei of teleosts do not occupy the same location as those of other fishes, it has been proposed that teleost electroreceptors may not be homologous with others, and that they arose at least twice during teleost evolution as new modalities. Holosteans have no electroreceptors.

As may be recalled from Chapter 4, neuromast organs arise from embryonic ectodermal placodes on the surface of the head and extending as a linear series along the side of the trunk and tail. One of the placodes at the level of the hindbrain in this linear series becomes the membranous labyrinth of the inner ear. The latter is therefore a complex of neuromast organs. Membranous labyrinths and the other neuromast organs of vertebrates constitute collectively an **acousticolateralis (octavolateralis)** system of sense organs innervated by cranial nerves VII, VIII, IX, and X.

Membranous Labyrinth

All vertebrates have a pair of fluid-filled **membranous labyrinths** or inner ears, each located within a similarly shaped **skeletal labyrinth** in the

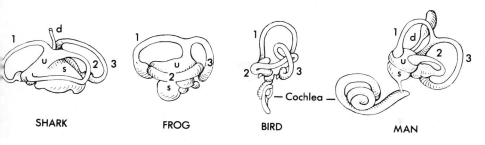

SHARK   FROG   BIRD   MAN

— Cochlea —

FIGURE 16-4

Left inner ears of representative vertebrates. *1,* Anterior semicircular duct with ampulla at base; *2,* lateral semicircular duct; *3,* posterior semicircular duct; *d,* endolymphatic duct and, in man, sac; *s,* sacculus; *u,* utriculus.

otic capsule of the skull (Fig. 16-4). The membranous labyrinth is not adherent to the walls of the skeletal labyrinth, being separated by a fluid-filled **perilymphatic space** (Fig. 16-12, *gray*) across which extend fine strands of connective tissue that stabilize the membranous labyrinth and anchor it in place. The fluid within the membranous labyrinths is **endolymph.** It resembles interstitial fluids elsewhere in the body. The fluid in the perilymphatic space, which cushions the membranous labyrinths in their bony canals, is **perilymph.** It consists chiefly of cerebrospinal fluid that seeps into the space by one route or another. The endolymph and perilymph are not confluent at any point. The membranous labyrinth is a highly specialized complex of neuromast mechanoreceptors.

Each membranous labyrinth of lower vertebrates other than cyclostomes consists of three **semicircular ducts** and two membranous sacs, the **utriculus** and **sacculus.** In the floor or wall of each sacculus in fishes and amphibians there is a small outpocketing, the **lagena** (Fig. 16-8, *A*). The lagena has expanded to become a **cochlea** in mammals (Fig. 16-8, *C*). The lumina of the semicircular ducts are continuous with the cavity within the utriculus, and each duct has a dilation, or **ampulla,** near the junction. The lumen of the utriculus is continuous with that of the sacculus (Figs. 16-4). The anterior and posterior vertical semicircular ducts are in vertical planes that are perpendicular to each other. The third, or lateral duct is in a horizontal plane. Thus there is a semicircular canal in each of the three planes of space. In lampreys, each labyrinth has only two semicircular ducts, an anterior and posterior vertical. Hagfishes have only the posterior vertical duct with, uniquely, an ampulla at each end. Also, in many agnathans there is little or no distinction between sacculus and utriculus. As mentioned earlier with reference to other cyclostome structures, one can only speculate whether their labyrinths are primitive.

Emerging dorsally from approximately the site of confluence of the sacculus and utriculus, except in a few teleosts, is an **endolymphatic duct** (Figs. 16-4, shark and man, and 16-12). It terminates in a blind **endolymphatic sac,** except in elasmobranchs, in which it opens to the exterior through two small endolymphatic pores on top of the head. In mammals the endolymphatic sacs lie in the subarachnoid space of the brain beneath the temporal bones. A pair of **perilymphatic ducts** emerge from perilymphatic spaces in most vertebrates and they, too, usually terminate in a small sac nearby. In sharks the perilymphatic ducts parallel the endolymphatic ones and terminate in a small sac in the endolymphatic fossa (a depression on the middorsal aspect of the neurocranium between the two otic capsules). In mammals the perilymphatic ducts empty directly into the subarachnoid space overlying the brain. Little is known of the role of the perilymphatic ducts of most vertebrates. In mammals they provide a passageway whereby cerebrospinal fluid can contribute directly to the perilymph.

The membranous labyrinth is a connective tissue structure lined by a simple squamous epithelium. In this epithelium, within each of the components of the labyrinth—sacculus, utriculus, ampullae, and lagena or cochlea—are one or more neuromast sites. Each site contains large numbers of hair cells (receptor cells) innervated by sensory fibers of the eighth cranial nerve. A neuromast site within an ampulla is a **crista**. Those elsewhere are **maculae** (Fig. 16-5). Cristae and maculae differ in structural details. A crista is a **papilla** of epithelial cells with hair cells near the surface. The kinocilia of the hair cells extend from the surface of the papilla embedded in a tall, gelatinous cupula that projects far out into the endolymph and is displaced as the inert endolymph responds to rotatory movements of the head. A macula is a flattened mound overlaid by a flat cupula. The upper gelatinous layer of cupulas overlying the utricular and saccular maculae, and sometimes those of the lagena, contains crystalline **otoliths** in anamniotes and a few reptiles. Otolithic membranes are not present in birds and mammals. The long macula-like neuromast organ of the crocodilian, avian, and mammalian cochlea is the **organ of Corti** (*1, 2,* and *3* in Fig. 16-10).

Otoliths are crystals of calcium carbonate combined with a protein. They grow by accretion. In many fishes a single large otolith nearly fills the sac-

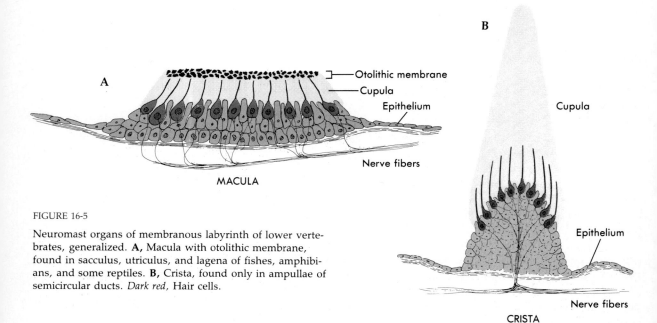

FIGURE 16-5

Neuromast organs of membranous labyrinth of lower vertebrates, generalized. **A,** Macula with otolithic membrane, found in sacculus, utriculus, and lagena of fishes, amphibians, and some reptiles. **B,** Crista, found only in ampullae of semicircular ducts. *Dark red,* Hair cells.

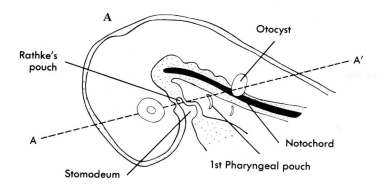

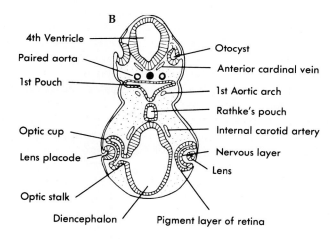

FIGURE 16-6

Origin of inner ear (from otocyst), retina (from optic cup), and lens (from lens placode). **A,** Head region of embryo. **B,** Cross section of head at level of A—Á. Left side of **B** is slightly earlier than right side.

culus. In others, including some sharks, otoliths are microscopic and abundant, and constitute an amorphous mass. The crystals spill out of the sacculus of *Squalus acanthias* if the latter is ruptured in the anatomy laboratory. They can be readily observed with a low-power lens, and their crystalline nature can be observed by transmitted light. Species differences in otolith morphology have made it possible to identify the fossils of some fishes to species level from large otoliths alone.

Membranous labyrinths arise during embryogenesis from surface ectoderm as a pair of **otic placodes** which, in fishes and amphibians, are in series with the placodes that give rise to the neuromasts of the lateral line system. Each otic placode sinks under the skin to become a fluid-filled vesicle, or **otocyst** (Fig. 16-6, *A* and *B*). Cartilage, then membrane bone (except in agnathans and cartilaginous fishes), is deposited around the membranous labyrinth, separated from the latter by the narrow perilymphatic space, thereby creating a skeletal labyrinth.

The sacculus, utriculus, and semicircular ducts are primarily organs of equilibration (balance, in a broad sense). However, in at least some fishes certain maculae function in hearing to one degree or another. The extent of the phenomenon of hearing in fishes remains to be explored; only general information for a small number of species is available. The cochlea, which is of more recent origin, is solely auditory.

EQUILIBRATORY FUNCTION OF THE LABYRINTH. The sacculus, utriculus, and semicircular ducts are receptors that monitor the position of the head in space, whether at rest (in static equilibrium) or in motion (in dynamic equilibrium). Information from these organs is employed reflexly by skeletal muscles to maintain a moment-to-moment posture of the body that favors survival. But what is the individual role of each of these components of the labyrinths in equilibration?

Answering this question with reference to the sacculus and utriculus in vertebrates as a group is possible only in the most general terms, largely for lack of sufficient experimental evidence from the several vertebrate classes. There are facts we are sure of. The maculae of the sacculus and utriculus lie in different planes; not all kinocilia in a single neuromast site are oriented in the same direction; the orientation of the kinocilia is affected by movements of the endolymph and of an otolithic or tectorial membrane, the latter having no otoliths; and displacement of kinocilia is translated into nerve impulses that are transmitted across synapses in the brain. Thus different maculae and different hair cells in the same macula are stimulated differentially as the endolymph is displaced by movements of the head. It has been concluded that the sacculus and utriculus respond to changes in linear velocity of a fish; that is, appropriately oriented hair cells fire when stimulated as a result of acceleration of the head when the fish is swimming in a straight line. One explanation proposes that linear acceleration tends to carry along the kinocilia but that the otolithic membrane, because of inertia, initially resists acceleration. As a result, certain kinocilia that are in contact with the overhead membrane are bent in a direction opposite that of acceleration, and this stimulates the sensory nerve endings at the base of hair cells. It is also thought that a tilted position of the head differentially stimulates one macula as opposed to another. The generalization is that the sacculus and utriculus are stimulated by *linear acceleration of the head,* and by the *orientation or position of the head,* but *not by angular movement* of the head.

Stimulation by angular acceleration of the head (rotation of the head) seems to be the province of the semicircular ducts. The cupula of a crista projects far out into the endolymph, where it is subject to swinging displacement by the shearing forces generated in the endolymph as the head turns in one plane or another. Consequent bending of the enclosed kinocilia is translated into a nerve impulse.

The sensory impulses generated in the semicircular ducts, sacculus, and utriculus by changes in linear and angular acceleration of the body and by the position of the head are transmitted to the brain stem and cerebellum where, as one result, reflex movements of the eyeballs are initiated so that the eyes of higher vertebrates, at least, are always looking in a direction compatible with the position of the head. Voluntary effort is required to turn your head swiftly to the side while continuing to gaze straight ahead.

(Try it!) That the labyrinth controls reflex eyeball movements can be demonstrated by spinning someone in a revolving chair rapidly a number of times and then stopping the chair abruptly. The individual will feel dizzy, and an observer will note that the eyeballs continue to exhibit rapid, jerky, side-to-front-to-side movements **(nystagmus)**, which cease as the inert endolymph ceases to apply shearing forces on the cristae of the semicircular ducts. The nystagmus is the cause of the dizzy feeling. Children, in fun, often turn round and round until, for the same reason, they become dizzy.

Input from the equilibratory components of the labyrinth also results in reflex motor activity that may alter the position of the trunk, appendages, and neck (if present), thereby maintaining the body and its parts in proper orientation (balance), with respect to gravity, as when a cat is turned upside down, dropped from a height, and lands on its paws. Maintenance of equilibrium is more readily *demonstrable* in tetrapods, since they have a larger number of independently movable members of the body and they are not suspended in a dense medium, but it is equally essential for survival of a fish in water. In its role of maintaining balance, the labyrinth is assisted by proprioceptors, to be discussed shortly.

AUDITORY FUNCTION OF THE LABYRINTH AND EVOLUTION OF THE COCHLEA. The labyrinth has an auditory function as far down the phylogenetic scale as those fishes in which maculae respond to longitudinal sinusoidal waves of the amplitude and frequency of sound (low amplitude, high frequency). In Ostariophysi (Order Cypriniformes or separate status), a large group of mostly freshwater teleosts that includes catfish, goldfish, and carp, sound waves in water evoke waves of similar frequency in the gas in the turgid swim bladder, and these are transmitted to the labyrinth by **weberian ossicles** (Fig. 16-7). These are a series of modified transverse pro-

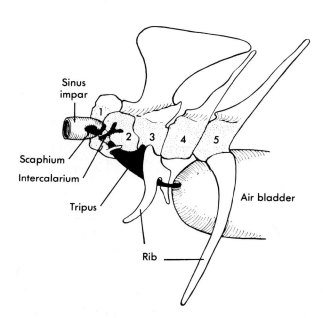

FIGURE 16-7

Weberian ossicles *(black)* of a teleost. *1* to *5*, Centra of first five vertebrae. The sinus impar is an extension of the perilymphatic space. The ossicles are connected by modified intervertebral ligaments.

cesses of the first three (occasionally four or five) trunk vertebrae that extend between the swim bladder and the **sinus impar,** an extension of the perilymphatic space. These fish can hear, although the mechanism provides no information as to the direction from which sounds are coming. The specific maculae that detect these sounds are not known. It is known, however, that the maculae that respond to sound in *lower tetrapods* are in the sacculus near the lagena. In Clupeiformes, an order of herringlike teleosts, an anterior extension of the swim bladder comes in direct contact with the labyrinth, but whether this is a route for transmitting sound waves is not known. Without doubt, there are other teleosts that have maculae with low enough threshold to detect sound waves. The extent of the phenomenon of hearing in fishes and the receptor sites involved is not yet known. Only general information on a small number of species is currently available.

An account of the evolution of the amniote cochlea begins with amphibians. In addition to the usual maculae associated with the sacculus and lagena of fishes, amphibians have two receptor sites that resemble maculae except that they have a **tectorial membrane** instead of a cupula. (A mammalian tectorial membrane is illustrated in Fig. 16-10.) One of them is the **amphibian papilla** of the sacculus, so named because it is not found in any other vertebrate (Fig. 16-8, *A*). The other is the **basilar papilla** in the basilar recess of the sacculus. These two papillae are receptors for sound.

The basilar papilla and the lagena were destined to become the chief components of the amniote hearing mechanism. In most reptiles the lagena has become a prominent sac suspended from the sacculus. It still houses the lagenar macula, whose function is unknown, and the basilar papilla, now at its entrance, serves for audition. In crocodilians, birds, and mono-

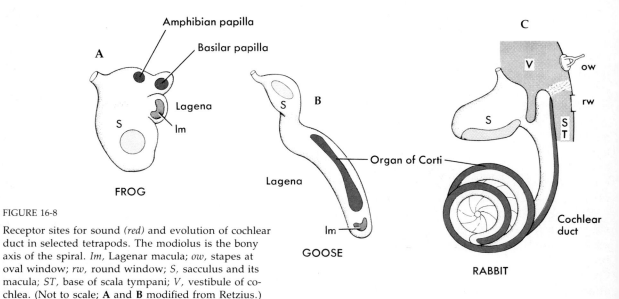

FIGURE 16-8

Receptor sites for sound *(red)* and evolution of cochlear duct in selected tetrapods. The modiolus is the bony axis of the spiral. *lm,* Lagenar macula; *ow,* stapes at oval window; *rw,* round window; *S,* sacculus and its macula; *ST,* base of scala tympani; *V,* vestibule of cochlea. (Not to scale; **A** and **B** modified from Retzius.)

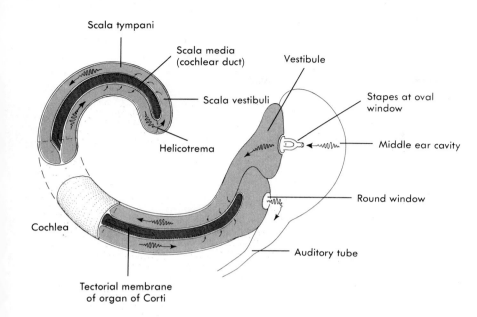

Scala tympani

Scala media
(cochlear duct)

Vestibule

Scala vestibuli

Stapes at oval
window

Helicotrema

Middle ear cavity

Cochlea

Round window

Auditory tube

Tectorial membrane
of organ of Corti

FIGURE 16-9

Pathway of sound waves from the middle ear through the scalae cavity and back to the middle ear at the oval window. *Light red,* Vestibule, scala vestibuli, and scala tympani; *dark red,* scala media. The tectorial membrane, of which only a short section is shown, extends the entire length of the cochlear duct.

tremes the lagena has become a long tube and the basilar papilla has become an elongated **organ of Corti** within the tube (Fig. 16-8, *B*). The term "cochlear duct" is often applied to this tube, but only in placental mammals is it part of a *spiral* hearing organ. Elongation of the lagena was accompanied by a corresponding extension of the perilymphatic space.

In placental mammals the duct housing the organ of Corti (the old lagena), now called the **cochlear duct,** has elongated still more and spirals upward around a bony pillar of the petrosal bone, the **modiolus** (Fig. 16-8, *C*), within which lies the spiral ganglion of the cochlear nerve. The cochlear duct or **scala media** is accompanied by perilymphatic spaces, the **scala vestibuli** and **scala tympani** (Fig. 16-9). The scala media, vestibuli, and tympani collectively constitute the **cochlea,** so named because in mammals the unit resembles a snail's shell. The spiral varies from one to five turns. The scala vestibuli begins at the vestibule, a portion of the perilymphatic space where sound waves enter the cochlea through the membrane stretched across the oval window **(fenestra ovalis)** of the otic capsule. It spirals to the apex of the cochlea where it is continuous with the scala tympani via a small aperture, the **helicotrema.** The scala tympani spirals down the helix and ends at the **secondary tympanic membrane** stretched across the round window **(fenestra rotundus).**

The organ of Corti of mammals is a long, macula-like neuromast receptor site resting on a spiraling, bony, shelflike projection of the modiolus, the **osseous spiral lamina,** which is continuous with a fibrous **basilar membrane** that separates the scala media from the scala tympani (Fig. 16-10, *3*). Overlying the length of the organ of Corti is a **tectorial membrane** that is attached to the inner wall of the cochlear duct. The distalmost tips of the hair cells of the organ of Corti are embedded in the tectorial membrane.

FIGURE 16-10

Cross section of one turn of a mammalian cochlea. *1, Cochlear duct (scala media) containing endolymph (red);* 2, tectorial membrane; *3, basilar membrane (attached at left to the osseous spiral lamina) and organ of Corti resting on basilar membrane; 4, Reissner's membrane; 5, spiral ganglion and cochlear nerve fibers. The scala vestibuli and scala tympani are perilymphatic spaces. The diameter of a similar section of the human cochlea would be less than 1 mm.*

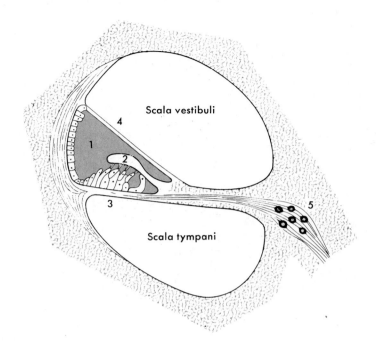

Separating the scala media from the scala vestibuli is a thin sheet of epithelium, **Reissner's membrane.**

Sound waves induced in the perilymph of the vestibule by vibrations of the footplate of the stapes are transmitted up the scala vestibuli, down the scala tympani, and, via Reissner's and the basilar membranes, into and through the endolymph of the scala media at all levels of the spiral (Fig. 16-9). Expansion and compression waves induced in the endolymph displace the organ of Corti toward the tectorial membrane at the same frequency, and the mechanical effect on the hair cells sets off volleys of nerve impulses in the sensory fibers that innervate the organ. It has been ascertained that hair cells in the first spiral respond maximally to vibrations of higher frequencies, hence higher pitches; those near the apex of the scala media respond only to low frequencies, and those in between exhibit graded responses. Vibrations in the perilymph are discharged instantaneously into the middle ear cavity at the round window. Lower tetrapods as well as mammals have oval and round windows in their otic capsules.

With expansion of the basilar papilla to become an organ of Corti, the relay nuclei on pathways for sound in the brain became increasingly larger. In reptiles for the first time the auditory nuclei in the tectum of the mesencephalon bulge above the surface to become auditory lobes (Figs. 15-13, snake, and 15-25). Differentiation of an auditory neocortex in the temporal lobes of the cerebral hemispheres (Fig. 16-24) was accompanied by a further increase in size and number of secondary sensory nuclei and of higher relay centers for sound.

THE MIDDLE EAR. The most common route of conduction of sound waves from air to the inner ear of tetrapods is via a **tympanic membrane (eardrum)** and one or more ossicles that extend across the **middle ear cavity (cavum tympanum)** between the drum and the secondary tympanic membrane at the oval window (Figs. 16-11 and 16-12). All tetrapods have a **stapes**—known also as a **columella** in lower tetrapods—unless the tympanic membrane is absent and the middle ear cavity remains rudimentary, which is not uncommon in some orders. The columella or stapes is the ancient hyomandibular cartilage of fishes.

A stage in the evolution of a columella is seen in the urodele *Ranodon* (Fig. 8-12, *B*), whose inner ear, like that of other urodeles, is rudimentary,

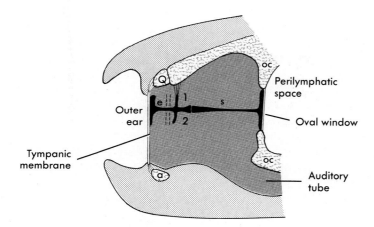

FIGURE 16-11

Stapes complex (*black*) and middle ear cavity (*red*) of a lizard, diagrammatic. *a,* Articular bone of lower jaw; *e,* extrastapes (extracolumella); *oc,* otic capsule; *Q,* quadrate bone of upper jaw; *s,* stapes (columella). *1* and *2,* Dorsal and ventral processes of extrastapes, the latter articulating with the quadrate bone (*broken lines*) beyond the plane of the section.

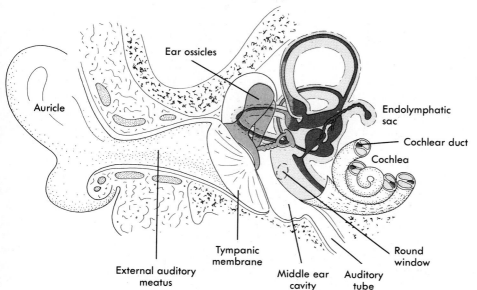

FIGURE 16-12

Mammalian ear complex. *Dark red,* membranous labyrinth; *light red,* middle ear ossicles; *gray,* perilymphatic space within bony labyrinth. The footplate of the stapes (see Fig. 16-13) rests against the oval window.

FIGURE 16-13

Human middle ear ossicles. The handle of the malleus is attached by ligaments to the eardrum, the anterior process is anchored by a ligament to the petrous portion of the temporal bone, and the footplate of the stapes rests against the membrane at the oval window. (From Schottelius, B.A., and Schottelius, D.D.: Textbook of physiology, ed. 17, St. Louis, 1973, The C.V. Mosby Co.)

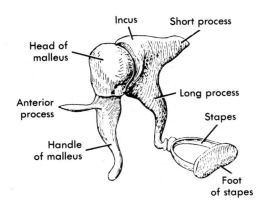

and who lacks a tympanic membrane. The columella still occupies its primitive (and embryonic) location as the dorsalmost segment of the hyoid arch, abutting against the otic capsule at the site where the oval window forms in other tetrapods. The columella of anurans, on the other hand, is a strong, cylindrical, mobile, replacement bone that is firmly attached to the large eardrum and traverses a broadly open middle ear cavity to reach the oval window in the prootic bone.

In reptiles and birds the columella, unless rudimentary, is composed of two segments, a cartilaginous **extrastapes** attached to the eardrum and a bony stapes in contact with the membrane of the oval window, both of hyomandibular origin (Fig. 16-11, *e*).

Mammals have two additional ossicles, a **malleus** and **incus** (Figs. 16-12 and 16-13) derived from the articular and quadrate bones of reptiles. Because the surface area of the eardrum of all tetrapods approaches twenty times that of the secondary tympanic membrane at the oval window (Fig. 16-12), vibratory stimuli applied to the eardrum have been amplified many times by the time they are delivered to the vestibule of the cochlea. The leverage achieved by the series of unaligned (nonlinear) articulations between the ear ossicles of mammals (Fig. 16-13) intensifies the effect.

There are two small muscles in the middle ear cavity of mammals. A **tensor tympani** innervated by the fifth cranial nerve inserts on the malleus and regulates the tension of the eardrum, and a **stapedial muscle** innervated by the seventh cranial nerve inserts on, and restrains, the movements of the stapes. Very loud sounds cause the tensor tympani and stapedial muscles to contract, which damps the vibrations of the ossicles and protects the hair cells of the organ of Corti. Relaxation of the muscles results in amplification of weak sounds, as when one "strains one's ears" to hear. The phylogenetic history of the ossicles and their muscles was discussed in earlier chapters.

The middle ear cavity arises ontogenetically as an evagination of the first pharyngeal (spiracular) pouch, which grows toward the developing ear ossicle or ossicles and partially surrounds it or them, isolating them from other tissues (Fig. 8-37). The cavitation process is completed by erosion of any remaining mesenchyme. The middle ear cavity remains in communication with the pharynx throughout life via an **auditory tube (eustachian tube),** which ensures that the air pressure on both sides of the eardrum will be the same. Anurans and some lizards have a wide, short, permanently open tube that opens into the buccopharyngeal cavity (Fig. 11-3, *A*). That of crocodilians, birds, and mammals is longer, of small diameter, and opens above the secondary palate (Fig. 11-3, *B*). The mammalian auditory tube remains closed at its nasopharyngeal exit except during the act of swallowing. Anyone who has suffered from increased pressure in the middle ear as a result of blockage of the auditory tube will recall the sensation and its effect on hearing. Frequent swallowing relieves the symptoms by opening the entrance to the tube, thereby equalizing the pressure on the two sides of the eardrum.

THE OUTER EAR. The **tympanic membrane** or eardrum is the usual portal to the ear ossicles of tetrapods. In anurans, turtles, and primitive lizards the eardrum is more or less flush with the surface of the head (Figs. 5-10 and 16-14). In more specialized lizards, crocodilians, birds, and mammals the eardrum lies at the end of an **external auditory meatus** or outer ear canal that may be short or long and has an entrance on the side of the head behind the angle of the jaws (Figs. 5-15, 16-11, and 16-12). In terrestrial mammals a fibrocartilaginous appendage, the **auricle (pinna)** collects sound waves and directs them into the outer ear canal. Cetaceans, pinnipeds, sirenians, and some moles have a very small auricle or none at all. The outer ear canal of whales is of very small diameter, but whales have excellent hearing, which is used to receive communications and for echolocation.

FIGURE 16-14

Head of an iguanid lizard showing slightly depressed eardrum just behind angle of jaws.

TETRAPOD HEARING IN THE ABSENCE OF AN EARDRUM. Although an eardrum and middle ear are the usual route to the auditory receptors of the inner ear of tetrapods, these are not always present. Urodeles, apodans, a few anurans (*Ascaphus*, for example), *Sphenodon*, and most limbless reptiles have no eardrum, the auditory tube and middle ear cavity are vestigial, and, except in snakes and amphisbaenians, the columella is rudimentary. Yet, most of these tetrapods are sensitive to either airborne or substrate-borne (seismic) sound waves, or both.

In amphisbaenians a long, narrow, thin process of the extrastapes extends forward on each side of the lower jaw just beneath the skin. It ends in a broad, flat, subcutaneous plate that functions like a tympanic membrane, vibrating in response to sound waves in the environment. The waves are transmitted to the oval window via the columella. Some other amphibians and reptiles whose lower jaw lies in contact with a substrate detect seismic waves in the frequency range of the inner ear by way of the mandible, the quadrate, and the columella, or some other bone that articulates with the otic capsule. In lizards a process of the extrastapes is attached to the quadrate bone. The route of airborne sound waves to the perilymph of snakes is not known, except that it involves the quadrate and the well-developed columella. Still other lower tetrapods detect seismic waves via the skeleton of the forelimbs, scapula, and either the extrastapes, which in some species is connected to the scapula by a ligament, or some other bone in the temporal region of the skull.

Anurans have an ossification, the operculum (not homologous with the operculum of fishes), in the membrane of the oval window. An opercular muscle connects the pectoral girdle with the operculum. Although the role of this **anuran opercular system** is not yet known, it seems likely that it receives seismic signals from the pectoral girdle. Thus there is opportunity for bone conduction of sound waves by one route or another from the environment to the perilymph of tetrapods in the absence of a drum and even of middle ear ossicles. Many prosthetic hearing devices for human beings employ bone conduction to bypass the middle ear.

Saccus Vasculosus

Suspended from the floor of the diencephalon just caudal to the pituitary in jawed fishes other than lungfishes is a thin-walled ventral evagination of the third ventricle which, because it is highly vascular, has been named the saccus vasculosus (Fig. 15-18). In some species it extends far caudad, underlying much of the brain stem. It is lined with hair cells that are surrounded by sustentacular cells and innervated by sensory nerve fibers that pass to the hypothalamus and other brain centers. The cilia of the hair cells project into the cerebrospinal fluid. It is thought that it may monitor the pressure of the cerebrospinal fluid (which varies with the depth of the fish), and that the information may be used by actinopterygians to regulate the volume of gas in the swim bladder, hence the animal's buoyancy at given depths. However, it will be recalled that elasmobranchs lack a swim bladder. The significance of the rich vascular supply is unknown. The sac is largest in deep sea fishes, not well developed in shallow freshwater species, rudimentary in cyclostomes, and absent in lungfishes.

Light Receptors

Many cold-blooded vertebrates have two sets of photoreceptors, lateral and median eyes. In lateral eyes, reflected light is translated into an image on a photosensitive epithelium (the retina), and the information is used in maintaining an appropriate orientation of the body in the environment. Median eyes form no image. The light stimulates neuroendocrine reflex arcs that maintain biological rhythms (Fig. 17-5). Lateral eyes are necessary for moment-to-moment survival of the individual. Median eyes participate in long-term survival of species that have them.

LATERAL EYES. The receptor site of the lateral eye is the **retina,** a membrane rich in nervous tissue and synapses, and located at the rear of a fluid-filled vitreous chamber of the eyeball (Fig. 16-15). The retina arises from the embryonic forebrain as lateral evaginations that soon become double-walled **optic cups** (Figs. 15-6 and 16-6, *B*). The optic cup retains an attachment to the brain via the **optic stalk.** The layer of the optic cup that light will first strike becomes the **nervous layer of the retina.** Some of the retinal cells become rods and cones, which are the photoreceptors. Others become bipolar association neurons, and still others become cell bodies of the optic nerve fibers, somewhat inappropriately called ganglion cells. The latter sprout long processes that grow along the optic stalk and into the brain. These processes constitute the optic nerve. From this description it can be seen that the retina arises from the brain and never becomes completely detached from it.

The rods and cones of the retina are **neurosensory cells,** that is, specialized neurons whose cell bodies lie in an epithelium. Rods are insensitive to color, but they have a low threshold for light, hence can function in poorly illuminated environments. Cones—green cones, blue cones, double cones,

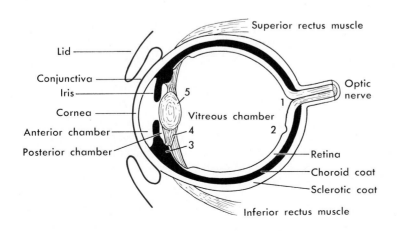

FIGURE 16-15

Vertebrate eyeball in sagittal section. *1,* Blind spot; *2,* fovea; *3,* ciliary body; *4,* suspensory ligament; *5,* lens. Red represents the bulbar conjunctiva (on the cornea) and the palpebral conjunctiva (on the under surface of the lids). The extrinsic muscles are shown in Fig. 10-14.

FIGURE 16-16

Nervous layer of retina. **A,** Histological appearance. **B,** Areas of synapse, diagrammatic. The light enters the retina at layer *10* (base of picture) and passes through the other layers to the rods and cones. *1* and *a,* Pigmented epithelium; *2,* layer of rods and cones; *3,* external limiting membrane; *4,* nuclei of the rods and cones; *5,* outer molecular layer; *6,* layer of bipolar cell bodies; *7,* inner molecular layer; *8,* layer of ganglion cells (cell bodies of optic nerve fibers); *9,* layer of optic nerve fibers; *10,* internal limiting membrane; *b,* rods and cones. (From Bevelander, G., and Ramaley, J.A.: Essentials of histology, ed. 8, St. Louis, 1978, The C.V. Mosby Co.)

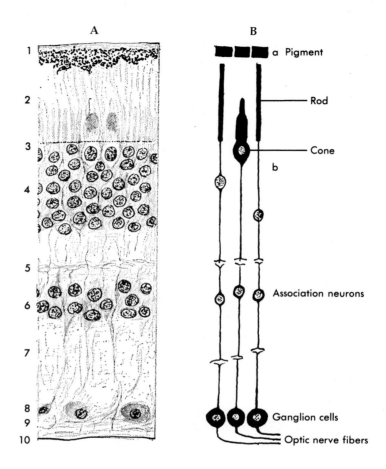

each sensitive to a specific range of wave lengths of light—function in color vision. Cones are lacking in a few vertebrates, including fishes and tetrapods that inhabit murky waters or dark habitats, and in snakes.

Toward the medial quadrant of the retina there is a blind spot, the **optic disc,** lacking rods and cones, where the optic nerve fibers leave the retina. Not far from the optic disc, directly behind the pupil, is a small, oval, yellowish spot, the **macula retinae (macula lutea),** which is composed predominantly of rods in nocturnal animals, cones in diurnal ones. Macular vision is *more acute.* In the center of the macula of a few mammals (chiefly the highest primates), most birds and lizards, and a small number of other vertebrates is a depressed area, the **fovea,** composed chiefly of cones. There vascularity is minimal, the layer of association cells and ganglion cells overlying it is thinnest, hence vision is *most acute.* Nocturnal animals lack a fovea. Raptorial birds (owls, eagles, hawks, other birds of prey) and some lizards have two foveas in each eye. The grand champion of all foveas is probably that of hawks, which have a man-sized eye with approximately 1 million cones per square millimeter in each fovea. The few other birds with two foveas in each eye are those that catch flying insects. The resolving power of a hawk's eye is said to be at least eight times that of the human eye.

The embryonic mesenchyme surrounding the optic cup forms a pigmented **choroid coat** and a dense **sclerotic coat** that are external to the retina (Fig. 16-15). The choroid coat is perforated by the **pupil.** Although usually circular, the pupil is slitlike in snakes and cats and rectangular in many ungulates. The part of the choroid coat surrounding the pupil is the **iris diaphragm.** This is the colored part of the eyeball that is seen encircling the pupil when the eye is open. In tetrapods the iris diaphragm contains smooth, radially arranged **dilator muscles** and circumferential **constrictor muscles** that together alter the diameter of the pupil in response to changes in light intensity. (The diameter of the pupil of fishes is usually fixed.) The effect of light may be demonstrated by shining a flashlight into a dark-adapted eye and watching the pupil constrict. The pupil appears black because it is dark within the eye.

The sclerotic coat **(sclera),** which is the white of the eye, is so named because it is often sclerotized (hardened) by cartilage or bone which, in reptiles and birds, takes the form of platelike ossicles (Fig. 16-17). In front of the iris diaphragm and pupil the sclerotic coat becomes transparent and more convex. This portion is the **cornea.** It is through the cornea that one sees the iris diaphragm and the pupil. Inserting on the outer surface of the sclerotic coat are rectus and oblique muscles that rotate the eyeball within the orbit (Fig. 10-14). These muscles are similar from fish to man except that in birds they are so poorly developed that they are practically useless. As a result, a bird must move its entire head to change the direction of its gaze.

The lens arises as a lens placode that sinks into position in front of the optic cup (Fig. 16-6, *B*). The placode is induced to form by inductor substances released by the embryonic retina. That the retina serves as the inductor may be demonstrated in an amphibian embryo by exchanging undifferentiated ectoderm from the region of the future thigh with that at the site where the lens will form. A lens is induced in the tissue transplanted to the head but not in the potential lens ectoderm that was removed from the influence of the retina. Removal of the optic cup results in no lens placode.

The chamber behind the lens, the **vitreous chamber,** is filled with a jellylike viscous refracting substance, the **vitreous humor.** The small chamber between the lens and iris diaphragm **(posterior aqueous chamber)** and that between the iris diaphragm and cornea **(anterior aqueous chamber)** are

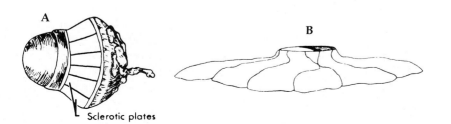

A

B

Sclerotic plates

FIGURE 16-17

Ossicles of the sclerotic coat of the eye. **A,** Owl's eye, showing sclerotic plates in place. **B,** Scleral ring of overlapping ossicles from a lizard's eye.

filled with **aqueous humor.** In teleosts, squamates, and birds, a pigmented and highly vascular conical or fan-shaped projection of the choroid coat extends into the vitreous chamber nearly to the lens from a position near the entrance of the optic nerve. It is known as the **falciform process** in fishes, **conus papillaris** in reptiles, and **pecten** in birds. It is thought that these may alert the animal to nearby moving objects by casting a shadow on the retina. Their intense vascularity has also led to the hypothesis that they participate in metabolic exchange between the blood and the interior of the eyeball.

Within the eyeball, encircling the periphery of the iris diaphragm at the level of the sclerocorneal junction is the **ciliary body,** a low, wedge-shaped shelf of the choroid coat in the form of a ring (Fig. 16-15, 3). Stretched between this ring and the lens is a **suspensory ligament** that inserts circumferentially on the lens capsule. Associated with the ciliary body in tetrapods is a ring of **ciliary muscle** fibers, most of which are arranged radially. The ciliary muscle in amniotes other than snakes regulates the curvature of the lens for near and far vision.

Accommodation of the eye for near or far vision is effected differently in different vertebrates. In lampreys a **cornealis muscle** *pulls on the cornea* from one side, flattening the cornea. In elasmobranchs and amphibians the lens of an eye at rest (emmetropic eye) is at a distance from the retina such that distant objects are focused on the retina. In accommodating for near objects the *lens is drawn forward* by protractor muscles. The resting eye of teleosts, on the other hand, is focused for near objects, and a retractor muscle, the **campanula,** extends from the falciform process to the lens and *pulls the lens backward* for distant vision. Changing the curvature of the lens during accommodation is uncommon in anamniotes.

In snakes, increased pressure in the vitreous humor generated by muscles near the iris *pushes the lens forward.* The lens returns passively to the resting position when the force that displaced it is removed. This is different from other amniotes, who change the shape of the lens in accommodating for near objects.

When the eye of an amniote other than a snake is focused at infinity, the ciliary muscle ring is relaxed. Because of the distance between the ring and the lens at this time, the suspensory ligaments extending radially between the lens and the ring are taut and the lens, being under tension, is flattened. Contraction of the muscles of the ciliary ring in adapting for near vision *pulls the ciliary body slightly toward the lens.* This relieves some of the tension within the suspensory ligament, and the lens, being resilient, becomes more convex. The ciliary muscle fibers are striated in reptiles and birds, smooth, hence slower to respond to motor stimuli, in amphibians and mammals. Therefore reptiles and birds can change from infinity to near vision, and reverse, more quickly than other tetrapods.

The surface of the eyeball that you can touch when the lids are open is covered with transparent skin, the **bulbar conjunctiva.** This is continuous with the **palpebral conjunctiva** on the under surface of the lids (Fig. 16-15). In snakes and many lizards the lids are permanently closed, but they are transparent, forming a **spectacle.** Each time a snake molts the surface of the spectacle is shed, and this carries away any scratches.

Epidermal glands in the orbit of terrestrial vertebrates keep the conjunctiva moist and clean. The **lacrimal gland** secretes lachryma (tears). The gland is poorly developed in some reptiles and birds, but enormous in marine turtles, in which it secretes salt. In many mammals a **harderian gland** secretes a more viscous fluid. It is absent in some mammals, especially permanently aquatic species. Some mammals also have an **infraorbital gland.** The fluids secreted by orbital glands and emptying into the palpebral fissure (the space enclosed by the eyelids) usually drain into a nasolacrimal duct that leads to the nasal cavity or to the vomeronasal organ, to be discussed shortly.

Vertebrates that live in caves or other dark recesses (some fishes, cave salamanders, caecilians, and moles, for example) are frequently blind and the eyes may even be vestigial. Frequently the lids fail to open. In hagfishes no eyeball whatsoever differentiates.

MEDIAN EYES. Many vertebrates below birds have a functional third eye on the top of the head (Fig. 16-18). Among these are lampreys, ganoid fishes, a few teleosts (especially larvae), anuran larvae, some adult anurans, *Sphenodon*, and some lizards. The median eye is an evagination of

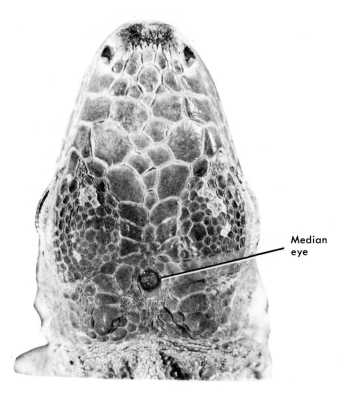

Median
eye

FIGURE 16-18
Parapineal eye of an iguana.

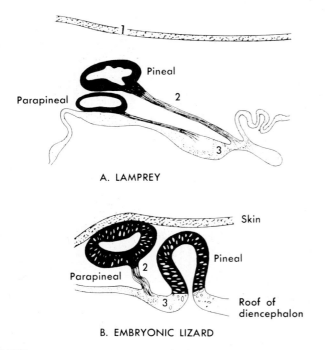

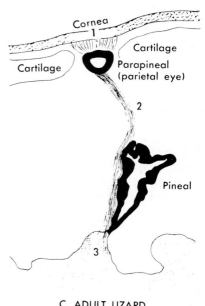

FIGURE 16-19

Epiphyseal complex of lamprey and embryonic and adult lizard. *1*, Cornea; *2*, fiber tract; *3*, habenular nuclei. The parapineal of the lizard lies within the parietal foramen of the skull. (**B** and **C** modified from Nowikoff.)

the roof of the diencephalon. Primitively, it was one of two evaginations constituting an **epiphyseal complex** (Fig. 16-19). The more anterior evagination is the parapineal, the posterior one is the pineal. It is usually the parapineal that is photosensitive, but in lampreys both are.

In lampreys (Fig. 16-19, *A*) the pineal ends as a hollow knob beneath the cornea, an area of skin devoid of pigment between the lateral eyes. The upper wall of the knob consists of several layers of cells that form a lens. The lower wall contains photosensitive cells and, beneath these, ganglion cells with long processes that pass down the stalk to sensory nuclei in the *right* side of the diencephalon. The parapineal of lampreys is similar, but its fibers terminate on the *left*. The pineal is dominant.

The parapineal of lizards, or **parietal eye,** lies in the parietal foramen immediately under a single, translucent epidermal scale in the midline of the head (Fig. 16-19, C). It consists of a cornea, lens, retina with photoreceptive cells resembling those of the vertebrate retina, ganglion cells, and a sensory fiber tract that extends down the epiphyseal stalk and enters the roof of the diencephalon.

In larval frogs the third eye is called the **frontal organ,** or **stirnorgan.** Authorities are uncertain whether this is the pineal or parapineal part of the epiphyseal complex. At metamorphosis the photoreceptive part of this organ regresses, leaving only the glandular component that produced the hormone melatonin in the larva. In at least one tree frog, however, the third eye persists throughout life.

Median eyes, unlike lateral eyes, form no retinal image. Instead, they monitor the duration of photoperiods, and the input affects internal biological rhythms such as daily spontaneous motor activity and seasonal changes in the gonads. They probably monitor also the intensity of solar radiation. The rate of metabolic processes is correlated with body temperature, which, in turn, depends on how much solar energy has been received. Third eyes were present in all major groups of Devonian fishes and in early amphibians and reptiles before being lost in most later vertebrates.

Infrared Receptors of Snakes

Rattlesnakes, copperheads, and water moccasins (family Crotalidae) have a deep pit receptor on each side of the head between the eye and the nostril. Within the pit are hair cells that are sensitive to infrared radiation (thermal stimuli). The tissue that underlies the epithelial receptor organ is highly vascular. Because of their location in the lore, which is the region in front of the eye of a reptile or bird, they are called **loreal pits.** They are at the caudal border of the loreal scale. (The loreal scale of a milk snake is shown in Fig. 5-16.) Loreal pits are readily visible, being several millimeters wide, twice as deep, and directed forward. Because of these pits, crotalid snakes are called pit vipers.

Physiological and behavioral studies have shown that loreal pits can detect temperature changes of as little as 0.001° C at a distance of several feet. This enables pit vipers to detect the presence of warm-blooded vertebrates such as mice or small birds at night, or in daytime when the prey is hidden under the substrate or in a nest or burrow, so long as the environmental temperature is lower than that of the prey. The receptors also enable the viper to strike accurately in the dark. Experimentally, they can also detect objects that are colder than the surrounding environment.

Pythons (family Pythonidae) and some boas (family Boidae) have a series of similar but smaller and less sensitive **labial pits** with slitlike openings between the scales surrounding the mouth. Some pythons have up to 30 of these infrared receptors. All thermal receptors on the head of snakes are innervated by sensory fibers from the fifth cranial nerve. The fibers terminate in a cranial nucleus that is separate from, but closely associated with, the trigeminal nucleus that receives mechanoreceptive input.

Evolution of reptilian pit organs, which lie above the maxillary bones, entailed restructuring of the maxillae in such manner that they could accommodate the pit organs while at the same time continuing to support and move the fangs. Pit organs supplement vision, mechanoreception, and chemoreception in the search for food.

SPECIAL VISCERAL RECEPTORS

There are two varieties of special visceral receptors, **olfactory** (for smell) and **gustatory** (for taste). Both are *chemoreceptors* that are sensitive to certain amino acids. Olfactory receptors respond to less specific chemical configurations than taste receptors. Differences in sensitivity apparently are related to side chains of the acids. That taste and smell are related phenomena is evident to anyone who has been unable to savor food because of a head cold.

Olfactory Organs

The olfactory organs of gnathostomes arise as a pair of ectodermal olfactory placodes that form just anterior to the embryonic stomodeum. The placodes sink into the head to form a pair of **nasal pits,** the epithelial lining of which differentiates into olfactory, supportive, and mucous cells (Figs. 1-1 and 12-8, *B*). The olfactory cells, which like rods and cones are neurosensory cells sprout processes that grow toward and penetrate the olfactory bulb of the forebrain (Fig. 16-20). These processes are the fibers of the olfactory nerve. There are an estimated 50 million olfactory cells in humans. For a substance to act as an odorant, it must have a chemical configuration compatible with that of a binding substance on the olfactory cell membrane, and it must be in solution. Characteristic branched, tubular, mucus-secreting **Bowman's glands** keep the epithelium moist in tetrapods. They are not present in fishes.

In fishes other than lobe-fins the differentiating olfactory epithelium becomes surrounded by connective tissue that forms a blind olfactory sac. A current of water into and out of the sac is ensured because each external naris is partitioned into incurrent and excurrent apertures so situated that the forward motion of the fish propels a stream of water into one aperture and out the other. The mucosa containing the olfactory epithelium may exhibit folds that increase the surface area. The olfactory cells monitor the water stream and are stimulated by odorants that may have their source in potential food, mates, or enemies. Probably the most primitive response to olfactory stimuli is reflex contraction of locomotor muscles, which propel the fish closer to, or farther from, the source of the odorant.

In lungfishes and tetrapods the olfactory pits push deep into the head to acquire an opening into the oral cavity or pharynx. The openings are internal nares. When this occurs, the olfactory epithelium is confined to a portion of the lining of this newly established nasal canal, so that it is appropriate to distinguish the olfactory epithelium from the epithelium of the remainder of the nasal passageway. The olfactory epithelium contains olfactory cell bodies just as in fishes, but in tetrapods it monitors an airstream instead of a water stream. Odorants in the airstream dissolve on the moist olfactory epithelium and stimulate the olfactory cells.

An olfactory epithelium is well developed in fishes, least developed in birds, which therefore have a poor sense of smell. The olfactory nerves of some species of whales disappear during embryonic life. However, a mammal trying to inhale under water would drown, so loss of the olfactory nerves was no disadvantage to whales.

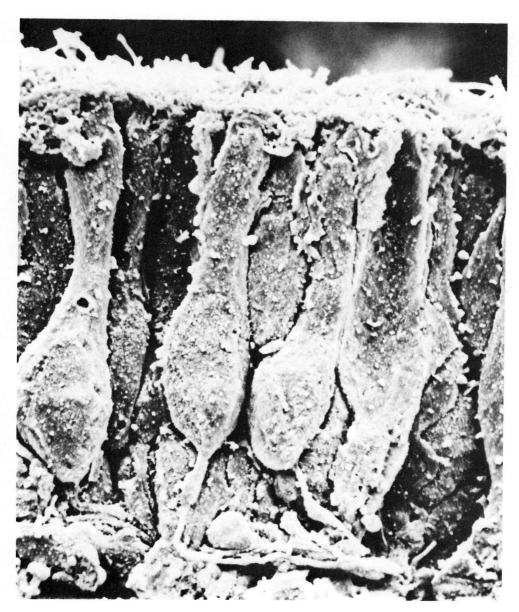

FIGURE 16-20

Section of olfactory epithelium from the channel catfish, *Icta-lurus punctatus*. The axons of three olfactory cells are readily identifiable. Height of the section was approximately 20 microns. (From Caprio, J., and Raderman-Little, R.: Tissue and Cell **10**[1]:1, 1978.)

FIGURE 16-21

Anatomical relationships of
the vomeronasal organ in a
generalized lizard. Only the
left organ is shown. Diagonal lines designate a parasagittal section. The lacrimal
duct is also called the nasolacrimal canal.

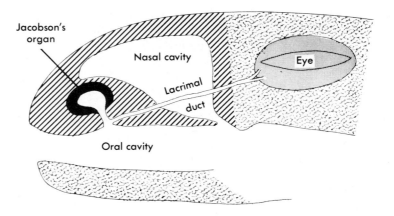

FIGURE 16-21

Anatomical relationships of
the vomeronasal organ in a
generalized lizard. Only the
left organ is shown. Diagonal lines designate a parasagittal section. The lacrimal
duct is also called the nasolacrimal canal.

Vomeronasal (Jacobson's) Organs

There has been a tendency among tetrapods for a ventral segment of the
olfactory epithelium to become more or less isolated from the nasal passageway, to the extent of becoming an accessory olfactory organ in some
species. The isolated olfactory area is called a vomeronasal organ because of
its usual location above the vomer bone. It is also known as **Jacobson's
organ.** The epithelium lacks Bowman's glands.

In urodeles the vomeronasal organs are a pair of deep grooves in the
ventromedial floor of the nasal canal. In anurans and apodans they are
blind sacs that open into the nasal canal. In squamates they lose their connection with the nasal canals and open into the anterior roof of the oral
cavity (Fig. 16-21), becoming two moist pockets that receive the tips of the
forked tongue each time it darts out of the mouth and back as it monitors
chemicals in the environment. Garter snakes have no tongue, but autora-
diographs have shown that chemicals on the lips, derived from potential
foodstuffs, reach the vomeronasal epithelium by diffusion. The organs are
more prominent in snakes than in other reptiles. Vomeronasal organs
appear in embryonic crocodilians and birds and then regress—mute testi-
mony to an ancestral function. They are undifferentiated in turtles.

Most mammals have vomeronasal organs just above the hard palate.
They open either in the floor of the nasal canals, as in many rodents, or
onto the hard palate via nasopalatine ducts that pass through incisive
foramina, as in cats. They are especially well developed in monotremes,
marsupials, and generalized insectivores, and many carnivores. They are
absent in cetaceans (where they would be as useless as other olfactory
epithelia), some bats, and adult higher primates. They and their nerves
develop in human embryos, reach maximum size about the fifth month of
gestation, and then regress. Those that open into the oral cavity are probably stimulated by food in solution in the cavity.

Vomeronasal organs are supplied by a **vomeronasal nerve,** a subdivision
of the olfactory nerve. The axons of the vomeronasal nerve in tetrapods
with a large vomeronasal organ pass to a small but distinct **accessory olfac-**

tory bulb. The size and differentiation of the latter is proportional to that of the vomeronasal organ, being absent when the organ is absent. In those mammalian embryos in which a vomeronasal organ appears transitorily, the neuronal and synaptic architecture of the transitory accessory bulb resembles that of the main olfactory bulb of amphibians and reptiles, being less complicated than that of the chief mammalian bulb.

Organs of Taste

Taste buds are barrel-shaped clusters of elongated receptor cells (taste cells) and supportive cells. The apices of the receptor cells have a prominent apical process that protrudes into a tiny **taste pore** in the moist epithelium (Fig. 16-22). The apical processes are covered with microvilli. Entwined about the base of each receptor cell are sensory nerve endings. The functional life of a taste receptor cell is only about 10 days, at which time it dies from "wear and tear." They are replaced by supportive cells, which are reserve receptor cells. The supply of supportive cells is continually being replenished from the basal layer of the epithelium.

In fishes, taste buds are widely distributed in the roof, side walls, and floor of the pharynx, where they monitor the incoming stream of respiratory water. In bottom feeders or scavengers, such as catfish, carp, and suckers, taste buds are distributed over the surface of the body to the tip of the tail. They are abundant on the barbels ("whiskers") of catfish. The exaggerated size of the sensory nucleus that receives incoming taste fibers in such fishes is illustrated in Fig. 15-14.

In most tetrapods taste buds are restricted to the tongue, posterior palate, and pharynx. There are fewer on the tongue in reptiles and birds than in mammals, and human embryos have more taste buds than children 7 years old as a result of failure to replace some that are lost.

FIGURE 16-22

Taste bud on the tongue of a monkey. Taste cells are *light red*. Their innervation is not shown.

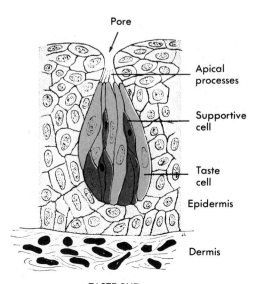

Pore

Apical processes

Supportive cell

Taste cell

Epidermis

Dermis

TASTE BUD

Taste buds from fish to man are supplied by the seventh, ninth, and tenth cranial nerves *in that sequence* from mouth to the end of the pharynx. Therefore, in man, taste buds on the anterior surface of the tongue are supplied by nerve VII, those on the posterior surface of the tongue by nerve IX, and those in the vicinity of the glottis by nerve X. Nerve VII also supplies all taste buds in the skin of fishes all the way to the tip of the tail.

GENERAL SOMATIC RECEPTORS

There are two categories of general somatic receptors: (1) general cutaneous receptors for light touch, pain, temperature, and pressure (touch sufficient to indent the surface) and (2) proprioceptors, found in striated muscles, joints, and tendons.

General Cutaneous Receptors

NAKED AND MISCELLANEOUS UNENCAPSULATED ENDINGS. The skin of all vertebrates contains free, or naked, endings of sensory nerve fibers that ramify among the epidermal cells and are stimulated by contact (Fig. 16-23, naked endings in epidermis). These are the most primitive cutaneous endings in vertebrates, and in cyclostomes they are the only ones. They give rise to what has been called **protopathic** sensation. This is a crude, poorly localized, phylogenetically ancient sensation. It is purely protective. It is not necessary that a fish know the texture of whatever touches it or whether the object is warm or cool. The mere fact of contact indicates possible danger. The impulses reach the thalamus, where reflex motor activity is initiated to reorient the animal so that it avoids the stimulus. The thermoreceptors of snakes are naked nerve endings. In mammals these endings give rise to vague, poorly localized, and sometimes unpleasant sensations including pain, as in toothache or when the skin is pricked. Free nerve endings for touch entwine around the base of each mammalian hair. Displace a single hair on your arm and note the sensation. Localization of the stimulus is a function of the cerebral cortex.

Lizards have mechanoreceptors and thermoreceptors that open to the surface between epidermal scales. These organs are of several morphological and functional varieties, the most numerous being **apical pits.** They are distributed widely over the body in groups of one to seven and, as their name implies, they lie at the apices—the posterior free borders—of scales. In some varieties a filamentous, hairlike bristle **(protothrix)** projects to the surface and transmits tactile stimuli to a receptor cell in the pit. Apical pits provide sites for input of exteroceptive stimuli in a scale-covered epidermis.

Some snakes—black snakes, for example—have mechanoreceptors between scales over the entire trunk and tail except on the venter, the majority of which have a high threshold, are fast-adapting, and are sensitive to vibrational stimuli in the same frequency range (up to 800 Hz) that induce cochlear potentials in these species.

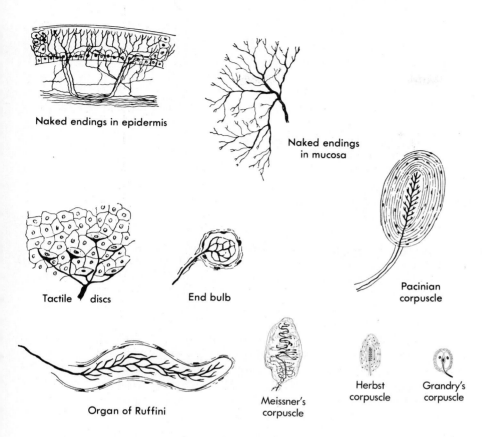

Naked endings in epidermis

Naked endings in mucosa

Tactile discs

End bulb

Pacinian corpuscle

Organ of Ruffini

Meissner's corpuscle

Herbst corpuscle

Grandry's corpuscle

FIGURE 16-23

Some general receptors. All are somatic except the naked endings in the mucosa and any pacinian corpuscles that occur in visceral locations.

ENCAPSULATED ENDINGS. In addition to naked nerve endings that ramify in the epidermis, tetrapods have acquired bulbous encapsulated endings that add to the repertoire of sensations (Fig. 16-23). These consist of nerve endings in association with epithelial-like cells within a connective tissue capsule. The simplest such endings are **tactile discs** that consist merely of a sensory ending in the epidermis in intimate association with a specialized epithelial cell. They are abundant in the snout of moles and pigs, where they are referred to as **Merkel's corpuscles.** Presumably they are responsible for the extreme sensitivity of the pig snout, which is employed in routing in the soil for truffles and other fungi where available. Similar discs are associated with the vibrissae of mice. **Grandry's corpuscles** along the margins of the beaks of many shore birds such as sandpipers, and of freshwater fowl such as ducks and geese consist of a nerve ending and two epithelial cells surrounded by a thin capsule. **Corpuscles of Herbst,** found on the beak, tongue, and palate of water birds, have a thick lamellar connective tissue capsule and a central core of epithelial-like cells. The **end bulbs of Krause** and the **organs of Ruffini** of mammals may be thermal receptors. **Meissner's corpuscles,** found only in primates and chiefly in hairless areas—toes, soles, palms, and especially the fingertips—probably are tactile receptors, but specific modalities of sensation cannot at this time be ascribed to any corpuscular receptors with certainty. All the foregoing varieties of sensory corpuscles other than tactile discs are found immediately beneath and partially surrounded by the epidermis in dermal papillae (Fig. 5-11, touch corpuscle).

Pacinian corpuscles, the largest encapsulated receptors, are found deep in the dermis or in the subepidermal connective tissue where they are stimulated by compression and, when an object is grasped by two fingers, participate in stereognosis (see next paragraph). Pacinian corpuscles are found also in the bursae of diarthroses where they subserve proprioception. They are also common in coelomic mesenteries, in the connective tissue stroma of some viscera, in periostea and in the peritoneum, including the pericardial peritoneum, where they function as *visceral* receptors. In the latter situations they are not known to evoke conscious sensations. A few simple exteroceptive corpuscles are found as far down the taxonomic scale as amphibians, but birds and mammals have the largest number and variety, including large **genital corpuscles** on the external genitalia of mammals and, in man, on other erogenous areas of the body.

Encapsulated exteroceptive endings in mammals are employed in **epicritic sensations** such as discrimination between small differences in warmth or coolness of objects, recognizing texture, perceiving as separate sites two points on the surface of the skin that are very close together and being touched simultaneously, and appreciation of the shape and weight of an object as a result of handling, the latter being stereognosis. After synapsing in the thalamus, the impulses are projected to the **somesthetic** (somatic sensory) **cortex** of the mammalian brain (Fig. 16-24), which is the center for epicritic sensibilities.

FIGURE 16-24

Topographical lobes of the human cerebral hemisphere and cortical loci for voluntary motor control (pyramidal area), hearing, sight, and somesthetic sensibilities. Most of the visual and auditory cortices are on infolded gyri. Cortical centers for taste and olfaction are less localized.

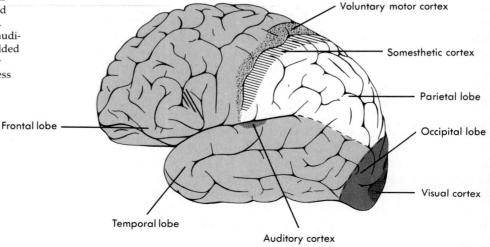

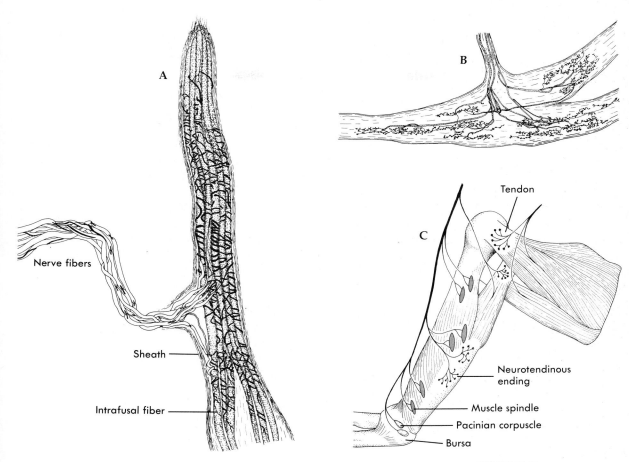

Tendon

Neurotendinous
ending

Muscle spindle

Pacinian corpuscle

Bursa

Nerve fibers

Sheath

Intrafusal fiber

A

B

C

FIGURE 16-25

Proprioceptors from muscle, tendon, and bursae. **A,** Portion of a muscle spindle of a dog; **B,** neurotendinous endings; **C,** representative proprioceptor sites (exaggerated). (**A** and **B,** modified from Huber and De Witt.)

Proprioceptors

Skeletal muscles, their tendons, and the bursae of joints are supplied with sensory endings that monitor continuously the activity status of muscles and joints. These proprioceptors are distinguished from exteroceptors, which monitor the external environment, and enteroceptors, which monitor the viscera. Proprioceptors maintain tonus reflexly in a resting muscle and assure synergy in the activity of functional groups of muscles when the latter are working against a load.

The sensory endings in most mammalian skeletal muscles—particularly appendicular muscles—are found in **muscle spindles** (Fig. 16-25, *A*). These are tiny fusiform bundles of fewer than a dozen specialized striated **intrafusal muscle fibers** that are wrapped in a connective tissue sheath and interspersed among and parallel to the **extrafusal fibers** (muscle fibers not in the spindle) near the extremities of the muscle. These receptors are extremely sensitive to very small changes in the length of a muscle, hence are sometimes referred to as **stretch receptors.** Each muscle fiber in the spindle is supplied with at least one proprioceptive nerve fiber (Fig. 16-26).

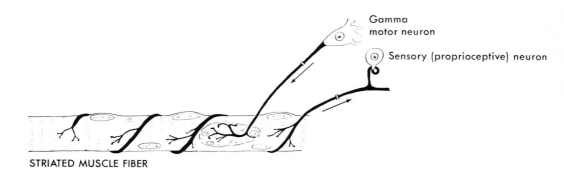

STRIATED MUSCLE FIBER

FIGURE 16-26

Innervation of an intrafusal muscle fiber of a muscle spindle. Tension within the muscle fiber is monitored by the proprioceptive neuron. The axon of the proprioceptive neuron synapses in the central nervous system with cell bodies of other neurons that stimulate extrafusal muscle fibers. The mechanism is essential for muscle tonus and posture.

Motor neurons supplying intrafusal fibers are called **gamma motor neurons** to distinguish them from **alpha motor neurons** that innervate extrafusal fibers.

In maintaining muscle tonus, which is a dynamic state, muscle spindles function as follows. The connective tissue envelope of the spindle is continuous with the endomysium that surrounds extrafusal fibers beyond the spindle, so that any change in the length of a muscle affects the spindle. When a muscle is at rest it is elongated, the spindle is slightly stretched, and proprioceptive nerve fibers are firing. Since proprioceptive fibers synapse in the central nervous system with alpha motor neurons, discharge from the spindle triggers contraction of a limited group of extrafusal fibers, and this increases the tonus of the muscle, shortening the muscle slightly. Shortening relieves some of the tension within the spindle, and as a result, firing of the proprioceptive fibers slows. Thereupon the muscle tends to elongate once again, and this triggers new discharges from the spindle. Thus feedback from muscles to central nervous system reflexly maintains the muscle in a state of dynamic tonicity while the muscle is at rest.

The role of spindles when muscles are working against a load is more complicated. Alpha and gamma motor neurons in motor nuclei of the spinal cord are being stimulated by motor nerve fibers whose cell bodies are in the brain. Stimulation of large numbers of gamma motor neurons results in contraction of intrafusal fibers in more than one muscle and consequent discharge of many proprioceptive impulses at a high frequency. The resulting steady reflexive firing of alpha motor neurons maintains a steady contraction of large numbers of extrafusal muscle fibers, and this continues, subject to brain control, whether the vertebrate is locomoting or playing a piano.

Appendicular muscles have the largest number of spindles per unit volume. Some mammalian muscles of the axial series have few spindles and some, such as the extrinsic eyeball muscles, have none. In muscles having few or no spindles proprioceptive fibers simply spiral around ordinary muscle fibers (**annulospiral endings**), or they terminate at one site on a regular muscle fiber in an ending that resembles a flower spray (**flower spray endings**). Muscle spindles and simple annulospiral endings have been described in tetrapods as low on the taxonomic scale as urodeles, but they are much less common than a variety of simpler neuromuscular sensory endings seen not only in tetrapods but in fishes as well.

Proprioceptive endings on tendons (neurotendinous endings) are flower spray endings that respond to forces generated in the tendon when its muscle contracts (Fig. 16-25, *B* and *C*). Proprioceptive endings in bursae are corpuscular, pacinian corpuscles being the predominant variety in mammals (Fig. 16-25, *C*). Naked sensory endings in myosepta, presumably for proprioceptive input, have been described in lampreys.

Most proprioceptive stimuli do not reach the centers of consciousness but are shunted into the cerebellum and elsewhere, where they make reflex connections. To demonstrate conscious proprioception, first close your eyes, then extend your arm or leg. Your awareness of the change in position is an example of conscious proprioception, also called kinesthesia and muscle sensibility. That the act was performed smoothly is a result of reflex responses to volleys of proprioceptive stimuli that did not reach the level of consciousness.

GENERAL VISCERAL RECEPTORS

General visceral receptors are mostly naked endings in the mucosa of the tubes, vessels, and organs of the body, in cardiac muscle, in smooth muscle including that of the blood vessels, and in the capsules, mesenteries, and meninges of the viscera (Fig. 16-23, naked endings in mucosa). Pacinian corpuscles subserve a visceral function in mesenteries and other coelomic locations, as mentioned earlier.

General visceral receptors are chiefly stretch and chemoreceptors, but they include a few baroreceptors and osmoreceptors. Among other roles, chemoreceptors monitor the pH of the blood (including oxygen and carbon dioxide content), which affects cardiorespiratory functions; and they monitor the pH of the contents of the stomach and proximal intestine, which affects digestive functions. Baroreceptors monitor blood pressure, and osmoreceptors in the hypothalamus and elsewhere monitor the concentration of certain solutes in the blood stream. General visceral receptors are stimulated by tactile and thermal stimuli in the pharynx but not beyond. Such input would have no survival value. Most visceral pain from healthy hollow visceral organs results from distension (stretch) and is detected by naked endings.

Three grossly observable vascular monitors are present in tetrapods. These are the carotid bodies, carotid sinuses, and aortic body. The **carotid body** lies close to, or embedded in, the wall of the common carotid or internal carotid artery, from which it receives its blood supply. It is richly supplied with sensory endings of the ninth cranial nerve and monitors the oxygen and, perhaps, carbon dioxide content of blood passing through the organ. Anoxia evokes a discharge of sensory impulses that reflexly increase the respiratory rate. The **carotid sinus** is a bulbous enlargement on the internal carotid artery at its origin from the common carotid. It is a baroreceptor that monitors arterial blood pressure. Low pressure evokes sensory discharges that provide input to the cardiovascular regulatory center of the medulla. The sinus is supplied by the ninth cranial nerve and also receives vagal and sympathetic fibers. An **aortic body,** described only in mammals, lies on the arch of the aorta. It is a baroreceptor innervated by sensory fibers in the vagal nerve.

Most of the sensory input from general visceral receptors gives rise to no conscious sensation, but to reflex control of smooth and cardiac muscles and glands. Because the same kind of monitoring of the internal environment is necessary in all vertebrates whether in water or on land, general visceral receptors are not subject to so many selective pressures as are somatic receptors. Therefore visceral receptors vary little from fish to man.

Chapter Summary

1. Receptors are transducers of mechanical, electrical, thermal, chemical, or radiant energy. General receptors are widely distributed in the body. Special receptors have a limited distribution and mostly are paired.

2. Somatic receptors provide information about the external environment (exteroceptors) and about skeletal muscle activity (proprioceptors). Visceral receptors monitor the internal environment (enteroceptors).

3. Neuromast organs consist of hair cells, sensory endings, and supporting cells. They are most common in fishes and aquatic amphibians and occupy shallow pits, grooves, ampullae, or canals and function principally in mechanoreception, less commonly in electroreception, thermoreception, and chemoreception. The lateral-line and cephalic canals are neuromast systems.

4. The membranous labyrinth or inner ear is a neuromast system consisting of semicircular ducts, sacculus, utriculus, endolymphatic and perilymphatic ducts and sacs, and either a lagena or a cochlear duct. It is filled with endolymph and surrounded by perilymph.

5. The primitive function of the labyrinth was equilibratory. An auditory function culminating in a cochlear duct was a later development.

6. Cristae are papillary receptor sites in the ampullae of the semicircular ducts. Receptor sites elsewhere in the labyrinth are flattened maculae. They often have otolithic membranes in lower tetrapods, tectorial membranes in higher ones. The organ of Corti is the most specialized neuromast complex.

7. Reception by fishes of vibrations of the frequency of sound waves in water is limited, but air bladders and weberian ossicles transmit such waves to the inner ear in some fishes.

8. Airborne sound waves in tetrapods are usually transmitted from tympanic membranes via middle ear ossicles. Seismic waves are transmitted via the mandible or forelimb skeleton to the columella or directly to the otic capsule. An opercular system in anurans and a specialized extrastapes in amphisbaenians provide alternative routes.

9. Middle ears and columella are found in most tetrapods, are rudimentary in a few. Mammals alone have a malleus, incus, and associated striated tensor tympani and stapedial muscles.

10. In fishes a saccus vasculosus evaginates from the floor of the third ventricle and contains hair cells and supporting cells. Its role is not clear.

11. Lateral eyes are evaginations of the embryonic forebrain. They consist of a retina, choroid, and sclerotic coat. The corneal surface is covered by a bulbar conjunctivum. Accommodation for distance and light intensity is chiefly by ciliary muscles that insert on the lens and by iridial dilators and constrictor muscles.

12. Median eyes have a cornea, lens, and retina but form no retinal image. They were present in all Devonian fishes and early amphibians and reptiles. They have persisted in cyclostomes, some bony fishes, larval and a few adult anurans, and many lizards. The parapineal (parietal) eye serves as a light receptor more frequently than the pineal.

13. Pit receptors stimulated by infrared radiations are found in pit vipers (loreal pits) and in pythons and some boas (labial pits). They supplement other receptors in the search for warm-blooded prey. The apical pits of lizards house mechanoreceptors.

14. Olfactory and vomeronasal epithelia are special visceral chemoreceptors that arise from embryonic nasal pits.

15. Taste buds are special visceral chemoreceptors on the surface of the head, trunk, and tail of some fishes, confined to the head in others, and restricted to the oral cavity and pharynx in tetrapods. They are supplied by cranial nerves VII, IX, and X in all tetrapods.

16. General somatic receptors include cutaneous receptors for protopathic and epicritic sensibilities and proprioceptors.

17. General cutaneous receptors include naked endings in the epidermis of all vertebrates, encapsulated endings in the dermis of tetrapods, and interscale reptilian pits. Proprioceptors are muscle spindles, naked endings in tendons, and pacinian corpuscles in tendons and bursae.

18. General visceral receptors are chiefly naked endings for stretch and chemoreception in visceral sites, and pacinian corpuscles in coelomic mesenteries and viscera. Carotid bodies, carotid sinus, and aortic bodies serve as baro- and osmoreceptors. Hypothalamic receptors subserve homeostasis by monitoring certain solutes in the circulation.

SELECTED READINGS

Adams, W.E.: The comparative morphology of the carotid body and carotid sinus, Springfield, Illinois, 1958, Charles C Thomas, Publisher.

Ariëns Kappers, C.U., Huber, G.C., and Crosby, E.C.: The comparative anatomy of the nervous system of vertebrates, including man, New York, 1936, The Macmillan Co. (Republished by Hafner Publishing Co., 1960.)

Botterman, B.R., Binder, M.D., and Stuart, D.G.: Functional anatomy of the association between motor units and muscle receptors, American Zoologist **18**:135, 1978.

Bullock, T.H., Northcutt, R.G., and Bodznick, D.A.: Evolution of electroreception, Trends in Neurosciences **5**:50, 1982.

Buning, T. d'C.: Thermal sensitivity as a specialization for prey capture and feeding in snakes, American Zoologist **23**:363, 1983.

Dodt, E.: The parietal eye (pineal and parietal organs) of lower vertebrates. In Jung, R., editor: Handbook of sensory physiology, part B, vol. 7/3B, West Berlin, 1973, Springer-Verlag.

Fein, A., and Szuts, E.Z.: Photoreceptors: their role in vision, London, 1982, Cambridge University Press.

Johns, P.R.: Growth of fish retinas, American Zoologist **21**:447, 1981.

King, A.S., and McLelland, J.: Form and function in birds, vol. 3, New York, 1985, Academic Press.

Parker, D.E.: The vestibular apparatus, Scientific American **243**:98, 1980.

Popper, A.N.: Comparative studies of hearing in vertebrates, New York, 1980, Springer-Verlag.

Popper, A.N., and Coombs, S.: The morphology and evolution of the ear of actinopterygian fishes, American Zoologist **22**:311, 1982.

Purves, E., and Pilleri, G.E.: Echolocation in whales and dolphins, New York, 1983, Academic Press.

Quay, W.B.: The parietal eye-pineal complex. In Gans, C., Northcutt, R.G., and Ulinski, P.: Biology of the reptilia, vol. 9, New York, 1979, Academic Press.

Shaw, E.: Schooling fishes, American Scientist **66**:166, 1978.

Stebbins, W.C.: The acoustic sense of animals, Cambridge, Massachusetts, 1983, Harvard University Press.

Tavolga, W.N., Popper, A.N., and Fay, R.R., editors: Hearing and sound communication in fishes, New York, 1981, Springer-Verlag.

Ulinski, P.S.: Design features in vertebrate sensory systems, American Zoologist **24**:717, 1984.

Weaver, E.G.: The reptile ear: its structure and function, Princeton New Jersey, 1978, Princeton University Press.

Wilczynski, W.: Central neural systems subserving a homoplasous periphery, American Zoologist **24**:755, 1984.

Symposia in American Zoologist

Vertebrate sound production, **13**:1137, 1973.

Recent advances in the biology of Sharks. Section III. Cranial nerves and sense organs, **17**:411, 1977.

17

ENDOCRINE ORGANS

*In this chapter we will study organs whose secretions in most instances
adapt vertebrates over a period of time to more or less gradual fluctuations in
the internal and external environments. We will learn that the brain is one
of these organs, translating cyclical environmental input into hormonal mes-
sages. We will find that vertebrates from fish to humans have the same array
of such organs, that they form from the same embryonic precursors, and that
they synthesize the same molecules. We will also note some of the molecular
and anatomical mutations that have occurred during phylogeny.*

Endocrine organs synthesize hormones. Hormones are products of specific groups of cells that have a regulatory effect on other cells, often
referred to as "target" cells. Target cells may be right next to the cells that
regulate them, but more often they are remote and the hormone is transported by the bloodstream.

The term "target cell" might lead to the mistaken notion that a hormone
passes only to the "target," as though aimed in that direction. This is not
the case. Most hormones are transported in the bloodstream, and the fluid
constituents of the stream escape from capillaries everywhere in the body
and bathe every living cell. Only those cells having an appropriate molecular "receptor site" on their membranes can be affected by the hormone.
For example, thyrotropic hormone from the pituitary gland bathes the cells
of the big toe in the same concentration that it bathes the cells of the thyroid
gland, which is its target; but only certain cells of the thyroid gland are able
to respond. A few hormones, such as insulin, have receptor sites on every
cell of the organism. These are **general metabolic hormones.**

Some hormones are produced by organs with additional functions. For
example, insulin in most vertebrates is produced by a gland that also
produces digestive enzymes. There are endocrine cells in the liver, kidney,
epithelium of the digestive tract, and many other locations. In this chapter
we will confine the discussion to organs that are solely endocrine, or that
produce hormones as one of their major functions.

Some endocrine organs appear to have been derived *phylogenetically* from
clusters of hormonal cells that at one time were confined to an epithelium.
Thyroid cells arise from the epithelium of the pharyngeal floor, where they
remain in protochordates and also during the long larval life of cyclostomes. In some fishes the epithelial thyroid tissue retains a duct that empties onto the floor of the pharynx throughout life. The glandular lobe of the
pituitary gland evaginates from the epithelial roof of the embryonic stomodeum, and it too retains a duct to the oropharyngeal cavity in some adult
fishes. Cells that secrete insulin in elasmobranchs are in the epithelium of

the smaller pancreatic ducts, not having escaped from the epithelium during ontogeny as they do in most other vertebrates. Insulin-secreting cells are still in the epithelium of the gut in lampreys, and immediately adjacent to it in hagfishes and some dipnoans.

Several **gastrointestinal hormones** that induce secretion of digestive enzymes are synthesized in the epithelium of the digestive tract. These include **gastrin,** which is produced by the mucosa of the stomach and evokes the secretion of gastric juice (an acid), and **secretin,** which is secreted by the intestinal epithelium when the acid contents of the stomach **(chyme)** enter the duodenum. Secretin, which has been found in all vertebrate classes, is transported by the bloodstream to the exocrine cells of the pancreas and liver, inducing these cells to secrete their respective digestive enzymes and, with respect to the pancreas, perhaps to synthesize them.

The earliest hormones probably had only local action, affecting nearby cells as they diffused from their source, rather than being transported by a closed circulatory system to distant organs. This is a necessity in coelenterates, because they lack a circulatory system; but hormones that act locally by diffusion are not uncommon in living vertebrates. Certain gonadal hormones, for example, have both local and remote effects.

Most hormones are combinations of amino acids—chiefly polypeptides, proteins, and glycoproteins—or they are steroids; a few are amines. All these compounds are common products of biochemical synthesis not only among vertebrates but also among invertebrates and plants. They represent chemical taxa that have evolved along with the organisms that produce them. Steroids, for example, are found in yeast and in many green plants and are universal in animals. There is a greater variety of steroids in lower than in higher animals. It is as though many analogues evolved, some developed a functional relationship that had survival value, and others disappeared. A number are still produced in the mammalian adrenal cortex for which no role has yet been found.

We will begin with the endocrine roles of the nervous system, a highly important source of regulatory secretions. Next the endocrine organs derived from ectoderm, mesoderm, and endoderm are discussed briefly and in that order, touching only on generalities as we examine their roles in the lives of vertebrates.

ENDOCRINE ROLES OF THE NERVOUS SYSTEM

The hormones of the nervous system are referred to as **neurohormones.** They are produced by **neurosecretory neurons.** Neurohormones should not be confused with neurotransmitters. The latter are amines, the former are small polypeptides.* Neurohormones are found as low in the animal kingdom as coelenterates.

---

*Some amines, epinephrine and melatonin, for example, serve as hormones in certain locations, and there is evidence that polypeptides sometimes are able to serve as neurotransmitters.

FIGURE 17-1

Neurosecretory neurons in a functional neurosecretory unit. The neurohemal organ contains sinusoidal vascular channels. The chief neurohemal organs in vertebrates are urophysis, posterior lobe of the pituitary, and median eminence of the diencephalic floor.

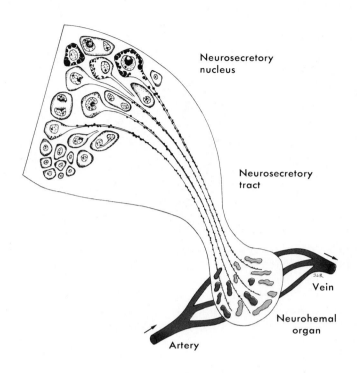

FIGURE 17-2

Hypothalamic neurosecretory neurons. Cell bodies in the hypothalamus manufacture neurohormones (black granules) that flow along the axons (neurosecretory fibers) and are discharged into the hypophyseal portal vein or into vascular channels in the posterior lobe or the pituitary. Neurohormones released into the portal vein help regulate the hormone-producing cells (*dark red*) of the anterior lobe. Those released in the posterior lobe affect tissue remote from the pituitary.

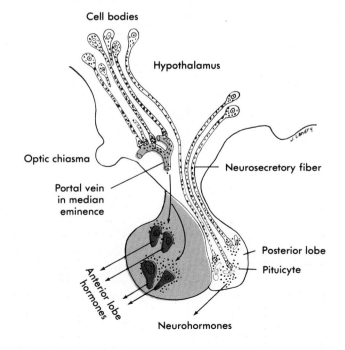

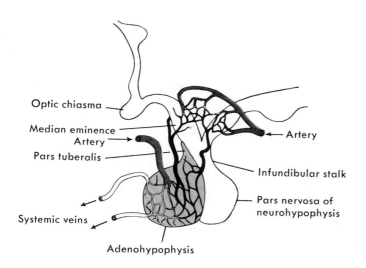

FIGURE 17-3

Schema of the hypophyseal portal system *(black)* of mammals. *Arrows* indicate direction of blood flow.

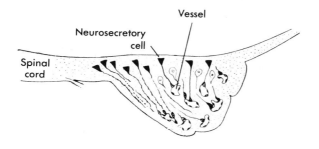

FIGURE 17-4

Urophysis (caudal neurohemal organ) of a carp. In some species this organ is more pendulous.

Neurohormones are synthesized in the cell bodies of neurosecretory neurons located in neurosecretory nuclei (Fig. 17-1). The secretions move along the axon, or neurosecretory fiber, chemically bound to stable proteins **(neurophysins),** and accumulate in the axon terminals until released into blood sinusoids by reflex nerve impulses. *The axon terminals plus the sinusoids constitute a neurohemal organ.* The major neurohemal organs of vertebrates are the **posterior lobe of the pituitary gland** (Fig. 17-2), the **median eminence** of the diencephalic floor immediately behind the optic chiasma (Figs. 17-2 and 17-3), and the **urophysis, or caudal neurosecretory organ,** at the end of the spinal cord of elasmobranchs, chondrosteans, and teleosts (Fig. 17-4).

Neuron cell bodies with stainable neurosecretory droplets were discovered by Dahlgren in the caudal region of the spinal cord of fishes early in the twentieth century, but their significance was unknown. In 1928 E. Scharrer drew attention to the presence of **"Dahlgren cells"** in the hypothalamus and suggested that they might be concerned with secretion. But it was not until 1949 that Gomori designed staining techniques suitable to demonstrate the validity of this concept. Thus, study of neurons in the spinal cord of a fish led to one of the most important discoveries in the field of endocrinology, the neurosecretory neurons of the brain of all vertebrate animals, including mankind.

As we now know, most vertebrate neurohormones are produced in the hypothalamus. Some of these are released into a hypophyseal portal vein in the median eminence and transported to the anterior lobe of the pituitary, where the target cells are located (Fig. 17-2). Synthesis or release of these hypothalamic neurohormones is regulated in part by the cyclic external environment and partly by feedback from endocrine glands. The obvious environmental influences include, among other variables, the circadian cycles of light and darkness, and seasonal changes in temperature, day length, rainfall, and salinity (in an aquatic environment). These are monitored by sense organs. One of the many effects of this input is promotion of gametogenesis and appropriate reproductive behavior (territorial defense,

FIGURE 17-5

Regulatory effects of the environment on reproduction.

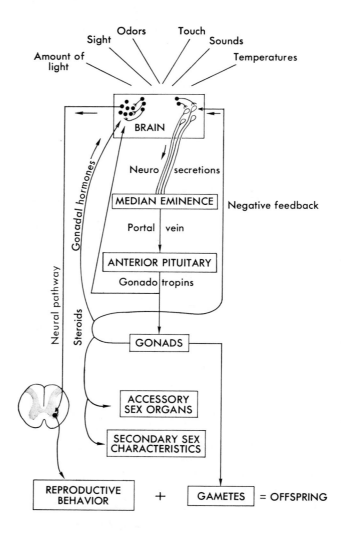

mating behavior, nest building, care of eggs and young) at the precise time of year when environmental conditions are most suitable for survival of offspring (Fig. 17-5). Such adaptive neuroendocrine reflex arcs (receptor–sensory nerve–hypothalamus–neurosecretions–median eminence–hypophyseal portal system–effector) are evidently the result of natural selection. Other hypothalamic neurosecretions are released into the posterior lobe (Fig. 17-2), where they enter the general circulation. Their role is discussed shortly.

Fishes other than cyclostomes have neurosecretory cells (Dahlgren cells) in the spinal cord at the base of the tail (Fig. 17-4). Their axons terminate in the **urophysis,** a neurohemal organ that varies from a mere swelling of the spinal cord at that level to a rather pendulous organ, depending on the species. The neurohormones of this **caudal neurosecretory system,** termed **urotensins,** raise the blood pressure of teleost fishes (vasopressor effect) and make a major contribution to osmoregulation in fishes in general. The unique roles of urotensins in these physiological adaptations have not yet been separated from those of other hormones having similar regulatory functions in fishes.

ENDOCRINE ORGANS DERIVED FROM ECTODERM
Pituitary Gland

The pituitary gland, or hypophysis, is suspended beneath the diencephalon, cradled in gnathostomes in the sella turcica. The pituitary consists of two major components with very different embryonic origins, a **neurohypophysis** that arises from the floor of the diencephalon, and an **adenohypophysis,** which, except in hagfishes, arises from the roof of the embryonic stomodeum (Fig. 17-6). The subdivisions of these lobes are as follows:

    Neurohypophysis
        Median eminence
        Infundibular stalk
        Posterior lobe (pars nervosa)
    Adenohypophysis
        Pars intermedia
        Pars tuberalis
        Anterior lobe (pars distalis)
        Ventral lobe of elasmobranchs

NEUROHYPOPHYSIS. The neurohypophysis forms from the floor of the diencephalon (Fig. 17-6, *dark gray*). It contains a recess of the third ventricle that is shallow in lower vertebrates, deepest in mammals, in which the diencephalic floor is drawn out into a long **infundibular stalk** with the **posterior lobe,** a neurohemal organ, at the end (Figs. 15-20 and 17-7). Just behind the optic chiasm the neurohypophysis has another neurohemal area, the **median eminence** (Figs. 17-2 and 17-3).

The posterior lobe *synthesizes* no known hormones. The hypothalamic neurohormones *released* there are carried to the heart for distribution

FIGURE 17-6

Embryogenesis of amniote pituitary. **A,** Rathke's pouch stage. **B,** Isolation of adeno-hypophyseal anlage *(light gray)* in contact with the floor of the diencephalon. **C,** Young pituitary consisting of adenohypophysis *(light gray)* and neurohypophysis *(dark gray)*. *1,* Median eminence; *2,* infundibular stalk; *3,* pars nervosa (posterior lobe). The subdivisions of the adenohypophysis are labeled.

FIGURE 17-7

Pituitary glands of representative vertebrates, sagittal sections, anterior to the left. *Gray,* neurohypophysis and contiguous brainstem; *red,* pars intermedia; *stippled area,* pars distalis (anterior lobe); *black,* pars tuberalis. The pars distalis of teleosts, as exemplified by the trout, exhibits two cytological regions, a rostral part *(coarse stipple)* and a proximal part *(fine stipple)*.

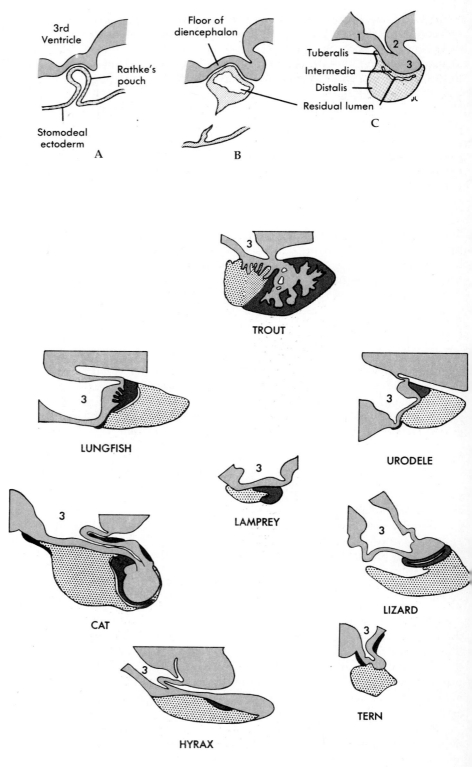

throughout the body. The most ubiquitous, and probably the oldest, "posterior lobe hormone" is arginine vasotocin, found in cyclostomes and all higher vertebrate classes, although in mammals it is secreted only during fetal life. A number of mutants have appeared during vertebrate phylogeny by the substitution of different amino acids at one or more positions on the peptide molecule (Fig. 17-8).

Arginine vasotocin is the only neurohormone found in cyclostomes. Its function in fishes other than lungfishes is unknown, but in terrestrial vertebrates it prevents dehydration by causing the kidney tubules to reclaim water from glomerular filtrate and by causing resorption of water stored in the urinary bladder. In lungfishes and amphibians that inhabit moist burrows during arid conditions it also stimulates absorption of water from the soil through the skin.

Arginine vasopressin regulates water and salt excretion in mammals and is therefore a human antidiuretic hormone **(ADH).** Another mutant, **oxytocin,** causes oviducal contractions in turtles, induces strong uterine contractions during birth in mammals, and causes "letdown" of milk into nipples during nursing by stimulating the smooth muscle fibers surrounding the milk-secreting alveoli of the lactating mammary gland. What these and other mutants are able to do for any species depends on whether some of their cells, either contractile or secretory, have receptor sites that are able to be occupied by the circulating mutant. A mutant molecule can have no effect in a species in which no cells are competent to respond.

Hypothalamic neurohormones released into the median eminence enter hypophyseal portal vessels that transport them to the adenohypophysis, where they stimulate or inhibit the release of hormones synthesized in the anterior lobe. The portal vessels in tetrapods and some fishes are discrete portal veins (Fig. 17-3). In teleosts many capillary-like vessels form a plexus that does not organize into a discrete portal vein, yet performs the same function by emptying into the capillary beds of the anterior lobe. This vascular arrangement is probably correlated with the unusually intimate anatomical relationship between the neurohypophysis and adenohypophysis in teleosts (Fig. 17-7, trout).

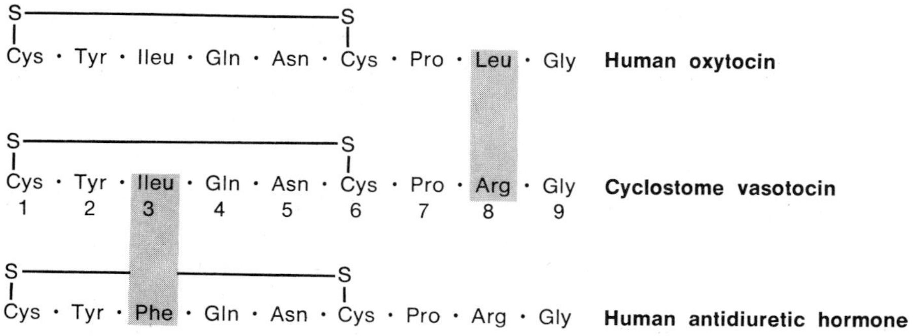

FIGURE 17-8

Two of many evolutionary mutants of arginine vasotocin (cyclostome vasotocin). *Gray bars* designate the only differences between cyclostome and human posterior lobe neurohormones. The human antidiuretic hormone is arginine vasopressin.

ADENOHYPOPHYSIS. The adenohypophysis of all vertebrates other than hagfishes arises as a bud of ectodermal cells from the roof of the stomodeum. In amniotes, sharks, and some lower bony fishes the bud is hollow and is known as **Rathke's pouch** (Fig. 17-6, *A*). In other fishes and in amphibians the bud is solid. When the anlage of the adenohypophysis has made intimate contact over a broad area with the floor of the brain, the connection between the stomodeum and the bud usually disappears (Fig. 17-6, *B*). However, it remains as an open ciliated duct leading to the oropharyngeal cavity in *Calamoichthyes, Polypterus,* and some primitive teleosts. Rathke's pouch may remain in the adult gland as a residual lumen (Fig. 17-6, *C*). According to Gorbman, the adenohypophysis of hagfishes arises from the endoderm of the roof of the nasohypophyseal duct (Fig. 12-2).

The origin of the adenohypophysis from the stomodeum suggests that primitively the adenohypophysis may have secreted into the oropharyngeal cavity. It was once thought that the pituitary (*pitua,* phlegm) was the source of phlegm that falls into the throat. Although they were wrong about the source of the phlegm, they may have been correct about the primitive site of secretion of the pituitary!

The adenohypophysis typically exhibits three regions: pars intermedia, pars distalis, and pars tuberalis. The **pars tuberalis** consists of a pair of narrow thin cephalic extensions of the pars distalis along the ventral surface of the diencephalon or up the infundibular stalk. It seems to be a functionally insignificant part of the adenohypophysis. No pars tuberalis forms in most fishes or squamates. In elasmobranchs the adenohypophysis has a fourth subdivision, the **ventral lobe** (Fig. 15-18). It is an expanded portion of the pars distalis that probably secretes gonadotropic and thyrotropic hormones.

The **pars intermedia** lies in intimate contact with the neurohypophysis. However, in birds and in cetaceans, manatees, and a few other mammals no discreet, histologically recognizable pars intermedia develops. In apes and humans the intermedia grows smaller in size with age, after being relatively large in the embryo. Size differences may be related to the function of the hormone **intermedin (melanophore stimulating hormone, MSH).** MSH causes pigment granules in some chromatophores of ectotherms to disperse, thus darkening the skin. It is not present in birds. In mammals it initiates melanogenesis in hair follicles.

The **pars distalis** synthesizes the following hormones:

Somatotropin (STH)
Thyrotropin (TSH)
Adrenocorticotropin (ACTH)
Prolactin—also known as luteotropic hormone (LTH)
Gonadotropins
    Follicle-stimulating hormone (FSH)
    Luteinizing hormone (LH)—in males, known as interstitial cell–stimulating hormone (ICSH)

Somatotropin stimulates synthesis of proteins from amino acids and is therefore a growth hormone. **Thyrotropin** (thyroid-stimulating hormone) stimulates the thyroid gland to accumulate iodine, to synthesize thyroid hormone, and to release it into the circulation. **Adrenocorticotropin** affects the zones of the adrenal cortex that secrete glucocorticoids. **Follicle-stimulating hormone** stimulates the growth of ovarian follicles and their secretion of estrogen. **Luteinizing hormone** induces ovulation of a mature ovum and subsequent conversion of a graafian follicle into a corpus luteum. Because it also induces the interstitial cells of testes to synthesize androgens, it is sometimes referred to as **interstitial cell–stimulating hormone.**

Prolactin is a general metabolic hormone with wide-ranging effects. It is the major freshwater-adapting hormone of euryhaline fishes (fishes that adapt to a wide range of salinity, such as saltwater fishes that migrate into freshwater streams to spawn). When preceded or accompanied by other hormones prolactin causes red efts to migrate to ponds at the approach of sexual maturity, stimulates production of pigeon milk by the avian crop sac and mammalian milk by mammary glands, and stimulates secretion of parental skin mucus that nourishes hatchlings of some teleost species. In all vertebrates prolactin has a role in inducing certain parental behavior such as nest building, turning and incubation of eggs, and protection of the young. Because it has a luteotropic effect, inducing the secretion of progesterone by the corpus luteum in some species, prolactin is known also as **luteotropic hormone.**

Pineal Body

The pineal organ, so named because in mammals it is shaped like a pine cone, is an evagination from the roof of the diencephalon that produces the hormone **melatonin.** Mammalian melatonin causes melanin granules in dermal melanophores of *larval amphibians* and many fishes to aggregate, thereby blanching the skin. The effect is opposite that of MSH. Experimental evidence indicates that in some species light impedes the synthesis of melatonin by inhibiting the activity of the enzyme serotonin *N*-acetyltransferase, which catalyzes the rate-limiting step in melatonin synthesis. The effect of light on the pineal body is direct when the pineal is located under translucent skin. Otherwise it is mediated by way of the optic nerves, relay nuclei in the brainstem, spinal cord, thoracic spinal nerves, sympathetic trunk, superior cervical sympathetic ganglion, and the conarial nerves (**nervi conarii),** which pass from the ganglion to the pineal organ. Darkness facilitates the synthesis of melatonin. Amphibian larvae with intact pineal bodies become pale when placed in darkness, and pinealectomy abolishes this response.

Melatonin has an inhibitory effect on ovarian and testicular weights in some mammals, and it inhibits sperm formation in these species. Failure of golden hamster pairs to produce young when exposed to long winter nights in higher northern latitudes has been shown to be caused by high melatonin levels and the consequent inadequacy of the male sperm count, not to lack of ovulation of viable eggs. This has survival value, preventing the birth of offspring under conditions unfavorable for the survival of neonates. The pineal body is discussed further in Chapters 15 and 16.

Adrenal Medulla and Aminogenic Tissue

The adrenal gland in most vertebrates is on the ventral surface of the kidney or near its cephalic pole. In mammals it consists of two components, a peripheral cortex and a central medulla (Fig. 17-9). However, cortex and medulla are two entirely different glands, separated from each other in many fishes and more or less interspersed among one another in other vertebrates below mammals (Fig. 17-10). Therefore we must discuss the adrenal as two separate glands, which it really is.

One component of the adrenal complex arises from neural crests. This tissue, homologous with the adrenal medulla of mammals, synthesizes two catecholamines: **norepinephrine (noradrenaline)** and **epinephrine (adrenaline)**. Because of its staining reaction, this aminogenic component is called **chromaffin tissue.***

In lampreys, elasmobranchs, and some teleosts, the aminogenic cells lie in clusters *scattered* along the postcardinal vein, and in lungfishes they are scattered along the dorsal aorta. In most teleosts they are associated with vestiges of the pronephroi (Fig. 17-10, teleost). In this location they are more or less interspersed among cells of the other adrenal component (**steroidogenic tissue**).

In most tetrapods the two components are *interspersed.* However, in some lizards and snakes the aminogenic tissue tends to *aggregate*, forming an almost complete capsule around the steroidogenic tissue (Fig. 17-10, lizard). This is the reverse of the condition in mammals. But even in mammals there are species in which the steroidogenic tissue does not form a complete cortex. In sea lions, for example, cortical tissue is scattered in the medulla, and medullary tissue is scattered in the cortex.

The adrenal glands of anurans are flattened, elongated masses on the ventral surface of the kidneys (Fig. 14-16). In urodeles they form small bright flecks and nodules along the postcava and are difficult to locate without a lens. In amniotes the adrenals are at or near the cephalic pole of the kidney. For this reason, in erect mammals they are also called **suprarenal** glands.

The aminogenic cells of the adrenal complex and the cell bodies of postganglionic neurons of the sympathetic nervous system are of the same lineage. Both arise from neural crests, both are innervated by preganglionic fibers of the sympathetic nervous system, and both synthesize norepinephrine and epinephrine. In fact, the aminogenic cells of the adrenal medulla

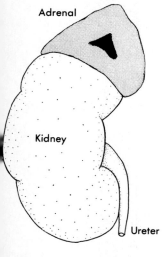

FIGURE 17-9

Adrenal gland of humans. The cortex (steroidogenic tissue, *gray)* surrounds the medulla (aminogenic tissue, *black).*

---

*Chromaffin tissue, in addition to that of the adrenal complex, is widely distributed in the trunk of most vertebrates. Masses occur in close association with sympathetic ganglia, where they are called paraganglia, and in the gonads, kidneys, heart, and other viscera. However, not all chromaffin tissue is aminogenic.

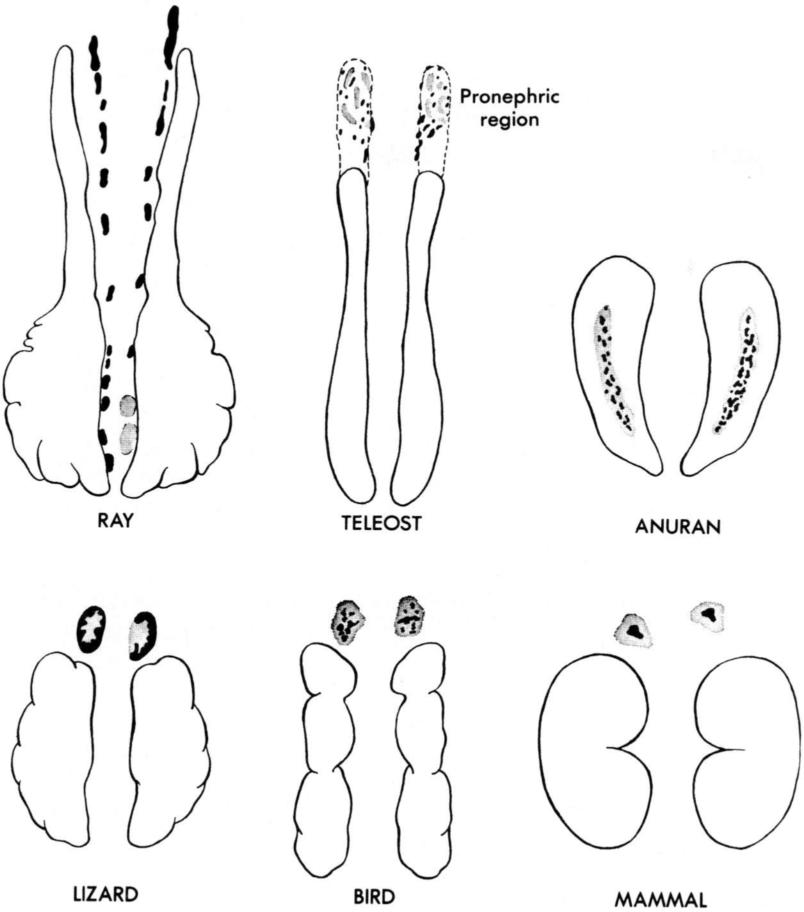

FIGURE 17-10

Adrenal components in selected vertebrates. Aminogenic tissue (medulla in mammals) is shown in *black*, steroidogenic tissue (cortex in mammals) in *gray*. The kidneys are shown in outline. Note the reversed location of the two components in lizards and mammals. The two masses of steroidogenic tissue between the caudal ends of the kidneys in the ray are referred to as interrenal bodies.

can be thought of as postganglionic cell bodies of the sympathetic nervous system that fail to sprout processes. Norepinephrine is a precursor of epinephrine. It is converted to epinephrine by the addition of a methyl group. This conversion takes place to a greater degree in the adrenal medulla than in most postganglionic cell bodies. Consequently, epinephrine is the predominant end product of the adrenal complex, whereas norepinephrine is the predominant end product of postganglionic neurones of the sympathetic system. The proportions of each in each tissue are fairly constant and characteristic of the species.

Among numerous roles, epinephrine released at times of exceptional stress stimulates glycogenolysis in the liver and skeletal muscles. The effect in the liver is to immediately release glucose-6-phosphate into the circulation. In striated muscles, lactic acid is released, it is reconverted to glycogen in the liver (gluconeogenesis), then released as additional glucose. The elevated glucose levels of the blood satisfy the increased metabolic demand of the heart and skeletal muscles for glucose during stress. Epinephrine and norepinephrine have similar actions at some receptor sites, but their relative potency differs with the site. Because norepinephrine is a potent vasoconstrictor, it has a significant role in the day-by-day management of the peripheral circulation. Both epinephrine and norepinephrine relax the smooth musculature of the bronchi, thereby enhancing the delivery of oxygen to the lungs. The preganglionic fibers supplying the adrenal medulla arrive via branches of the splanchnic nerves. Metabolic control centers in the hypothalamus initiate these incoming impulses.

ENDOCRINE ORGANS DERIVED FROM MESODERM
Adrenal Cortex and Steroidogenic Tissue

We have just seen that the adrenal complex of vertebrates is composed of two entirely different components, aminogenic and steroidogenic, which may be intimately associated or spatially separated. The aminogenic component becomes the adrenal medulla in mammals, and the steroidogenic component becomes the cortex.

The steroidogenic component is derived from mesodermal cells that arise from the coelomic mesothelium of the gonadal ridge (Fig. 14-17) and from the underlying nephrogenic mesoderm. Therefore steroidogenic cells are closely associated with kidneys. In sharks and rays the steroidogenic cells usually form one or more compact masses (**interrenal bodies**) between the caudal ends of the kidneys (Fig. 17-10). In teleosts the steroidogenic cells most frequently collect in the pronephric region near the aminogenic cells, and in tetrapods they are interspersed among the aminogenic cells to form a more or less discrete adrenal gland on or near the kidney (Fig. 17-10). All such steroidogenic masses are homologous with the adrenal cortex of mammals. Although the term "cortical tissue" is appropriate for these masses in mammals only, it is sometimes used to designate any steroidogenic mass that produces "corticoids," that is, steroids similar to those of the mammalian adrenal cortex. Most of the steroids isolated from the adrenal cortex are intermediate metabolic products having varying degrees of corticoid activity depending on the species on which they are tested.

Cortisone, cortisol, and **corticosterone** are glucocorticoids, so named because they stimulate the conversion of noncarbohydrates into glucose (gluconeogenesis). Cortisol is the major saltwater-adapting hormone of euryhaline teleosts, and the chief glucocorticoid in human beings, monkeys, hamsters, and certain other mammals. Corticosterone, on the other hand, is the chief glucocorticoid in rats and rabbits. Some mammals, such as cats, secrete more or less equal measures of both. Cortisone is not an important mammalian glucocorticoid, but it is important in some lower vertebrates. During prolonged stress glucocorticoids enhance the synthesis of epinephrine in the adrenal medulla.

Aldosterone is a mineralocorticoid and is the most potent sodium-regulating steroid in tetrapods. Its action on the ascending arm of the mammalian loop of Henle has been mentioned earlier.

Unlike aminogenic tissue, steroidogenic tissue receives no motor innervation. The region of the mammalian adrenal cortex that secretes glucocorticoids (zona fasciculata and zona reticularis) is under the control of adrenocorticotropic hormone. The region that secretes aldosterone (zona glomerulosa, the outermost zone of the cortex) is under a more complicated control mechanism involving angiotension, a compound formed in the bloodstream under the catalyzing effect of a renal enzyme (renin). Note, however, with respect to adrenocorticotropin, that this anterior pituitary hormone is in turn under the influence of neurohormonal releasing factors from the hypothalamus. Thus the hypothalamus exerts a regulatory effect on aminogenesis via the nervous system, and on the synthesis of glucocorticoids via hormones (Fig. 17-11).

The origin of steroidogenic tissue from the same mesothelium that gives rise to gonads is interesting because only corticoid tissue and gonads produce steroid hormones in vertebrates. The mammalian adrenal cortex produces some 50 different steroids, including small amounts of male and female sex hormones. The bearded lady is an example of what may happen when the adrenal cortex of a female produces excessive quantities of male hormones.

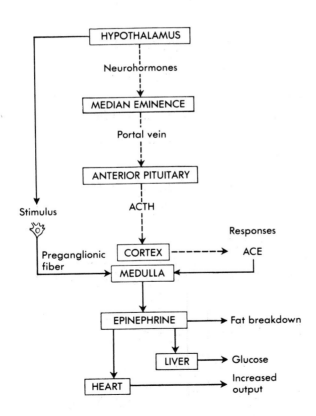

FIGURE 17-11

Some regulatory functions of the adrenal medulla. When presented with a suitable neural stimulus (left center), a preganglionic neuron of the sympathetic nervous system stimulates the medulla to release epinephrine. The latter elicits many responses, three of which are indicated at the right. *ACE*, Adrenal corticoids.

Gonads as Endocrine Organs

Gonads arise from the coelomic mesothelium as a pair of gonadal ridges medial to the kidneys (Fig. 14-17). Ovaries and testes of most vertebrates produce three varieties of steroid hormones: **estrogens, androgens,** and **progestogens.** Collectively, these hormones are essential for reproduction.

Androgens produced by the interstitial cells of the testes (Fig. 17-12) and estrogens produced by ovarian follicles (Fig. 14-19, *A*) are primarily responsible for the maintenance of male and female organs, of secondary sex characteristics such as combs and wattles in roosters and mammary glands in mammals, and for reproductive behavior patterns from fishes to mammals, including cyclic changes. Most accessory sex organs (ducts and glands other than the gonads) and some secondary sex characters (characteristics associated with one sex) atrophy on ablation of the gonads (the primary organs), but atrophy can be prevented by administration of appropriate androgens or estrogens. Androgens are the precursors of estrogens, and the pathways in the biosynthesis of sex steroids are common to all vertebrates from fishes to mammals, although there are quantitative and qualitative differences between species. In mammals testosterone is the most potent androgen and 17β-estradiol is the most important estrogen. As mentioned earlier, androgens are also synthesized in the adrenal cortex, and their overproduction may result in sexual abnormalities.

Progesterone, one of the progestogens, is a precursor in the synthesis of androgens, estrogens, and corticosteroids throughout the vertebrate

FIGURE 17-12

Section of mammalian testis showing interstitial tissue *(light red)* and seminiferous tubules. *Dark red,* capillaries.

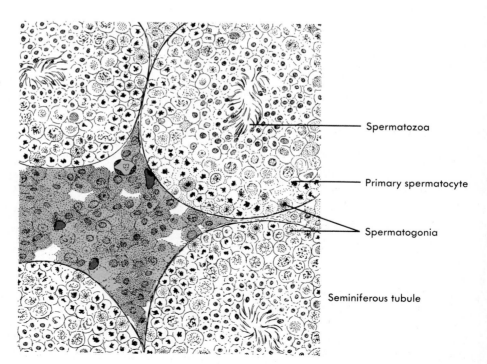

Spermatozoa

Primary spermatocyte

Spermatogonia

Seminiferous tubule

classes, but its role below mammals, if any, has not been ascertained. In female mammals progesterone has achieved an independent role: preparation of the uterine lining for implantation, and maintenance of the lining in a progestational state, a state that will support pregnancy, after implantation. This role inspired the name "progesterone." It has also become essential for ovulation in mammals.

The progesterone used in the maintenance of pregnancy comes from corpora lutea (Fig. 14-19, *A*), which is the name given to ovarian follicles after the cells have undergone luteinization (chemical and morphological change) under the influence of luteinizing hormone from the pituitary. The source of the very small quantities necessary for ovulation is the maturing graafian follicle. By negative feedback to the hypothalamus, progesterone inhibits the formation of a new wave of ovarian follicles after ovulation "in anticipation" of pregnancy in human females; and during pregnancy it inhibits the maturation of new ova in all mammals. Pregnancy is the anticipated outcome of mating in feral mammals.

In addition to steroid hormones, the mammalian ovary produces **relaxin,** a peptide. Relaxin produced during pregnancy softens the ligaments of the pubic symphysis and sacroiliac joints of the mother before birth, thus enlarging the birth canal for easier delivery of the fetus.

Corpuscles of Stannius

Embedded in the posterior part of the mesonephric kidneys or attached to the mesonephric ducts of ray-finned fishes are spherical epithelioid bodies, the corpuscles of Stannius. They are easily mistaken for interrenal bodies, but their embryonic origin is different because they arise as evaginations of the pronephric duct. In most teleosts there are two, but in *Amia* there are 40 to 50. In large salmon the corpuscles may reach 0.5 cm in diameter.

Ablation of the corpuscles has been followed by increases in the concentration of calcium in tissue fluids of bony fishes, indicating that the corpuscles may produce a hypocalcemic factor. There also seems to be a functional relationship, direct or indirect, between the corpuscles and steroidogenic tissue, inasmuch as ablation of the latter stimulates the corpuscles. The significance of these findings remains to be ascertained.

ENDOCRINE ORGANS DERIVED FROM ENDODERM

The endodermal linings of the embryonic pharyngeal pouches develop thickenings that become parathyroids, thymus, and ultimobranchial bodies, and the pharyngeal floor evaginates to form thyroid tissue. Except in a few lower vertebrates, these organs separate from the pharynx, sink into the surrounding mesenchyme, and migrate some distance from their origin. All have endocrine functions. The endocrine role of the gastric and intestinal mucosa, which are also of endodermal origin, has been discussed. The endocrine portion of the pancreas is also endodermal, and is discussed next.

FIGURE 17-13

Island of Langerhans *(light red)* and surrounding acini of a mammalian pancreas. *Dark red,* blood vessels.

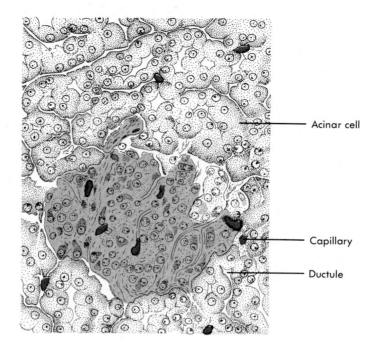

Acinar cell

Capillary

Ductule

Endocrine Pancreas

In addition to acini that secrete digestive enzymes, the pancreas of lower bony fishes and tetrapods contains many microscopic islands of endocrine cells **(islands of Langerhans, pancreatic islets)** that secrete **insulin** and **glucagon** (Fig. 17-13). However, the exocrine and endocrine components are not always spatially associated. In lampreys the insulin-secreting cells are in the mucosa and submucosa of the intestinal wall, and in Myxiniformes they form a small lobe, or "islet organ," around the bile duct completely separated from the diffusely scattered exocrine cells. In elasmobranchs the endocrine cells lie in the epithelium of the pancreatic ductules. In most teleosts they are aggregated into a few compact macroscopic nodules, often two or three, in the same mesenteries that support the diffuse exocrine pancreatic tissue.

The seemingly random location of the endocrine tissue in fishes is understandable once their embryogenesis is known. Both the exocrine (acinar) and endocrine (islet) cells arise from the epithelial lining of the foregut, which, however, is usually displaced by the formation of pancreatic buds (Fig. 11-24). Later, potential endocrine cells bud off from the lining of the developing pancreatic ductules to become isolated islands interspersed among the acini. If the endocrine cells fail to bud off, they will be in the lining of the ductules, as in elasmobranchs. No discrete exocrine-endocrine pancreatic bud forms in cyclostomes, and the endocrine cells migrate independently to their final sites.

Insulin stimulates the uptake and regulates the utilization of glucose in all the tissues, and promotes glycogenesis—the conversion of dietary sugar into glycogen—in the skeletal muscles and liver. It also performs other roles associated with carbohydrate and fat metabolism and is the most potent of the lipogenic hormones. Glucagon, sometimes referred to as a hyperglycemic hormone, has a role essentially opposite that of insulin, elevating the blood sugar level by promoting glycogenolysis in the liver, but not in the striated muscles, where the glycogen is protected. The two hormones together maintain a stable level of blood sugar. In times of stress epinephrine adds its own hyperglycemic effect. Phylogenetically, insulin and glucagon are among the oldest polypeptide hormones. Mammalian insulin causes a decrease in blood sugar levels when administered to cyclostomes and gnathostome fishes.

Thyroid Gland

Early in vertebrate evolution some of the cells of the epithelium of the pharyngeal floor had the capacity to accumulate iodine and to bind it to tyrosine, the first step in the biosynthesis of thyroid hormone. Cells of the hypobranchial groove of the amphioxus perform this synthesis, as does the subpharyngeal gland of larval lampreys (Figs. 2-8, C, and 17-14). In both organisms the iodinated protein is secreted into the pharynx and absorbed farther along the digestive tract. When the duct of a lamprey's subpharyngeal gland closes at metamorphosis, the iodinated protein is thereafter secreted into blood vessels from thyroid follicles that have become isolated

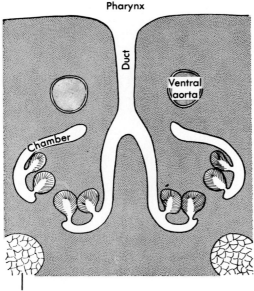

Pharynx

Duct

Ventral aorta

Chamber

Branchial cartilage

FIGURE 17-14

Cross section of subpharyngeal gland (endostyle) of a larval lamprey (ammocoete) at site of its duct. All chambers empty into the duct. At metamorphosis the duct closes and some of the isolated glandular cells form thyroid follicles.

beneath the pharyngeal floor (Fig. 17-15). A **thyroid follicle** consists of cuboidal epithelium surrounding a colloid-filled cavity that stores, temporarily, the iodinated protein (Fig. 17-16). The follicles in hagfishes are similarly located, but hagfish larvae have no subpharyngeal gland.

The thyroid glands of gnathostomes, like endostyles, arise as an unpaired evagination from the pharyngeal floor near the second pharyngeal pouches (Fig. 17-17). After the evagination has reached its adult location, thyroid follicles organize. The embryonic stalk connecting the thyroid to the pharynx usually disappears, leaving a ductless gland. However, a thyroid duct remains in the elasmobranch *Chlamydoselache*. In mammals a small pit on the surface of the tongue near its root marks the site where the thyroid evagination took place; and in human beings short sections of the stalk often persist as a cystlike **thyroglossal duct** that occasionally requires surgical removal.

FIGURE 17-15

Subpharyngeal gland of the larval lamprey *(black)* and thyroid follicles in the adult.

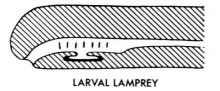

LARVAL LAMPREY

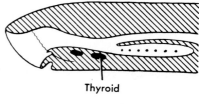

Thyroid

ADULT LAMPREY

FIGURE 17-16

Actively secreting thyroid follicles.

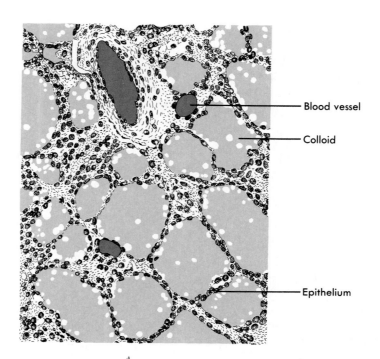

Blood vessel

Colloid

Epithelium

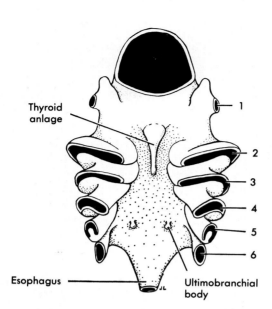

FIGURE 17-17

Pharynx of shark embryo, viewed from below. *1,* Spiracle; *2 to 6,* gill slits.

In teleosts, thyroid follicles are usually scattered singly or in small groups along the ventral aorta under the pharyngeal floor. They may accompany some of the afferent branchial arteries into the gill arches. In a few teleosts they follow the dorsal aorta *caudad* and even invade the kidneys. In still others they form one or two compact masses between the bases of the first gill pouches.

Except in cyclostomes and teleosts, the unpaired thyroid evagination develops into either a single gland or a pair. Sharks have a median thyroid just behind the mandibular symphysis near the insertion of the coracomandibular muscles. A median thyroid is also characteristic of snakes, turtles, a few lizards, and *Echidna*. Most other adult vertebrates have paired thyroid glands.

In amphibians the two glands lie in the floor of the pharynx under cover of the mylohyoid muscle (Fig. 17-18, *A*). In amniotes the glands migrate caudad for varying distances, taking a position close to the trachea and common carotid arteries from which they receive a rich arterial supply (Fig. 17-18, *B* to *D*). The gland was named thyroid because of its location close to the thyroid cartilage in mammals.

The ability to combine iodine into an organic molecule is not restricted to vertebrates or even animals, and **thyroxine,** the hormone of the vertebrate thyroid, has been identified in various species of segmented worms, insects, and mollusks. In vertebrates the synthesis of thyroxine (T_4) is stimulated by thyrotropic hormone from the adenohypophysis. In peripheral tissues thyroxine is converted to **triiodothyronine** (T_3), the biologically active form of the hormone. Thyroid hormone regulates the rate of cellular respiration, stimulates thermogenesis in homeotherms, induces metamorphosis of larval amphibians, and is generally permissive to the action of many other hormones on their target tissues.

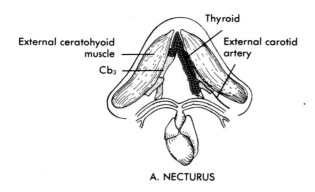

A. NECTURUS

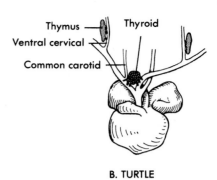

B. TURTLE

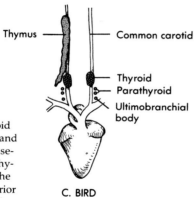

C. BIRD

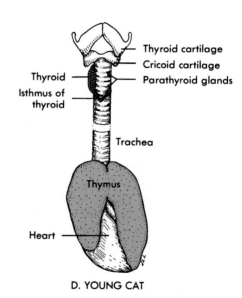

D. YOUNG CAT

FIGURE 17-18

Thymus *(light red)*, thyroid *(dark red)*, parathyroids, and ultimobranchial body in selected vertebrates. The thymus of necturus lies in the angle between the posterior ends of the masseter and external ceratohyoid muscles and is not illustrated. Cb_3, The ceratobranchial cartilage of the third pharyngeal arch. In turtles the parathyroids are embedded in the thymus.

The mammalian thyroid produces another hormone, **calcitonin,** which impedes removal of calcium from bone. It is a product of **parafollicular cells (C cells),** which lie between thyroid follicles, having migrated into the mammalian thyroid from the embryonic ultimobranchial bodies.

Parathyroid Glands

Parathyroid glands arise as evaginations from pharyngeal pouches and are so named because they usually lie beside or embedded in the thyroid gland (Fig. 17-18, *C* and *D*). They are not found in fishes or larval or neotenous amphibians. A few reptiles have three pairs of parathyroid glands, which arise as endodermal outgrowths from the second, third, and fourth pharyngeal pouches, but most tetrapods have only two pairs, since second pouch anlagen usually fail to mature. When there is only one pair, as in a few urodeles, crocodilians, some domestic fowl, and some mammals, the glands develop from either the third or fourth pouch, depending on the species; and occasionally the single gland on each side has contributions from both pouches.

Parathyroid glands produce **parathyroid hormone.** Levels of serum calcium below a given range, which varies with the species, evoke release of parathyroid hormone. Parathyroid hormone causes calcium to be released from bone and other storage sites, restoring serum calcium levels to their normal range.

STORAGE AND RELEASE OF CALCIUM. No hormone is required to cause calcium to be deposited in bone or other sites, at least in tetrapods. Only its release is hormonally regulated. It is protected from release by calcitonin from the mammalian thyroid and from the ultimobranchial bodies of lower vertebrates, except cyclostomes. Parathyroid hormone overrides this effect in tetrapods, and a parathyroid hormone–like peptide may be released from the pituitary gland of teleosts. As mentioned, the corpuscles of Stannius may produce a hypocalcemic factor, but how this effect may be mediated is not known.

Ultimobranchial Glands

Ultimobranchial glands develop from the epithelium of the last pair of pharyngeal pouches in all vertebrates except cyclostomes, and they produce **calcitonin** throughout life, except in mammals (Figs. 17-17, 17-18, C, and 17-19). In elasmobranchs the left gland alone matures and is situated between the pharynx and the pericardium, on the left, just anterior to the esophagus, very close to its embryonic origin. In teleosts it is median or paired, depending on the species, and they lie in the septum transversum below the esophagus. In reptiles, in which it is often unpaired, and in birds the glands lie close to the thyroid.

Adult mammals have no ultimobranchial glands, with the possible exception of scaly anteaters. The potential calcitonin cells (C cells) in mam-

FIGURE 17-19

Pharyngeal derivatives of sharks and mammals (diagrammatic frontal sections looking down onto pharyngeal floor). Left sides are earlier in ontogeny than the right. *Numbers* identify ectodermal grooves or pharyngeal pouches. *Arrows* indicate caudal growth of anlagen. Whether the ultimobranchial gland of mammals is from pouch 4 or from a vestige of pouch 5 is not certain. Endocrine anlagen, except thyroid, are *red.*

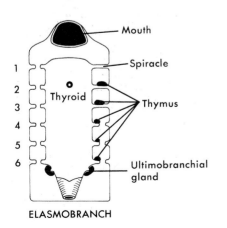

ELASMOBRANCH

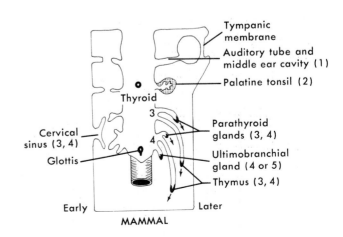

MAMMAL

mals migrate from the thickened epithelium of the last pair of pouches to the thyroid anlagen, where they become parafollicular cells. The fourth pharyngeal pouch of mammals is unusual in that it has an extra pouch evaginating from its caudal wall. Whether or not the latter, from which the ultimobranchial gland arises, is homologous with the fifth pouches of lower vertebrates is unknown. No ultimobranchial glands have been found in cyclostomes, and no calcitonin has been demonstrated in their serum.

Thymus and Bursa of Fabricius

Thymus glands arise as thickenings of the linings of several pharyngeal pouches (Fig. 17-19). The epithelia become invaded by cells that migrate from the hemopoietic regions of the embryonic yolk sac, and later from the fetal liver. The thymic anlagen migrate to their adult location, which has become increasingly farther down the neck in amniotes. The thymus is a lymphoid organ that, in young mammals at least, secretes a hormone that plays an important role in securing full expression of host immunity.

Primitively, thymus probably developed from all pouches. In lampreys it is said to differentiate from all seven. In jawed fishes and tailed amphibians it usually develops from all pouches except the first, and transient thymus tissue has been described from the first. In caecilians the first six pouches participate, and in salamanders, thymus arises from the third, fourth, and fifth pouches. In amniotes the third and fourth pouches are the sole contributors, and in a few mammals the third pouch is the sole source. In frogs the second pouch is frequently the sole source.

There is a tendency for successive thymus anlagen to unite during development, so there may be fewer adult thymus masses than embryonic anlagen. In fishes, except elasmobranchs, all anlagen more or less fuse to form a single, elongated gland lying above the branchial chambers. In amphibians and reptiles they may fuse or remain separate. The single gland on each side in frogs usually lies just behind the tympanic membrane. In reptiles and birds the thymus may consist of a series of large nodes in the neck extending caudad as far as the thyroid (Fig. 17-18, bird). In young mammals the thymus is a large bilobed mass in the thoracic cavity between the sternum and anterior portion of the pericardium. After puberty it undergoes fatty degeneration. In the larvae of at least one elasmobranch (*Heptanchus cinereus*), ducts lead from the first six lobes of the thymus into the pharynx. These ducts may persist in young adults.

Interest in the thymus as an endocrine organ alternately waxed and waned over the years as bits of evidence for an endocrine role were reported in the literature, only to fail the test of reproducibility. Among the facts that had to be taken into account is that the thymus of mammals is largest in fetal, neonatal, and juvenile animals, and undergoes fatty infiltration as the young attain sexual maturity. Recent studies by immunologists have provided part of the explanation. The thymus is large just before birth and early in life, when it is housing and processing **lymphocytic stem cells** that are released into the general circulation and which, along with their descendants, produce antibodies. A polypeptide hormone, **thymosin**, has also been isolated from the thymus of mammals and appears to be an important factor in imparting immune competence to the stem cells, particularly those related to killing tumor cells.

The role of the thymus in most young birds is supplemented by the **bursa of Fabricius,** a lymphoid organ that arises as a middorsal evagination of the embryonic cloaca and extends into the pelvic cavity, where it is sandwiched between the large intestine and the synsacrum. It resembles the thymus in structure, and regresses completely at sexual maturity. Its role, like that of thymus tissue, appears to be to impart antibody capability to the lymphatic stem cells that it produces.

HORMONAL CONTROL OF BIOLOGICAL RHYTHMS

All living organisms exhibit rhythms of metabolism and behavior. These vary from annual cycles in which gonads usually regress and recrudesce and reproductive behavior waxes and wanes, to daily (circadian) cycles in which metabolism exhibits phases of elevated and depressed activity and behavior is rhythmically altered. The daily photoperiod (period of light) is the principal environmental time setter, or synchronizer, of circadian rhythms, but other environmental factors such as temperature and more subtle stimuli may also be effective (Fig. 17-5). Environmental rhythms are detected by sense organs, especially the parapineal and retina. The information is then transmitted to a small number of hypothalamic nuclei that serve as pacemakers to regulate and synchronize biorhythms by elaborating and releasing neurohormones. The hormones that transmit circadian information may be divided into three categories: synchronizers, modifiers, and inducers.

Synchronizing hormones set the phases of circadian rhythms. That is, through their synchronizing effects on many different tissues, including other endocrine glands, they determine the onset, duration, and termination of daily cycles. Through their effects specific systems become "go" and a specific metabolic activity develops. Adrenal corticoids are probably the most important synchronizing hormones. The corticosteroid rhythm is under direct control of the daily light-dark cycle in those vertebrates exposed to one. A phase of the corticosteroid rhythm determines, for example, the time of day or night that the liver is able to respond to the fat-synthesizing effect of prolactin. This has been demonstrated in fishes, several species of birds, and in mammals. In mammals that have been tested thus far, fat is synthesized by the liver at the onset of light. Seasonal changes in the daily photoperiod, which are a result of the elliptical path of the earth around the sun during the course of 365¼ days, are accompanied by changes in the phases of the corticosteroid rhythm. Consequently, the rhythms of fat synthesis are different in different seasons.

Modifying hormones regulate the amplitude of circadian rhythms. Thyroid hormones are important modifiers that must be present to either drive or permit the expression of rhythms of other hormones. Depressed levels of serum thyroxin result in lowered sensitivity of all tissues, including other endocrine organs and the neurosecretory cells of the hypothalamus. Thyroid hormone therefore is able to modify circadian and annual rhythms of metabolism and behavior.

Inducing hormones act in conjunction with synchronizing hormones. These, too, are synthesized, stored, and ultimately released rhythmically. For example, pituitary prolactin levels in male and female golden hamsters in Baton Rouge, Louisiana, are higher at 8:00 PM than at 8:00AM in January and August. Their release into the circulatory system induces those metabolic processes that have been rendered inductible by the earlier release of synchronizing hormone. *The time interval between release of corticosteroids and the subsequent release of prolactin is critical,* and is not necessarily the same for each different function of prolactin. Prolactin and the adrenal corticoids "get out of sync" seasonally for certain functions. The result is an annual rhythm superimposed on the circadian rhythm. The phenomenon is adaptive, because prolactin has many roles. The reproductive hormones and insulin also act in an inducing capacity. The effect of seasonal environmental changes on reproductive structure, physiology, and behavior were noted earlier in the chapter.

The importance of the time-interval factor in the function of a synchronizing-inducing hormone complex is illustrated by the direction in which certain species of migratory birds take off after an experimental regimen of corticosteroids (synchronizers) and prolactin (inducing agent for take-off). Birds administered corticosteroids at a specific hour in the lower Mississippi River Valley and then administered prolactin 4 hours later, for 9 days, will at the end of that time, if released, set off on a southbound migratory route. Other birds of the same flock, if concurrently administered corticosteroids followed by prolactin 12 hours later will, if released, set off on a northbound migratory route. The responses reflect the behavior of these species of birds in nature. The ambient environment at each terminus of a migratory route, operating via the hypothalamus, provides a time-interval relationship between synergizing and inducing hormone of such duration as to ensure that the migratory route from that location will have survival value.

The vertebrate body, as the saying goes, is "fearfully and wonderfully made." But of all living species, man alone has the ability to understand his own physical nature and to analyze himself intellectually and morally. The extent to which the human species utilizes these unique insights for the common good may determine the fate of the species on the planet Earth.

Chapter Summary

1. Hormones are products of specific groups of cells that regulate nearby or remote cells of a different nature. They are mostly polypeptides, proteins, glycoproteins, or steroids. A few are amines.

2. Neurohormones are small polypeptides produced by neurosecretory neurons and released into circulatory channels of neurohemal organs.

3. Known neurohemal organs of vertebrates are the median eminence, posterior lobe of pituitary, and urophysis of elasmobranchs, chondrosteans, and teleosts.

4. The hypothalamus is the major source of vertebrate neurohormones. It is regulated partly by the external environment and hormonal feedback. Its secretions are released in the median eminence and posterior lobe of the pituitary.

5. The pituitary consists of a neurohypophysis derived from the floor of the third ventricle and an adenohypophysis derived from the roof of the stomodeum.

6. The neurohypophysis consists of a median eminence, an infundibular stalk, and a posterior lobe (pars nervosa). It synthesizes no known hormones. Hypothalamic neurohormones released from the posterior lobe prevent dehydration and stimulate smooth muscles in restricted locations. Those released from the median eminence reach the pituitary via hypophyseal portal vessels.

7. The adenohypophysis usually consists of pars distalis, pars intermedia, pars tuberalis, and, in elasmobranchs, a ventral lobe. The pars distalis secretes STH, TSH, ACTH, FSH, LH (ICSH), and prolactin. The pars intermedia produces intermedin, a melanophore-stimulating hormone that darkens some skin.

8. The urophysis secretes urotensins, which have vasopressor and osmoregulatory effects.

9. The pineal produces melatonin during periods of darkness. It blanches the skin of some fishes and of larval amphibians and has a gonad-inhibiting influence in some mammals.

10. Aminogenic tissue, which includes the mammalian adrenal medulla, synthesizes hormones, including epinephrine, that are amines. In some fishes it is spatially separate from the steroidogenic component of the adrenal complex.

11. Steroidogenic tissue arises from the mesothelial covering of the gonadal ridge and produces adrenal and gonadal steroids.

12. Adrenal steroidogenic tissues include the interrenal bodies of fishes and adrenal cortex of mammals. Aldosterone regulates sodium excretion, and glucocorticoids elevate blood sugar levels by gluconeogenesis.

13. The gonads of both sexes produce steroids that include androgens and estrogens. The ovary also produces progesterone and relaxin, which affect the female reproductive tract.

14. Corpuscles of Stannius are derivatives of the pronephric duct of ray-finned fishes. They have a hypocalcemic effect on serum calcium.

15. The endocrine pancreas produces insulin, which stimulates glucose uptake and utilization, and glycogenesis; and glucagon, which elevates blood sugar by glycogenolysis. Its relationship to the exocrine pancreas is less intimate in some fishes than in tetrapods.

16. Thyroid follicles arise from a median evagination of the pharyngeal floor and form discrete median or paired glands except in cyclostomes and teleosts. Thyroxine and triiodothyronine increase metabolic rates. C cells of the thyroid produce calcitonin in mammals.

17. Parathyroid glands are absent in fishes and larval and neotenic amphibians. They are derivatives of pharyngeal pouches II to IV in some reptiles, III and IV in birds and mammals. They produce parathyroid hormone, which elevates serum calcium, and calcitonin, which reduces it.

18. Ultimobranchial glands develop from the last pharyngeal pouches and produce calcitonin. They are absent in cyclostomes and adult mammals.

19. The thymus is a lymphoid organ that arises from the epithelium of one or more pharyngeal pouches. The bursa of Fabricius is a similar organ that evaginates from the cloaca of birds. Thymus and bursa produce lymphocytic stem cells, whose descendants participate in immune reactions. The thymus of mammals also produces the hormone thymocresin, which imparts antibody capability to the stem cells it produces.

SELECTED READINGS

Amoroso, E.C., Heap, R.B., and Renfree, M.B.: Hormones and the evolution of viviparity. In Barrington, E.J.W., editor: Hormones and evolution, vol. 2, New York, 1980, Academic Press.

Barrington, E.J.W.: Hormones and evolution, 2 vols., New York, 1980, Academic Press.

Bentley, P.J.: Comparative vertebrate endocrinology, ed. 2, New York, 1982, Cambridge University Press.

Clark, N.B.: Evolution of calcium regulation in lower vertebrates, American Zoologist **23**:719, 1983.

DeVlaming, V.L.: Actions of prolactin among vertebrates. In Barrington, E.J.W., editor: Comparative endocrinology, vol. 2, New York, 1980, Academic Press.

Glick, B.: The thymus and bursa of Fabricius: Endocrine organs? In Epple, A., and Stetson, M.H.: Avian endocrinology, New York, 1980, Academic Press.

Gorbman, A.: Early development of the hagfish pituitary gland: evidence for the endodermal origin of the adenohypophysis, American Zoologist **23**:639, 1983.

Gorbman, A., and others: Comparative endocrinology, New York, 1983, John Wiley and Sons.

Hadley, M.E.: Endocrinology, Englewood Cliffs, N.J., 1984, Prentice-Hall.

Harris, G.W., and Donovan, B.T., editors: The pituitary gland, 3 vols., Berkeley, California, 1966, University of California Press.

Holmes, R.L., and Ball, J.N.: The pituitary gland: a comparative account, New York, 1974, Cambridge University Press.

Ishii, S., Hirano, T., and Wada, M., editors: Hormones, adaptation, and evolution, New York, 1980, Springer-Verlag.

Loretz, C.A., Bern, H.A., Foskett, J.K., and Mainoya, J.R.: The caudal neurosecretory system and osmoregulation in fish. In Farner, D.S., and Lederis, K., editors: Neurosecretion: molecules, cells, systems, New York, 1982, Plenum Press.

Ralph, C.L., editor: Comparative endocrinology: developments and directions. Progress in clinical and biomedical research, vol. 205, New York, 1986, Alan R. Liss.

Setchell, B.P.: The mammalian testis, Ithaca, New York, 1978, Cornell University Press.

Stoddart, D.M.: Mammalian odours and pheromones, London, 1976, Edward Arnold Publishers, Ltd.

Symposia in American Zoologist

Comparative aspects of the endocrine pancreas, **13**:565, 1973.

Endocrine role of the pineal gland, **16**:1, 1976.

Evolution of endocrine systems in lower vertebrates, a symposium honoring Professor Aubrey Gorbman, **23**:593, 1983. Thirteen papers covering a broad range of endocrine organs and functions, with extensive bibliographies.

Photoperiodism in the marine environment, **26**:386, 1986.

APPENDIX I

SYNOPTIC CLASSIFICATION OF VERTEBRATES

This classification has been provided so that students may readily ascertain the kinships of animals they meet in the text. If it is consulted whenever an unfamiliar group is first mentioned, the student will find that he or she is acquiring a familiarity with vertebrate taxonomy with minimal effort.

This is a conventional *natural order* classification. As discussed under "Taxonomy and Systematics" in Chapter 1, not all classification schemes are alike, and all are tentative. For example, agnathans, here given the status of a class, are sometimes classified as a superclass, or even as a subphylum; and cartilaginous fishes (here assigned class status) are sometimes assigned subclass status in a class Elasmobranchiomorphii, which includes the bony placoderms. There is no universally accepted classification of agnathans and cartilaginous jawed fishes. These and other differences between this and other classifications of vertebrates are often due to substitution of cladistic taxonomic theory in part, but not in all parts, of the scheme. The primary consideration in adopting the present scheme was to achieve maximal utility for the user.

All commonly recognized classes are included, but subclasses have been omitted when all members are extinct and no representative has been mentioned in the text. All widely recognized orders containing living taxa are included, except those of bony fishes and birds. Inclusion of the many subgroups in those classes would not be in accordance with the purpose of the synopsis. Totally extinct taxa are indicated by an asterisk.

Subphylum Hemichordata. An invertebrate taxon of doubtful status.
    Class Enteropneusta. Acorn worms.

Subphylum Urochordata (Tunicata)

    Class Ascidiacea
    Class Larvacea (Appendicularia)
    Class Thaliacea

Subphylum Cephalochordata. *Branchiostoma, Asymmetron,* sole genera

Subphylum Vertebrata (Craniata)

    Class Agnatha. Jawless fishes.
        *Order Heterostraci
        *Order Osteostraci
        *Order Anaspida   } Ostracoderms
        *Order Thelodonti (Coelolepida)

Order Petromyzontiformes. Lampreys. ⎱ Cyclostomes
Order Myxiniformes. Hagfishes. ⎰

***Class Acanthodii.** Armored Paleozoic jawed fishes.
***Class Placodermi.** Armored Paleozoic jawed fishes.
  ***Order Arthrodira.** Arthrodires.
  ***Order Antiarchi.** Antiarchs.
  Additional extinct orders.
Class Chondrichthyes. Cartilaginous fishes.
  Subclass Elasmobranchii. Naked gill slits.
   ***Order Cladoselachii.** Primitive Paleozoic sharks.
   ***Order Pleuracanthodii.** Freshwater Paleozoic sharks with lobed fins.
   Order Squaliformes. Sharks.
   Order Rajiformes. Sawfish, skates, rays.
  Subclass Holocephali. Gill slits covered by an operculum.
   Order Chimaeriformes. Chimaeras.
Class Osteichthyes. Higher bony fishes.
  Subclass Sarcopterygii (Choanichthyes). Lobe-finned fishes, many with internal nares.
   Order Crossopterygii. Chiefly Paleozoic.
    ***Suborder Rhipidistia.** Probable ancestors of amphibians.
    Suborder Coelacanthiformes. Specialized crossopterygians, internal nares absent; *Latimeria* sole living crossopterygian.
   Order Dipnoi. True lungfishes; *Lepidosiren, Neoceratodus, Protopterus* sole living genera.
  Subclass Actinopterygii. Ray-finned fishes.
   Superorder Chondrostei. Chiefly Paleozoic.
    ***Order Palaeonisciformes.** Paleozoic ganoids.
    Order Polypteriformes. Two genera of African lungfishes—*Polypterus* and *Calamoichthyes;* ganoid scales.
    Order Acipenseriformes. Sturgeons, several genera, including *Acipenser;* and spoonbills—*Polyodon* and *Psephurus;* ganoin lacking.
   Superorder Holostei. Dominant Mesozoic fishes with ganoid scales; two living genera—*Amia calva* (bowfin) and *Lepisosteus* (gars).
   Superorder Teleostei. Recent bony fishes; 95% of all living fishes.
    Order Clupeiformes. Herringlike fishes, salmonids, others.
    Order Cypriniformes. Goldfish, carp, North American catfish, buffalo fish; minnows; exhibit weberian apparatus.
    Order Anguilliformes. Eels.
    Order Gadiformes. Codfishes, others.
    Order Perciformes. Perchlike fishes; largest order of teleosts.
    And up to 35 additional living orders.
Class Amphibia. Highest anamniotes.
  ***Subclass Labyrinthodontia.** Earliest tetrapods; abundant, diverse; centra of vertebrae formed by intercentra and pleurocentra.
   ***Order Ichthyostegalia.** Late Devonian; the earliest amphibians; *Ichthyostega.*
   ***Order Temnospondyli.** Permian; *Archegosaurus.*
   ***Order Anthracosauria.** Paleozoic; probably ancestral to reptiles.
  ***Subclass Lepospondyli.** Of debatable phylogeny.
  Subclass Lissamphibia. Triassic to modern amphibians.
   ***Order Proanura.** Triassic precursors of Anura.
   Order Anura. Frogs, toads, and tree toads.

Order Urodela (Caudata). Tailed amphibians.
Order Apoda (Gymnophiona). Caecilians.
Class Reptilia. Lowest amniotes, mostly extinct.
  Subclass Anapsida. No temporal fossae.
    ***Order Cotylosauria.** Stem reptiles.
    Order Testudinata (Chelonia). Turtles and tortoises.
  ***Subclass Euryapsida (Synaptosauria; Parapsida).** Large marine reptiles, including plesiosaurs and ichthyosaurs; one dorsal temporal fossa.
  Subclass Lepidosauria. Diapsid (two lateral temporal fossae) or modified diapsid skull).
    Order Rhynchocephalia. *Sphenodon punctatum* sole living species.
    Order Squamata.
      Suborder Lacertilia. Lizards.
      Suborder Amphisbaenia. Amphisbaenians.
      Suborder Serpentes (Ophidia). Snakes.
  Subclass Archosauria. Diapsids; includes bird stock.
    ***Order Thecodontia.** Stem archosaurs.
    ***Order Pterosauria.** Flying reptiles (pterodactyls).
    ***Order Saurischia.** Dinosaurs with reptilelike pelvis.
    ***Order Ornithischia.** Dinosaurs with birdlike pelvis.
    Order Crocodilia. Crocodiles, alligators, caimans, gavials.
  ***Subclass Synapsida.** Mammal-like reptiles. One lateral temporal fossa.
    ***Order Pelycosauria.** Early synapsids.
    ***Order Therapsida.** Late synapsids; mammalian precursors.
Class Aves. Feathered vertebrates.
  ***Subclass Archaeornithes.** Earliest birds, derived from bipedal archosaur; *Proavis, Archaeopteryx,* sole known species.
  Subclass Neornithes. All other extinct and living birds.
    ***Superorder Odontognathae.** Toothed Cretaceous marine birds; *Hesperornis,* sole species.
    Superorder Neognathae. Ratites and carinates.
      Order Columbiformes. Doves.
      Order Pelecaniformes. Pelicans, cormorants, etc.
      Order Anseriformes. Ducks, geese, other waterfowl.
      Order Falconiformes. Hawks, eagles, vultures.
      Order Galliformes. Grouse, quail, domestic foul.
      Order Psittaciformes. Parrots, paroquets.
      Order Passeriformes. Perching birds—up to 64 families, including songbirds.
    And about 15 other living orders.
Class Mammalia. Vertebrates with hair.
  Subclass Prototheria. Egg-laying mammals.
    Order Monotremata. Duckbilled platypuses and echidnas.
  Subclass Metatheria. Yolk sac serves as placenta.
    Order Marsupialia. Koala bear, opossum, kangaroo, and other pouched mammals.
  Subclass Eutheria. True placental mammals.
    Order Insectivora. Moles, shrews, hedgehogs, flying lemurs, others.
    Order Chiroptera. Bats.
    Order Primates
      Suborder Lemuroidea. Lemurs, lorises, pottos, bush babies.
      Suborder Tarsioidea. Tarsiers.

Suborder Platyrrhini. South American monkeys and marmosets. Nostrils open to the side.

Suborder Catarrhini. Anthropoids with nostrils opening downward.

Superfamily Cercopithecoidea. Old World monkeys.

Superfamily Hominoidea. Apes and man.

Family Pongidae. Gorillas, chimps, other apes.

Family Hominidae. Man and extinct prehuman hominids.

*Australopithecus africanus, A. afarensis.

*Homo erectus. Java man.

*Homo neanderthalensis. Sometimes cited as *Homo sapiens neanderthalensis.*

Homo sapiens. Modern and *Cro-Magnon man.

Order Carnivora

Suborder Fissipedia. Terrestrial carnivores. Felines, canines, hyenas, bears, raccoons and pandas, civets, mustelids, others.

Suborder Pinnipedia. Marine carnivores. Seals, sea lions, walruses.

Order Cetacea. Whales, dolphins, porpoises.

Order Edentata. Sloths, armadillos, South American anteaters.

Order Tubulidentata (Xenarthra). Aardvarks. Insectivorous. One species.

Order Pholidota. Pangolins. Toothless, insectivorous: *Manis,* sole genus.

Order Rodentia. Gnawing mammals other than lagomorphs.

Suborder Sciuromorpha. Squirrels, woodchucks, others.

Suborder Myomorpha. Mice-like rodents.

Suborder Caviomorpha (Hystricomorpha). Porcupines, chinchillas, cavys, others.

Order Lagomorpha. Pikas and rabbits.

Order Perissodactyla. Ungulates with mesaxonic foot; usually odd-toed. Horses, tapirs, rhinoceros.

Order Artiodactyla. Ungulates with paraxonic foot; usually even-toed.

Suborder Suina. Pigs, hippopotami, peccaries—relatively primitive artiodactyls.

Suborder Ruminantia. Cud chewers with complex stomachs.

Family Camelidae. Camels, llamas.

Family Cervidae. Deer, caribou (reindeer).

Family Giraffidae. Giraffes.

Family Antilocapridae. American pronghorn antelope sole species.

Family Bovidae. Oxen, sheep, goats, true antelopes.

Family Tragulidae. Chevrotains.

Order Proboscidea. Elephants, *mastodons. Subungulates.

Order Hyracoidea. Two genera of hoofed hyraxes. Subungulates.

Order Sirenia. Dugongs, manatees. "Sea cows" from ungulate stock.

And additional extinct orders.

APPENDIX II

SELECTED PREFIXES, SUFFIXES, ROOTS, AND STEMS EMPLOYED IN TEXT*

Following are some classical components of terms used in the text, with examples. Familiarity with the entries should enable a reader to deduce, within useful limits, the meanings of many additional words, such as *hemangioepithelioblastoma*, which otherwise may be a meaningless jumble of letters. The list is no substitute for a general unabridged or standard medical dictionary, but it is sufficiently long (800 entries) and varied to motivate a reader toward habitual use of those standard works. Such a practice will extend the reader's intellectual horizons far beyond the boundaries of comparative anatomy. Pragmatically, it will result in better recognition and recall and also in more accurate spelling of technical terms. The list makes no pretense to completeness.

Meanings are those relevant to the subject matter of the text. Headings that are stems **(acanth-)** are usually the smallest combinations of letters that are common to the derivatives. The symbol > means *hence* and separates a classical from a derived meaning.

---

*This is not a glossary. Meanings of terms not included here can be located in the text by referring to the index.

a- lacking, without; as in *acelous, Agnatha.*
ab- from, away from.
    abducens a nerve that abducts (q.v.) the eyeball.
    abduct to move a part away from the longitudinal axis, as in raising the arm laterally.
acanth- spine, spiny.
acelous lacking a cavity.
acetabulum a cup for holding vinegar; the socket in the innominate bone.
acinus a grape.
    acinar resembling a cluster of grapes.
acr- extremity, highest.
    acrodont tooth on the summit of the jaw.
    acromion process at proximal extremity of the shoulder (*-omo*).
actin- ray.
    Actinopterygii ray-finned fish (see *pter-*).
ad- to, toward, upon.
    adduct to draw toward.
    adrenal a gland on the kidney.
aden- gland.
    adenohypophysis glandular part of the pituitary.
    adenoid resembling a gland; a nasal tonsil.
-ae nominative plural ending, as in chordae tendineae (tendinous chords); genitive singular ending as in radix aortae (root of the aorta).
aestivate to spend summer in a state of lowered metabolism.
af- same as *ad-* (to, toward); the *ad-* is changed to *af-* to make the word easier to pronounce.
    afferent carrying something to or toward something else (see *-ferent*).
Agnatha *a-* (lacking) + *gnath* (jaws).
ala- wing; see also *ali-*.
    alar winglike.
alba white.
alecithal lacking yolk.
ali- wing.

alisphenoid wing of the sphenoid.
alveolus a small chamber or sac.
amel- enamel.
    ameloblast a cell that produces enamel.
amphi- on both sides; in two ways.
    amphiarthrosis a joint with severely limited movement.
    amphicelous having concavities at both ends.
    amphioxus an animal with both ends pointed (*-oxy*).
ampulla a flask > a small dilation.
an- without.
    anamniote an animal lacking an amnion.
    anapsid lacking an arch.
    anura lacking a tail.
ana- up, again.
    anadromous *ana-* + *-dromos* (course) > able to migrate from the sea up into freshwater streams.
    anastomose to unite end to end.
    anatomy *ana-* + *-tome* (q.v.).
anch- gill.
andr- male.
    androgen a hormone that induces maleness.
angi- vessel.
ankyl- a growing together of parts.
    ankylose to fuse in an immovable articulation.
anlage (pl., **anlagen**) an embryonic rudiment or precursor of a developing structure.
annulus a ring.
    annulus tympanicus bony ring to which the eardrum is attached.
ante- before.
    antebrachium the forearm; the part before the brachium.
anthrop- refers to human beings.
anti- against, opposite.
    antidiuretic inhibiting loss of water (diuresis) via kidneys.
antrum a cavernous space.
Anura *an-* + *uro* (q.v.); tailless amphibians.

apical at the apex.
Apo- from, away from.
Apoda *a* + *pod-*; without legs.
aponeurosis *apo-* + *neuron* (a tendon); a broad, flat, tendinous sheet.
apophysis an outgrowth or process.
apsid refers to an arch.
arch- first, primary, ancient.
    archenteron primitive gut.
    archetype an early model.
    archinephros hypothetical primitive kidney.
    archipallium first roof of the telencephalon.
arcuate arched.
arrector pili (pl., **arrectores pilorum**) muscle that erects a hair.
arthro- joint, articulation.
    arthrodire placoderm with joints, because of dermal plates, in the neck (*-dire*).
artio- an even number.
    artiodactyl having an even number of digits.
arytenoid resembling a ladle.
Ascaphus genus of anuran lacking a drum (*scapha*).
ataxia *a-* (lacking) + *taxia* (order); a disorder of the neuromuscular system.
-ate having the property of, as septate: with septa.
atlas the vertebra that supports the head like the mythical Atlas holds up the earth.
atrium the courtyard of a Roman home > a cavity that has entrances and exits.
auricle an ear or earlike flap.
auto- self.
    autostyly a condition in which the upper jaw braces (*-styly*) itself against the skull.
    autotomy cutting one's self, as when a lizard breaks off the end of its tail.
axial in the longitudinal axis.
azygos *a-* (lacking + *zyg-* (a yoke) > on one side only.

baro- pressure.

baroreceptor a sense organ that monitors pressure.

basi- most ventral; pertaining to a basal location.

basihyal ventral element of hyoid skeleton.

bi- double, twice.

bicornuate having two horns (*cornua*).

bicuspid having two cusps.

bipartite having two parts.

bio- life.

blast- an embryonic precursor; a germ of something.

blastema an embryonic concentration of mesenchyme.

blastocoel cavity of the blastula.

blastocyst the mammalian blastula.

blastula a little (*-ula*) embryo.

bovine pertaining to oxen or cows.

brachi- arm.

brachiocephalic associated with the arm and head.

branchi- gill.

branchiomeric referring to parts of the branchial arches.

bucco- cheek.

bulbus bulb.

bulbus arteriosus muscular swelling on ventral aorta.

bulbus cordis term sometimes applied to conus arteriosus in lungfishes, amphibians, and mammalian embryos; part of the heart.

bulla, a bubble > a bubblelike part such as the tympanic bulla.

buno a hill or mound.

bunodont a tooth with low cusps.

bursa a sac or pouch.

caecum a blind pouch.

calamus a stem or reed; the stem of a feather.

calcaneo- pertaining to the heel bone.

canaliculus a little canal.

capitulum a little head.

caput head.

cardi- heart.

cardinal of basic importance; chief.

carina a keel.

carn- flesh.

carnivore a flesh-eating animal.

carotid from a word meaning heavy sleep; compression of the carotid artery cuts off blood to the brain.

carpo- wrist.

cata- down, complete.

catarrhine having a nose (*-rhin*) with nares directed downward.

cauda tail.

caudad toward (*-ad*) the tail.

cecum see *caecum*.

cel- see *-coel-*.

cen- new.

cenogenic of recent origin.

cephal- head.

cera wax.

cerat- horn.

ceratohyal horn of the hyoid.

ceratotrich a horny, hairlike fin support.

-cercal tail.

cerumen secretion of mammalian external ear glands; see *cera*.

cervical pertaining to the *cervix* (q.v.).

cervix neck.

cheir-, chir- hand.

Chiroptera mammals in which the hand is modified as a wing; bats.

Chelys tortoise.

chiasma shaped like the Greek letter chi (X) > a crossing.

choana a funnel-shaped opening.

chole- bile.

chondr- cartilage.

chorion a membrane.

choroid, chorioid resembling a membrane (see *chorion*).

chrom-, chromato- color.

circa, circum around, about.

circadian about a day in length.

cladistic pertaining to phylogeny.

clava club.

clavo-, cleido- clavicle.

cleidomastoid a neck muscle attached to clavicle.

cleidoic closed, locked up > a reptilian, bird, or monotreme egg with much yolk and a shell.

cloaca a sewer > common terminus for digestive and urinary tracts.

cochlea a snail with a spiral shell > spiral labyrinth of inner ear.

-coel- hollow, a cavity.

coelom body cavity.

collagen gelatinous, gluelike material.

columella a little (*-ella*) pillar or column.

com-, con-, cor- with, together.

conari of the pineal (q.v.).

conch- shell.

concha the pinna of the ear, shaped like a clamshell.

contra- opposite.

contralateral on the opposite side.

conus a cone.

conus arteriosus chamber of heart after the ventricle.

copr- feces.

coprodeum fecal passage derived from cloaca.

cor- see *com-*.

coracoid shaped like a crow's beak.

corn- horn.

cornified changed to horn by keratinization.

cornu (pl., **cornua**) horn of the hyoid.

corona a wreath or crown.

coronary sinus forms a "wreath" around the heart.

corpus (pl., **corpora**) body.

corpora quadrigemina the two pairs of twins (*-gemini*) of the roof of the mesencephalon of amniotes.

corpus luteum a yellow body of the ovary.

corpus spongiosum spongy body.

costa rib.

costal cartilage a cartilage at the ventral end of a rib.

cotyledonary cup shaped.

coxa hip.
cribriform sievelike.
cricoid resembling a ring.
crista a ridge or crest.
crotalum a rattle > *Crotalus*, a rattle-snake.
crus leg.
cten- comb.
ctenoid resembling a comb.
cucullaris from a word meaning a hood; the two cucullaris (trapezius) muscles resemble collectively a hood or shawl.
cuneiform wedge shaped.
cusp a peak or point.
cutaneous referring to skin.
cyclo- circular.
cyclostome agnathan with round mouthlike funnel.
cyno- dog.
cynodont dog-toothed.
cysto- sac, bladder.
cyt-, cyto-, -cyte, cell.

dactyl- finger, toe.
de- away from.
decidu- fall off or be shed.
deciduous placenta one in which the uterine lining is partly shed at parturition.
deltoid resembling the Greek letter delta (Δ).
demi- half.
demibranch gill on one face of a gill arch.
dendro- tree.
dendrite neuronal arborization.
dent- tooth.
dentin bone like that in teeth.
derm- skin.
dermatome layer of a somite giving rise to skin.
Dermoptera mammals in which the skin forms a wing membrane (see *-pter*).
dermato- referring to skin.
dermatocranium skull bones phylogenetically derived from skin.
-deum, -daeum a passageway.
deuter- two.
deuterostome an animal that uses

the blastopore as an anus and forms a second mouth.
di- twice, double.
diapophysis one of two lateral processes.
diapsid having two arches.
diarthrosis freely movable joint between two bones or cartilages.
Dipnoi fish with two breathing apertures (external and internal).
dia- through, apart, completely, between.
diaphragm a separation *(phragma)* between two parts.
diaphysis ossifying shaft of a long bone.
-didym- twins > the testes.
digiti- fingers or toes.
digitigrade walking *(-grade)* on the digits.
dino- fearful, terrible.
dinosaur a fear-inspiring reptile.
diphy- double.
diphyodont having two successive sets of teeth.
diplo- double, two.
diplospondyly two vertebrae in each body segment.
dis- separation, apart from.
dissect to disassemble.
diverticulum an outpocketing.
dorsum the back.
dorsad toward the back.
duodenum 12; the length of the human duodenum is about the breadth of 12 fingers.
dura tough, hard.
dys- faulty, painful, difficult.

e- without.
Edentata an order of mammals lacking teeth.
ect- outer.
ectopterygoid the outer pterygoid bone.
ectotherm an animal whose temperature varies with the environment.
-ectomy *ex-* (out of) + *tome* (cut), as in appendectomy.

ectopic *ex-* (out of) + *topo-* (place).
ectopic pregnancy a pregnancy in which the fetus is implanted elsewhere than in the uterus.
ef- variant of *ex-*.
efferent that which carries away from; efferent branchial arteries carry blood out of the gills.
elasmo- plate.
elasmobranch with gills composed of flat plates.
-ella a diminutive, as in columella.
en- in, into.
encephalon the brain, a structure in the head.
endo- within, inner; see also *ento-*.
endochondral within cartilage.
endolymph the lymph within the membranous labyrinth.
endotherm an animal that maintains a relatively constant body temperature regardless of environmental fluctuations.
enteron the gut.
ento- within, inner; see also *endo-*.
entoglossal within the tongue.
ep-,·epi- upon, above, over.
epaxial above an axis.
ependyma an outer garment > the membrane covering the part of the central nervous system exposed to the neurocoel.
epididymis (pl., **epididymides**) a structure lying on the testis *(didym-)*.
epiglottis a skeletal flap over the glottis.
epimere dorsal mesoderm.
epiphysis *epi* + *physis* (q.v.), pineal complex.
epiploic relating to greater omentum (the **epiploon**).
epithelium a surface layer of cells.
epitrichium *epi* + *trich-* (hair), a temporary layer of epidermal cells overlying the hair of fetal mammals. The mammalian periderm.
erythro- red.
eso- carrier.
esophagus carrier of substances that have been eaten *(phag-)*.

estr- female.

ethmoid sievelike.

eu- true.

eury- wide.

euryhaline able to live in waters with a wide range of salinity (*halo-*).

ex-, exo- out, out of, away from, outer.

excurrent pore a pore for the exit of a current of water.

exoskeleton a skeleton in the skin.

extra- beyond, outside of.

extraembryonic outside of the embryo.

falciform shaped like a sickle.

fauces throat.

fenestra a window > an aperture.

-ferent, -ferous carrying, as in *afferent* (q.v.).

fil thread.

fimbria fringe.

foramen a small opening, usually transmits something.

foramen magnum the large foramen in the occipital region.

fore- before, in front.

forearm the part of the arm before the upper arm.

fossa a pit, cavity, depression, vacuity.

fovea a small pit; site of most acute vision in retina.

frenulum a little bridle > the membrane that bridles (ties) the tongue to the floor of the oral cavity.

frug- fruit.

frugivore a fruit eater.

fundus the bottom of a cavity.

fusiform spindle shaped.

gan- bright.

ganglion a swelling > a group of cell bodies outside the central nervous system.

gastr- a belly, a stomach; the digastric muscle has two bellies.

gastralia ventral abdominal ribs.

gastrula a little stomach > a stomachlike embryo.

gen- origin.

genio- chin.

genotype the specific assortment of genes.

geniculate ganglion a kneelike bend of facial nerve; see *genu*.

genu knee.

geo- earth.

glans acorn > the tip of the penis.

glenoid resembling a socket.

glia glue.

glomerulus a little glomus (ball or skein) > a tiny plexus of blood vessels.

gloss- tongue.

gnath- jaw.

gon- seed > generative, as in glucagon (giving rise to glucose).

gonad the source of gametes.

gubernaculum a rudder > a governor, as the ligament that (partly) governs the position of the testis.

gula throat.

gular fold a fold at the throat of some tetrapods.

gustatory related to gustation (taste).

gymn- naked.

gyrus a ridge between grooves.

haem-, hem- blood.

hallux great toe.

hamate having a hook.

hamulus a little hook.

hemi- half; equal to *demi-*.

hemo see *haem-*.

hepat- liver.

hept- seven.

herbivore an animal that devours grasses (herbs).

hetero- other, different; opposite to *homo-*.

heterodont having different kinds of teeth.

hex- six.

hilum a small notch.

hipp- horse.

hist- tissue.

histogenesis the formation of a tissue.

holo- entire, whole.

holonephros a kidney extending the length of the coelom.

hom-, homeo-, homo-, homoio- like, similar.

homeostasis maintenance of a constant internal environment.

homeotherm an animal that maintains a steady body temperature despite ambient (surrounding) temperature; an endotherm.

homodont teeth all alike.

hyaline clear, glassy.

hyoid shaped like the capital Greek letter upsilon (γ).

hyostyly see *-styly*.

hyp-, hypo-, under, below, less than ordinary.

hypaxial below a given axis.

hypophysis a growth under the brain; the pituitary body.

hypural a bone below the urostyle of fishes.

hyper- above, beyond the ordinary.

ichthy- fish.

-iform having the shape of.

ileo- pertaining to the ileum of the intestine.

ilio- pertaining to the ilium of the pelvis.

impar unpaired.

in- not.

innominate not named.

incus an anvil.

infra- beneath, under.

infundibulum a little funnel.

inguen the groin.

inguinal in the region of the groin.

inter- between.

intercalary plate part of neural arch between neural plates in fishes.

interrenal steroidogenic tissue between the kidneys.

intra- within.

intramembranous within a membrane.

intrasegmental within a segment.

ipsi- the same.
   ipsilateral on the same side.
irid- rainbow.
   irides pl. for irises.
   iridophore a pigment cell containing refractory bodies that result in iridescence.
ischi- hip, pelvis.
iso- equal, alike.
   isolecithal egg an egg having even distribution of yolk.
-issimus a superlative ending; the longissimus dorsi is the longest muscle of the back.
-itis inflammation of.

jejunum empty; part of intestine that is often empty at death.
juga- yoke > something that joins; the **jugal bone** in mammals is a yoke uniting maxilla and temporal bones.
jugular pertaining to the neck.
juxta- next to, near.
   juxtaglomerular near glomeruli.

kat- down; same as *cat-*.
keratin from a Greek word meaning horn; a relatively insoluble substance in cornified cells.
kinetic capable of moving.

labium lip.
labyrinth a maze.
   labyrinthodont an early tetrapod with greatly folded dentin in the teeth.
lac-, lact- milk.
lacrimal pertaining to tears; from lachryma (a teardrop).
lacuna a lake.
lag- hare.
lagena a flask > flask-shaped part of inner ear.
lambdoidal having the shape of the Greek letter lambda (λ).
lamella a thin plate.
lamina a thin sheet, plate, or layer.
laryng- larynx.
latissimus the broadest; see *-issimus*.

lecith- yolk.
lemmo- sheath or envelope.
lemur from a word meaning a nocturnal being or ghost.
lepid-, lepis-, lepo- scale, husk.
   Lepisosteus a ganoid fish with bony scales.
lepto weak, thin, delicate.
leuco-, leuko- white, colorless.
levator that which elevates or raises.
lien- spleen.
lingua the tongue.
lip- fat.
liss- smooth.
longissimus the longest; see *-issimus*.
lumbar pertain to the loins; the region of the vertebral column just cephalad to the sacrum.
lumen light > an opening that light can pass through, the cavity in a tube.
lunar, lunate moon shaped.
luteo- yellow.

macula a spot.
magnus, -a, -um large.
malleus a hammer.
mandibula a jaw.
manu- hand.
manubrium a handle.
marsupium a pouch.
mastoid like a breast (*mast-*).
mater mother.
maximus, -a, -um largest.
meatus a canal or passageway.
medulla bone marrow.
   medulla spinalis the marrow of the backbone > spinal cord.
meg- great, very large.
melan- dark, black.
   melanophore bearing (*-phore*) dark pigment.
meninx (pl., **meninges**) a membrane.
mento- chin.
   mental foramen foramen on mandible near chin.
-mer- a segment, a part, one of a series.
mes- middle, midway, intermediate.

mesaxonic foot one in which the weight-bearing axis passes through the middle toe.
mesenchyme a tissue (*enchyme*) that is not yet differentiated.
mesentery associated with the midline of the enteron.
mesonephros an intermediate kidney.
met- after, succession, change, behind.
   metacarpal a bone distal to a carpal.
   metamere one of a series of segments.
   metamorphosis change in structure.
   metanephros hindmost kidney.
mimetic capable of mimicking; having the characteristics of a mime.
mitral refers to a bishop's miter or headdress; the mitral valve is the bicuspid valve of the mammalian heart.
mono- one.
   monotreme a mammal with one caudal opening.
morph- shape, structure, form.
   morphogenesis development of form.
   morphology study of form; anatomy.
morula a mulberry.
myel- marrow.
   myelencephalon the marrow inside the skull > the medulla.
   myelin fatty material.
mylo- from a word meaning a millstone.
   mylohyoid a muscle attached near the grinding teeth.
myo- muscle.
   myocardium muscle of heart.
   myomere one of a series of muscle segments.
   myotome part of somite giving rise to muscle.

neo- new, recent.
nephr- kidney.
   nephrogenic giving rise to kidney.

nephron a functional kidney unit.

nomen- (pl., **nomina**) name.

noto- back.

notochord cordlike skeleton of the back.

nuchal refers to the nape of the neck.

occiput part of the head surrounding the foramen magnum; in mammals, the back of the head.

octo- eight.

ocul- eye.

odon-, odont- tooth.

odontognath a bird with teeth on the jaws.

odontoid resembling a tooth.

-oid like, having a resemblance to, as in hominoid (humanlike).

-ole small, as in arteriole.

olecranon the head (cranium) of the ulna (olēnē) at the elbow joint.

oligo- few, small.

-oma swelling.

omentum a free fold of peritoneum.

omni- all.

omnivore an animal that eats plants and animals.

omo- shoulder.

omphalo navel.

ontogenesis onto (individual) + genesis (origin); the development of an individual.

oö- egg (pronounced oh-oh).

oöcyte egg cell.

oöphoron oö + -phore- (q.v.); ovary.

operculum a cover or lid.

ophthalm- eye.

opisth- at the rear, behind.

opisthocelous with a cavity at the posterior end.

opisthonephros mesonephric kidney of adult anamniotes.

orb- a circle.

orbit cavity for the eyeball.

ornith- bird.

oro- mouth.

ortho straight.

-orum of the, as in branchiorum (of the gills).

os bone.

ossicle a small bone.

ossify to become bony.

os mouth.

os uteri entrance to uterus from vagina.

osmoregulation electrolyte homeostasis.

oste- bone.

Osteichthyes bony fish.

osteon unit of bone in concentric layers.

ostium an entranceway > a mouth.

ostraco- shell.

oto- ear.

otic refers to the ear.

otocyst a vesicle that becomes the inner ear.

-ous having the characteristic of.

ovale shaped like a hen's egg; oval.

ovi-, ovo- egg.

oviparous egg laying.

ovipositor a structure for laying eggs.

ovine pertaining to sheep.

-oxy- sharp, acute, acid.

pachy- thick.

paed-, ped- child.

paedogenesis reproduction in larval state.

palae- see pale-.

palatine referring to the palate.

paleo- old, ancient.

pallium a cloak > a roof.

panniculus a small piece of cloth > a layer of tissue.

papilla a nipple > a nipple-shaped structure.

par-, para- beside, near.

parasphenoid parallel to the sphenoid.

parotid near the ear.

parie- (pl., **parietes**) pertaining to the body wall.

-parous bearing, giving birth to.

pars (pl., **partes**) part.

pectoral refers to the chest.

pedicel a slender stalk.

pelvis a basin.

penta- five.

perennibranchiate having permanent gills.

peri- around.

perilymph fluid surrounding membranous labyrinth.

periss- odd.

peritoneum something that stretches over or around > the coelomic lining.

pes (pl., **pedes**) foot.

petro- stone, rock.

petrosal bone the bone surrounding the inner ear, which resembles a steep rugged rock.

phag- eat.

phalanx (pl., **phalanges**) a line of soldiers > a bone of a digit.

pharyng- pharynx.

pheno- visible.

phenetic pertaining to phenotypes.

phenotype a hereditary character or set of characters expressed in the soma as opposed to genotype.

-phil loving > having an affinity for.

-phore- bearing, one that bears, as in photophore (an organ that emits light).

phrenic refers to the diaphragm.

phylo- tribe.

phylogenesis creation of new taxa.

phylogeny evolutionary history of a group.

-physis that which grows.

physo- bellows > lung or air bladder.

physoclistous lacking a duct from air bladder.

physostome a fish that can get air to the swim bladder via the mouth.

pia tender, kind.

pia mater the delicate meninx of the brain.

pilo- hair.

pineal resembling a pine cone.

pisci- fish.

pisiform shaped like a pea.

placo- thick, flat, platelike.
  placode in embryology, an ectodermal thickening that gives rise to something.
  placoderm fish with (bony) plates in the skin.
planta- the sole of the foot.
  plantigrade a flat-footed stance.
platy- flat, wide, broad.
plesio- close (to) > resembling.
  plesiosaur a marine animal that resembles, and is, a reptile.
pleur- rib, side.
  pleural refers to ribs.
plexus a network.
pneumato- air.
pneumo- lung.
pod- foot.
poikilotherm an ectotherm (q.v.).
pollex the thumb.
poly- many, much, variegated.
pons a bridge.
porcine pertaining to swine.
post- after, behind.
  posttrematic behind a trema or slit.
pre- before, in front of; see also *pro-*.
  pretrematic in front of a trema or slit.
prim- first, earliest.
  primate one first in rank.
pro- favoring, on behalf of.
  prolactin hormone necessary for milk production.
pro- in front of, before, preceding.
  procelous with a cavity at the cephalic end.
  prostate a gland standing (*stat-*) at the beginning of the urethra.
proboscis a feeding tube > an elephant's trunk.
procto- anus.
proprio- one's own.
  proprioception reception of stimuli from muscles, joints, tendons.
pros- toward, near.
  prosencephalon the anterior end of the embryonic brain.
proto- early, first.
pseudo- false.

psoas loin.
pter-, pteryg- wing, feather.
  pterosaur winged reptile.
  pterotic wing of the otic complex.
  pterygoid resembling a wing.
pulmo- lung.
pyg- rump.

quadrate square, four sided.
quint- five.

rachi- referring to an axial support.
  rachis spine of a feather.
  rachitomous specifies a vertebra consisting of several pieces; see *-tome*.
radix (pl., **radices**) a root.
ramus a branch.
rectus, rectum straight.
rete (pl., **retia**) a network.
  rete mirabile a remarkable (*mirabilis*) network of vessels.
  reticulum a little network.
retro- behind, backward.
rheo- current, flow.
rhin- nose.
  rhinencephalon an olfactory part of the brain.
  rhinoceros an animal with a horn (*cerato-*) on the nose.
rhomb- rhomboid.
rhynch- snout.
rostrum a beak or platform.
ruga (pl., **rugae**) a wrinkle.

sacculus a little sac.
sagittal from a word meaning an arrow.
sangui- blood.
sarco- flesh.
  Sarcopterygii fish with fleshy lobe at base of fin.
saur- lizard > reptile.
scala a ladder.
scalene a triangle with sides and angles unequal.
scler- hard, skeletal.
  sclerotome part of somite giving rise to skeletal components.

sebum grease, wax.
  sebaceous having an oily secretion.
seco- refers to cutting.
  secodont shearing molariform teeth of carnivorous mammals.
-sect- cut, divide.
selen referring to the moon.
  selenodont teeth with enamel disposed in crescentic folds.
sella turcica a seat (*sella*) shaped like a Turkish saddle.
semi- half.
  semilunar shaped like a half-moon.
seminal pertaining to seed > to semen.
  seminiferous carrying sperm.
serrate notched or toothed along the edge; the serratus muscle is serrate.
sex- six.
sigmoid S-shaped.
sinus a cavity.
  sinusoid a thin-walled, sinuslike vascular channel.
soma-, somato- body.
  somite a body segment.
sphenoid wedge shaped.
spiracle a breathing hole.
splanchn- viscera.
splen- spleen.
spondyl- vertebra.
squam- scale.
  squamous scalelike, flattened; squamate.
stapes a stirrup.
  stapedial associated with the stapes.
stato- standing, fixed.
steg- refers to covering plates; also to a roof.
stellate star shaped.
stereo- solid > shape; stereognosis is knowledge (*-gnosis*) of form or weight acquired by feeling or lifting.
stom- mouth.
stratum layer.
strept- twisted, curved.
stria a stripe.

-style pillar > a process such as the urostyle.

    styloid having an elongated shape.

-styly braced; in hyostyly the jaws are braced against the hyoid.

sub- under, below.

    subclavian under the clavicle.

    subunguis under the nail.

    subungulate not quite an ungulate.

sudor sweat.

sulcus a groove.

super-, supra- over, above, in addition.

    suprarenal a gland above the mammalian kidney.

sur- over, above; equivalent to *super-*.

    surangular bone a bone above the angular.

sym-, syn- together.

    symphysis a growing together; see *-physis*.

    synapse a junction.

    synarthrosis an immovable suture-like joint.

    synsacrum sacrum united with other vertebrae.

tarsus ankle; also, a connective tissue plate in the eyelid.

tax- arrangement.

    taxon a taxonomic unit such as a phylum or species.

    taxonomy the orderly arrangement of taxons; classification.

tectum a roof.

tegmentum a covering.

tel-, teleo-, telo- end, complete.

    telencephalon anterior end of the brain.

    teleology the use of design or purpose to explain natural phenomena.

tela a web > a thin weblike membrane.

temporal refers to the temple or the side of the skull behind the eye.

teres round.

tetra- four.

thalamo- a chamber.

theco- a case or sheath.

    thecodont having socketed teeth.

therio- an animal with hair, a beast.

thyroid shield shaped.

-tome cut; also, the result of cutting, as a section or sheet.

trabecula a little beam > a strand, ridge, rod, or bundle.

trans- across.

    transect to cut across.

trapezoid a four-sided plane with two parallel sides.

    trapezius muscle named for its shape in humans.

-trema a slit.

tri- having three parts.

    trigeminal from word meaning triplets; a nerve with three primary branches.

trochlea a pulley; the trochlear nerve of humans passes through a pulley at its attachment.

troph- nourishment.

truncus trunk.

    truncus arteriosus ventral aorta.

tuber a swelling or knob > a tuberosity, tubercle, or protuberance.

    tuberculum a little tubercle.

tunic a coat or wrap.

tympanum a drum > eardrum.

ulna elbow > the bone at the elbow.

ultimobranchial a gland derived from the last (*ultimo-*) branchial pouch.

ultra- beyond.

-ulus, -ula, -ulum diminutive endings denoting tiny.

uncus a hook.

    uncinate hooked.

unguis nail, claw, hoof.

ungula hoof.

    ungulate hooved.

uro- from Greek **ouro,** meaning tail or urine.

    urogenital = urinogenital.

    urophysis a growth at the base of the tail.

    uropygium the rumplike tail of a bird

utricle a little sac or vesicle.

vagina a sheath.

vagus wandering.

vas (pl., **vasa**) vessel.

    vasa vasorum vessels of the blood vessels.

velum a veil > a thin membrane.

venter abdomen; the part opposite the dorsum or back.

ventricle a cavity in an organ.

vesica a bladder or vesicle.

vestibule an antechamber or entrance way.

vitelli- yolk.

vitreous having a glassy appearance.

vivi- alive.

volar pertaining to palm or sole.

vomer a plowshare; the mammalian vomer bone resembles a plowshare.

-vorous eating, devouring, as in insectivore.

Xanth- yellow.

xiph- sword.

    xiphoid process a swordlike process of the sternum opposite the manubrium.

ypsiloid shaped like the Greek letter upsilon (Y).

zyg- yoke > something that links two things.

    zygapophysis a vertebral process that articulates with a more anterior or posterior one.

zygomatic arched.

zygote result of union of gametes.

COMPREHENSIVE
REFERENCES

The following are mostly multivolume works relevant to many chapters. *A single volume work of the same nature—a book on the biology of fishes, for example—will be found at the end of Chapter 3.* The latter and the following list supplement the Selected Readings at the ends of many chapters. Other entries below have been arbitrarily included.

Alexander, R.M.: The chordates, ed. 2, Cambridge, England, 1981, Cambridge University Press.

Bellairs, D.A., and Cox, C.B., editors: Morphology and biology of reptiles, Linnaean Society Symposium Series, New York, 1977, Academic Press.

Bloom, W., and Fawcett, D.: Textbook of histology, ed. 10, Philadelphia, 1975, W.B. Saunders Co.

Cole, J.F.: A history of comparative anatomy, London, 1944, The Macmillan Co., Ltd.

Farner, D.S., King, J.R., and Parkes, K.C., editors: Avian biology, 7 vols., New York, 1971-1983, Academic Press.

Gans, C.: Biomechanics: An approach to vertebrate biology, Philadelphia, 1974, J.B. Lippincott Co.

Gans, C., and others, editors: Biology of the reptilia, 15 vols. Vols. 1-13, New York, 1969-1982, Academic Press. Vols. 14 and 15, New York, 1985, John Wiley and Sons, Inc.

Getty, R., editor: Sisson and Grossman's the anatomy of the domestic animals, ed. 5, 2 vols., Philadelphia, 1975, W.B. Saunders Co.

Goodrich, E.S.: Studies on the structure and development of vertebrates, London, 1930, The Macmillan Co., Ltd. (Reprinted by The University of Chicago Press, Chicago, 1986.) *Some outdated theory but excellent morphology of fossil and extant species.*

Grassé, P.P., editor: Traité de zoologie, anatomie, systématique, biologie, vols. 12-16, Paris, 1954-1983, Masson et Cie.

Hardesty, M.W., and Potter, I.C., editors: The biology of lampreys, New York, 4 vols. through 1983, Academic Press.

Hildebrand, M., Bramble, D.M., Liem, K.F., and Wake, D.B., editors: Functional vertebrate morphology, Cambridge, Massachusetts, 1985, Harvard University Press.

Hoar, W.S., and Randall, D.J., editors: Fish physiology, 10 vols., New York, 1969-1984, Academic Press.

Jarvik, E.: Basic structure and evolution of vertebrates, 2 vols., New York, 1980, Academic Press.

King, A.S., and McLelland, J., editors: Form and function in birds, 3 vols., New York, 1979-1985, Academic Press.

Lofts, B., editor: Physiology of the amphibia, 3 vols., New York, 1974-1976, Academic Press.

Nickel, R., and others: The viscera of the domestic mammals, New York, 1973, Springer-Verlag.

Prosser, C.L., editor: Comparative animal physiology, ed. 3, vol. 2, Philadelphia, 1973, W.B. Saunders Co.

Schmidt-Nielsen, K.: Animal physiology, adaptation, and environment, ed. 3, Cambridge, 1983, Cambridge University Press.

Weiss, L., editor: Histology: cell and tissue biology, ed. 5, New York, 1983, Elsevier Science Publishing Co.

Young, J.Z.: The life of vertebrates, ed. 3, New York, 1981, Oxford University Press.

MEDICAL DICTIONARIES AND ANATOMICAL TERMINOLOGY

Baumel, J., and others, editors: Nomina anatomica avium, New York, 1980, Academic Press.

Dorland's illustrated medical dictionary, ed. 26, Philadelphia, 1981, W.B. Saunders Co.

Nomina anatomica, ed. 5, Baltimore, 1983, The Williams & Wilkins Co.

Nomina anatomica veterinaria, ed. 3, Ithaca, N.Y., 1984, Habel.

Stedman's medical dictionary, ed. 24, Baltimore, 1982, The Williams & Wilkins Co.

INDEX

Pages cited in **boldface** contain an illustration only. Other cited pages frequently contain illustrations. Tables are designated by "t."